Progress and Trends in Rheology

Progress and Trends in Rheology

Edited by
Hanswalter Giesekus · Kurt Kirschke · Joseph Schurz

Proceedings of the
First Conference of European Rheologists
Graz (Austria), April 14 – 16, 1982

Special Edition from Rheologica Acta, Vol. 21, No. 4/5

Springer-Verlag Berlin Heidelberg GmbH 1982

ISBN 978-3-7985-0612-1 ISBN 978-3-662-12809-1 (eBook)
DOI 10.1007/978-3-662-12809-1

Originally published by Dr. Dietrich Steinkopff Verlag, GmbH & Co. KG, Darmstadt in 1982

Contents

(Page numbers refer to the numbers at the lower margin)

Papers Submitted but not Presented

Proceedings of the First Conference of European Rheologists, Graz (Austria), April 14 – 16, 1982

INTRODUCTION

The First Conference of European Rheologists, which was held at the University of Graz from $14^{th} - 16^{th}$ April 1982, was originally intended to be only an Annual Conference of the German Society of Rheology organized as a joint meeting together with the Working Group of Rheology in the Austrian Chemical Society. The process by which it became a real Conference of European Rheologists in which more than two hundred scientists from about twenty countries participated was outlined in the inaugural address of the President of the German Society of Rheology as follows:

"As is well known there are too few rheologists in most countries for them to be able to restrict their scientific communication to colleagues from only their own country. In fact, I think there is a common feeling, at least among older researchers in this field, that rheologists continue to think of themselves as belonging to a worldwide scientific family. Every International Congress shows this anew. However, it is not possible for everyone to participate in these big international events and even the occasional participation in the conference of a neighbouring national society does not seem to be sufficient to satisfy to a desirable degree the need for international personal contacts for the stimulation and exchange of ideas. This communication is now even more important, not only because of the specialization occurring with the progress of knowledge but also because of new interrelationships between research fields previously considered to be independent.

Therefore about ten years ago some European societies began to organize joint meetings. The first one was the British-Dutch Symposium at Lunteren (Holland) in 1973, followed by the British-Italien Joint Conference at Shrivenham (England) in 1975, the British-Dutch-Italien Joint Conference at Pisa (Italy) in 1977, the Dutch-Italian-British Joint Conference at Amsterdam (Holland) in 1979, and the Belgian-British Joint Meeting at Liège (Belgium) in 1981. There were also two Conferences on Chemical Engineering Rheology, organized jointly by British and German groups, the first taking place at Salford (England) in 1974, the second at Aachen (West-Germany) in 1979, and a third one is being organized for London in 1983.

The experience of these joint conferences stimulated some national societies, in particular the British Society of Rheology, to explore the possibilities for a closer liaison and cooperation between the European societies and groups. After some correspondence between the then secretary of the British Society, Dr. Chris Petrie, and the representatives of the other societies (including the answering of a questionnaire), delegates of nine European groups met a Naples during the VIII. International Congress in 1980 to discuss the proposed activities.

Although there was not a majority in favour of the idea of a European Society of Rheology, two decisions were reached. The first was to establish a 'Provisional Committee of European Delegates to the International Committee on Rheology' with Dr. Petrie as the secretary, its role to be that of an 'intelligent post-box'. The somewhat cumbersome title of this committee indicates that its delegates should be the same as the delegates of the respective societies in the International Committee on Rheology (which is responsible for deciding on the location of International Congresses), so that there would be an opportunity for them to become better acquainted and to co-operate more closely. 'Intelligent post-box' indicates that the first duty of the committee should be to disseminate information and to steer joint activities.

The second decision at Naples referred particulary to the Graz Conference. It was agreed that the German Society and the Austrian Group invite all other European societies of rheology to join them in organizing this meeting.

It should not be concealed that there were some doubts about the usefulness of a European Confer-

ence. It was feared that such a meeting would become too big and therefore loose the informality which was understood to be one of the valued attributes of nationally organized conferences and necessary for the cross-fertilization of ideas and interdisciplinary discussion".

However, at the end of the conference there was far reaching agreement not only that these apprehensions had proved to be unfounded but that this meeting had really aided the progress of rheology. So many opportunities were available for establishing contacts and exchanging ideas and experiences. The only disappointment was that in spite of the neutral status of the host country and its relative proximity to Eastern and South-Eastern Europe and the fact that many scientists from these countries had expressed interest in taking part, only very few of these were given the opportunity to participate. In particular there were no visitors from the DDR or the USSR.

Delegates from the various national societies and groups met during the conference and agreed to pass on the following recommendations: Subsequent Conferences of European Rheologists should not be held more often than every four years and preferably midway between the International Congresses. The Czechoslovakian delegates also used this opportunity to offer their country for the next Conference in 1986. As there were no objections to this proposal the Second Conference of European Rheologists is likely to be organized by the CSSR Group of Rheology.

The preparatory work for the present conference was undertaken jointly by Prof. K. Kirschke, Secretary of the German Society of Rheology, and Prof. J. Schurz, Chairman of the Austrian Group of Rheology. The latter was also responsible for the local organization.

The conference began on Wednesday morning with a performance of two movements of a string quartett ("Lerchenquartett") from Joseph Haydn by members of the Hochschule für Musik und Darstellende Kunst Graz. It was formally opened by Prof. J. Schurz, followed by some words of welcome from Prof. K. Freisitzer, Rector Magnificus of the Karl-Franzens Universität Graz. The inaugural address was given by Prof. H. Giesekus, President of the German Society of Rheology, and the opening lecture by Prof. E. Broda from Vienna. Four invited plenary papers followed the opening ceremony. The 81 contributed papers were then presented in five parallel sessions on Wednesday afternoon, Thursday and Friday. On Wednesday evening a banquet was given by the Governor (Landeshauptmann) of Styria and on Thursday evening by the Mayor (Bürgermeister) of the City of Graz. A ladies program offered a walk through the "Old City of Graz" and excursions to neighbouring places of interest. The closing address on Friday afternoon was given by Prof. G. W. Becker, Past President of the German Society of Rheology and currently its delegate to the International Committee on Rheology.

These Proceedings contain the opening lecture and the invited papers, which are printed in full, as well as short versions of 63 contributed papers and 11 papers which were submitted but not presented. A list of papers presented but not included herein is given at the end.

Dr. M. Hibberd, presently at the University of Dortmund, assisted with the editorial work by checking and improving the English text of some authors whose native tongue is not English.

Prof. Dr. Hanswalter Giesekus, Dortmund
President of the German Society of Rheology

Prof. Dr.-Ing. Kurt Kirschke, Berlin
Secretary of the German Society of Rheology

Prof. Dr. Joseph Schurz, Graz
Chairman of the Working Group of Rheology
in the Austrian Chemical Society

Rheologica Acta Rheol. Acta **21**, 357–359 (1982)

OPENING LECTURE

Ludwig Boltzmann – Man, physicist, philosopher, biologist

E. Broda

Institute of Physical Chemistry, Vienna University

Key words: Boltzmann, thermodynamics, statistics, philosophy, Darwin

It is not so generally known that Ludwig Boltzmann was not only one of the greatest theoretical physicists of all time, but also a magnificent philosopher, a great leader in theoretical biology, and a most interesting and attractive person. It was here at Graz that Boltzmann spent his most productive and happy years. He was here Professor of Mathematical Physics from 1869 to 1873, and of Experimental Physics from 1876 to 1890. He was, by the way, an excellent experimenter.

Boltzmann was born at Vienna in 1844 and spent much of his youth in Upper Austria. His teacher at Vienna University was Josef Stefan, a son of illiterate peasants of Slovene mother tongue in Carinthia (Kärnten). A modest and lovable person, Stefan had great merits. Together with Helmholtz, he was the prophet of Maxwell's theory of the electromagnetic field in Central Europe, where it had found much resistance. Further, together with Josef Loschmidt he developed the Physics Department of Vienna University into a stronghold of atomism. At that time, many physicists still rejected the atom. For instance, in France none of the major physicists of the 19th century was a supporter of the atom. Loschmidt was the first after nearly two and a half millenia of atomism to venture, successfully, an estimate of the size of the atom.

In physics, Boltzmann's name appears in the Boltzmann theorem (the equipartition law: the expression is Maxwell's), the Maxwell-Boltzmann distribution law, the Boltzmann factor, Boltzmann statistics, the Boltzmann transport equation, the Boltzmann constant (named by Planck) and the Stefan-Boltzmann radiation law. Boltzmann's derivation of this law was called a true pearl of theoretical physics by Hendrik Antoon Lorentz in his moving obituary to Boltzmann. The expression "Boltzmann principle" was introduced by Einstein for the famous equation $S = k \log W$ that is engraved on Boltzmann's beautiful grave of honour in the Central Cemetery of Vienna. Along with Maxwell and J. Willard Gibbs, Boltzmann was the founder of Statistical Mechanics. These achievements of Boltzmann are common knowledge among physicists and chemists and need not be further discussed here.

In his later period Boltzmann took increasing interest in philosophy. His starting point were the problems of epistemology, the origin meaning of the ultimate categories in physics: Space, time, law of nature. Boltzmann was opposed to the idealist, aprioristic, views of Immanuel Kant, though he did not deny his greatness. He cherished others less. A lecture in 1905 to the Philosophical Society at Vienna University was at first called "Proof that Schopenhauer was a mindless ignorant philosophaster who scribbles nonsense and through empty word trash debases the heads fundamentally and for ever". These were words that originally had been directed by Schopenhauer himself against Hegel. However, this idealist philosopher was likewise not appreciated by Boltzmann.

Berkeley's form of idealism was called by Boltzmann the greatest folly ever hatched by the human brain. It was pointed out by him that subjective idealism must in the end lead to solipsism. This also applies to the positivism that was promoted by Boltz-

777

mann's colleague at Vienna since 1895, Ernst Mach. Mach refused to accept the objective reality of an external world. He denied that science can give information about it, and limited its task to the ordering of "complexes" or "elements" of sensations, by application of "economy of thought" (Denkökonomie).

Boltzmann was polite with Mach and acknowledged his spirit, but he firmly rejected his ideas. This attitude was clearly strengthened by the fact that Mach denied atomism. He maintained this negative view on atoms to the end of his days in 1916, and the same was true, by the way, of his negative view on Einstein's theory of relativity. It is an ironic fact that precisely Boltzmann became successor to Mach as Professor of Philosophy at Vienna University when the latter had to withdraw prematurely in 1901 in consequence of a stroke. At the same time Boltzmann remained Professor of Theoretical Physics.

Boltzmann's philosophy was realism, or, as at the end he occasionally called it, materialism. He derived his conviction of the reality of the external world as an object of science from what may be called "concordance". The information obtained by different senses, with the help of different instruments, by different persons, and at different times, always lead to similar conclusions about the external, real, world. This could not happen if the independent world had no objective existence.

Boltzmann's passionate interest in biology and Darwinism was closely linked to his philosophy. He called the 19th century Darwin's century, and he expected "all salvation for philosophy" from Darwin's teachings. Already in an address to Graz University in 1887 he compared the "obscurantists" who still struggled against Darwin's ideas to the enemies of the theorem of Pythagoras after this had been established. Their horrible noises were compared to those of oxen Pythagoras was going to sacrifice to the gods in gratitude.

Boltzmann put forward ideas on the origin of life on the basis of struggle for survival and selection that are quite consistent with present views. He assumed what is now known as chemical evolution in the primeval ocean. Moreover, Boltzmann explained photosynthesis as having arisen in the tendency of plants to harness the energy of light to their requirements.

This thought had been expressed by Julius Robert Mayer in 1845 already. However, Boltzmann showed that the utilization of light energy is bound up with the degradation of its value. The energy of the light of the Sun (short-wave) has low entropy, the energy of the radiation emitted by the Earth (long-wave) has high entropy. In their struggle for survival, the plants exploit the difference. This is, of course, the concept held in our time.

In Boltzmann's view the nervous and, later, the intellectual functions likewise arose in evolution. The brain evolved in man to extraordinary size, as in the stork the beak, in the giraffe the neck. Memory and subsequently consciousness are products of gradual improvement of nervous functions among animals. Also the rules and laws of thought, based on acceptance of space, time and causality, were formed in evolution and are now inborn in mankind. But the laws are not necessarily faultless, as Boltzmann strongly insisted against Heinrich Hertz. They must be further adjusted in the light of more experience.

According to Boltzmann the morality and the striving for beauty arose among animals and in mankind on the same basis. Happiness indicates circumstances favourable for survival of the species, unhappiness the opposite. This Darwinian view was strongly, indeed violently, argued in Boltzmann's powerful rejoinder in 1904 to Wilhelm Ostwald's lecture "On Happiness". Here this remarkable man, in many ways a scientific opponent of Boltzmann, while he was his personal friend, had put forward a simple algebraic equation to express happiness in quantitative terms.

At Graz Boltzmann derived the transport equation. Here he made probability the sole basis of Statistical Mechanics, and here he set up Boltzmann's principle. Here he explained the Stefan radiation law, and here he interpreted photosynthesis on the basis of entropy. But here he also married – the niece of the Mayor of Graz, the cousin of his friend, the composer Wilhelm Kienzl. Here, at Kroisbach, he also had his famous country house with a fine wide view where he discussed the greatest problems of physics and life with his students, among them Svante Arrhenius and Walther Nernst.

Boltzmann enjoyed nature, poetry and music, he had a lively sense of humour, and his friendliness to people knew no limit. So he had also many grateful students, among whom Fritz Hasenöhrl, Marian von Smoluchowski, Paul Ehrenfest und Lise Meitner may be mentioned in addition to Arrhenius and Nernst. Moreover, Boltzmann's influence on Max Planck and Albert Einstein was enormous. So it is a tragic fact that at the end of his life when his health failed, he had become the target of attacks by scientific and philosophical adversaries, and felt isolated, Boltzmann succumbed. He committed suicide at Duino, near Trieste, in September, 1906.

References

Boltzmann, L., Populäre Schriften, edited and introduced by E. Broda, Vieweg (Braunschweig 1979).
Broda, E., Ludwig Boltzmann, Mensch, Physiker, Philosoph, Deuticke (Vienna 1955). Also in Japanese and Romanian. American edition to appear in 1982.
Broda, E., Ludwig Boltzmann, Phys. i. u. Z. **10**, 79 (1979).
Broda, E., Boltzmann, Einstein, natural law and evolution, Comp. Biochem. Physiol. **67 B**, 373 (1980).
Broda, E., Ludwig Boltzmann, Der Mann und das Werk, Mitt. Verein Gesch. Naturwiss. **1**, 21 (1981).
Broda, E., Ethics and evolution in Boltzmann's and Einstein's thought, in: M. Kageyama et al., eds., Science and Scientists, Reidel (Dordrecht 1981).
Broda, E., Boltzmann and Darwin, lecture, to be published (Vienna 1981), manuscript available.
Broda, E., Darwin and Boltzmann, lecture, Kühlungsborn 1981, to be published, manuscript available.

(Received May 12, 1982)

Author's address:

Prof. Dr. E. Broda
Institute of Physical Chemistry
Vienna University
Währinger Straße 42
A-1090 Wien

Rheologica Acta Rheol. Acta **21**, 360–365 (1982)

INVITED PAPERS

Topics in thermodynamics of rheological behaviours*)

C. Huet

Centre d'Enseignement et de Recherches en Matériaux de Construction (CERMaC),
Ecole Nationale des Ponts et Chaussées, Paris

Abstract: Survey of results recently obtained by the author about hereditary behaviour, Gibbs equation, caloric equation of state, local equilibrium, reversible accessibility, use of internal variables, behaviour classification. Toward a new and unified approach: thermodynamics in natural variables.

Key words: Constitutive equation, thermodynamics, local state, natural variable, hereditary behaviour

1. Introduction

Rheologists are very well acquainted with the complexity of rheological behaviour. It is well known (cf. for instance [1]), that the problem of a complete, general and unified theory of thermodynamics of materials with hereditary behaviour is not yet available, and that the subject remains controversial.

Actually, one usually recognises two main schools of thought that developed, during the XXth century, quite independently, from the same roots of the XIXth century, i.e. Rumford [2], Carnot [3], Meyer [4], Clausius [5], Massieu [6], Boltzmann [7], Gibbs [8], Helmholtz [9], Duhem [10].

The first school, we might name here, with European chauvinism, "the European school", is represented mainly by physicists and chemists such as De Donder [11], Onsager [12], Defay [13], Prigogine [14] and Meixner [15]. In the forties, it developed a theory that has since appeared to be successful in wide branches of Physics and Chemistry. This theory is now well known as the "Thermodynamics of Irreversible Processes" (TIP).

The second school, that for sake of simplicity and symmetry we might name perhaps "the American school", was developed mainly after the Second World War by mechanicists and mathematicians as Truesdell [16, 17], Noll [16], Coleman [18] and Day [19].

This second theory was alleged by its main author (quoting Gibbs himself, 1901) to be a "Rational Thermodynamics" [17], and is now widely known under this name. Quite ununderstandingly, the presentation of the theory by this author was usually accompanied, probably for pedagogic purposes, by a somewhat irrational aggressivity against the other schools. This was still perceptible in the opening lecture given by him at the VIIIth International Congress on Rheology in Naples, in September 1980 [20]. From this lecture, rheologists could learn that one doesn't need rheology, but only a constitutive equation, and that one doesn't need rheologists, as rheology does not exist.

The VIIIth Congress itself, attended by more than six hundred people, demonstrated, as the present Meeting seems about to, that this claim is too bold: it is apparent that industry, that physics, that chemistry, that biology need rheology.

As a matter of fact we don't need "a" constitutive equation. What is needed is the right one. That is, we need the constitutive equation able to represent the, generally very complicated, behaviour of the real material we are actually dealing with. This real material can be a polymer foam or mass, some rocky material, a mixture of sand and clay, a fiber composite, a cement or bituminous concrete. It can even be blood or something else. But we cannot decide arbitrarily what its properties are: we do not deal, generally, with school problems.

Thus we need experimental methods to investigate the behaviour of this real material. And this represents a large – probably the largest – part of rheological activities.

*) Invited paper, presented at the First Conference of European Rheologists at Graz (Austria) April 14–16, 1982.

Of course, an experimental method cannot be elaborated without a theoretical basis. A theory will be good if it provides a framework large enough to contain most of the behaviours exhibited by real materials when we are still in the situation where we don't know what their properties are. Further, this framework must be precise enough to give us procedures by which we can recognize the main features of these behaviours, which can be thus classified, and then by which we can determine, by appropriate experiments, the magnitude of behaviour coefficients and explicit expressions of behaviour functions.

A good theory must also tell us how we can optimize the use of available experimental data to obtain as much information as possible, or optimize the experimentation itself. For instance it must tell us when the information obtained from a rheological experiment can be sufficient for our purpose, and when further information, perhaps of non-rheological nature, will be necessary.

It seems to us that we cannot remain in a situation where each of the two schools claims it is the only valid one. We have to put the two theories flat on the same table, to confront them, and to use from both what seems to be logically consistent and best for our needs.

We want to present here a summary of a few results of our own recent attempts towards that purpose. Therefrom, it would appear that finding a way to reconcile the two approaches is not hopeless.

We presented the basic ideas of these results [21] at the 12th National Colloquium of the French Group of Rheology (Paris, 1977), specially devoted to the Thermodynamics of Rheological behaviour.

In this lecture, we shall recall these basic ideas and briefly present a few developments of the theory obtained since that time.

2. Universal equations, constitutive equations and the thermodynamics of behaviour

To begin with, thermomechanics is based upon a set of balance equations of universal character. Expressing universal principles, they stand for whatever material we are dealing with. Usually this set involves more unknowns than equations, and a further set of equations has to be built on. These are the constitutive equations. They can differ from one material to another.

Mathematicians postulate them, thus defining what they call a behaviour or a material. Physical theoreticians try to construct them from microscopic constitution. Engineers have to determine them from experiment.

Rheologists are involved in all three approaches.

The balance equations that are generally considered in the thermomechanics of continua are the five conservation equations for mass, momentum, moment of momentum and energy.

The moment of momentum equation reduces to stress tensor symmetry, leaving only 5 scalar equations for many more scalar unknowns.

Thus, for the system of equations to be closed, we must add, to the universal conservation equations, a further set of equations, in number equal to the difference between the number of unknowns and the number of universal equations.

The further equations are called constitutive equations. We learned from Truesdell in Naples [20] that the name was coined by Bateman as early as 1910.

Since we are restricting ourselves here to the rheological behaviour of materials, without consideration to their electric, or magnetic behaviour for instance, we are used to calling them rheological constitutive equations.

Continuum mechanicists, in order to make the models of material behaviours we try to use physically realistic, introduced a condition expressing the second principle of thermodynamics.

Truesdell [17] called this condition Clausius-Duhem inequality, as it results from Duhem's attempt [22] to apply the Carnot-Clausius principle [3, 5] to continua in non-homogeneous conditions.

The introduction of this inequality into the mechanical formalism constitutes the corner-stone of rational thermodynamics.

Whatever the state of the polemic is at the present time, we cannot help but recognize how useful and powerful this modeling of a physical principle into the formalism of continuum mechanics is, especially for rheological purposes.

For instance, Coleman, in a celebrated paper [18], and using this formalism, has shown that, for viscoelastic behaviour, which essentially is an out-of-equilibrium behaviour, this condition provides a set of properties, e. g. potentiel properties, usually devoted to thermodynamics of equilibrium or to reversible processes.

This seems to confirm the idea that, by some local state or local equilibrium principle, one can extend the equilibrium formalism to out-of-equilibrium situations.

As we have shown extensively in our study of the local state concept [21], this can be done to some extent and in a few instances, but never completely.

When it can be, it must be done in a way depending on the behaviour exhibited by the material in specific rheological experiments.

Hence we have to be cautious, and to learn. For what kind of behaviours and for what kind of experiments, can or must this be done?

3. The local state concept

Just as Clausius-Duhem inequality is the cornerstone for Rational Thermodynamics, the local state concept or principle, or hypothesis, and especially the explicit use of a Gibbs equation are the cornerstones for the Thermodynamics of Irreversible Processes (TIP).

According to Prigogine himself, this local state principle, or local equilibrium hypothesis, states (cf. [23]) that "explicit evaluation of entropy production is possible for the class of phenomena for which the local entropy can be expressed by the same law and in terms of the same independent variables, as if the system were at equilibrium". The author adds cautiously that "situations considered in rheology are excluded" and quotes attempts made by the American school to go further. On the other hand, the local state principle is widely used in textbooks on continuum mechanics (see for instance [24, 25]).

So, what is the situation for rheological behaviour?

For Newtonian fluids, the above statement is generally accepted as being true. For more complicated behaviour, it is generally accepted as hypothesis of specific form. Actually, it can be shown that it is not true even for viscous behaviour if the fluid exhibits volume viscosity.

So we have to be cautious, especially with the set of variables we are dealing with. As we want experimental checking, it is best to use the set of macroscopic observables.

In situations with uniform temperatures, this set is $(u, \varepsilon; T, \rho_0^{-1}\sigma)$, or $(u, \varepsilon; T^{-1}, \rho_0^{-1}T^{-1}\sigma)$, where u is the massic internal energy, ε the strain, T the temperature, σ the stress, and ρ_0 the volumic mass in initial configuration. The second set can be recognised as the set of conjugate observables in entropic representation of equilibrium thermodynamics. The first is the one used in the energetic representation. Here, we restrict ourselves to small strains. We know that the corresponding formalism can be extended to finite deformations by use of Lagrangian representation, with Green tensor Δ in place of ε, and Piola-Kirchhoff stress Π in place of σ.

Rheological constitutive equations relates, in each set, the second pair of variables to the first one or vice versa, or mixed combinations of these pairs.

We know (cf. for instance [26]) that these four variables can be controlled from the boundary of a specimen in homogeneous condition, justifying their name of observables. In the general formalism, only two of these four variables can be taken as independent variables.

At equilibrium, we know that the whole behaviour is completely determined by the explicit knowledge of a fundamental equation, expressing massic entropy in terms of u and ε.

These equilibrium properties are numerous and well known. They can be split into the following main properties:

(i) Entropy is a state function of u and ε, that is, possible states in Gibbs space (s, u, ε) lie on a surface of this space (equilibrium surface property).
(ii) Entropy is a potential for T^{-1} and $\rho_0^{-1}T^{-1}\sigma$, involving reciprocity relations.
(iii) A Gibbs equation holds, that is: $\dot{s}$ is a total differential form in terms of $\dot{u}$ and $\dot{\varepsilon}$, with T^{-1} and $\rho_0^{-1}T^{-1}\sigma$ as coefficients.
(iv) Reversible accessibility from one state to another state holds.
(v) Second differential of s in terms of u and ε is negative definite.
(vi) Sets of independent variables can be interchanged and give equivalent formalisms by mean of duality, involving various Legendre transformations.

If we consider the out-of-equilibrium case, the local equilibrium hypothesis could be understood, strictly speaking, as the preservation of the whole set of the above properties.

But, by use of Clausius-Duhem inequality, it can be shown that this is not the case, and that at least one part of the above properties is necessarily lost.

As a matter of fact, this could be predicted, since we have only two possibilities:

– either we must release at least one of fundamental equation properties when we are in an out-of-equilibrium situation;
– or the fundamental equation cannot fully characterize equilibrium behaviour.

Now the problem is, given a specific real material, what properties, if any, of equilibrium are preserved, and what properties are lost. We shall see that we can answer this question, if we make specific experiments.

The answer can depend on the material itself, and on the set of observable variables we take as independent ones. From the kind of answer we get from a given experiment, we can establish a behaviour classification.

4. A classification based on adiabatic relaxation

At first, we make a powerful choice, the Gibbs set (u, ε) involved in Gibbs space (s, u, ε). The corresponding specific experiment is adiabatic relaxation after a history of u and ε achieved at a finite rate. In this experiment we suddenly fix u and ε at their values attained at a time t_0, and we measure temperature and stress response of the material.

There are only four possibilities:

a – There is a sudden change of temperature and/or stress, followed by a delayed change: the material is of general behaviour; T^{-1} and $T^{-1}\sigma$ depend explicitly on current rates $\dot{u}$ and/or $\dot{\varepsilon}$, and the previous history. A trivial example is the Maxwell model in parallel with a dashpot.

b – There is a delayed change in T^{-1} and $T^{-1}\sigma$, but no sudden change: the material is what we have called colemanian [21]. In this case, T and $T^{-1}\sigma$ depend on history, but not on present rates $\dot{u}$ and $\dot{\varepsilon}$. A trivial example is the Maxwell model.

c – There is a sudden change in T and $T^{-1}\sigma$, but no delayed change. In this case, T and $T^{-1}\sigma$ depend on current rates $\dot{u}$ and $\dot{\varepsilon}$, but not on previous history. The material is of what we have called markovian behaviour [21]. It is called of rate-type behaviour by other authors. Trivial examples are the Kelvin model and Newtonian fluids (with or without volume viscosity).

d – There is no change at all during the experiment (and no permanent instantaneous deformation as assumed from the beginning): the behaviour is reversible (thermoelastic).

In cases a and b, when delayed response appears, it is clear that entropy s cannot be expressed in terms of u and ε by the same function as if the material were in equilibrium. This because s varies at ε and u fixed, as does not occur in equilibrium case. Actually, it cannot be expressed by a function at all, but only by a functional. Thus the local state or local equilibrium hypothesis, as quoted from Prigogine in par. 2, does not hold. But we can get further information by application of Clausius-Duhem inequality.

The first main result provided by Clausius-Duhem inequality is that in none of these exhausting cases, entropy can depend on the rates $\dot{u}$ and $\dot{\varepsilon}$ of u and ε, even when there are sudden changes in T and $T^{-1}\sigma$. Thus instantaneous relaxation (in the large, i.e. including T response) is isentropic. Further, delayed relaxation is not isentropic: entropy is growing at fixed u and ε during the experiment till equilibrium is established (i.e. till the time when T and σ attain values remaining constant in time).

The second main result is that, for Colemanian behaviour (i.e. without instantaneous relaxation), entropy has potential properties. But equilibrium surface property is lost: entropy, if it can be defined, is under the equilibrium surface as long as delayed relaxation is not stabilized.

The third main result is that for Markovian behaviour (i.e. with no delayed relaxation), entropy has equilibrium surface properties: it is a state function of u and ε, the same as in equilibrium. This because it doesn't have delayed evolution, and, further, no instantaneous one (as it cannot depend on $\dot{u}$ and $\dot{\varepsilon}$). But potential properties are lost.

In the general case a, there is no surface equilibrium property, nor potential property.

In case d of reversible behaviour, both properties hold, and also other equilibrium properties.

Thus we see that there are cases when one of the two first equilibrium properties of entropy is preserved. But they are peculiar cases, and when one property is preserved, the other is not. Further, we cannot assume arbitrarily that this one or the other property is preserved. The indication of it is provided by experiment.

5. A TIP formalism without Gibbs equation for Markovian behaviour

For none of the cases a to c, Gibbs equation holds in terms of observable current variables u, ε, T, σ because these behaviours are irreversible behaviours for which entropy production γ_i, i.e. the deviation of $\dot{s}$ from the Gibbs differential form, is strictly positive. Entropy rate $\dot{s}$ involves further terms coming from departure of T and σ from their equilibrium values, and/or a term $\dot{s}_{u,\varepsilon}$, a continuation functional expressing evolution rate of s at fixed u and ε. But for markovian behaviour this functional vanishes. Entropy production is then a bilinear form in terms of $\dot{u}$ and $\dot{\varepsilon}$ on one hand, $(T^{\mathrm{eq}-1} - T^{-1})$, $(T^{-1}\sigma - T^{\mathrm{eq}-1}\sigma^{\mathrm{eq}})$ on the other hand. We get thus a formalism completely analogous to classical TIP, but based on Clausius-Duhem inequality, and not on Gibbs equation, invalid here as previously said.

Of course, Gibbs equation still remains valid when expressed in terms of equilibrium values T^{eq} and σ^{eq} of T and σ. We have shown more recently [27] that, when taken in terms of other variables (for instance T and σ) the associated Legendre transform γ exhibits delayed evolution involving a continuation functional $\dot{\gamma}_{T,\sigma}$. But this functional degenerates into a function explicitly known. Thus the behaviour is explicitly known for every kind of experiment (isothermal creep and so on) from the response to adiabatic relaxation.

6. The use of internal variables

We have also shown [27] more recently from the above formalism that equilibrium properties and the Gibbs equation can be restored in full in out-of-equilibrium situations by use of internal variables, that is variables not involved in universal balance equations. Of course this result is not completely new (cf. for instance [28, 29]). But the conditions of its validity are now explicit, and the number of assumptions to be made (cf. for instance [28]) is reduced. For instance, Gibbs equation is no more an assumption or a mathematical trick, but a consequence of classical first and second principles of thermodynamics applied to a concrete physical system of known constitution.

As a matter of fact, all internal variables must be Markovian of order one. Thus, the reason for which the method holds is obvious since, for Markovian behaviour, entropy is a state function of the Gibbs variables for each subsystem and its entropy production rate can hence be explicitly calculated. In experiments such as adiabatic relaxation, internal variables involve spontaneous change because we are not able to control them from the boundary of the specimen even in homogeneous conditions.

7. Toward a new approach: thermodynamics in natural variables

When studying Rational Thermodynamics, one may be struck by one fact: the thermodynamic variables, such as s and T, that are introduced as arguments or functions in the set of constitutive equations necessary for closure of the set of universal conservation equations, are completely heterogeneous to these conservation equations themselves. Thus this introduction looks rather artificial, and almost irrational, since it is made only heuristically in the way generally used in other branches of physical science, and involves unavoidably prejudicial assumptions (as a matter of fact, need for constitutive relations prevents continuum mechanics to be built on a completely rational basis).

Hence, in the classical formalism of thermomechanics, implicit choices are made by the very choice of arguments that are used in constitutive equations. This restricts the number of classes of behaviours that can be represented by the formalism resulting from a given choice. We have proposed recently [30] to avoid these restrictions by introducing explicitly (as in TIP cf. [23]) the entropy balance equation in the set of universal equations.

Thus we get one more independent variable. We proposed to take the natural set (ρ_0, ε, u, s) of variables to be the fundamental set, at least as a point of departure. We call it the natural set (or natural combination) of arguments because each argument corresponds to one balance equation: ρ_0 to mass, ε to momentum (by sym(grad v) as required by objectivity, the velocity v being interpreted as the massic momentum), u to energy and s to entropy. In this approach, entropy production rate must be given as a constitutive equation in terms of (ρ_0, ε, u, s).

By use of this unusual set of independent variables we have shown [30] that the whole classical formalism and preceeding results can be rebuilt, by elimination of one of the variables u or s.

But it seems to us that the formalism can be developed for itself and could exhibit interesting properties.

For instance we can see very easily in this new formalism that, in this set of variables, T, σ and γ_i exhibit no delayed response: when the set (ρ_0, ε, u, s) is fixed, so is the set (T, σ, γ_i), instantaneously. The problem is thus reduced to a dynamic one, without functionals coming from constitutive equations.

For a few features, this approach has some resemblance with one proposed independently by Green and Naghdi (cf. [31]). But it is still a heuristical proposal and we have to go further in the study to know if it can truly be useful.

8. Concluding remarks

We have seen that, for the application of thermodynamics to rheological behaviour, we don't need any new principle such as the local state principle or the local equilibrium principle, or an out-of-equilibrium Gibbs equation. We need only the classical principles of mechanics and thermodynamics, and experimental evidence.

In the formalism we have developed, TIP and derived theories take place in classical thermomechanics, either as specific to a peculiar class of behaviour that can be identified experimentally, or as resulting from the introduction of internal variables, that must exhibit, themselves, specific properties that can be used as criteria.

For Markovian behaviour, reversible accessibility of out-of-equilibrium state holds. But it must be understood in a restrictive fashion – only for the Gibbs space coordinates, and not for stress and temperature. It is impossible to establish reversibly the out-of-equilibrium stress of a viscous fluid, a result that leads us to prefer our analysis to the formalism developed, for instance, by Lambermont and Lebon in [32]. As one can easily see, it is only the Markovian character of the subsystems involved in [32] that provides their approach with consistency. This consistency cannot be granted unless this specific condition is not explicitly stated. Actually, in the model considered in [32], the dashpot is unloaded, and thus the system is not in an out-of-equilibrium situation.

In the preceding development, we had to consider if the material response to our experiment is instantaneous or delayed. Of course, the conclusion is only relative to the choice of a time scale, and with regard to the dynamic characteristics of the apparatus involved in the experiment.

This could be the beginning, if needed, of the rehabilitation of the Deborah number, this pillar of rheological folklore knocked down by Truesdell in his Naple's speech to world rheologists.

References

1. Domingos, J. J., M. N. R. Nina, J. H. Whitelaw (eds.), Foundations of continuum thermodynamics (BUSSACO International Symposium, Portugal, 1973), Mac Millan Press (London 1974).
2. Comte Rumford. Mémoires sur la chaleur, Firmin-Didot (Paris 1804).
3. Carnot, S., Réflexions sur la puissance motrice du feu, Bachelier (Paris 1824), Gauthier-Villars (Paris 1878).
4. Meyer, J. R., Bemerkungen über die Kräfte der unbelebten Natur, Liebigs Ann. Chem. **42**, 233–240 (1842).
5. Clausius, R., Théorie Mécanique de la chaleur, 2 volumes, Lacroix (Paris 1854–1868).
6. Massieu, F., C. R. Acad. Sc., Paris **69**, 858 (1869).
7. Boltzmann, L., Sitzgsber. Akad. Wiss. **63** (2), (1871).
8. Gibbs, J. W., On the equilibrium of heterogeneous substances, Trans. Connecticut Acad. **3**, 108 (1875); Cf. also the french translation by Le Chatelier, Paris, and: Diagrammes et surfaces thermodynamiques, trad. Roy. Scientia, Phys. Math. n° 22 (Paris 1903).
9. Helmholtz, H., Sitzgsber. Akad. Wiss., Berlin 1.22 (1882).
10. Duhem, P., Le potentiel thermodynamique et ses applications (Paris 1886).
11. De Donder, Th., Leçons de thermodynamique et de chimie physique; réédition Gauthier-Villars (Paris 1920).
12. Onsager, L., Phys. Rev. **37**, 405 (1931); **38**, 2265 (1931).
13. Prigogine, I., R. Defay, Thermodynamique chimique, Desoer (Liège 1944).
14. Prigogine, I., Thermodynamique des phénomènes irréversibles, Desoer, (Liège 1947).
15. Meixner, J., H. G. Reik, Thermodynamik der irreversiblen Prozesse, in: S. Flügge (ed.), Encyclopedia of Physics, Volume III/2, Principle of thermodynamics and statistics. Springer-Verlag (Berlin 1959).
16. Truesdell, C., W. Noll, The non-linear field theories of Mechanics, Springer-Verlag (Berlin 1965).
17. Truesdell, C., Rational thermodynamics, MacGraw-Hill (New York 1969).
18. Coleman, B. D., Arch. Rational Mech. Anal. **17**, 1–46 (1964).
19. Day, W. A., The thermodynamics of simple materials with fading memory. Springer-Verlag (Paris 1972).
20. Truesdell, C., Sketch for a history of constitutive relations; in: G. Astarita, G. Marruci, L. Nicolais (eds.), Rheology (VIII. Intern. Congr. Rheology, Naples, September 1–5, 1980), Volume 1, p. 1–27, Plenum Press (New York 1980).
21. Huet, C., Sur la notion d'état local en Rhéologie (12e Colloque du Groupe Francais de Rhéologie, Thermodynamique des Comportements rhéologiques, Paris, 7–8 Decembre 1977, Cahiers du Groupe Français de Rhéologie, n° 4, 1979, p. 611–651).
22. Duhem, P., Traité d'énergétique, Gauthier-Villars (Paris 1911).
23. Glansdorff, P., I. Prigogine, Structure, stabilité et fluctuations, Masson (Paris 1971).
24. Germain, P., Cours de Mécanique des Milieux continus, p. 143–145, Masson (Paris 1973).
25. Mandel, J., Introduction à la mécanique des milieux continus déformables, p. 97, p. 104, Editions scientifiques de Pologne (Varsovie 1974).
26. Huet, C., Remarques sur la procédure d'assimilation d'un matériau hétérogène à un milieu continu équivalent (15e Colloque Annuel du Groupe Francais de Rhéologie, Paris 3–4–5 Décembre 1980, Editions Anciens ENPC, Paris, 1981).
27. Huet, C. (unpublished result, 1979).
28. Meixner, J., Entropy and entropy production, in [1], p. 129–142.
29. Bataille, J., J. Kestin, J. Mécanique **14**, 365–384 (1975).
30. Huet, C., Thermodynamique des comportements rhéologiques en variables naturelles (1980–1981, submitted for publication).
31. Naghdi, P. M., Energy **5**, 771–781 (1980).
32. Lambermont, J., G. Lebon, Int. J. Non-linear Mechanics **9**, 55–74 (1974).

(Received May 17, 1982)

Author's address:

Prof. Dr. C. Huet
Centre d'Enseignement et de Recherche
en Matériaux de Construction
Ecole Nationale des Ponts et Chaussées
28 rue des Saints-Pères
F-75007 Paris

Rheologica Acta Rheol. Acta **21**, 366–375 (1982)

A unified approach to a variety of constitutive models for polymer fluids based on the concept of configuration-dependent molecular mobility*)

H. Giesekus

Abteilung Chemietechnik der Universität Dortmund

Abstract: The concept of a configuration-dependent tensorial drag (or mobility) coefficient developed some sixteen years ago is presented in a more generalized manner. In particular, the special case of a one-mode model is discussed. This is shown to be equivalent to the one-mode Leonov model as well as to a recently published model of Dashner and VanArsdale.

For an isotropic mobility tensor, the model is shown to include not only the simple upper convected Maxwell model but also those generalizations in which the shear modulus and the relaxation time are introduced as functions of either the present configuration or the configuration history, the latter case covering the model of Acierno et al. The introduction of a non-isotropic mobility, which is linearly related to the configuration tensor, enables at least a simple qualitative description of a number of well-known flow properties to be given; namely, finite extensional viscosity, shear thinning and non-disappearing second normal-stress difference, non-linear relaxation and stress-overshoot in transient shear flows. It is further shown that models in which the shear modulus and the relaxation time are introduced as functions of the rate of strain tensor, or in which objective time derivatives other than the upper convected derivative are used, are not compatible with this approach.

Finally, a short discussion of multi-mode models is given in which emphasis is placed on the problem of whether the different modes can be considered to be independent, as is the case in nearly all existing models, or whether the concept of entangled network structures requires coupling of these modes. The latter would mean that mobility tensors and shear moduli of every configurational mode would have to be considered to depend on the full set of configuration tensors.

Key words: Constitutive model, polymer fluid, entangled network structure, configuration tensor, tensorial mobility

1. Introduction

The aim of this paper is to present a unified approach, which seems to be relatively unknown, by which a wide variety of constitutive models for polymer fluids can be classified, in spite of the fact that these models were originally derived in quite different ways. This approach is based on the concept of a configuration-dependent tensorial drag coefficient or (the reciprocal of this quantity) a configuration-dependent tensorial mobility, which governs the motion of the structural elements of the molecules relative to their surroundings under the influence of elastic tractions. This approach was first presented sixteen years ago [1] but is given here in a somewhat modified and more generalized manner.

*) Invited paper, presented at the First Conference of European Rheologists at Graz (Austria), April 14–16, 1982.

814

2. The polymer fluid model

Polymer molecules are idealized by unbranched or branched chains of structural elements called "beads" which are joined either by elastic or by rigid connections (often called "springs" and "rods", respectively) and which are additionally subjected to "Brownian motion" forces. Regardless of whether these molecules are dissolved in a liquid or form a melt, the motion of the beads relative to the surrounding beads of the same or neighbouring molecules may occur by entanglement loss and regeneration processes.

Deformation of a polymer fluid results in a macroscopic motion which may be described in the Eulerian form by the motion of a continuum with local velocity $\boldsymbol{v}(\boldsymbol{x})$ and local deformation gradient

$$\boldsymbol{L} = (\nabla \boldsymbol{v})^{\dagger} = \boldsymbol{D} + \boldsymbol{W} \tag{1}$$

where the symmetric part $\boldsymbol{D}$ designates the rate of strain tensor and the antisymmetric part $\boldsymbol{W}$ the rate of rotation tensor. Using this quantity, the material time derivative $d\boldsymbol{v}$ of an infinitesimal vector $d\boldsymbol{x}$ embedded in the continuum is given as

$$d\boldsymbol{v} = \frac{D}{Dt} d\boldsymbol{x} = d\boldsymbol{x} \cdot \nabla \boldsymbol{v} = \boldsymbol{L} \cdot d\boldsymbol{x} \,. \tag{2}$$

This can be formulated as

$$\frac{\mathfrak{D}}{\mathfrak{D}t} d\boldsymbol{x} = \frac{D}{Dt} d\boldsymbol{x} - \boldsymbol{L} \cdot d\boldsymbol{x} = 0 \,, \tag{2a}$$

$\mathfrak{D}/\mathfrak{D}t$ being called the "upper (or contravariant) convected time derivative", which is thus seen to signify the rate of a vectorial quantity relative to the motion of the continuum.

Now consider a polymer molecule in which the instantaneous loci of the beads are given by position vectors $\boldsymbol{r}_i$ taken from a point moving with the continuum, preferably from the centre of resistance (fig. 1). Because of the action of the above-mentioned intermolecular and Brownian forces, the motion of the beads is expected to differ systematically from the motion of the continuum, i.e.

$$\frac{\mathfrak{D}\boldsymbol{r}_i}{\mathfrak{D}t} = \frac{D\boldsymbol{r}_i}{Dt} - \boldsymbol{L} \cdot \boldsymbol{r}_i \neq 0 \,. \tag{3}$$

Here the $\boldsymbol{r}_i$ are treated as "infinitesimal" vectors like $d\boldsymbol{x}$; in other words, the local deformation gradient varies only macroscopically but is understood to be constant over molecular distances.

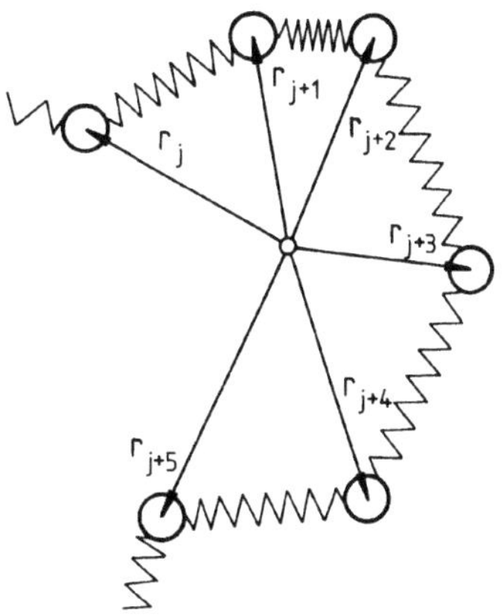

Fig. 1. Portion of a polymer molecule (bead-spring model)

The relationship between the relative motion and the generating force $\boldsymbol{f}_i$ may be quite generally described by a tensorial drag coefficient ζ_i as follows:

$$\boldsymbol{f}_i = \zeta_i \cdot \frac{\mathfrak{D}\boldsymbol{r}_i}{\mathfrak{D}t} = \frac{\mathfrak{D}\boldsymbol{r}_i}{\mathfrak{D}t} \cdot \zeta_i^{\dagger} \,. \tag{4}$$

This is a straight-forward generalization of Stokes' law. In the special case of the motion of a sphere with radius a_i in an isotropic fluid with viscosity η we would obtain $\zeta_i = 6\pi\eta a_i \mathbf{1}$.

In contrast to the action of intermolecular forces between the beads in connection with the undisturbed motion in a solvent which results in a continuous process, entanglement loss and regeneration as well as Brownian motion are stochastic processes resulting in irregular fluctuations. Therefore $\boldsymbol{f}_i$ and $\mathfrak{D}\boldsymbol{r}_i/\mathfrak{D}t$ should not be understood to be actual instantaneous quantities but rather expectation values taken over a large ensemble of equivalent beads or, in short, as "smoothed-out" forces and velocities.

As is well known, the forces $\boldsymbol{f}_i$ acting on the beads must be compensated by macroscopic stresses which can be described by a set of not necessarily symmetric excess stress tensors

$$\boldsymbol{S}_i = -n_i \langle \boldsymbol{r}_i \boldsymbol{f}_i \rangle = -n_i \left\langle \boldsymbol{r}_i \frac{\mathfrak{D}\boldsymbol{r}_i}{\mathfrak{D}t} \cdot \zeta_i^{\dagger} \right\rangle \,. \tag{5}$$

Here n_i is the number of beads of the i-th kind in the unit volume, and the arrow-shaped brackets designate averages over the configuration space of vectors $\boldsymbol{r}_i$ determined by a distribution function $\psi(\boldsymbol{r}_1, \ldots, \boldsymbol{r}_N)$

which is subjected to some "diffusion equation" defining the evolution of the molecular configuration, cf. e.g. [2].

3. Configuration dependence of the drag coefficient

Now the question arises: On which quantities does the drag tensor ζ_i depend? Because the structural elements are assumed to be spherical beads there is no dependence on their orientation as would be the case with non-spherical elements, but only on the *surrounding* beads. In dilute solutions this dependence is restricted only to the beads belonging to the respective molecule, but in concentrated systems there exists an interaction with other molecules as well. This makes a rigorous treatment of the problem extremely difficult, if not impossible. Therefore hypotheses are used to define some more idealized systems for which the problem under consideration is tractable.

3.1. Hypotheses of reptational motion

A first class of hypotheses starts from an idea of de Gennes [3] that the motion of a polymer molecule essentially occurs only along a tube around its actual configuration, therefore figuratively called "reptational motion". This concept was further elaborated by Doi and Edwards [4]. However, recently Curtiss and Bird [5] showed that this hypothesis can be reformulated in such a way that a drag tensor ζ_i is introduced, which depends (just as in dilute systems) on the position vectors $\boldsymbol{r}_i$ of the respective molecule only and, to a first approximation, on the connectors between the two neighbouring spheres, i.e. $\boldsymbol{R}_i = \boldsymbol{r}_i - \boldsymbol{r}_{i-1}$ and $\boldsymbol{R}_{i+1} = \boldsymbol{r}_{i+1} - \boldsymbol{r}_i$. With some additional simplifications, for example, that all beads are of equal size and the joints between them are rigid and of equal length (i.e. the Kramers freely jointed bead-rod chain model), the so called "mild curvature assumption", and the hypothesis that the dependence of ζ_i on the connector orientation is linear, this concept could be used to give definite predictions with respect to different flow properties which look quite reliable. But the calculations, in which not only averages of the type of second-order tensors $\langle \boldsymbol{R}_i\boldsymbol{R}_i \rangle$ but also of fourth-order tensors $\langle \boldsymbol{R}_i\boldsymbol{R}_i\boldsymbol{R}_i\boldsymbol{R}_i \rangle$ have to be evaluated by means of the distribution function $\psi(\boldsymbol{R}_i)$, which therefore must be explicitly known, are very laborious and as a consequence only quite simple types of flow can be analysed.

3.2. Hypotheses of a drag coefficient dependent only on the average molecular configuration

A second class of hypotheses postulates, in contrast to the first one, that the drag tensor does not depend on the *actual* configuration of the molecule under consideration but only on the *average* configuration of all molecules. In this case ζ_i can be taken outside the brackets of eq. (5):

$$\boldsymbol{S}_i = -n_i \left\langle \boldsymbol{r}_i \frac{\mathfrak{D} \boldsymbol{r}_i}{\mathfrak{D} t} \right\rangle \cdot \zeta_i^{\dagger} . \tag{6}$$

If this equation is multiplied by ζ_i^{-1} and added to its transpose the following expression results:

$$\zeta_i^{-1} \cdot \boldsymbol{S}_i^{\dagger} + \boldsymbol{S}_i \cdot \zeta_i^{-1\dagger}$$

$$= -n_i \left\{ \left\langle \frac{\mathfrak{D} \boldsymbol{r}_i}{\mathfrak{D} t} \boldsymbol{r}_i \right\rangle + \left\langle \boldsymbol{r}_i \frac{\mathfrak{D} \boldsymbol{r}_i}{\mathfrak{D} t} \right\rangle \right\} = -n_i \frac{\mathfrak{D}}{\mathfrak{D} t} \langle \boldsymbol{r}_i \boldsymbol{r}_i \rangle \tag{7}$$

where

$$\frac{\mathfrak{D}}{\mathfrak{D} t} \langle \boldsymbol{r}_i \boldsymbol{r}_i \rangle = \frac{D}{D t} \langle \boldsymbol{r}_i \boldsymbol{r}_i \rangle - \{ \boldsymbol{L} \cdot \langle \boldsymbol{r}_i \boldsymbol{r}_i \rangle + \langle \boldsymbol{r}_i \boldsymbol{r}_i \rangle \cdot \boldsymbol{L}^{\dagger} \} \tag{8}$$

denotes the upper convected time derivative of a second-order tensor which is not identical with the respective formula for a vector, cf. eq. (3).

To every molecular configuration there can be associated an isotropic equilibrium configuration characterizing the state of no deformational motion at all:

$$\langle \mathring{\boldsymbol{r}}_i \mathring{\boldsymbol{r}}_i \rangle = \tfrac{1}{3} \langle \mathring{\boldsymbol{r}}_i \cdot \mathring{\boldsymbol{r}}_i \rangle \, \boldsymbol{1} . \tag{9}$$

Then the actual configuration can be described as a mapping from the equilibrium configuration by means of a "configuration tensor" $\boldsymbol{b}_i$ as follows:

$$\langle \boldsymbol{r}_i \boldsymbol{r}_i \rangle = \boldsymbol{b}_i \cdot \langle \mathring{\boldsymbol{r}}_i \mathring{\boldsymbol{r}}_i \rangle = \tfrac{1}{3} \langle \mathring{\boldsymbol{r}}_i \cdot \mathring{\boldsymbol{r}}_i \rangle \boldsymbol{b}_i . \tag{10}$$

This tensor may be understood to be a measure of deformation of an elastic continuum, though not in the strict sense of a *material* continuum but in a statistical sense, the respective continua representing only configurational states of a polymer chain. This does not imply the restriction that every individual vector $\mathring{\boldsymbol{r}}_i$

has to be mapped into an associated vector $\boldsymbol{r}_i$ by an affine transformation[1]).

With this interpretation we need no longer assume that there has to be an individual configuration tensor $\boldsymbol{b}_i$ and drag tensor ζ_i associated to every position vector $\boldsymbol{r}_i$. Instead the whole set of beads ($i = 1,\ldots,N$) can be classified into classes ($k = 1,\ldots,K$) with n_k beads per unit volume with a common configuration tensor $\boldsymbol{b}_k$ and drag tensor ζ_k for each class.

Now it is appropriate to define a relative mobility tensor $\boldsymbol{\beta}_k$ as follows:

$$\zeta_k^{-1} = \boldsymbol{\beta}_k \mathring{\zeta}_k^{-1} \tag{13}$$

which means that in the state of equilibrium $\mathring{\boldsymbol{\beta}}_k = \mathbf{1}$. With this definition and the above mentioned interpretation, eq. (7) simplifies to give

$$\frac{1}{2}[\boldsymbol{\beta}_k \cdot \boldsymbol{S}_k^{\dagger} + \boldsymbol{S}_k \cdot \boldsymbol{\beta}_k^{\dagger}] + \mathring{\eta}_k \frac{\mathfrak{D}\boldsymbol{b}_k}{\mathfrak{D}t} = 0 \tag{14}$$

where the "equilibrium viscosity" $\mathring{\eta}_k$ is

$$\mathring{\eta}_k = \tfrac{1}{6} n_k \overline{\langle \mathring{\boldsymbol{r}}_i \cdot \mathring{\boldsymbol{r}}_i \rangle}_k \mathring{\zeta}_k \tag{15}$$

and $\overline{\langle \mathring{\boldsymbol{r}}_i \cdot \mathring{\boldsymbol{r}}_i \rangle}_k$ is the average taken over the respective class of beads.

4. One-mode model

Classification of the total number of structure elements into K classes is intended to cover systems with a wider molecular weight distribution, different kinds of molecules, block or graft copolymers, etc. However, before investigating such rather complicated types of models it seems appropriate to analyse first the simplest type with $K = 1$, called the one-mode model, in which the index k in eqs. (13 – 15) can be removed. Despite this considerable simplification there is some hope that the essential properties of the respective models will be retained.

[1]) If this were the case, i.e. if

$$\boldsymbol{r}_i = \chi_i \cdot \mathring{\boldsymbol{r}}_i = \mathring{\boldsymbol{r}}_i \cdot \chi_i^{\dagger}, \tag{11}$$

then

$$\langle \boldsymbol{r}_i \boldsymbol{r}_i \rangle = \chi_i \cdot \langle \mathring{\boldsymbol{r}}_i \mathring{\boldsymbol{r}}_i \rangle \cdot \chi_i^{\dagger} = \tfrac{1}{3} \langle \mathring{\boldsymbol{r}}_i \cdot \mathring{\boldsymbol{r}}_i \rangle \chi_i \cdot \chi_i^{\dagger} \tag{10a}$$

would result, allowing one to identify

$$\boldsymbol{b}_i = \chi_i \cdot \chi_i^{\dagger} \tag{12}$$

as the Finger deformation tensor associated with this mapping.

The question now arises as to the way in which the tensors $\boldsymbol{s}$, $\boldsymbol{\beta}$ and $\boldsymbol{b}$ would depend on the configuration of the molecules and how they would be connected with one another. In principle this problem would have to be solved by means of a detailed kinetic analysis based on the evaluation of the distribution function. However, the concept of an elastic fluid in connection with the theory of elastic polymer networks suggests that the configuration tensor $\boldsymbol{b}$ would give a complete representation of the configurational state (often called the "recoverable strain") and could therefore be understood as the only independent inner variable of the deformational motion[2]), whereas $\boldsymbol{S}$ and $\boldsymbol{\beta}$ depend on $\boldsymbol{b}$ in a way characterizing the special material.

If this hypothesis is accepted, $\boldsymbol{S}$ and $\boldsymbol{\beta}$ have to be symmetric tensors, which are commutable with $\boldsymbol{b}$ and, correspondingly, with one another, and then eq. (14) can be further simplified to

$$\boldsymbol{\beta}(\boldsymbol{b}) \cdot \boldsymbol{S}(\boldsymbol{b}) + \mathring{\eta} \frac{\mathfrak{D}\boldsymbol{b}}{\mathfrak{D}t} = 0\,, \tag{16}$$

first published in [1].

If a new function

$$\boldsymbol{\varphi}(\boldsymbol{b}) = -\frac{1}{\mathring{\eta}} \boldsymbol{\beta}(\boldsymbol{b}) \cdot \boldsymbol{S}(\boldsymbol{b}) \tag{17}$$

is introduced into this equation, it can be written concisely as

$$\frac{\mathfrak{D}\boldsymbol{b}}{\mathfrak{D}t} = \boldsymbol{\varphi}(\boldsymbol{b})\,. \tag{18}$$

This is equivalent to the basic equation of a phenomenological theory for elastic fluids recently derived by Dashner and VanArsdale [6] using methods of rational mechanics[3]).

At the same time, eqs. (16) and (18) are equivalent to a one-mode specialization of Leonov's constitutive equation [7], which this author obtained with the aid

[2]) In this hypothesis it is implicit that higher moments of configuration vectors, e.g. $\langle \boldsymbol{r}_i \boldsymbol{r}_i \boldsymbol{r}_i \boldsymbol{r}_i \rangle$, are of no importance.

[3]) These authors use $\boldsymbol{c} = \boldsymbol{b}^{-1}$ instead of $\boldsymbol{b}$ as the configuration tensor. Correspondingly, the *lower* convected derivative of this tensor

$$\frac{\mathfrak{d}\boldsymbol{c}}{\mathfrak{d}t} = \frac{D\boldsymbol{c}}{Dt} + [\boldsymbol{L}^{\dagger} \cdot \boldsymbol{c} + \boldsymbol{c} \cdot \boldsymbol{L}] \tag{19}$$

is found to be a function $\boldsymbol{f}(\boldsymbol{c})$. However, it is easily verified that both formulae are completely equivalent with

$$\boldsymbol{f}(\boldsymbol{c}) = -\boldsymbol{c} \cdot \boldsymbol{\varphi}(\boldsymbol{c}^{-1}) \cdot \boldsymbol{c}\,.$$

of some arguments from the thermodynamics of irreversible processes. Correspondingly, Leonov decomposes the velocity gradient $\boldsymbol{L}$ into a reversible part $\boldsymbol{L}_{(e)}$ and an irreversible part $\boldsymbol{L}_{(p)}$. The first part is defined by a condition analogous to the general condition to which the Finger deformation tensor $\boldsymbol{B}$ of any (not necessarily elastic) continuum is subjected, i.e. $\mathfrak{D}\boldsymbol{B}/\mathfrak{D}t = 0$:

$$\left[\frac{\mathfrak{D}\boldsymbol{b}}{\mathfrak{D}t}\right]_{(e)} = \frac{D\boldsymbol{b}}{Dt} - [\boldsymbol{L}_{(e)} \cdot \boldsymbol{b} + \boldsymbol{b} \cdot \boldsymbol{L}_{(e)}^{\dagger}] = 0\,. \quad (20)$$

Then

$$\frac{\mathfrak{D}\boldsymbol{b}}{\mathfrak{D}t} = \left[\frac{\mathfrak{D}\boldsymbol{b}}{\mathfrak{D}t}\right]_{(e)} + \left[\frac{\mathfrak{D}\boldsymbol{b}}{\mathfrak{D}t}\right]_{(p)} = -[\boldsymbol{L}_{(p)} \cdot \boldsymbol{b} + \boldsymbol{b} \cdot \boldsymbol{L}_{(p)}^{\dagger}], \quad (21)$$

where $\boldsymbol{L}_{(p)}$ is understood to be an inner variable of the configurational state and therefore a function of $\boldsymbol{b}$ only. This implies that this tensor must be a symmetric tensor $\boldsymbol{D}_{(p)}$, commutable with $\boldsymbol{b}$, whereas there is no irreversible part associated with the rate of rotation tensor, i.e. $\boldsymbol{W}_{(p)} = \boldsymbol{0}$[4]. Comparison of eq. (21) with eqs. (16) and (18) leads to the equivalence relation

$$\boldsymbol{D}_{(p)} = \frac{1}{2\mathring{\eta}}\boldsymbol{\beta} \cdot \boldsymbol{S} \cdot \boldsymbol{b}^{-1} = -\frac{1}{2}\boldsymbol{b}^{-1} \cdot \boldsymbol{\varphi}(\boldsymbol{b})\,. \quad (22)$$

There is an additional, quite plausible hypothesis which was first mentioned in [1] and is elaborated in more detail by Leonov as well as by Dashner and VanArsdale, namely that the tensor of excess stress

$$\boldsymbol{S}(\boldsymbol{b}) = s_1\boldsymbol{b} + s_0\boldsymbol{1} + s_{-1}\boldsymbol{b}^{-1} \quad (23)$$

is associated with a Helmholtz free-energy function or elastic potential $W(\boldsymbol{b}) = W(\mathrm{I}_b, \mathrm{II}_b, \mathrm{III}_b)$, where $\mathrm{I}_b, \mathrm{II}_b, \mathrm{III}_b$ are the principal invariants of $\boldsymbol{b}$, in the following way:

$$s_1 = 2\,\mathrm{III}_b^{-1/2}\frac{\partial W}{\partial \mathrm{I}_b}, \quad s_0 = 2\,\mathrm{III}_b^{-1/2}\left[\mathrm{II}_b\frac{\partial W}{\partial \mathrm{II}_b} + \mathrm{III}_b\frac{\partial W}{\partial \mathrm{III}_b}\right],$$

$$s_{-1} = -2\,\mathrm{III}_b^{1/2}\frac{\partial W}{\partial \mathrm{II}_b}\,. \quad (24)$$

This corresponds to the constitutive model of a so-called hyperelastic material, which is thus postulated to hold for the configuration tensor of the fluid under consideration in the same way as for the deformation tensor of a solid elastic polymer network.

It has been tacitly assumed that the polymer fluid under consideration represents an incompressible material. This leads some authors to conclude that the configuration volume characterized by the third invariant III_b of the configuration tensor should also have this property, i.e. $\mathrm{III}_b = 1$[5], but no intelligible reasoning for this can be seen. In contrast to the material volume, the configuration volume does not have the property of impenetrability, as is evident from the underlying model of a transient network of interpenetrating entangled molecular chains. Of course, there might be special reasons for introducing such a constraint because of special model considerations but it cannot be assumed a priori.

In connection with this, the excess stress tensor $\boldsymbol{S}(\boldsymbol{b})$ is sometimes understood to necessarily represent a deviatoric tensor, i.e. $\mathrm{tr}\boldsymbol{S} = 0$, but this also seems to be a misconception. Of course, instead of $\boldsymbol{S}$ one can always introduce $\mathrm{dev}\,\boldsymbol{S} = \boldsymbol{S} - \frac{1}{3}(\mathrm{tr}\,\boldsymbol{S})\boldsymbol{1}$ into the complete constitutive equation without doing any harm. But if $\boldsymbol{S}$ itself is subjected to such a restriction the one-to-one correspondence, otherwise expected to exist between $\boldsymbol{b}$ and $\boldsymbol{S}$, is lost. As a consequence it is then no longer possible to consider $\boldsymbol{S}$ rather than $\boldsymbol{b}$ to be the independent inner variable and correspondingly to interpret $\boldsymbol{b} = \boldsymbol{b}(\boldsymbol{S})$ and $\boldsymbol{\beta} = \boldsymbol{\beta}(\boldsymbol{S})$ which is sometimes advantageous, cf. sect. 5.

5. Some special one-mode models

To proceed to a really applicable model some more stringent specializations have to be introduced. The theory of rubber elasticity suggests that the so-called neo-Hookean material is also a reasonable first approximation for a transient polymer network:

$$\boldsymbol{S} = \mathring{G}(\boldsymbol{b} - \boldsymbol{1}) \quad (25)$$

where $\mathring{G}$ is a constant shear modulus. The simplest hypothesis with respect to the mobility is, of course, $\boldsymbol{\beta} = \boldsymbol{1}$, i.e. an isotropic constant drag. Introducing this expression together with eq. (25) in eq. (16) results in

$$\boldsymbol{b} + \mathring{\lambda}\frac{\mathfrak{D}\boldsymbol{b}}{\mathfrak{D}t} = \boldsymbol{1}\,, \quad \mathring{\lambda} = \mathring{\eta}/\mathring{G}\,. \quad (26)$$

[4]) Leonov introduces this as a matter of course although it can only be derived easily from the aforementioned dependence on the symmetric tensor $\boldsymbol{b}$. If, in a more general case, each member of a sequence of tensors $\boldsymbol{L}_{(p)k}$ were to depend on the full sequence of tensors $\boldsymbol{b}_k$, this problem would be more complicated, cf. sect. 6.

[5]) Corresponding to this hypothesis the expressions (24) are given in [7] in a more specialized form, derived from these by setting $\mathrm{III}_b = 1$ and $s_0 = -p^*$, respectively.

If, by inversion of eq. (25), S is substituted as the variable, one immediately obtains (with $\mathfrak{D}\mathbf{1}/\mathfrak{D}t = -2\boldsymbol{D}$) the constitutive equation of an upper convected Maxwell fluid:

$$S + \mathring{\lambda}\frac{\mathfrak{D}S}{\mathfrak{D}t} = 2\mathring{\eta}\boldsymbol{D}\,. \tag{26a}$$

This is also the one-mode special case of Lodge's network model [8].

The merits of this model are as well known as its shortcomings. The most important of the latter are: in steady viscometric flows a constant shear viscosity and first normal-stress coefficient together with a disappearing second normal-stress difference, and in uni- and bi-axial extensional flow an unrestrictedly growing extensional viscosity with infinite limiting values at finite rates of extension. If these insufficiencies are to be overcome appropriate generalizations must be introduced.

5.1. *Models with isotropic mobility*

The most obvious generalization is the substitution of the constant mobility and constant shear modulus by configuration dependent quantities, but retaining the isotropic character of $\boldsymbol{\beta}$:

$$\boldsymbol{\beta} = \beta(\boldsymbol{b})\mathbf{1}\,, \quad G = G(\boldsymbol{b}) \tag{27}$$

where in these scalar functions $\boldsymbol{b}$ stands for the invariants I_b, II_b and III_b of the configuration tensor. Then eq. (16) can be written in two different ways:

$$\boldsymbol{b} + \lambda(\boldsymbol{b})\frac{\mathfrak{D}\boldsymbol{b}}{\mathfrak{D}t} = \mathbf{1}\,, \quad \lambda(\boldsymbol{b}) = \mathring{\eta}/\beta(\boldsymbol{b})G(\boldsymbol{b}) \tag{28}$$

and

$$[S/G(S)] + \lambda(S)\frac{\mathfrak{D}[S/G(S)]}{\mathfrak{D}t} = 2\lambda(S)\boldsymbol{D}\,, \tag{28a}$$

respectively. For practical use some simple specializations for $\lambda(\boldsymbol{b})$ and $G(\boldsymbol{b})$ or $\lambda(S)$ and $G(S)$ must, of course, be introduced. This generalized Maxwell fluid is also well-known. In particular, it is intimately related with the Kaye model based on the idea of stress-dependent network properties [9].

In this modified model the idea is introduced that the drag coefficient, as well as the shear modulus, depend on the entanglement density, itself being a function of the configuration tensor. But a more detailed view of entanglement loss and regeneration processes leads to the concept that the entanglement density is not only a function of the instantaneous configuration but also depends on the configuration history. In the simplest case quasi-linear differential equations are proposed:

$$\frac{d\lambda}{dt} = \Lambda(\lambda, \boldsymbol{b})\,, \quad \frac{dG}{dt} = \Gamma(G, \boldsymbol{b})\,. \tag{29}$$

Furthermore, the time dependences of both quantities may be coupled by means of a "structural parameter" x to give

$$\lambda = \lambda_0\Phi(x)\,, \quad G = G_0\Psi(x)\,, \quad \frac{dx}{dt} = \Xi(x, \boldsymbol{b})\,. \tag{30}$$

The model of Acierno, La Mantia, Marrucci and Titomanlio [10] is of this type.

5.2. *Models with non-isotropic mobility*

By the foregoing generalizations some essential shortcomings of the upper convected Maxwell model are overcome, i.e. these fluids are able to describe shear-thinning effects, a finite limiting behaviour in extensional flow and non-linear relaxation. Moreover, the last-mentioned model also includes stress-overshoot in start-up flows. But all these fluids continue to be Weissenberg fluids, i.e. they show, besides less manifest restrictions, a disappearing second normal-stress difference in steady viscometric flows. To overcome this unrealistic feature a *non-isotropic* mobility tensor $\boldsymbol{\beta}(\boldsymbol{b})$ has to be introduced.

As has been described elsewhere [11, 12], under quite general conditions the second normal-stress difference N_2 is proportional to $[(\beta_{\mathrm{I}}/\beta_{\mathrm{II}}) - 1]$ where β_{I} and β_{II} designate the proper values of $\boldsymbol{\beta}$ in the plane of the shearing motion. But to obtain models which are useful for making predictions in more general flow situations, a special hypothesis about the quantitative dependence of $\boldsymbol{\beta}$ on $\boldsymbol{b}$ must be introduced. The simplest, non-trivial model of this kind is given by the linear relation

$$\boldsymbol{\beta} = \mathbf{1} + \alpha(\boldsymbol{b} - \mathbf{1})\,, \quad 0 \leqslant \alpha \leqslant 1 \tag{31}$$

in addition to a linear dependence of S on $\boldsymbol{b}$ given in eq. (25). Thus we arrive at a constitutive equation:

$$[1 + \alpha(b - 1)] \cdot (b - 1) + \overset{\circ}{\lambda} \frac{\mathfrak{D} b}{\mathfrak{D} t} = 0\,, \qquad (32)$$

including the upper convected Maxwell equation as the limiting case when $\alpha = 0$.

Notwithstanding the fact that this represents a non-linear equation, it has analytic solutions, for instance, in the case of steady shear and extensional flows and also for relaxation and extensional start-up flow. Additionally, start-up in simple shear flow can be evaluated by a routine computer program. A detailed analysis of these flows is given in [12]. With $\alpha > 0$ the extensional (or Trouton) viscosity tends to finite limiting values for positive as well as negative rates of extension, i.e. uniaxial and biaxial extensional flows, the limiting values as well as the gradients of these curves being related in a simple way to the parameter α. But it is this parameter which also determines the initial value of the ratio of the second to the first normal-stress differences in simple shear flow. For this type of flow the model predicts shear-thinning for both the shear-stress and the normal-stress differences which depend in a characteristic way on α. Indeed, these effects are most pronounced with $\alpha = 1$, i.e. $\beta = b$, but deviations from the upper convected Maxwell model ($\alpha = 0$) are already quite marked for small values of α. This is remarkable as such models can be expected to describe the real behaviour of polymer fluids best, because the small values of $N_2/N_1 \approx 0.1$ which have always been found experimentally, correspond to $\alpha \approx \frac{1}{5}$.

Furthermore, this model gives non-linear relaxation curves, but the predictions with respect to start-up follows seem to be the most interesting ones: In simple extensional flows monotonically increasing curves are always found, whereas in simple shear flow stress-overshoot followed by a small undershoot is predicted to occur when a critical value of the normalized rate of shear is exceeded, the character of which is quite different for the shear-stress and the normal-stress differences [13].

In principle it is not difficult to generalize this model in a way similar to the preceding ones, i.e. by consideration of a configuration-dependent entanglement density, leading to a shear modulus and a relaxation time which are either functions of b or of the configuration history. Also more general types of dependences of S on b according to eqs. (23) and (24) may be introduced, as has been discussed in particular by Leonov and Prokunin [14]. But, of course, the simplicity of the model is then lost and only numerical evaluation would be possible, even for the simplest flows.

5.3. Models not included in the concept of a configuration-dependent mobility

As has been shown, a wide class of models are included in the above approach. But there are also well-known models which are obviously not covered by it at all. This refers, for example, to models in which G and λ are understood to be functions of D, instead of S or b, and also to models in which the upper convected derivative of S is substituted by a different objective time derivative

$$\frac{\mathbb{D}_{(\xi)} S}{\mathbb{D} t} = \frac{\mathfrak{D} S}{\mathfrak{D} t} + \xi [D \cdot S + S \cdot D]\,, \qquad (33)$$
$$0 \leqslant \xi \leqslant 2$$

which includes the well-known special cases of the lower convected time derivative ($\xi = 2$) and the co-rotational or Jauman time derivative ($\xi = 1$). With these derivatives different types of Maxwell models have been constructed, of which the classical model of Zaremba-Fromm-DeWitt and Oldroyd's model A (being only a slight generalization) are the best known, but the Oldroyd eight-constant model [15] and the model recently proposed by Phan-Tien and Tanner [16] are also of this type. This can be understood by the fact that in our approach the beads are caused to move relative to the continuum so that the relative velocity is determined not by D but rather by its irreversible part $D_{(p)}$. From the relation

$$(D - D_{(p)}) \cdot b + b \cdot (D - D_{(p)}) = \frac{Db}{Dt} - [W \cdot b - b \cdot W] \qquad (34)$$

which can easily be derived from eq. (20), it can be seen at once that D and $D_{(p)}$ differ not only in transient flows ($Db/Dt \neq 0$) but also in flows with constant stretch history, provided that the material rotates under the deformational motion according to the rate of rotation tensor W.

Our constitutive equation (32) can be reformulated in terms of the time derivative (33) as follows:

$$S + \overset{\circ}{\lambda} \frac{\mathbb{D}_{(\alpha)} S}{\mathbb{D} t} + \frac{\alpha}{2\overset{\circ}{G}} [S \cdot (S - 2\overset{\circ}{\eta} D) + (S - 2\overset{\circ}{\eta} D) \cdot S] = 2\overset{\circ}{\eta} D\,. \qquad (35)$$

Thus it is obvious that this material cannot be distinguished from a Maxwell model constructed with the respective objective time derivative if the tensor product $S \cdot (S - 2\mathring{\eta}D)$ has no symmetric part. As has been shown in [12], this is the case with $\alpha = 1$ for simple extensional as well as for simple shear flow, i.e. in these flows the above fluid with $\beta = b$ behaves in the same way as one obeying the *co-rotational* Maxwell model. In contrast, for $0 < \alpha < 1$, this condition is not fulfilled exactly for the above-mentioned flows, but as in every steady flow the respective term in eq. (35) is at least of third order in D and W, the difference may be detectable only with sufficiently high velocity gradients.

If, on the other hand, in a Maxwell fluid of the type discussed above, the time derivative (33) is modified in such a way that D is substituted by $D_{(p)}$ in the additional term on the right-hand side, the following equation is obtained:

$$S + \mathring{\lambda}\left[\frac{\mathfrak{D}S}{\mathfrak{D}t} + \xi(S \cdot D_{(p)} + D_{(p)} \cdot S)\right] = 2\mathring{\eta}D\,. \quad (36)$$

According to eqs. (18) and (22) we have

$$D_{(p)} = -\frac{1}{2}b^{-1} \cdot \frac{\mathfrak{D}b}{\mathfrak{D}t}\,. \quad (37)$$

If this expression is introduced into eq. (36) and S in the term $\mathfrak{D}S/\mathfrak{D}t$ is substituted by b using eq. (25), one obtains the constitutive equation:

$$[1 - \xi(1 - b^{-1})]^{-1} \cdot S + \mathring{\eta}\frac{\mathfrak{D}b}{\mathfrak{D}t} = 0\,. \quad (38)$$

Comparison with eq. (16) results in

$$\beta = [1 - \xi(1 - b^{-1})]^{-1}\,. \quad (39)$$

In the special case $\xi = 1$ this is identical with $\beta = b$, i.e. the above model (32) with $\alpha = 1$.

But because this expression can be rewritten as

$$\beta = 1 + \xi(b - 1) \cdot [1 + (1 - \xi)(b - 1)]^{-1} \quad (39\,a)$$

it is seen at once that also for $0 < \xi < 1$ it differs only slightly from the linear relation (31) (if ξ is identified with α) provided that the configuration tensor b does not deviate too much from the unit tensor.

Correspondingly, more complicated constitutive equations using time derivatives of the type (33) will fit into our approach after introducing the respective substitution.

6. Multi-mode models

Although one-mode models are capable of describing essential features of real polymer systems, for finer details multi-mode models, as defined by eq. (14), are needed. If the hypothesis is made that all modes are independent from one another, i.e.

$$\beta_k = \beta_k(b_k)\,, \quad S_k = S_k(b_k) \quad (40)$$

the respective system of differential equations is uncoupled and each can be treated as a one-mode model, at least in those cases where the deformation history is given. It is this hypothesis which indeed underlies all models mentioned above, as far as they are introduced as multi-mode models at all.

But if we take our concept seriously it must be doubted whether this is really an appropriate supposition. Certainly, this would be adequate in the case of a multi-mode convected Maxwell model corresponding to Lodge's network theory (i.e. $\beta_k = 1$, $G_k = \mathring{G}_k$), but when β_k and G_k depend on the configuration at all, they should in principle depend on the full set of configuration tensors, (as was pointed out in [1]), because entanglement loss and regeneration processes are interaction processes occurring between all species of molecules. As a consequence, this should also apply to the non-isotropic part of the drag tensor. Therefore, it has to be assumed that

$$\beta_k = \beta_k(b_1, \ldots, b_K)\,, \quad G_k = G_k(b_1, \ldots, b_K)\,. \quad (41)$$

This leads to a coupling of all differential equations (14), and so S_k and β_k are no longer commutable[6]).

A special consequence of the Leonov formulation of the theory occurs if the modes are not independent, contrary to Leonov's assumption. From the multi-mode generalization of eq. (20) and eq. (14) there results an equivalence relation (with $L_{(p)k}$ understood to be dependent on $b_1, \ldots, b_K$):

$$L_{(p)k} \cdot b_k + b_k \cdot L_{(p)k}^{\dagger} = \frac{1}{2\mathring{\eta}_k}(\beta_k \cdot S_k^{\dagger} + S_k \cdot \beta_k^{\dagger})\,. \quad (42)$$

[6]) This hypothesis retains the symmetric character of the excess-stress tensors S_k. If, instead, the more general hypothesis

$$S_k = S_k(b_1, \ldots, b_K) \quad (41\,a)$$

is introduced this is no longer the case. Physically the distinction is as follows: in the first case there is no average torque acting on any of the structures of type k, whereas in the second case these structures may exert torques on one another although they will cancel to give a resultant stress $S = \sum_{k=1}^{K} S_k$ which is a symmetric tensor.

A solution of it of the same type as eq. (22) is easily seen to be

$$L_{(p)k} = \frac{1}{2\mathring{\eta}_k} \beta_k \cdot S_k^{\dagger} \cdot b_k^{-1}. \tag{43}$$

This is generally not a symmetric tensor. However, as eq. (42) represents a relation between symmetric linear combinations, this is not a unique solution as every solution $\Lambda_{(p)k}$ of the equation

$$\Lambda_{(p)k} \cdot b_k + b_k \cdot \Lambda_{(p)k} = 0 \tag{44}$$

can be added to it. Because this homogeneous equation has three linearly independent solutions $\Lambda_{(p)k}^{(\nu)}$ ($\nu = 1,\ldots,3$), a symmetric tensor

$$\bar{D}_{(p)k} = L_{(p)k} + \sum_{\nu=1}^{3} l_\nu \Lambda_{(p)k}^{(\nu)} \tag{45}$$

could be constructed by an appropriate choice of l_ν which would also be a solution of eq. (42). Therefore the irreversible parts of the velocity gradients $L_{(p)k}$ can always be considered to be symmetric tensors as was conjectured by Leonov under less general conditions[7]).

Even in the linear model sketched in sect. 5.2, a matrix $\alpha_{kk'}$ has to be introduced instead of the one constant α, leading to

$$\beta_k = 1 + \sum_{k'=1}^{K} \alpha_{kk'}(b_{k'} - 1). \tag{46}$$

Of course, the practical application of such a model would be much more complicated, but it seems likely that simplifications could be introduced to reduce the number of parameters to such an extent that the model could be used for practical purposes.

In the foregoing, the model of a material with a drag coefficient dependent only on the *average* molecular configuration was considered to be a theory in its own right. However, it can also be considered to be a simplified reptation model derived from the Curtiss-Bird concept by preaveraging. If this view were held it would be sensible to propose that the different modes be independent, according to eqs. (40), and in the special linear case

$$\beta_k = 1 - \alpha_k(b_k - 1) \tag{46a}$$

could be introduced instead of eq. (46).

In special circumstances, for example for homopolymers with a relatively narrow molecular-weight distribution, it may be possible to substitute a single parameter α for α_k, independent of k, but under more general conditions a variation of the α_k (e.g. dependent on molecular weight, degree of branching) has to be expected.

As is well known, it is possible, in some cases, to transform multi-mode models of the rate-type into a single equation of the integral-type in which a memory function describes the material properties. In the simplest case, i.e. Lodge's network model, this function does not depend on the stress or strain history, whereas in all more general models such a dependence has to be introduced. However, it is easy to verify that none of the generalizations of our multi-mode models is compatible with any single-integral model. In particular, no such model with explicit introduction of strain or rate-of-strain history in the memory function is compatible with our approach, whereas introduction of the configuration or stress history, as even in the simplest case of a Kaye model, leads to an implicit model which can serve only for successive approximation purposes. Therefore there is no reason for going into more detail here.

7. Conclusions

It has been shown that the introduction of a configuration-dependent mobility does indeed enable one to classify a wide variety of different constitutive models under a unifying approach which elucidates their interrelationships. At the same time, this approach definitely excluded some other well-known models, making it obvious that their basic assumptions were not compatible with this concept. However, it was shown that such models were nevertheless able to deliver predictions in certain flow situations (particularly steady flows) which were more or less indistinguishable from those of models compatible with the above approach, whereas in others (e.g., transient flows) they differed considerably. This reminds one of the well-known fact that some constitutive models fit experimental data fairly well in some cases but in others only poorly. Therefore, the approach presented here and the critical comments in connection with it, including those with respect to an adequate description of the configuration dependence in multi-mode models, may stimulate ideas for pheno-

---

[7]) If, however, $L_{(p)k}$ is given by the defining relation

$$\frac{\mathfrak{D} r_k}{\mathfrak{D} t} = -L_{(p)k} \cdot r_k = -r_k \cdot L_{(p)k}^{\dagger} \tag{42a}$$

which is equivalent to

$$\frac{D r_k}{D t} = L_{(e)k} \cdot r_k \tag{42b}$$

and is a somewhat more specific assumption than the corresponding multi-mode generalization of eqs. (20) and (21), it can be easily shown that eq. (43) represents a unique solution. A more detailed analysis is needed to clarify the structural background of these alternatives.

menological models of polymer fluids which are based on sound molecular-kinetic concepts and represent a reasonable compromise between more rigorous but impracticable theories on the one hand and practicable but unrealistic models on the other.

Acknowledgement

The author is indebted to Dr. M. Hibberd for this help in improving the English text.

References

1. Giesekus, H., Rheol. Acta **5**, 29–35 (1966).
2. Bird, R. B., O. Hassager, R. C. Armstrong, C. F. Curtiss, Dynamics of polymeric liquids, Vol. 2: Kinetic theory, Wiley (New York 1977).
3. De Gennes, P. G., J. Chem. Phys. **55**, 572–578 (1971).
4. Doi, M., S. F. Edwards, J. Chem. Soc., Faraday Trans. II, **74**, 1789–1801, 1802–1817, 1819–1832 (1978); **75**, 38–54 (1979).
5. Curtiss, C. F., R. B. Bird, J. Chem. Phys. **74**, 2016–2025, 2026–2033 (1981).
6. Dashner, P. A., W. E. VanArsdale, J. Non-Newtonian Fluid Mech. **8**, 59–67 (1981).
7. Leonov, A. I., Rheol. Acta **15**, 85–98 (1976).
8. Lodge, A. S., Trans. Faraday Soc. **52**, 120–130 (1956); Elastic liquids, Academic Press (New York 1964).
9. Kaye, A., Brit. J. Appl. Phys. **17**, 803–806 (1966).
10. Acierno, D., F. P. La Mantia, G. Marrucci, G. Titomanlio, J. Non-Newtonian Fluid Mech. **1**, 125–146 (1976).
11. Giesekus, H., The physical meaning of Weissenberg's hypothesis with regard to the second normal-stress difference, in: J. Harris (ed.), The Karl Weissenberg 80th Birthday Celebration Essays, East African Literature Bureau (Campala, Nairobi, Dar Es Salaam 1973) pp. 103–112.
12. Giesekus, H., J. Non-Newtonian Fluid Mech. **11**, 69–109 (1982).
13. Giesekus, H., J. Non-Newtonian Fluid Mech. (submitted).
14. Leonov, A. I., A. N. Prokunin, Rheol. Acta **19**, 393–403 (1980).
15. Oldroyd, J. G., Proc. Royal Soc. London, **A 245**, 278–297 (1958).
16. Phan-Tien, N., R. I. Tanner, J. Non-Newtonian Fluid Mech. **2**, 353–365 (1977), N. Phan-Tien, J. Rheol **22**, 259–283 (1978).

(Received May 21, 1982)

Author's address:

Prof. Dr. H. Giesekus
Abteilung Chemietechnik
Universität Dortmund
Postfach 500500
D-4600 Dortmund 50

Rheologica Acta Rheol. Acta **21**, 376 – 387 (1982)

Some aspects of linear and nonlinear viscoelastic behaviour of polymer melts in shear*)

H. C. Booij and J. H. M. Palmen

DSM, Central Research, Geleen (The Netherlands)

Abstract: In linear viscoelastic investigations the frequency dependence of the phase shift between stress and strain appears to be very characteristic of the molecular structure of the material. This function is also a good approximation of the slope of the double logarithmic plot of the absolute value of the shear modulus G_d vs. the angular frequency ω. The product $(G_d/\pi)\sin 2\delta$ comes very close to the relaxation spectrum $H(\tau)$, with $\tau = 1/\omega$, in all physical states of the material.

The experimentally observed separability of time and strain effects in nonlinear viscoelasticity of highly viscous isotropic polymer fluids imposes restraints to the form of the constitutive equation. A single integral superposition equation of the Boltzmann type containing the product of a time function and a nonlinear strain function gives good results in describing experimental data in shear as well as in elongation. The molecular structure affects both functions in a different way. A universal definition of the nonlinear tensorial strain measure has not yet been developed. There are some indications that a definition on the basis of the principal stretch ratios may be fruitful.

Key words: Linear viscoelasticity, thermorheological simplicity, material-specific function, nonlinear constitutive equation, strain measure

1. Introduction

In the rheology of polymer fluids the stationary flows in capillary, coaxial cylinder or cone-and-plate devices have drawn much attention. However, flows which occur in plastic processing machines often involve highly time-dependent strain histories as well as extensional modes of deformation. It is, therefore, important to study these types of flow and to examine nonequilibrium situations, the more so because these can also provide information on the rates of loss and reformation of molecular entanglements which play a central role in modern theories of viscoelastic behaviour of polymer fluids. It is not the intention of this paper to give a review of all aspects of linear and nonlinear viscoelasticity. An excellent survey of current understanding of the time-dependent nonlinear behaviour of polymer fluids has recently been presented by Soong in Rubber Chemistry and Technology [1]. In addition to steady-state properties, entangled polymers also exhibit a wide array of complicated time-dependent stress qualities under transient flow conditions. Soong dealt with stress growth and overshoot phenomena upon the inception of steady shear flow, stress relaxation after cessation of that flow, stress relaxation subsequent to a sudden shear strain, elastic recovery after steady shear flow, reduction-in-shear-rate and flow reversal experiments. Yet, his enumeration can be readily extended by e.g. large amplitude oscillatory motion [2 – 5], nonlinear creep response to known stress programs [6], small oscillations superimposed parallel or orthogonal to steady shear flow [7 – 10, 19], double- or triple-step strain [11 – 15] and spike strain [13] experiments.

The insight that these transient responses are manifestations of true material properties brings a new dimension in our understanding of the behaviour of macromolecules, and provides very powerful

*) Invited paper, presented at the First Conference of European Rheologists at Graz (Austria), April 14 – 16, 1982.

means in testing the various theoretical approaches, because they play a more discriminating role than steady state flow results (for recent inquiries see [16–23]). However, transient investigations are much more difficult to perform experimentally than steady state ones, because most of the transients take place on a relatively short time scale, so that the instruments must be mechanically very stiff [22–25]. Probably the best way to evaluate the stress components in time-dependent flow fields is to measure the birefringence and the extinction angle, making use of the fact that the stress-optical coefficient is independent of strain and time [23, 26–30].

All the abovementioned transient experiments can also be performed in the tensile mode [30]. However, experimental difficulties in obtaining reliable data in that mode are even much severer than in shear flow experiments. Especially the need for a homogeneous strain throughout the whole sample puts sometimes too high demands on the nonlinear viscoelastic behaviour of polymer melts [40].

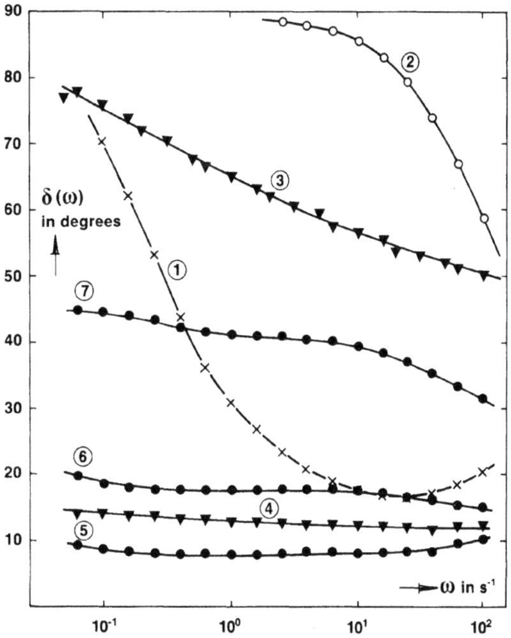

Fig. 1. Phase shift δ between stress and strain in oscillatory experiments as a function of the angular frequency ω for ① PS, $M = 2 \times 10^5$, 433 K; ② PS, $M = 3.7 \times 10^4$, 433 K; ③ PE, $M_w = 1 \times 10^5$, 453 K; ④ PE, $M_w = 1.5 \times 10^6$, 463 K; ⑤ PVC, 403 K; ⑥ PVC, 440 K; ⑦ PVC, 478 K

2. Some remarks on linear viscoelasticity

2.1. Interrelations among dynamic functions

In an oscillation experiment, in which the deformation is varied sinusoidally with an angular frequency ω, linear viscoelastic materials yield a stress which is periodically in ω, but lags behind the strain in phase over δ radians. The quotient of stress and strain gives the complex dynamic modulus which in case of shear is of the form $G^*(i\omega) = G_d(\omega)\exp[i\delta(\omega)] = G'(\omega) + iG''(\omega)$. It has become clear that the function $\delta(\omega)$ is very characteristic of structure parameters of the material, such as molar mass (distribution), network structures due to crosslinking, crystallinity or reinforcement with fine fillers, etc. Figure 1 shows an illustration of this specificity. The steep curve ① presents the phase shift as a function of the frequency at 433 K for a melt of an anionic polystyrene (PS) with a narrow molar mass distribution. At low frequencies the viscoelastic behaviour tends towards that of a simple liquid ($\delta \to \pi/2$) and at high frequencies towards that of an elastic rubber ($\delta \to 0$).

This curve is shifted to higher frequencies (curve ②) for a PS-sample of lower molar mass. If we look at a polyethylene (PE) with a wide molar mass distribution (Marlex 6050, $M_w/M_n \simeq 7.5$) and compare the $\delta(\omega)$-curve ③ of this with the one of a much higher molar mass Hostalen GUR sample (curve ④), then especially the large vertical shift of the flat curves is striking. Also supermolecular structures are very well discernible in the shape of the $\delta(\omega)$-curves. A continuous network of polymer chains is present in the melt of polyvinylchloride (PVC), in which small crystallites serve as physical crosslinks. For that reason a PVC melt is very elastic at 403 K (curve ⑤). A part of the crystalline structure melts at higher temperatures leading to relatively more viscous liquids when going to 440 K (curve ⑥) and to 478 K (curve ⑦).

Normally we are more interested in the shape of $\delta(\omega)$-curves than in those of the $G_d(\omega)$-functions. The reason for this is that the latter functions do not yield essential additional information. It can be demonstrated in a very general way [31] that the so-called Kramers-Kronig relations hold between $\ln G_d(\omega)$ and $\delta(\omega)$, and that the nearly local relationship

$$\delta(\omega) \simeq \tfrac{1}{2}\pi[d\ln G_d(u)/d\ln u]_{u=\omega} \tag{1}$$

can be used as a good approximation of this interrelation [32]. This is demonstrated in figure 2, where δ as well as $\pi d \ln G_d/2d \ln \omega a_T$ are plotted against the

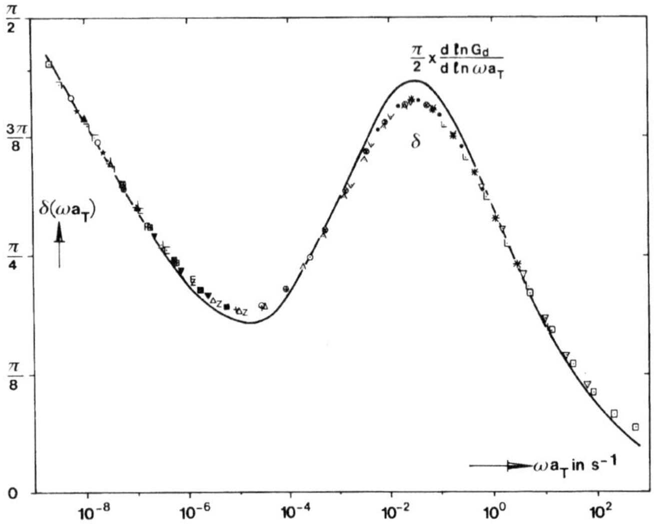

Fig. 2. Experimental values of $\delta(\omega a_T)$ compared with calculated curve of $\frac{\pi}{2}[d\ \ln G_d(u)/d\ \ln u]_{u=\omega a_T}$ for PVAc, at a reference temperature of 308 K

(reduced) frequency at 308 K for a sample of polyvinylacetate (PVAc). At high frequencies this material behaves like a glassy solid, around $\omega a_T = 0.1$ the glass-rubber transition manifests itself; and at still lower frequencies the material goes through the rubber plateau region and approaches second-order Newtonian liquid behaviour. Throughout all these regions the approximation (1) invariably holds within deviations of only some percent.

So the course of the $\delta(\omega)$ curve is very characteristic of the structure of a melt on molecular scale. On the other hand it proves to be particularly useful to specify the linear viscoelastic behaviour of a material with the distribution function of mechanical relaxation times. This spectrum $H(\tau)$ is very closely related to the functions $\delta(\omega)$ and $G_d(\omega)$. According to Williams and Ferry [33; 34, Ch. 4, Sec. A], a second approximation of this spectrum, using the experimentally measured shape of the dynamic storage modulus function $G'(\omega) = G_d\ \cos\delta$ and the estimated negative slope m of $H(\tau)$ on a double logarithmic plot, gives the equation

$$H(\tau) = \frac{\sin m\pi/2}{m\pi/2}\left[G'\ \frac{d\ln G'}{d\ln\omega}\right]_{\omega=1/\tau}. \quad (2)$$

Applying eq. (1) we obtained from eq. (2):

$$H(\tau) \approx \frac{\sin 2\delta}{2\delta}\,G_d\left[\frac{2\delta}{\pi}+\frac{d\ln\cos\delta}{d\ln\omega}\right]_{\omega=1/\tau} \quad (3)$$

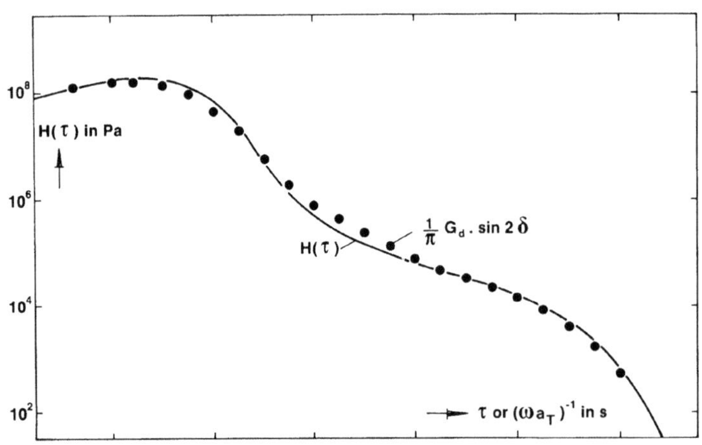

Fig. 3. Comparison between experimental values of $\frac{1}{\pi}G_d(\omega a_T)\sin 2\delta(\omega a_T)$ and calculated curve of relaxation spectrum $H(\tau)$

and as a further approximation

$$H(\tau) \approx \tfrac{1}{\pi}[G_d \cdot \sin 2\delta]_{\omega = 1/\tau}. \qquad (4)$$

Eq. (3) turns out to be remarkably well met by the experimental results, but even the very simple eq. (4) gives already a reasonable first approximation. This is demonstrated in figure 3, where this approximation is compared with the best approximation of $H(\tau)$ obtained by means of an iteration procedure from the master curves at 308 K of the moduli of the above-mentioned sample of PVAc.

2.2. *Thermorheological simplicity*

In the above-mentioned example use has been made of a reduced frequency scale, which has been obtained by application of the time-temperature superposition principle of thermorheologically simple materials. According to Schwarzl and Staverman [35], a linear viscoelastic material exhibits a thermorheologically simple behaviour if all its characteristic functions, like retardation or relaxation moduli, spectra, etc., meet the same time-temperature shifts. Backed by molecular theories this idea has been further evolved into the formulation of Ferry [34, Ch. 11]. It is based on two principles: the moduli are proportional to the product of temperature and density (or to the product of temperature, square density and mean square undisturbed molecular end-to-end distance [36]), while all relaxation times depend on one and the same monomeric friction coefficient. The first condition is hard to prove, because these products are practically constant so that the experimental accuracy is insufficient to confirm these forms. However, it implies that the quotient of G'' and G', that means $\tan\delta$, is independent of temperature and density. So the best way to start a superposition procedure is to superpose the $\delta(\omega)$-curves at various temperatures and to use the frequency-temperature shift factors a_T obtained in this way to shift the $G_d(\omega)$-curves, after which the vertical shift factors can be deduced.

In order to prove the second condition the shapes of the original curves at different temperatures must match over a substantial range of frequencies and each of them should specify at least parts of the glass-rubber transition, the rubber plateau region and the terminal zones of the time scale. Plazek found in very precise investigations on the creep and recovery behaviour of PS [37] and of PVAc [38] that the temperature dependences in the softening region are significantly greater than those in the terminal zone. In his textbook Ferry [34, Ch. 11, Sec. F2] treats some further anomalies in the application of reduces variables, and a very recent report is from Yanovsky [39] on 1,2-polybutadiene. Plazek therefore very carefully formulates: "It seems safe to assume in general that groups of viscoelastic mechanisms contributing to different maxima in $G''(\omega)$ will a priori have different temperature dependences". Normally, the experimentally employed frequency or time scale is so limited that the results cannot disprove this statement.

3. Nonlinear constitutive equations

The type of constitutive equations we are dealing with in this paper is related to the theories of temporary networks due largely to Lodge [41 – 43] and to Yamamoto [44]. The molecular mechanisms involved in the creation and destruction of physical junctions are not specified in these original theories (but they are assumed not to be affected by flow), nor was the precise nature of the entanglements made clear. The performance of this rubberlike liquid or upper convected generalized Maxwell liquid model in shear and elongation is not consistent with that of real polymer fluids, and for that reason the theory has been extended in various ways.

In order to be able to describe quantitatively the linear viscoelastic behaviour of polymer melts one replaces the memory functions by integrals of exponentials using empirical relaxation spectra to account for polydispersity in molecular mass. Neither the rheological constitutive equations derived from the temporary network models, nor the ones obtained in the past several years from kinetic theories using the Kramers freely joined bead-rod chain model (as did Doi and Edwards [17, 45] or Curtiss and Bird [46 – 48]) or using computer simulations of the dynamics of highly entangled polymers (as did Evans and Edwards [49 – 51]) incorporated polydispersity or long-chain branching. Only the kinetic transient network model of Shen et al. [52, 53] gives very good results in correlating the viscoelastic properties with molecular mass distribution. In this model entanglements not only produce an enhanced friction at certain sites along the chain, but they also give rise to extra elastic forces between each pair of entanglement points of the same chain, which are transmitted through the entangled network. Besides interchain entanglements, the model also takes into account intrachain entanglements. Until now this theory has only been worked out for shear stresses and has not yet yielded a complete tensorial constitutive equation.

To accommodate the nonlinearities (e.g. non-Newtonian viscosity, deviations from second-order normal stresses) one may allow the memory function to depend on the history of the strain rate [6, 19, 54–59], of the strain [44, 60–65, 13, 45], of the stress [66] or of the stored elastic energy [67, 68, 5], as is commonly practiced, or one may make the relaxation times themselves functionals of these quantities. However, all these modifications are subject to stringent limitations, for the theory of Doi and Edwards [69] and stress relaxation experiments of Osaki et al. [70, 18, 23] on concentrated polystyrene solutions make it evident that two ranges of nonlinearity can be distinguished. At short times the degree of nonlinearity increases with time, but the nonlinearity of the material becomes time-independent at times sufficiently longer than the equilibration time for fluctuations of contour length of a polymer chain confined in a tube. The broadness of this time domain is proportional to $(Mc)^{1.5}$ [71], where c is the concentration and M the molar mass of the polymer. Since we are mainly interested in high-molar mass polymer melts, we shall direct our attention especially to this range. Here the effects of strain and time (or temperature) are separable in stress relaxation experiments up to moderately large strains. This means that the shape of the relaxation spectrum does not change with the magnitude of strain (the time-shift invariance) and that any nonlinear behaviour observed arises solely from a nonlinearity in the stress-strain relationship.

The separability or factorizability of strain and time effects was discovered and developed in the field of crosslinked rubbers [72, 73]. A summary of further evidence on other materials is given by Chang et al. [74], and this separability also proves to hold for polymer solutions and melts, provided that $Mc < 10^6$ g/cm^3 [75–79, 18, 22].

Experimentally, the traditional and easiest way to determine separately the effects of time and strain is by means of step-strain stress relaxation tests with varying step height [73, 80–82, 70, 76, 11, 22]. Of course, also different kinds of transient measurements may be employed to obtain the time and strain functions separately (see e.g. [19] or [78]). For instance, Yen and McIntire [63] tried to derive the nonlinear strain function by inversion of the integral equation for the steady state non-Newtonian viscosity function. Their and also our own experience, as well as numerical analysis [83], reveal that the function acquired in this way is not identical to the one obtained by inversion of the integral describing the shear rate dependence of the steady state first normal-stress difference. On the other hand, the nonlinear strain function which fits the normal-stress data usually also gives a rather good approximation of the shear stresses. The most critical function in this procedure, however, is the relaxation spectrum.

The above references concern data obtained in tensile or in shear experiments. There are only a few publications on the comparison of the nonlinear viscoelastic behaviour of the same polymer fluid in both modes of deformation. Phillips [84] as well as Wagner [85] analysed the experimental data on a melt of LDPE of Meissner [24]. Wagner obtained different nonlinear strain measures for shear and uniaxial elongation and found these to be identical if plotted versus a more or less arbitrary combination of the first and second strain invariants.

All the considerations of this section lead to a single integral constitutive equation of the form

$$P(t) = -PI + \int_{-\infty}^{t} \int_{-\infty}^{+\infty} \frac{H(\tau)}{\tau} \cdot \exp\left[-\left(\frac{t-t'}{\tau}\right)\right] 2S(t-t')\, d(\ln \tau)\, dt' \quad (5)$$

where $2S(t - t')$ is the generalized tensorial nonlinear measure of the strain between t' and t. This type of Boltzmann superposition equation with a product of a time function and a strain function originates from Leaderman [86]; it is discussed by e.g. White and Tokita [61] and one of its advocates is Phillips [87, 88].

In imitation of Tanner and Williams [83] some authors [18, 76, 77, 82, 85] prefer to write the nonlinear strain measure as the product of the Finger strain tensor and the so-called damping function [64]. This function has been interpreted in terms of the network theories as a deformation-dependent survival probability for a given temporary physical network strand [85]. Hence the number of entanglements decreases with increasing strain: in a melt of LDPE 90% of all entanglements are destroyed after a shear strain of about 13 shear units or after a uniaxial extension of about 4 Hencky strain units, i.e. an elongation of more than 50 [85].

This disentangling due to strain and the subsequent nonlinearity in stress should be distinguished strictly from the relaxation of the stress. Tsang and Dealy [21, 89] deduced from a series of transient shear data that the time scale for the reformation of entanglements is significantly longer than that for stress relaxation. This implies that in describing experiments in which the strain passes through a maximum the nonlinear strain measure should be taken to be a functional operator of the previous strain history in the sense that the maximum strain in the time interval $t - t'$ after the creation time t' determines the strain measure $2S(t - t')$. This is called the irreversibility assumption [13, 65], which states that the strands lost during a non-decreasing deformation are not reformed during a subsequent decreasing deformation. Of course this is closely related to the discussion on the double-step-strain findings and the independent alignment assumption in the Doi-Edwards theory [12–18].

4. A generalized nonlinear strain measure

A remarkable result of the theory of Doi and Edwards [45] is that the strain function $2S(t - t')$ in eq. (5) should be determined only by the magnitude of strain and therefore must be independent of molecular structure parameters. This is certainly not supported by experimental evidence. Wagner [85] summarizes some published empirical nonlinearity functions for different materials, and an even more extended survey of strain measures $2S_{xy}$, obtained experimentally on various polymer solutions and melts, is given by Petrie [90]. We can add to these other measures, e.g. the one obtained for ABS melts for shear strains up to 6 [77] or for polybutadiene up to shear strains of 3.5 [30]. A selection of these functions is collected in table 1.

Numerous algebraic forms have been proposed to fit these functions. One of them attracted attention recently, namely the Seth strain measure. In 1962 Seth [91] introduced a generalized measure of strain based on an arbitrary power of the principal stretch ratio. This has been used for various types of materials. Blatz et al. [92, 93] and Ogden [94] applied this strain measure for the isothermal large deformation elasticity of incompressible rubberlike solids. Since for these highly elastic materials the isotropic, undistorted state is mostly assumed to be the reference state for subsequent deformations, the strain is described by means of the Green strain tensor $E_i = \frac{1}{2}(\lambda_i^2 - 1)$, where λ_i $(i = 1, 2, 3)$ denotes the stretch ratios in the principal strain directions. If we are dealing with liquids, which by definition do not have a preferred natural configuration, it is conventional and logical to consider the present configuration as the reference state. Then we use the Almansi strain measure $e_i = \frac{1}{2}(1 - \lambda_i^{-2})$. Seth [91], now, suggested a more general measure to characterize the deformation field, namely

$$S_i = \frac{1}{n}(1 - \lambda_i^{-n}) \tag{6}$$

where n is called the coefficient of strain measure and may assume positive or negative, integral or nonintegral values.

All the known strain measures can be represented by this generalized measure (e.g. $n = -2$: Green; $n = 2$: Almansi; $n = -1$: Cauchy; $n = +1$: Swainger; $n = 0$ ($\ln \lambda_i$): Hencky). From a number of investigations [93, 95, 96] it appeared that n is an essentially temperature-independent material parameter, implying that the generalized strain measure represents a material as well as kinematic quantity. We now want to state explicitly the form of this measure for two different deformation fields:

Table 1. Some published nonlinear shear strain measures

Author	Shear measure $2S_{xy}$	Fluid	Graph in figure 4
Zapas [80]	$\frac{\gamma}{1 + \frac{2}{9}\gamma^2}$	PIB sol.	5
Phillips [88]	$\gamma \exp(-0.133\,\gamma)$	LDPE melt	1
Phillips [87, 88]	$\frac{\gamma}{1 + (0.188\,\gamma)^3}$	PIB sol.	3
Osaki et al. [75]	$\gamma\,[0.913 \exp(-0.368\,\gamma) + 0.087 \exp(-0.084\,\gamma)]$	PS sol.	4
Wagner and Laun [76]	$\gamma\,[0.57 \exp(-0.31\,\gamma) + 0.43 \exp(-0.106\,\gamma)]$	LDPE melt	2
Booij and Palmen [77]	$\gamma\,[0.88 \exp(-0.7\,\gamma) + 0.12 \exp(-0.15\,\gamma)]$	ABS melt	6

i) Simple tension in 1-direction [93]:

$$\lambda_1 = \lambda, \quad \lambda_2 = \lambda_3 = \lambda^{-1/2},$$

$$S_1 = S(\lambda) = \tfrac{1}{n}(1 - \lambda^{-n}),$$

$$S_2 = S_3 = S(\lambda^{-1/2}) = \tfrac{1}{n}(1 - \lambda^{n/2}).$$

ii) Simple shear [94]:

$$\lambda_1 = \frac{\sqrt{\gamma^2 + 4} + \gamma}{2} = \lambda,$$

$$\lambda_2 = \lambda^{-1} = \frac{\sqrt{\gamma^2 + 4} - \gamma}{2},$$

$$\lambda_3 = 1,$$

$$S_1 = S(\lambda) = \tfrac{1}{n}(1 - \lambda^{-n}),$$

$$S_2 = S(\lambda^{-1}) = \tfrac{1}{n}(1 - \lambda^{n}),$$

$$S_3 = 0.$$

These are the strains in the principal directions. If the simple shear γ is in the x-direction, then the physical components of the strain in x, y-coordinates are obtained by transformations:

$$S_{xx} - S_{yy} = (S_1 - S_2)\cos 2\theta = [S(\lambda) - S(\lambda^{-1})]\frac{\lambda^2 - 1}{\lambda^2 + 1}, \tag{7}$$

$$S_{xy} = \frac{1}{2}(S_1 - S_2)\sin 2\theta = [S(\lambda) - S(\lambda^{-1})]\frac{\lambda}{\lambda^2 + 1} \tag{8}$$

with $\tan 2\theta = 2/\gamma$ and $S_{xx} - S_{yy} = \gamma S_{xy} = (\lambda - \lambda^{-1})S_{xy}$.

If λ tends towards 1, $S_1 - S_2 = \frac{3}{2}\varepsilon$ in tension, and if $n = \pm 2$, $S_{xx} - S_{yy} = \frac{1}{2}\gamma^2$ and $S_{xy} = \frac{1}{2}\gamma$ in shear. For these reasons normally the factor 2 is used in eq. (5) when the Seth measure is applied [90].

The nonlinear shear strain measures, tabulated in table 1, can be converted into the difference of the strains in the principal directions, $S(\lambda) - S(\lambda^{-1})$, by use of eq. (8). The resulting functions are represented in figure 4, disclosing great differences in nonlinearity between the various fluids. The shape of these strain difference measures as functions of the principal stretch ratio λ can be compared with the ones calculated with the Seth measure (6). These are reproduced in figure 5 for the indicated values of n. Even qualitatively the shape of these functions is different. The experimentally observed degree of nonlinearity at large values of λ cannot be described by the Seth formula, no more than the strong nonlinearity of the curves 4, 5 and 6 in figure 4.

The nonlinear principal strains can also be calculated from uniaxial elongation experiments. Then $S_1 - S_2 = S(\lambda) - S(\lambda^{-1/2})$, and this is equal to $(\lambda^2 - \lambda^{-1})h(\lambda)$ in the Wagner-Laun notation. Wagner [85] reports the equation $h(\lambda) = [0.0025\,\lambda^2 + 0.9975\,\lambda^{0.3}]^{-1}$ for the damping function obtained from experiments of the Meissner-Laun-Münsted group on

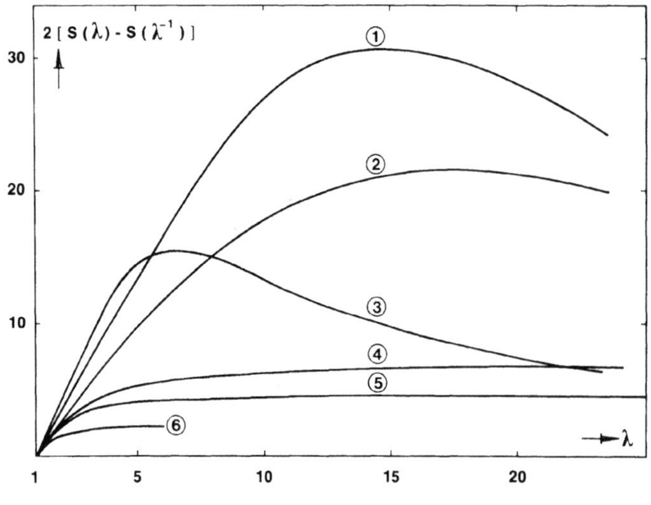

Fig. 4. Difference of principal strain measures in shear as a function of the principal stretch ratio λ, calculated from the experimental functions identified in table 1

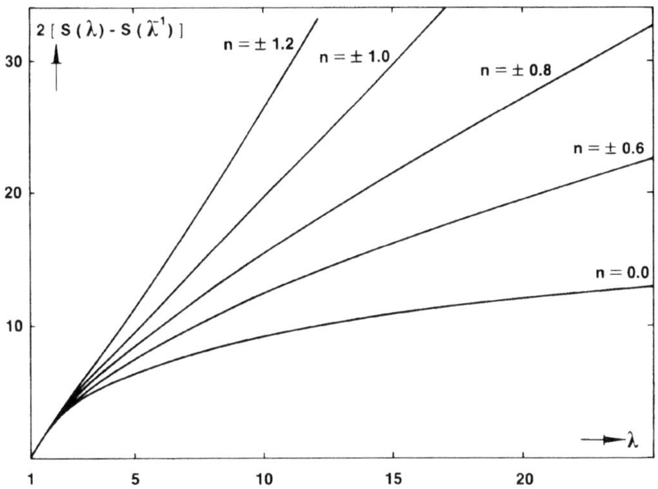

Fig. 5. Difference of principal strain measures in shear calculated from Seth's measure eq. (6) for the indicated values of n

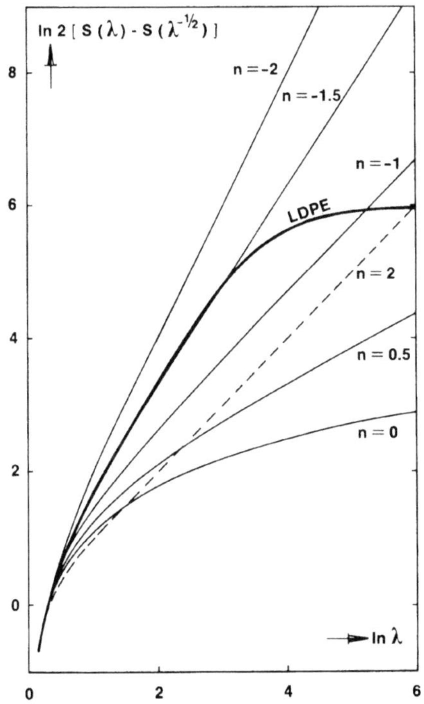

Fig. 6. Difference of principal strain measures in tension as a function of the principal stretch ratio λ, calculated from Seth's measure eq. (6) for the indicated values of n, and compared with the curve for LDPE-melt, obtained from the empirical formula of Wagner [85]

a long-chain branched LDPE melt. From this we calculated the form of the function $S(\lambda) - S(\lambda^{-1/2})$, represented in figure 6. The same figure also shows the curves of the nonlinear Seth strain measures for this type of deformation, namely $(\lambda^{n/2} - \lambda^{-n})/n$, for some values of n. Up to stretch ratios of about 20 the results on LDPE can be described remarkably well with $n = -1.5$. However, in shear this value gives much too linear strain curves (see fig. 5). Both conclusions are consistent with the results of the analysis Shaw [96] did on the same data. Summarizing, it must be put forward that the Seth strain measure, although flexible and handy, is not the proper quantity to use in nonlinear viscoelastic constitutive equations for polymer fluids.

Lodge and Meissner [97] have suggested that various transient flows should be run with the same principal elongation ratios λ, and the principal stresses compared. In extension the principal elongation ratio and principal stress are measured directly, whereas in shear they can be calculated by $\sigma_1 - \sigma_2 = \sqrt{N_1^2 + 4P_{xy}^2}$ and $\lambda = \frac{1}{2}(\sqrt{\gamma^2 + 4} + \gamma)$, where P_{xy} and N_1 are respectively the shear stress and the first normal-stress difference. However, if the largest principal strain in tension and in shear tests are equal, the other principal strains are not, because $\lambda_2 = \lambda_3 = \lambda^{-1/2}$ in tension and $\lambda_2 = \lambda^{-1}$, $\lambda_3 = 1$ in shear. This implies that under these circumstances only the principal stresses σ_1 are equal and not the measurable stress differences $\sigma_1 - \sigma_2$.

5. A particular type of experiment

With the aim to demonstrate one of the potentialities of application of the nonlinear constitutive equation (5) we will fall back on some by now classical data obtained in parallel superposition experiments [8] at room temperature on a 5% (*w*) solution in decalin of a high molecular weight ethylene/propylene copolymer containing 41% (*w*) propylene. In this type of experiment the total shear rate equals the sum of a steady shear rate q and an oscillatory component $\alpha\omega\cos\omega t$. For the calculation of the shear stress we use the Wagner approach for $2S_{xy}(t-t')$, namely $h(\gamma(t-t'))\cdot F_{xy}(t-t')$, where F_{xy}, the shear component of the Finger strain tensor, is equal to $\gamma(t-t') = q(t-t') + \alpha\sin\omega t - \alpha\sin\omega t'$ for this case.

Furthermore, it is supposed that the damping function $h(\gamma)$ can be fitted with a sum of exponentials $\sum_i f_i \exp(-n_i\gamma)$, with $\sum_i f_i = 1$. Introduction of this strain measure into eq. (5) yields for the steady-state shear stress in parallel superposition experiments:

$$P_{xy}(q,\omega,t) = P_{xy}(q) + \alpha G'(\omega,q)\sin\omega t + \alpha G''(\omega,q)\cos\omega t$$

with

$$P_{xy}(q) = \int_{-\infty}^{+\infty} \sum_i f_i \frac{H(\tau)\tau q}{(1+n_i\tau q)^2}\, d\ln\tau\,, \tag{9}$$

$$G'(\omega,q) = \int_{-\infty}^{+\infty} \sum_i f_i \frac{H(\tau)\omega^2\tau^2[1+\omega^2\tau^2 - n_i^2\tau^2q^2(3+2n_i\tau q)]}{[(1+n_i\tau q)^2+\omega^2\tau^2]^2(1+n_i\tau q)^2} \cdot d\ln\tau\,, \tag{10}$$

$$G''(\omega,q) = \int_{-\infty}^{+\infty} \sum_i f_i \frac{H(\tau)\omega\tau[1+\omega^2\tau^2 - n_i^2\tau^2q^2]}{[(1+n_i\tau q)^2+\omega^2\tau^2]^2}\, d\ln\tau\,. \tag{11}$$

The relaxation spectrum $H(\tau)$ was derived by an iteration procedure from data obtained in pure oscillatory experiments and the nonlinearity function could be represented by $h(\gamma) = 0.88\exp(-0.35\,\gamma) + 0.12\exp(-0.08\,\gamma)$. With the use of these functions eq. (9) gives a good fit of the shear rate dependence of the shear stress. By means of eqs. (10) and (11) the absolute value of the complex shear modulus $G_d(\omega,q) = (G'^2 + G''^2)^{1/2}$ and the phase shift $\delta(\omega,q) = \arctan(G''/G')$ were calculated as functions of the frequency at various values of the superimposed shear rate q.

A comparison with experimental data is shown in figures 7 and 8. A very reasonable fit is obtained for these dynamic shear functions. This can not be taken as an exclusive proof of the correctness of the applied constitutive equation. A similar degree of agreement can be attained by other models (see e. g. [67]). Moreover, the set of equations used does not give a satisfactory description of the normal stress components in parallel superposition on the same solution, no more than other models did [8, 67]. Even for these steady-state shear experiments an adequate explanation of the three-dimensional stress-strain relations has not yet been achieved.

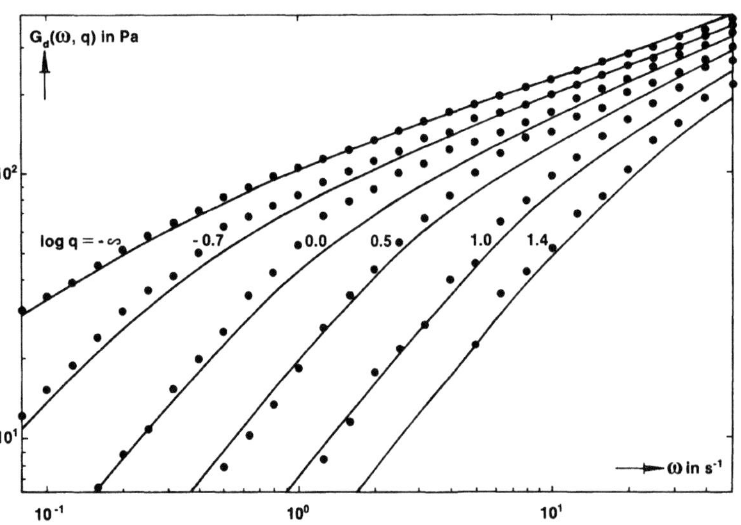

Fig. 7. Calculated frequency dependence of the dynamic modulus $G_d(\omega,q)$ (drawn lines) compared with the experimental data (●) for the indicated values of $\log q$ for a 5% solution of an ethylene/propylene copolymer in decalin at room temperature

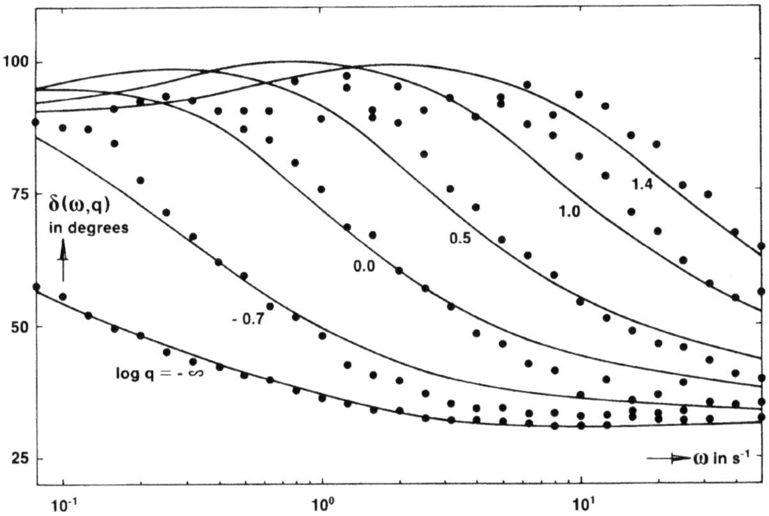

Fig. 8. Calculated effect of the indicated values of $\log q$ on the frequency dependence of the phase shift $\delta(\omega, q)$, compared with the experimentally measured effect (●)

6. Conclusions

Considerable progress has been made in the characterization of the steady state and transient nonlinear viscoelastic behaviour of polymer fluids. Especially in simple shear a good conformity of experimental data to theoretical predictions can be achieved by using a single integral constitutive equation of the Boltzmann superposition type with factorized time and strain functions. Both functions appear to be material specific, whereas the molecular theory of Doi and Edwards leads to a form of nonlinearity which is only dependent on the magnitude of the strain. Also in elongation the same constitutive equation is applicable, but until now no strain measure has been published that fits simultaneously the data obtained in both types of deformation. The concept that this measure should be a nonlinear function of the stretch ratios, or maybe of the recoverable stretches [98], in the principal strain directions may prove to be very prolific in future rheological research [74].

Acknowledgement

The authors thank their colleagues J. Sleijpen and P. Leblans for their contributions in experiments, calculations and discussions.

References

1. Soong, D. S., Rubber Chem. Techn. **54**, 641 (1981).
2. Macdonald, I. F., B. D. Marsh, E. Ashare, Chem. Eng. Sci. **24**, 1615 (1969).
3. Tee, T.-T., J. M. Dealy, Trans. Soc. Rheol. **19**, 595 (1975).
4. Vinogradov, G. V., Yu. Yanovsky, A. I. Isayev, J. Polym. Sci. **A2,8**, 1239 (1970).
5. Thurston, G. B., J. Non-Newtonian Fluid Mech. **9**, 57 (1981).
6. Liu, T., D. S. Soong, M. C. Williams, Polym. Eng. Sci. **21**, 675 (1981).
7. Osaki, K., M. Tamura, M. Kurata, M. Kotaka, J. Phys. Chem. **69**, 4183 (1965).
8. Booij, H. C., Rheol. Acta **5**, 215 (1966); **7**, 202 (1968).
9. Simmons, J. M., Rheol. Acta **7**, 184 (1968).
10. Laufer, Z., H. L. Jalink, A. J. Staverman, Rheol. Acta **14**, 641 (1975).
11. Zapas, L. J., T. Craft, J. Res. Nat. Bur. Stand. **69A**, 541 (1965).
12. Osaki, K., Y. Einaga, M. Kurata, N. Yamada, M. Tamura, Polymer J. **5**, 283 (1973).
13. Wagner, M. H., S. E. Stephenson, Rheol. Acta **18**, 463 (1979).
14. McKenna, G. B., L. J. Zapas, J. Rheol. **23**, 151 (1979); **24**, 367 (1980).
15. Marrucci, G., F. P. LaMantia, G. Spadaro, G. Titomanlio, Rheology 2, Proc. VIII. Intern. Congr. Rheol., p. 443, Plenum Press (New York 1980).
16. Osaki, K., S. Kimura, M. Kurata, J. Rheol. **25**, 549 (1981).
17. Doi, M., J. Polym. Sci., Polym. Phys. Ed. **18**, 1891, 2055 (1980).
18. Osaki, K., M. Kurata, Macromolecules **13**, 671 (1980).
19. Shroff, R. N., M. Shida, J. Appl. Polym. Sci. **26**, 1847 (1981).
20. Hayashi, S., Polym. J. **13**, 693 (1981).
21. Tsang, W. K.-W., J. M. Dealy, J. Non-Newtonian Fluid Mech. **9**, 203 (1981).
22. Vrentas, C. M., W. W. Graessley, J. Non-Newtonian Fluid Mech. **9**, 339 (1981).
23. Osaki, K., S. Kimura, M. Kurata, J. Polym. Sci., Polym. Phys. Ed. **19**, 517 (1981); **20**, 623 (1982).

24. Meissner, J., J. Appl. Polym. Sci. **16**, 2877 (1972).
25. Attané, P., P. Le Roy, G. Turrel, J. Non-Newtonian Fluid Mech. **6**, 269 (1980).
26. Gortemaker, F. H., M. G. Hansen, B. de Cindio, H. M. Laun, H. Janeschitz-Kriegl, Rheol. Acta **15**, 256 (1976).
27. Bessho, N., K. Osaki, M. Kurata, J. Soc. Rheol. Japan **5**, 68 (1977).
28. Osaki, K., N. Bessho, T. Kojimoto, M. Kurata, J. Rheol. **23**, 457, 617 (1979).
29. Takahashi, M., T. Masuda, N. Bessho, K. Osaki, J. Rheol. **24**, 517 (1980).
30. Kimura, S., K. Osaki, M. Kurata, J. Polym. Sci., Polym. Phys. Ed. **19**, 151 (1981).
31. Booij, H. C., G. P. J. M. Thoone, Rheol. Acta **21**, 15 (1982).
32. Booij, H. C., J. H. M. Palmen, B. J. R. Scholtens, Preprints 27th IUPAC Symp. on Macromolecules, p. 791 (Strasbourg 1981).
33. Williams, M. L., J. D. Ferry, J. Polym. Sci. **11**, 169 (1953).
34. Ferry, J. D., Viscoelastic Properties of Polymers, 3. ed., John Wiley and Sons (New York 1980).
35. Schwarzl, F., A. J. Staverman, J. Appl. Phys. **23**, 838 (1952).
36. Graessley, W. W., S. F. Edwards, Polymer **22**, 1329 (1981).
37. Plazek, D. J., J. Phys. Chem. **69**, 3480 (1965).
38. Plazek, D. J., Polymer J. **12**, 43 (1980).
39. Yanovsky, Yu. G., Intern. J. Polym. Mater. **8**, 187 (1980).
40. Meissner, J., T. Raible, S. E. Stephenson, J. Rheol. **25**, 1, 673 (1981).
41. Lodge, A. S., in: V. G. W. Harrison (ed.), Proc. Second Intern. Congr. Rheol., Oxford, 1953, p. 229, Butterworths (London 1954).
42. Lodge, A. S., Trans. Faraday Soc. **52**, 120 (1956).
43. Lodge, A. S., "Elastic Liquids", Acad. Press (London 1964).
44. Yamamoto, M., J. Phys. Soc. Japan **11**, 413 (1956); **12**, 1148 (1957); **13**, 1200 (1958).
45. Doi, M., S. F. Edwards, J. Chem. Soc., Faraday Trans. II, **74**, 1789, 1802, 1818 (1978); **75**, 38 (1979).
46. Bird, R. B., Chem. Eng. Commun. **9**, 321 (1981).
47. Curtiss, C. F., R. B. Bird, J. Chem. Phys. **74**, 2016, 2026 (1981).
48. Bird, R. B., H. H. Saab, C. F. Curtiss, J. Phys. Chem. **86**, 1102 (1982).
49. Evans, K. E., S. F. Edwards, J. Chem. Soc., Faraday Trans. II, **77**, 1891, 1929 (1981).
50. Edwards, S. F., K. E. Evans, J. Chem. Soc., Faraday Trans. II, **77**, 1913 (1981).
51. Evans, K. E., J. Chem. Soc., Faraday Trans. II, **77**, 2385 (1981).
52. Hansen, D. R., M. C. Williams, M. Shen, Macromol. **9**, 355 (1976).
53. Soong, D. S., M. C. Williams, M. Shen, in: A. G. Walton (ed.), Structure and properties of amorphous polymers, p. 1, Elsevier (Amsterdam 1980).
54. Bogue, D. C., Ind. Eng. Chem. Fund. **5**, 253 (1966).
55. Bird, R. B., P. J. Carreau, Chem. Eng. Sci. **23**, 427 (1968).
56. Carreau, P. J., Trans. Soc. Rheol. **16**, 99 (1972).
57. Yamamoto, M., Trans. Soc. Rheol. **15**, 331, 783 (1971).
58. Rokudai, M., R. Fujiki, Appl. Polym. Symp. **20**, 23 (1973).
59. Soong, D., M. Shen, J. Rheol. **25**, 259 (1981).
60. Bernstein, R., E. A. Kearsley, L. J. Zapas, Trans. Soc. Rheol. **7**, 391 (1963); **9**, 27 (1965).
61. White, J. L., N. Tokita, J. Phys. Soc. Japan **22**, 719 (1967).
62. Tanner, R. I., J. M. Simmons, Chem. Eng. Sci. **22**, 1803 (1968).
63. Yen, H.-C., L. V. McIntire, Trans. Soc. Rheol. **18**, 495 (1974).
64. Wagner, M. H., Rheol. Acta **15**, 136 (1976).
65. Wagner, M. H., S. E. Stephenson, J. Rheol. **23**, 489 (1979).
66. Kaye, A., Brit. J. Appl. Phys. **17**, 803 (1966).
67. Booij, H. C., Ph. D. Thesis, University Leiden (1970).
68. Acierno, D., F. P. LaMantia, G. Marrucci, G. Titomanlio, J. Non-Newtonian Fluid Mech. **1**, 125, 147 (1976).
69. Doi, M., J. Polym. Sci., Polym. Phys. Ed. **18**, 1005 (1980).
70. Fukuda, M., K. Osaki, M. Kurata, J. Polym. Sci., Polym. Phys. Ed. **13**, 1563 (1975).
71. Osaki, K., S. Kimura, K. Nishizawa, M. Kurata, Macromolecules **14**, 455 (1981).
72. Tobolsky, A. V., R. D. Andrews, J. Chem. Phys. **13**, 3 (1945).
73. Guth, E., P. E. Wack, R. L. Anthony, J. Appl. Phys. **17**, 347 (1946).
74. Chang, W. V., R. Bloch, N. W. Tschoegl, Rheol. Acta **15**, 367 (1976).
75. Osaki, K., S. Ohta, M. Fukuda, M. Kurata, J. Polym. Sci., Polym. Phys. Ed. **14**, 1701 (1976).
76. Wagner, M. H., H. M. Laun, Rheol. Acta **17**, 138 (1978).
77. Booij, H. C., J. H. M. Palmen, in: G. Astarita, G. Marrucci, L. Nicolais (eds.), Rheology 2, Proc. VIII. Intern. Congr. Rheol., p. 501, Plenum Press (New York 1980).
78. Zapas, L. J., J. C. Phillips, J. Rheol. **25**, 405 (1981).
79. Attané, P., P. LeRoy, J. M. Pierrard, G. Turrel, J. Non-Newtonian Fluid Mech. **9**, 13 (1981).
80. Zapas, L. J., J. Res. Nat. Bur. Stand. **70A**, 525 (1966).
81. Einaga, Y., K. Osaki, M. Kurata, S. Kimura, M. Tamura, Polym. J. **2**, 550 (1971).
82. Laun, H. M., Rheol. Acta **17**, 1 (1978).
83. Tanner, R. J., G. Williams, Trans. Soc. Rheol. **14**, 19 (1970).
84. Phillips, M. C., in: P. L. Clegg, F. N. Cogswell, D. E. Marshall, S. G. Maskell (eds.), Polymer Rheology and Plastics Processing, Plastics and Rubber Institute (London 1975).
85. Wagner, M. H., Rheol. Acta **18**, 33 (1979).
86. Leaderman, H., Trans. Soc. Rheol. **6**, 361 (1962).
87. Phillips, M. C., Trans. Soc. Rheol. **20**, 185 (1976).
88. Phillips, M. C., J. Non-Newtonian Fluid Mech. **2**, 109 (1977).
89. Dealy, J. M., W. K.-W. Tsang, J. Appl. Polym. Sci. **26**, 1149 (1981).
90. Petrie, C. J. S., J. Non-Newtonian Fluid Mech. **5**, 147 (1979).
91. Seth, B. J. in: M. Reiner and D. Abir (eds.), Second Order Effects in Elasticity, Plasticity and Fluid Dynamics, p. 162, McMillan (New York 1964).

92. Blatz, P. J., B. M. Chu, H. Wayland, Trans. Soc. Rheol. **13**, 83 (1969).
93. Blatz, P. J., S. C. Sharda, N. W. Tschoegl, Trans. Soc. Rheol. **18**, 145 (1974).
94. Ogden, R. W., Proc. Roy. Soc. **A326,** 565 (1972).
95. Chang, W. V., R. Bloch, N. W. Tschoegl, J. Polym. Sci., Polym. Phys. Ed. **15**, 923 (1977).
96. Shaw, M. T., Rheol. Acta **20**, 231 (1981).
97. Lodge, A. S., J. Meissner, Rheol. Acta **12**, 41 (1973).
98. Malkin, A. Ya., O. Yu. Sabsai, V. P. Beghishev, Intern. J. Polym. Mater. **9**, 1 (1981).

Authors' address:

Dr. H. C. Booij, J. H. M. Palmen
DSM – Central Research
P.O. Box 18
NL-6160 MD Geleen

Rheologica Acta Rheol. Acta 21, 388–393 (1982)

Some pending problems in polymer melt rheology, as seen from the point of view of Doi's slip-link model*)

H. Janeschitz-Kriegl

Johannes-Kepler-Universität Linz (Austria)

Abstract: An exposition is given of results, as obtained with the aid of Doi's slip-link model, being considered as the most simple version of the famous "reptation model". It turns out that this model which exhibits three distinct phases of relaxation (an extremely fast phase, an equilibration phase and a slow disengagement phase) is capable of explaining several peculiar features of polymer melt rheology:

a) The molecular mass dependence of the breadth of the rubber plateau in the storage modulus, of the zero-shear viscosity and of the normal-stress coefficients.

b) The molecular mass independence of the equilibrium (shear and tensile) compliances for monodisperse polymers (semi-quantitative prediction).

c) The seemingly contradictory sensitivity of these compliances for the breadth of the molecular mass distribution.

d) The critical value of the shear stress at which melt fracture occurs in capillary flow.

e) An equilibration phase in tensile experiments on unvolcanized rubber.

In this evaluation optical (flow birefringence) measurements are preferentially used.

Key words: Birefringence, flow, fracture, relaxation, reptation

1. Introduction

In a series of papers Doi and Edwards [1] and, very recently, Doi by himself [2] developed a perspicuously simple theory based on the reptation model originally proposed by De Gennes [3]. The present communication aims at a review of some experimental results which are mostly in favour of the indicated approach. First of all some very recent linear viscoelastic data, as obtained on polymers of a very narrow molecular mass distribution, are presented. In continuation, a series of literature data is evaluated, including a reference to an early discussion of the influence of polydispersity on polymer melt rheology [4], the melt fracture phenomenon [5] and the response of unvolcanized rubber to the tensile test [6]. A related and more elaborate theory, as meanwhile proposed by Curtiss and Bird [7], will not be included in the discussion, however. In some respects this latter theory furnishes very similar results, in fact. A discussion of results going beyond those of the simple theory has recently been started by the latter authors [8].

*) Invited paper, presented at the First Conference of European Rheologists at Graz (Austria), April 14–16, 1982.

2. Cyclic measurements on nearly monodisperse polystyrenes

In Doi's recent paper three relaxation phenomena are discerned, viz. a very rapid relaxation (phase A) in which chain sections between slip-links are engaged (Rouse spectrum), a considerably slower relaxation (phase B) in which the average contour length of the "primitive" chains is restored (after the rapid macroscopic deformation initiating the relaxation) and a rather slow relaxation (phase C) in which the chain molecules completely "disengage" by longitudinal diffusion from their deformed "tubes". The leading relaxation times of these phases are:

$$\tau_A = \zeta b^2 N_e^2 / 6\pi^2 kT,$$
$$\tau_B = 2\tau_A (N_0/N_e)^2, \quad (1)$$
$$\tau_C = 6\tau_A (N_0/N_e)^3,$$

where ζ is the (scalar) friction factor of the repeating unit, b is the length of this unit, N_e is the number of these units between neighbouring slip-links ("entanglement spacings") and N_0 is the number of repeating units in the whole macromolecule. Clearly, τ_A is independent of the molecular mass of the macromolecule and τ_B and τ_C are well separated if $N_0 \gg N_e$. In addition it should be mentioned that phase B does not play any role in linear viscoelasticity since the average contour length remains unaffected by small macroscopic deformations.

The relaxation time spectrum of phase A is identical with the one derived by Rouse [9] in his early paper with $\tau_1 = \tau_A$. The spectrum of phase C is given by the following equations:

$$\tau_p = \tau_C/p^2 \quad \text{with } p \text{ odd},$$
$$G_p = (3/5)\, G_0 (8/p^2\pi^2), \quad (2)$$
$$G_0 = \rho RT/M_e, \quad N_e = M_e/M_0.$$

After in the second eq. (2) the factor (3/5) is omitted for reasons discussed later, the storage and loss moduli G' and G'' are calculated as functions of circular frequency for a strictly monodisperse polystyrene of relative molecular mass 770000 with $M_e \simeq 18010$ (see Graessley [10]). These predictions are shown in figure 1 on double logarithmic scales. Obviously, the left side of this figure (the liquid range)

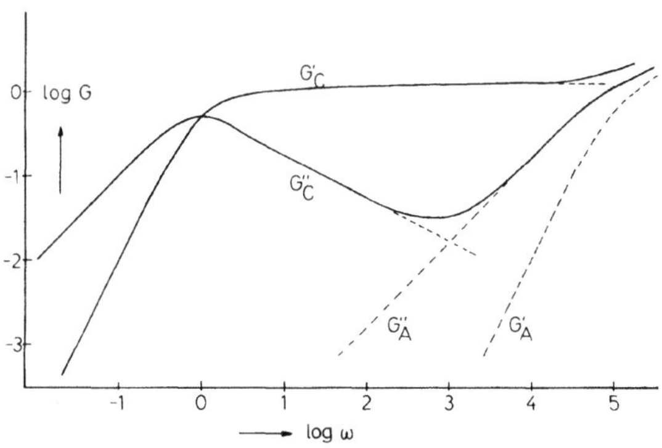

Fig. 1. Storage and loss moduli of a polystyrene of a sharp relative molecular mass of 770000, as calculated according to Doi's theory. (Normalization: $G_0 = 1$ and $\tau_C = 1$)

resembles very much a graph obtained for a single Maxwell element of $\tau = \tau_C = 1$ and $G_0 = 1$ (normalization of the graph). The explanation of this finding is given by the fact that the relaxation strengths decrease (according to the second eq. (2)) with $1/p^2$ so that the second one is only one ninth of the first one. For the Rouse model these relaxation strengths are all equal to G_0. (Moreover, also even p's are included). This makes the difference so that G' and G'' both finally increase with $\omega^{1/2}$ (right side of the graph).

In figure 2 the experimental results for an anionic polystyrene of the most frequent relative molecular mass of $M_p = 770000$ and a ratio $M_w/M_n = 1.09$ (Polymer Standard, Polymer Laboratories, Shrewsbury U.K.) are shown for a reference temperature of 180 °C. At first sight, this figure seems rather similar to figure 1. A closer inspection, however, reveals that the central parts of these figures are rather different. The "plateau" of the experimental curve for G' is not really flat and the corresponding loss moduli G'' are much higher than those of the theoretical curve. In this connection one may wonder whether the amplitudes of the oscillation of the shear were small enough for expected linear properties. But the independence of the results from the shear amplitude was specially checked, so that an influence of the non-linear phase B could be excluded.

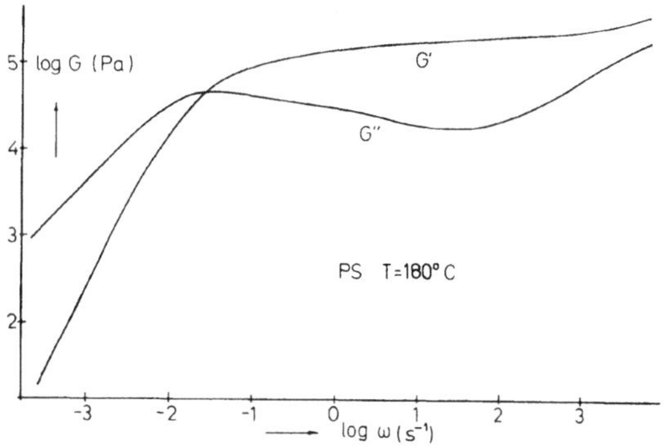

Fig. 2. Experimental storage and loss moduli of a polystyrene with very narrow molecular mass distribution of most frequent relative molecular mass $M_p = 770000$ at a reference temperature of 180 °C

For smaller molecular masses the discrepancy between theory and experiment becomes less perspicuous, due to the smaller distance between log τ_A and log τ_C.

A remarkable property of the reviewed theory is the additivity of the contributions of single chain molecules to stress (and birefringence). As a matter of fact, this theory is a mean field theory. As a consequence, the influence of polydispersity can easily be calculated. If M_k is the molecular mass of the k-th fraction, one simply obtains the total stress tensor:

$$\sigma = \sum_k \sigma_k . \tag{3}$$

On the basis of this equation, any influence of a possible polydispersity of the investigated sample can easily be estimated. For the above mentioned sample one finds, on the basis of the assumption of a logarithmic normal distribution (with $M_w/M_n = 1.09$), that the left side (in the liquid range) of figure 1 is modified (in the right direction) whereas the central part remains practically unchanged. From this experience one learns that not only – as has been known for a while – the exponent of the third eq. (1) is somewhat too low (from experience one expects 3.5 instead of 3 [11]), but also the damping is too low in the plateau region. Our experience is that especially this latter fact spoils the correct prediction of storage and loss moduli for polydisperse samples from their gel permeation chromatographic data. But if the linear viscoelastic data are not correctly described, one can all the less expect a quantitative prediction of non-linear (flow) properties.

Pertinent measurements and calculations were carried out by A. Schausberger and G. Schindlauer at Linz University. The dynamic shear viscometer, which has been used, is a commercial apparatus by Rodem (De Bilt, Holland) [12].

3. Equilibrium compliances and polydispersity

In spite of the merely semi-quantitative agreement between theory and experiment, as demonstrated in the previous section, the elucidative value of the theory is indisputable. About one decade ago it was concluded on the basis of extensive experimentation on polymers of very narrow molecular mass distributions that the equilibrium shear compliance J_e of monodisperse polymers is a temperature independent constant [13], in contrast to the prediction of the Rouse theory (modified for polymer melts), according to which J_e is expected to increase linearly with molecular mass [14]. This finding seems to spoil an early attempt to derive polydispersity parameters from polymer melt rheology [4].

A compilation of literature data, as given by Wales in his monograph will not be reproduced here. Instead of it, some data as obtained by Wales [15] for the optical equivalent J_{e0} of the equilibrium shear

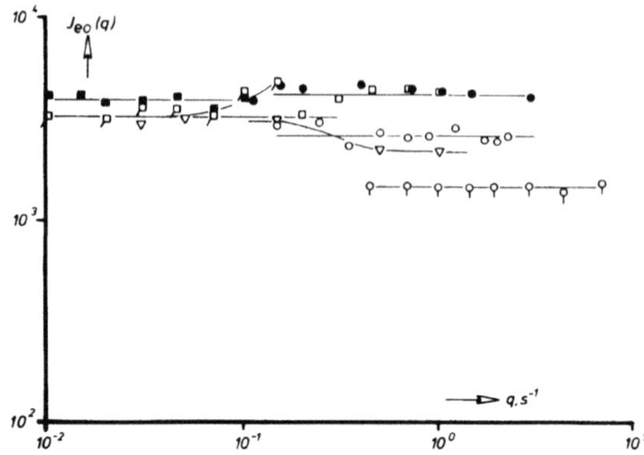

Fig. 3. Optical corollary J_{e0} to the equilibrium shear compliance J_e, plotted as function of the shear rate for "monodisperse" polystyrenes of indicated rel. mol masses according to Wales [15]): ♀ ... 51000 (158 °C), ○ ... 96000 (168 °C), ▽ ... 160000 (165 °C), ● ... 224000 (196 °C), □ ... 507000 (200 °C), ■ ... 860000 (214 °C), ⚐ ... 1700000 (242 °C). Temperature reduction has no influence on J_e

compliance, will be shown in figure 3. As one notices, the quantity

$$J_{e0} = \frac{1}{2}\,\frac{n_{11} - n_{22}}{n_{12}^2} = 2\,\frac{\Delta n \cos 2\chi}{(\Delta n \sin 2\chi)^2} \tag{4}$$

with

$$J_e = C J_{e0} \tag{4a}$$

is practically independent also of the steady-state shear rate q, so that an extrapolation to zero-shear rate is superfluous. (In these equations Δn is the flow birefringence in the 1,2-plane, χ is the corresponding extinction angle and C is the stress-optical coefficient, being $\simeq 5 \cdot 10^{-9}$ Pa^{-1} for the investigated polystyrene melts in the applied temperature range.)

As the relative molecular masses of the investigated practically monodisperse polystyrenes vary considerably (see the caption to the figure), and the results for all higher molecular mass samples cumulate at about $2.5 \cdot 10^3$, one can conclude that J_e is practically molecular mass independent and of the order of 1.3 $\cdot 10^{-5}$ Pa^{-1} (see eq. (4a)).

According to Doi's theory one should expect the following relation between J_e and the plateau modulus G_0:

$$J_e = (5/3)(6/5\ G_0) . \tag{5}$$

Since J_e is a characteristic parameter of the liquid range, the corresponding equation for a single Maxwell element should furnish a very similar result, viz.:

$$J_e = 1/G_0 . \qquad (6)$$

From a comparison of the latter two equations one can derive that the factor (5/3) of eq. (5) should be omitted. In fact, it has the same origin as factor (3/5) in the second eq. (2): Both factors can be shown to be due to a preaveraging step in Doi's calculations.

Since G_0 is of the order of $2 \cdot 10^5$ Pa (see fig. 2 or previous results by Onogi et al. [16]), one can expect that $J_e \simeq 0.5 \cdot 10^{-5}\ \mathrm{Pa}^{-1}$, which is a little lower than the above mentioned value of $1.3 \cdot 10^{-5}\ \mathrm{Pa}^{-1}$, but certainly of the correct order of magnitude.

It turns out that this little discrepancy can easily be explained in terms of polydispersity. Curiously enough, it is found that the influence of molecular mass just cancels out for a strictly monodisperse material. For a polydisperse sample one obtains (on the basis of the additivity of contributions to stress components – see eq. (3)) according to Doi's theory:

$$J_e = (5/3)(6/5\ G_0)\ \langle M^7 \rangle \langle M \rangle / \langle M^4 \rangle^2 \qquad (7)$$

where the angular brackets indicate averages over the molecular mass distribution.

Since the principle expressed by eq. (3) is of a more general nature than the details of Doi's theory, it is meaningful to calculate J_e also for an assembly of simple Maxwell elements, every element representing a sharp fraction of the polydisperse polymer. (In the liquid range of a strictly monodisperse polymer the influence of the leading relaxation time is predominant!) If, in addition, use is made of the experience that this leading relaxation time is proportional in practice to the 3.5-th power of the molecular mass, one obtains:

$$J_e = (1/G_0)\langle M^8 \rangle \langle M \rangle / \langle M^{4,5} \rangle^2 . \qquad (8)$$

It will be clear to the reader from eqs. (7) and (8) that the equilibrium shear compliance must be a quantity extremely sensitive to the breadth of the molecular mass distribution. In this way the results of figure 3 become quite understandable. The spread can easily be considered as the consequence of slightly varying molecular mass distributions. The average value $J_e = 1.3 \cdot 10^{-5}\ \mathrm{Pa}^{-1}$ of the high molecular mass samples is certainly influenced by the finite breadths of the distributions so that the theoretically predicted value of $J_e \simeq 0.5 \cdot 10^{-5}\ \mathrm{Pa}^{-1}$ remains very realistic for strictly monodisperse samples (of $M > M_e$!).

On the other hand, eqs. (7) and (8) may serve as useful starting points for a polydispersity investigation on polymer melts. In this way, part of the early expectations [4] appears to be preserved. Rather recently Münstedt [17] has measured the equilibrium tensile compliances D_e of melts of a series of polystyrenes of varying molecular mass distributions, characterized by the familiar ratio M_w/M_n. These

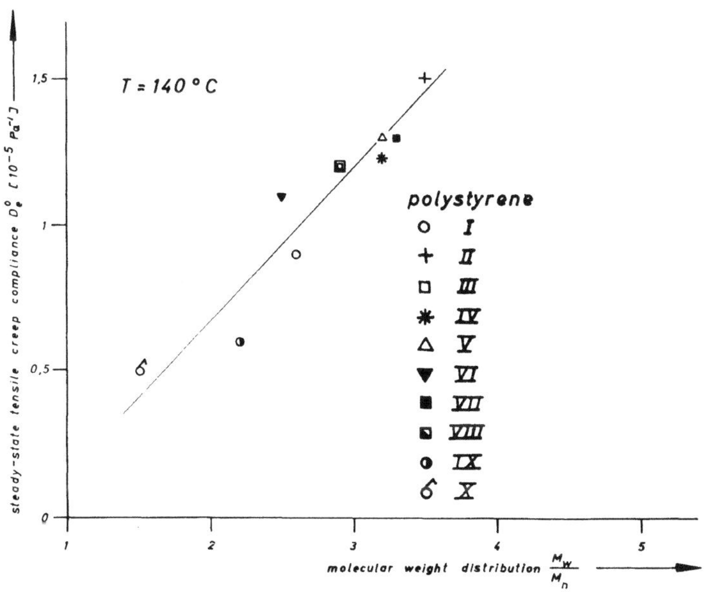

Fig. 4. Equilibrium tensile compliance D_e against the ratio M_w/M_n of weight to number average molecular mass for a series of polystyrenes according to Münstedt [17]

compliances can be measured directly as recoils with the aid of a proper tensile rheometer. Münstedt's results are presented in figure 4. The influence of M_w/M_n on D_e is quite obvious. If the drawn straight line is extrapolated tentatively to $M_w/M_n = 1$, one obtains for a monodisperse sample $D_e \simeq 0.15 \cdot 10^{-5}$ Pa^{-1}. Since $J_e = 3\,D_e$, the corresponding value of J_e would be $0.45 \cdot 10^{-5}\,\mathrm{Pa}^{-1}$ which is rather close to the theoretical one.

4. Melt fracture

One of the interesting predictions of the Doi-Edwards theory for the non-linear range of flow properties is concerned with the well-known melt fracture phenomenon. In figure 5 a dimensionless theoretical curve, as given by Doi and Edwards, is shown. It gives the shear stress according to the disengagement phase C as a function of the shear rate. This curve shows a maximum at $q\tau_C\,(= \kappa T_d)$ of about two. The height of this maximum is $\sigma_{12}/G_0 \simeq 0.36$ (G_0 in the papers of Doi and Edwards is three times our G_0). If one clings to the model, one has to expect instability for $q\tau_C > 2$. In reality, at high values of the shear rate phase B must become effective, which is ignored so far. However, if for a high molecular mass sample τ_C is well separated from τ_B (see eqs. (1)), the situation may occur which is sketched in figure 5 by the dashed line showing the possible influence of phase B. In this situation melt fracture is conjectured to occur at $q\tau_C \simeq 2$. For a polydisperse sample the location of this phenomenon on the shear rate axis is less clear. On the other hand, the height of the maximum, as being independent of molecular mass like G_0, should be rather invariant. For polystyrenes with $G_0 \simeq 2 \cdot 10^5$ Pa one should expect a critical shear stress maximum of about $0.7 \cdot 10^5$ Pa. Interestingly enough, Den Otter [18] quotes a range of critical shear stresses between 0.4 and $3.0 \cdot 10^5$ Pa for a great variety of industrial polymers.

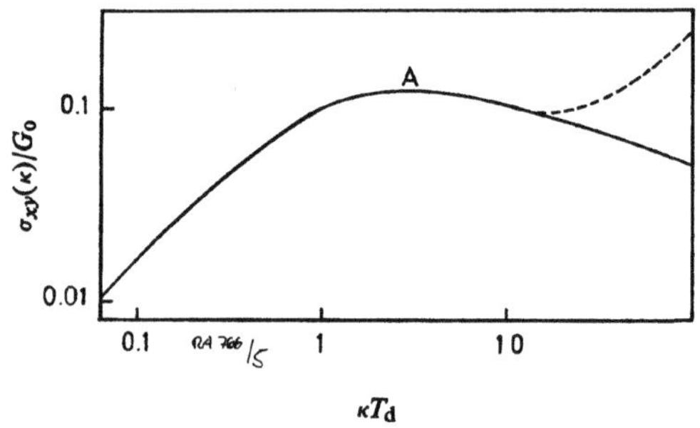

Fig. 5. Reduced shear stress vs. reduced shear rate according to Doi and Edwards [1]. (Courtesy of the Chemical Society)

5. Equilibration phase in rubber

As is well-known, the neo-Hookean elasticity modulus E_a for tensile experiments on rubbers (with λ being the stretching ratio) is given by

$$(E_a/3) = \sigma/(\lambda^2 - \lambda^{-1}) = 2[C_1 + C_2/\lambda]\,, \qquad (9)$$

if the Mooney-Rivlin approximation [19] is used. Quite recently, Noordermeer and Ferry [20] reported that, with stress relaxation measurements on 1,2-polybutadienes, C_1 relaxes much faster than C_2. This is clearly shown with the aid of Mooney-Rivlin plots,

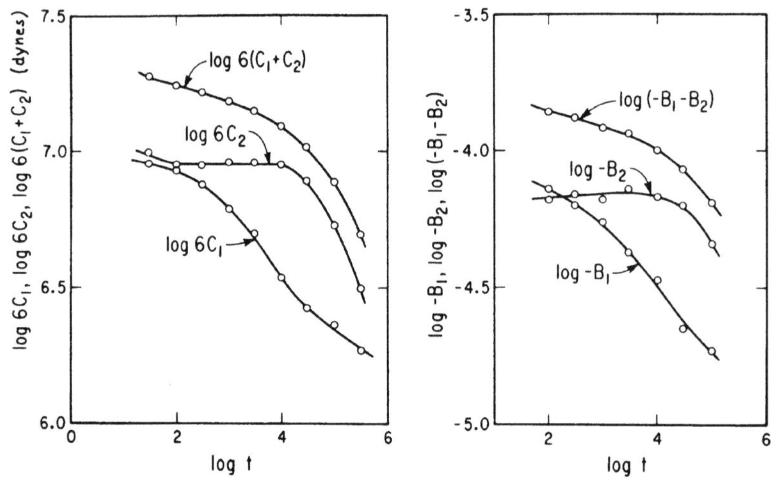

Fig. 6. Plots of C_1 and C_2 (and their optical equivalents) vs. time for a 1,2-polybutadiene sample (sample A) at 0°C according to Noordermeer and Ferry [20]. (Courtesy of John Wiley, New York)

where the intersection of the straight line (describing E_a vs. $1/\lambda$) with the ordinate axis comes down much faster than the slope of this line. This is shown in figure 6. In the left part the mechanical quantities are shown, whereas in the right part the equivalent optical quantities, as derived from the measured birefringence, are presented. Doi has indicated in the mentioned paper [2] that the decrease of C_1 should be ascribed to the equilibration phase B, whereas the decrease of C_2 should be due to the disengagement phase C. Interestingly enough, Hoffmann [21] has reported some time ago that unvolcanized rubber shows a neo-Hookean elasticity modulus just proportional to $(1/\lambda)$. It may be that in Hoffmann's experiments the equilibration occurred so fast that it was overlooked.

References

1. Doi, M., S. F. Edwards, J. Chem. Soc., Faraday Trans. II, **74**, 1789, 1802, 1818 (1978); **75**, 38 (1979).
2. Doi, M., J. Polym. Sci., Phys. Ed. **18**, 1005 (1980).
3. De Gennes, P. G., J. Chem. Phys. **55**, 572 (1971).
4. Peticolas, W. L., Rubber Chem. Tech. **36**, 1422 (1963).
5. Tordella, J. P., J. Appl. Phys. **27**, 454 (1956).
6. Ciferri, A., P. J. Flory, J. Appl. Phys. **30**, 1498 (1959).
7. Curtiss, C. F., R. B. Bird, J. Chem. Phys. **74**, 2016, 2026 (1981).
8. Bird, R. B., C. F. Curtiss, H. H. Saab, Proceedings of the IUPAC-Symposium on Macromolecules (Amherst 1982).
9. Rouse, P. E., J. Chem. Phys. **21**, 1272 (1953).
10. Graessley, W. W., Adv. Polym. Sci. **16**, 1 – 179 (1974).
11. Berry, G. C., T. G. Fox, Adv. Polym. Sci. **5**, 261 – 357 (1968).
12. Te Nijenhuis, K., "Dynamic Mechanical Studies on Thermo-Reversible Ageing Processes in Gels of Polyvinyl Chloride and Gelatine", Doctoral Thesis, Delft (1979).
13. Mieras, H. J. M. A., C. F. H. van Rijn, Nature **218**, 865 (1968).
14. Fox, T. G., J. Polym. Sci. **C9**, 35 (1965).
15. Wales, J. L. S., "The Application of Flow Birefringence to Rheological Studies of Polymer Melts", Doctoral Thesis, Delft (1976).
16. Onogi, S., T. Masuda, K. Kitagawa, Macromolecules **3**, 109 (1970).
17. Münstedt, H., Rheol. Acta **14**, 1077 (1975).
18. Den Otter, J. L., Plastics & Polym. **38**, 155 (1970).
19. Treloar, L. R. G., "The Physics of Rubber Elasticity", 3rd ed., Clarendon Press (Oxford 1975).
20. Noordermeer, J. W. M., J. D. Ferry, J. Polym. Sci., Phys. Ed. **14**, 509 (1976).
21. Hoffmann, M., in: M. Hoffmann, H. Krömer, R. Kuhn (eds.): "Polymeranalytik I", Thieme (Stuttgart 1977).

(Received April 8, 1982)

Author's address:

Prof. Dr. H. Janeschitz-Kriegl
Physikalische Chemie
Johannes-Kepler-Universität Linz
Altenberger Straße 69
A-4040 Linz

CONTRIBUTED PAPERS (SHORT VERSIONS)

Herbert Freundlich – Pioneer authority in surface, colloid and rheological science: Recollections on the centenary commemorations

D. W. Gillings

Department of Chemistry, University College of London

Key words: Herbert Freundlich, centenary commemorations, thixotropy

1. Introduction

For much of its history, science evolves in a mainly steady way, yet there have undoubtedly been times of marked advance. The later years of the last century saw the establishment of physical chemistry as a sector of science in its own right. Regarding this, the most concise way in which one can grasp the diversity of advances is to recall the originators – Gibbs, van der Waals, Arrhenius, van t'Hoff, to name but a very few in specific fields. Among others to remember is Wilhelm Ostwald, who presented much of this diversity in his expository work, underlining its character in the title, "Outlines of *General* chemistry". The work of his school at Leipzig attracted far-sighted collaborators, among whom was the widely-talented young Herbert Freundlich, who took the vocation of science from an ample choice of careers. Right from the first stage of original work, he took the first steps in what was to become an ever-broadening specialisation with his work "Die Ausfällung kolloidaler Lösungen durch Electrolyte" [1]. From this, at the turn of the century, he attained the status of authority in colloid science, ranging from its fundamentals in surface phenomena to widespread industrial application. "Kapillarchemie", destined to be foremost among the standard works, appeared in 1909, and in successive editions. Over the 25 years that followed, he founded the school of research at the Kaiser-Wilhelm-Institut in Berlin, and with a band of distinguished collaborators, published extensively on a great diversity of colloid chemical and related topics. The achievements of so many years prior to his untimely death in 1941 were narrated by the late Frederick Donnan, in his memoir [2], who wrote: "... the world of science (was deprived) of its most distinguished investigator and expounder of colloid and interfacial phenomena ... who had won respect, esteem, and affection ... of a host of friends in many lands ...". These years of such great endeavour with so many colleagues were tragically interrupted by his departure from the Institute in Berlin in 1933, under the discriminating pressures of the regime of the time. In Donnan's words again: "... rather than (comply) ... he resigned, and (accepted the invitation) coming to University College London as Honorary Research Associate ... supported with wise foresight by Imperial Chemical Industries, ... an honoured and welcome guest in the same college where Thomas Graham had worked."

The continuation of Freundlich's work in London gave some of us the opportunity for our post-graduate research with the most welcome benefit of his most unusual scientific leadership. I was among these fortunate few, and fortunate again that I was able to return to the College 40 years later, and together with several of his contemporaries of the earlier days, could assure a commemoration of Freundlich's life and work during 1980, the 100th year after his birth.

2. The centenary commemorations

Herbert Freundlich's personal concern for due honour to others for their life work was most characteristic. I well remember his speaking of contributions to the Festschriften, of Bodenstein, of Elster and Geitel and others, and we thus looked at the prospect for something in the same tradition and worthy of the occasion. Possibly under the shadow of "big science" – albeit in good company – interest in colloids has perhaps been a trifle muted, certainly not universal, even where relevant to applied science [3] and it was thus very welcome to enjoy practical support firstly from University College, and sustained by the specialist groups of the Society of Chemical Industry and the then Chemical Society.

The immense range of Freundlich's work set us a task in selecting topics which would provide for a wide review. The leading themes of "Kapillarchemie" alone would suffice several times, without even turning to continuing research papers. In the event, thanks to the knowledge, experience and standing of the authors, this wide review was closely approached in the five papers presented at the joint Societies' Groups Symposium of January 28, 1980 the birthday centenary [4], as will be apparent from short titles: "Colloid systems within colloid systems – the example of latex"; "Gas separation by adsorption"; "Nucleation and crystal growth"; "... Freundlich ... and colloidal dispersions ..."; "Surfactant layers ...".

As well as these "sector" papers, two of us [5] among the last of the school in Europe spoke in a more personal vein, rounding off an occasion well-supported and fitting, within its scope. Yet all this did leave much that one would have wished to achieve. It was therefore good news indeed that the Division of Colloid and Surface Chemistry of the American Chemical Society would present a session of Centennial Lectures at their 1981 Colloid Symposium. A splendid programme was prepared by Professor Wilfried Heller, who had worked with Freundlich in Germany before the departure, continued in the United States after Freundlich's appointment as Distinguished Service Professor at the University of Minnesota, and remains working well and truly in the field at Wayne State University, Detroit. The great success of this event, which drew a large attendance, redounds greatly to Professor Heller's credit, particularly as to the scope of the subject matter, and to the authority and world-wide affiliation of the contributors, as evident from the title-list: E. Matijevic, New York, on "Interactions of monodisperse colloidal metal oxides"; M. Kahlweit, Max-Planck-Institut Berlin, "Kinetics of formation of associated colloids"; T. Healy, Melbourne, "Electrokinetic phenomena ... (at) ... solid-aqueous solution interface(s)"; A. Vrij, van t'Hoff Laboratories, Utrecht, on "Structural properties ... of colloid particles ... by light scattering ...".

I would also like especially to mention Professor Heller's speech "Freundlich as a scientist", the extensive biographical work he has prepared, and his happy thought of presenting a collection of short memoirs from many past collaborators, from the Kaiser-Wilhelm times onward, at the banquet of celebration of the event.

3. The continuing influence of Freundlich's work

One might expect that the general standing of such a life's work would be familiar both to colloid specialists, and in outline, at least, to many others in science. However, over the past 40 years, increasing numbers and narrowing dimensions of so many specialisations have altered the prospects of overall views. The personal testimony of historical experience is significant, but of course subjective. The number and variety of relevant items in the literature means that the classical scan of abstracts in person is hardly practicable, and in a way, the perspective so lost is hardly to be replaced. However, we do have to-day's technology in retrieval, as realised in the Citation Index.

Any doubt about the extent of the Freundlich influence is dispelled at once by the briefest analysis of the citations. From the early 1960's onward, several hundred examples of reference to his publications are listed, having been referred to in over 100 journals, reviews and books, in relation to topics spreading from fundamental physical and life science to applied process technology and engineering. There is, moreover, much more than passing reference, and it is clear from following up specific citations that the past work still suggests starting points for new departures. It must be enough, here and now, to attempt a brief illustration by a single example – little more than a snapshot at random. This reports application of rheological measurement in research on the mechanism of aggregation of monomers illustrative of biochemical processes, as summarized in the table at the top of the next page.

There are 49 entries from a wide range of authors in the bibliography to this paper, and of these, the Freundlich reference is not only the earliest, but is quoted as among starting-points relevant in research of 40 years later. One can also read a certain general accord in the points raised in discussion with what might be imagined in a projection from the times of "Kapillarchemie", in the introduction of and appeal to the theme of molecular forces. By 1964, of course, it is valid to consider some very specific differentia-

Article carrying citation:		Article quoted in reference:	
Identification:	Subject matter:	Identification:	Subject matter quoted:
"Order-disorder transition in biocolloids" by W. L. Peticolas, J. Chem. Phys. **40** (6) p. 1463ff. March, 1964.	... theory of order-disorder transition tested by viscosity measurements at temperatures and concentrations of critical change...	"Streaming double-refraction in dye-stuff sols" ... H. Freundlich, C. Schuster, H. Zocher, Z. Phys. Chem. **105**, 119 (1923).	... classical colloid chemists observed ..., rod-shaped colloidal aggregates undergo *sharp* changes at *critical* rising temperature, and colloid properties of sol ... vanish suddenly.

tion between the nature of various forces determing the transformation from the monomeric state to rod-like aggregation. Compared with the present state of knowledge, the background in 1923 was sparse indeed. None the less, the references at that time to the nature, magnitude and range of molecular forces can scarcely be dismissed as superficial, and there is some encouragement to attempt a further constructive look into possible comparison, with the background which has been accumulating on the fundamentals. Indeed, attempting a short version of what Freundlich wrote on this soon shows that it was substantial, and in the present context, one can do little more than indicate only the first of possible points at which to start, namely the introduction of van der Waals' equation – primarily for the gaseous state – into the descriptive treatment of surface forces in condensed phases, as manifest by surface tension. To quote [6]: "... Although the theory of van der Waals does not in general accurately represent the behaviour of liquids, we may regard it ... (as a basis for indication of) ... order of magnitude of internal pressure ...".

Then, with cautious realism, Freundlich continues: "... It would doubtless be very desirable to start from a general theory of liquids, and base the capillary chemistry of liquids upon it. But if we consider how complicated van der Waals' theory of liquids already becomes for a mixture ... we can gauge the difficulty of the task. ... we can (hardly) make quantitative use of progress made (recently) from the kinetic point of view in the theory of the structure of crystals and atoms. We have, however ... a possibly better qualitative understanding ..."; and he concludes: "I have contented myself with connecting the related phenomena of capillary chemistry by a few ... (thermodynamic generalisations). This does not mean that molecular-kinetic considerations will not receive attention when ... successful...".

Since the time of "Kapillarchemie" there has, of course, being massive work on molecular forces, and on many lines, though many years did pass before the import of some of the background became generally appreciated. As with colloid science itself, this is a field of which the depth and diversity daunts any inclination towards any attempt either at synopsis or selection. We are, of course, never to know directly the extent of Freundlich's insight into the physics of these phenomena, whether at molecular level or as observed directly in colloid macro-systems, but he does refer to discussion with London [7] for example, and one could be sure that he would have kept abreast of the advanced developments which continue to emerge. I would have welcomed his view on the correspondence which might be postulated between "observable" liquid-gaseous interfaces, and molecular dynamic analysis of current theoretical work. Even if this had to be accepted as "forbiddingly indefinite", I think one could be certain of a constructive response, tolerant of the errors of speculation, a correction by the Good Shepherd of strays from the scientific path.

4. Thixotropy

There might be excessive risk of speculation were these lines to be extended further in comparing long range insights and actual evolution at such fundamental levels, but there is no doubt of the reality and significance of the major developments which could be ascribed to Freundlich's more immediate perceptions. Of these, there is none more major than that of the observational and experimental work in rheology, a field now mature, but only newly named in his time, and in particular, that which established the significance of a phenomenon, which itself also merited a new description, namely thixotropy.

This requires no line-by-line introduction to this, of all possible audiences. The work which had started almost tentatively in the early 1920's rapidly gathered real momentum, and by 1935, Freundlich wrote his most notable monograph, entitled simply "Thixotropy", one of a series on the Colloid State, with Frederick Donnan as general editor. It was considered a masterpiece, being a current exposition, a retrospective review and outlining alike the sectors of probable future advance. Although it has been quoted, as mentioned, in a number of citations, immediate projections have not seemed very accessible, as indeed is the case with the work itself, and the real breadth and depth of the treatment has to be inferred from later work which can be traced back in part to these foundations. Although not very long (around 20 000 words) it draws on many sources in its 83 references. It is hardly possible to summarise the variety and concentration of subject matter both synoptically and in a few paragraphs, and for the present, it must be enough to select only three suitably diverse points, more a skeleton than a summary:

– The basic questions of "equilibrium" sol/gel structure, mechanism of transformation, and conditions for stability are considered by reference to a "family diagram" of plots of the familiar potential energy/inter-particle distance relation for a range of system conditions from stable sol, through thixotropic state, to coagulation to gel. Much of this is consistent with how the knowledge has since developed, but at the time, he wrote: "... In this way, fairly general insight is gained, but ... (one can hardly regard investigation of thixotropy on these lines) ... as a method for (directly) testing a theory of van der Waals' forces." [7].
– There are projections of industrial application, two of the more important having arrived at the "household word" stage, namely the oil industry's "drilling muds" and "non-drip" paints.
– The discussion of biological phenomena, one of the themes of this European meeting, and one of the sources of Freundlich's original attraction to colloid science, even before 1900.

Where we may think that Freundlich's presentiments in this and other sectors have been borne out by the course of events and realisations, such a course is less like a defined track on well-surveyed land than an ocean passage. Oceans are for navigators, and perhaps, during these few years around his centenary, it may help to have his life of work as a reminder in setting the future courses. May I close by returning to Frederick Donnan, to whom we owe much for what it has been possible to write so far, with the last words of his memoir: "We, his many friends in many lands, will keep the memory of Herbert Freundlich ever green."

References

1. Freundlich, H., Z. Physikalische Chemie **44**, 129 (1903).
2. Donnan, F., Biographical Memoirs of Fellows of the Royal Society, p. 27 (1942).
3. Reports of a multi-disciplinary committee on colloid science: Science Research Council, London (1972, 1978).
4. Freundlich Centenary Symposium: Chemistry and Industry, London, **1980** (8) 360, (May 3, 1980).
5. Bondy, C., Symposium address, University College Chemical Society, London, January 28, 1980 (text in 1980 Newsletter, Association of University College Chemists, to be re-published). Gillings, D. W., see ref. [4], pp. 360 – 363.
6. Freundlich, H., "Colloid and Capillary Chemistry" (translated by H. S. Hatfield), p. 10 et seq., Methuen & Co. (London 1926).
7. Freundlich, H., "Thixotropy": Herman & Cie., Monograph Séries "Actualités Scientifiques ...": p. 12, line 1 et seq. (Paris 1935).

(Received May 26, 1982)

Authors' address:

Dr. D. W. Gillings
Department of Chemistry
University College of London
20, Gordon Street
London WC1H 0AJ (U.K.)

Rheologica Acta Rheol. Acta 21, 398 – 399 (1982)

Constitutive equations of the differential type as approximations

J. Meixner

Institute of Theoretical Physics, Rheinisch-Westfälische Technische Hochschule Aachen

Abstract: Constitutive equations of the differential type are often considered as approximations to constitutive equations of the after-effect type. In this case, conclusions drawn from them in conjunction with the Clausius-Duhem inequality are unfounded and untenable.

Key words: Constitutive equation, Clausius-Duhem inequality

Constitutive equations of the differential type (for the definition see [1], p. 93) have become very popular during the past two decades with a colossal discharge of literature. Combined with the Clausius-Duhem inequality they lead to some remarkable results.

The applied procedure shall here be demonstrated in one of the simplest possible examples which is a one-dimensional isothermal stress-strain relation of the differential type

$$\sigma = \sigma(\varepsilon, \dot{\varepsilon}), \tag{1}$$

where σ, ε are appropriate stress and strain measures. In addition a free energy is introduced which also depends on $\varepsilon, \dot{\varepsilon}$,

$$f = f(\varepsilon, \dot{\varepsilon}). \tag{2}$$

σ is assumed to be a continuous function of $\dot{\varepsilon}$ and f is assumed to have continuous partial derivatives f_ε and $f_{\dot{\varepsilon}}$. Then the Clausius-Duhem inequality can be written as (see [1], p. 361)

$$\sigma\dot{\varepsilon} - \rho\dot{f} \geqq 0 \tag{3}$$

with ρ = density, or

$$\sigma(\varepsilon,\dot{\varepsilon})\dot{\varepsilon} - \rho(\varepsilon)f_\varepsilon(\varepsilon,\dot{\varepsilon})\dot{\varepsilon} - \rho(\varepsilon)f_{\dot{\varepsilon}}(\varepsilon,\dot{\varepsilon})\ddot{\varepsilon} \geqq 0. \tag{4}$$

This inequality is required to hold for all values of ε, $\dot{\varepsilon}$, $\ddot{\varepsilon}$ (see [1], p. 361), in particular for arbitrarily large, positive or negative, values of $\ddot{\varepsilon}$. Then it follows immediately from (4) that

$$\partial f(\varepsilon,\dot{\varepsilon})/\partial\dot{\varepsilon} = 0, \quad \text{i.e.} \quad f = f(\varepsilon). \tag{5}$$

Moreover, if (4) is considered for very small values of $\dot{\varepsilon}$, one arrives at

$$\sigma(\varepsilon, 0) = \rho(\varepsilon) f_\varepsilon(\varepsilon). \tag{6}$$

Thus the thermostatic relation between stress and free energy is a special result of the general theory. There remains the dissipation inequality

$$[\sigma(\varepsilon,\dot{\varepsilon}) - \sigma(\varepsilon, 0)]\,\dot{\varepsilon} \geqq 0. \tag{7}$$

So far, so good, because this is all pure mathematics carried through under well formulated assumptions.

But there is considerable doubt whether materials exist which are exactly of the differential type (even if one includes any finite number of time and space derivatives of external thermodynamic variables in the constitutive equations) for arbitrary values of the various derivatives.

More generally, one sets up constitutive equations of the memory type, which are also called to be of after-effect character. In our one-dimensional and isothermal example they would read

$$\sigma(t) = \Sigma\{\varepsilon(t-s);\quad 0 \leqq s < \infty\}, \tag{8}$$

$$f(t) = \Phi\{\varepsilon(t-s);\quad 0 \leqq s < \infty\}. \tag{9}$$

That is to say, $\sigma(t)$ and $f(t)$ depend on the previous treatment (or strain history) of the material element, i.e. they are functionals of the history $\varepsilon(t - s)$, $s \geqq 0$.

Now the transition to constitutive equations of the differential type under suitable conditions seems to be quite easy. We quote Truesdell and Noll ([1], p. 93, see also Noll [2], p. 221) with an adaptation to the present example:

"It may happen that only a very short part of the strain history has an influence on the stress, i.e. that the value of $\sigma(t)$ depends only of the values of $\varepsilon(t - s)$ for s very near to zero. If the strain has sufficiently many time derivatives, then $\varepsilon(t - s)$ may be approximated, for s near zero, by its Taylor expansion about $s = 0$ up to some order r. This expansion is determined, for $r = 1$, by $\varepsilon(t)$ and $\dot{\varepsilon}(t)$. A material in which the stress depends only on a finite number of derivatives of $\varepsilon(t)$ is called a material of the differential type."

Similar considerations can be made for the free energy functional.

Of course, this restricts severely the class of permitted histories. They have, for $r = 1$, to be so that the term with $\ddot{\varepsilon}(t)$ in the Taylor expansion can be neglected, in other words, that at least $\ddot{\varepsilon}(t)$ be sufficiently small during the whole process.

A similar formulation is given by I. Müller [3]:

„Der Grundgedanke, der zur Unterscheidung zwischen Materialien mit verschieden gutem Gedächtnis führt, ist der folgende: Es sei $h_t(s) := h(t - s)$ mit $0 \leqq s < \infty$ eine analytische Funktion, die die Geschichte eines thermodynamischen Feldes repräsentiert. Wir approximieren diese Geschichte durch eine beim $(N + 1)$-ten Glied abgebrochene Taylorreihe

$$h_t(s) \approx \sum_{n=0}^{N} (-1)^n \frac{d^n h(t)}{dt^n} s^n/n! \,. \tag{10}$$

Es ist bekannt, daß diese Reihe die Funktion $h_t(s)$ in einer Umgebung T_N um $s = 0$ recht gut wiedergibt und zwar ist T_N um so größer, je größer N ist. Wenn nun das Gedächtnis eines Körpers nicht über die Zeit T_N hinausreicht, so werden die Materialgrößen zur Zeit t nicht vom Verhalten der Funktion $h_t(s)$ für $s > T_N$ abhängen und die Geschichte $h_t(s)$ kann dann in den Materialgleichungen durch die $N + 1$ Variablen

$$d^n h(t)/dt^n \quad (n = 0, 1, \ldots, N)$$

ersetzt werden."

If one looks, however, at the procedure with which the Clausius-Duhem inequality is exploited for such constitutive equations, one notices that arbitrarily large, positive and negative, values of $\ddot{\varepsilon}(t)$ or $d^{N+1}h(t)/dt^{N+1}$, respectively, are made use of (see the demonstration example given before), and therefore the condition under which the Taylor expansion can be truncated is violated.

Thus we are confronted with the following situation:

1. Materials with memory can possibly be approximated by materials of the differential type. But this is possible only for a suitably restricted class of histories.

2. An exploitation of the Clausius-Duhem inequality with such approximate constitutive equations does not allow consideration of arbitrarily large, positive and negative, values of those derivatives which are omitted in these constitutive equations.

Thus all results which are derived in this manner for approximate constitutive equations of the differential type are unfounded and untenable.

References

1. Truesdell, C., W. Noll, The Non-Linear Field Theories of Mechanics in: S. Flügge (ed.), Encyclopedia of Physics Vol. III/3, Springer-Verlag (Berlin-Heidelberg-New York 1965).
2. Noll, W., A Mathematical Theory of the Mechanical Behavior of Continuous Media, Arch. Rational Mech. Anal. **2**, 197–226 (1958).
3. Müller, I., Thermodynamik. Die Grundlagen der Materialtheorie (Düsseldorf 1973).

(Received May 19, 1982)

Author's address:

Prof. Dr. J. Meixner
Institut für Theoretische Physik
Rhein.-Westf. Technische Hochschule
D-5100 Aachen

An approximation method to relate the linear viscoelastic functions

F. Mainardi and G. Turchetti

Istituto di Fisica dell'Università di Bologna

Abstract: A method based on rational approximations is presented to interpolate the data from sinusoidal experiments in linear viscoelasticity. Bounds to the corresponding dynamical function and a discrete approximation to the spectrum are established. From this approximation the related viscoelastic functions can be computed. The method is checked by considering two theoretical models of physical interest and a satisfactory accuracy is achieved.

Key words: Linear viscoelasticity, creep compliance, rational approximation, Padé approximant

1. Introduction

The linear viscoelastic phenomena are described by various material functions whose interrelations are analytically known (see i.e. [1 – 3]). The experimental procedures allow to determine some of these functions at a finite number of points so that methods have to be introduced to interpolate the data and to compute approximately the other functions (see i.e. [2 – 4]).

The purpose of the present paper is to recall the attention of rheologists to a new mathematical method, introduced by the authors [5], which makes an extensive use of the peculiar nature of the viscoelastic functions, which are integral transforms of positive measure. A more detailed analysis is under preparation, taking also into account recently published results (see i.e. [6]).

Referring to one-dimensional compliances, the material functions $J(t)$ and $\hat{J}(i\omega)$, known as the creep compliance and the complex compliance respectively, are expressed by

$$J(t) = J_0 + \int_0^\infty [1 - \exp(-t/\tau)]\,\xi(\tau)\,d\tau\,, \tag{1}$$

$$\hat{J}(i\omega) = J_0 + \int_0^\infty \frac{\xi(\tau)}{1 + i\omega\tau}\,d\tau = J'(\omega) - iJ''(\omega) \tag{2}$$

where t is the time, ω the frequency, J_0 the instantaneous compliance and $\xi(\tau)$ the retardation spectrum.

Up to the not essential constant J_0 the previous functions are integral transforms of positive measure on $R^+ = [0, +\infty)$ given by

$$\Xi(\tau) = \int_0^\tau \xi(\tau')\,d\tau'\,. \tag{3}$$

In particular $\hat{J}$ turns out to be a so-called Stieltjes function, that is a Hilbert-like transform of positive measure. For such a class of functions relevant features of their approximation based on Padé approximants can be proved [7, 8].

In this note we assume a dynamic response, say $J'(\omega)$, to be given at N frequencies and we present a rational interpolation method which allows to obtain bounds and to approximate the related viscoelastic functions $\Xi(\tau)$, $J''(\omega)$ and $J(t)$.

2. The mathematical scheme

Let $x = \omega^2$ and $J'(x)$ be given in a sequence of N points $0 < x_1 < x_2 < \ldots < x_N$. A rational interpolation $J'_N(x)$ is easily obtained from the Thiele continued fraction and corresponds to a Padé approximant (PA) for $J'(x)$, that is, to the $[(N/2 - 1)/(N/2)]$ PA for N even, and $[\frac{1}{2}(N-1)/\frac{1}{2}(N+1)]$ PA for N odd.

Due to the positivity of the measure it can be shown that $J'_N(x)$ has the following bounding properties:

$$(-1)^k[J'(x) - J'_N(x)] > 0\,, \quad x_k < x < x_{k+1}\,, \qquad (4)$$

$k = 0,1,\ldots,N$ where $x_0 = 0$, $x_{N+1} = \infty$.

From $J'_N(x)$ and $J'_{N-1}(x)$ computed by dropping the first point x_1 one obtains upper and lower bounds for $x > x_2$.

Furthermore, expanding $J'_N(x)$ into simple fractions we obtain the following representation

$$J'_N(x) = \xi_{0,N} + \sum_{k=1}^{[N/2]} \frac{\xi_{k,N}}{1 + x\tau_{k,N}^2} \qquad (5)$$

where [.] denotes the integer part, $\xi_{0,N} = 0$ for N even, and the $\xi_{k,N}$ are all positive. We remind that eq. (5) is a consequence of the analyticity properties of $\hat{J}(i\omega)$ and allows to obtain a stepwise approximation to the measure $\Xi(\tau)$, that reads

$$\Xi_N(\tau) = \xi_{0,N}\theta(\tau) + \sum_{k=1}^{[N/2]} \xi_{k,N}\theta(\tau - \tau_{k,N}) \qquad (6)$$

where θ denotes the Heaviside function.

Rigorous upper and lower bounds to $\Xi(\tau)$ can be obtained by extending to the interpolation problem the procedure due to Krein for the moment problem.

Eq. (6) corresponds to a discrete approximation of the retardation spectrum $\xi(\tau)$ which enables us to obtain the corresponding approximate expressions for the viscoelastic functions $J''(\omega)$, $J(t)$ as

$$J''_N(\omega) = \omega\left[\xi_{0,N} + \sum_{k=1}^{[N/2]} \frac{\xi_{k,N}\tau_{k,N}}{1 + \omega^2\tau_{k,N}^2}\right], \qquad (7)$$

$$J_N(t) = \xi_{0,N} + \sum_{k=1}^{[N/2]} \xi_{k,N}[1 - \exp(-t/\tau_{k,N})]\,. \qquad (8)$$

When the actual spectrum is discrete we have $\hat{J}_N(i\omega) = \hat{J}(i\omega)$, $J_N(t) = J(t)$ for N large enough. In any other case $\hat{J}_N(i\omega)$ converges uniformly to $\hat{J}(i\omega)$ when $N \to \infty$.

In any practical application the data are affected by experimental errors and the positivity of $\xi_{k,N}, \tau_{k,N}$ gives a set of constraints to be necessarily fulfilled by any linear viscoelastic model.

3. Numerical results

In order to check the method we have considered the data for $J'(\omega)$ as given by two theoretical models of physical interest and we have compared our results with respect to the exact viscoelastic functions. The models, denoted by a), b), correspond to a discret and a continuous spectrum respectively. Model a) is characterized by

$$\hat{J}(i\omega) = \sum_{k=1}^{M} \frac{\xi_k}{1 + i\omega\tau_k}\,, \qquad J(t) = \sum_{k=1}^{M} \xi_k[1 - \exp(-t/\tau_k)] \qquad (9a)$$

where we assume $M = 4$, $\xi_k = 1$ and $\tau_k = 1/k^2$, according to the Mooney-Rouse theory of cross-linked polymers [2].

Model b) is defined by

$$\hat{J}(i\omega) = \frac{1}{1 + (i\omega)^{1/2}}\,; \quad J(t) = 1 - E_{1/2}[-t^{1/2}] \qquad (9b)$$

where $E_{1/2}(.)$ denotes the Mittag-Leffler function of order $\frac{1}{2}$ (see e.g. [9]).

For the model a) we have checked that with $N = 2M = 8$ data for $J'(\omega)$, that is with a $PA[\frac{3}{4}]$, the viscoelastic functions are exactly reproduced.

For the model b) we have assumed $N = 12$ data for $J'(\omega)$ at the frequencies ω_k given by $\log(\omega_k) = (k - 6)/4$, $k = 1, 2, \ldots, 12$. The maximum relative error for $J'(\omega)$ computed from the upper and lower bounds was found to be below 0.5‰. The results for the creep compliance $J_N(t)$ given by eq. (8) were compared with the collocation approximation developed by Schapery [10]. The collocation method allows to obtain $J(t)$ from $J'(\omega)$ and explicitly reads

$$J_N^c(t) = \sum_{k=1}^{N} c_k[1 - \exp(-\omega_k t)] \qquad (10)$$

where the coefficients c_k are determined by minimizing the L^2 norm of $[J_N^c(t) - J(t)]$. The results are summarized in table 1, where the proposed approximation appears to be slightly better than the collocation results for a wide range of times.

Table 1. Exact values of $J(t)$ for the model (9b) and relative errors for the Padé and the collocation methods: $\Delta P = |1 - J_N(t)/J(t)|$, $\Delta C = |1 - J_N^c(t)/J(t)|$ for $N = 12$ input data

t	0.01	0.05	0.1	0.5	1	5	10
$J(t)$	0.1035	0.2096	0.2764	0.4768	0.5724	0.7676	0.8294
ΔP	0.19	0.012	0.0002	0.001	0.0013	0.0052	0.0094
ΔC	0.29	0.02	0.0035	0.0035	0.0021	0.013	0.022

References

1. Gross, B., "Mathematical Structure of the Theories of Linear Viscoelasticity", Hermann (Paris 1953).
2. Ferry, J. D., "Viscoelastic Properties of Polymers", 2. ed., Wiley (New York 1970).
3. Nowick, A. S., B. S. Berry, "Anelastic Relaxation in Crystalline Solids", Academic Press (New York 1972).
4. Booij, H. C., G. P. J. M. Thoone, Rheol. Acta **21**, 15 (1982).
5. Mainardi, F., G. Turchetti, Lett. Nuovo Cimento **26**, 38 (1979).
6. Schwarzl, F. R., "On the mechanical properties of polystyrene" First Conference of European Rheologists, Graz, April 14–16, 1982.
7. Baker, G. A., J. L. Gammel (eds.), "The Padé Approximants in Theoretical Physics", Academic Press (New York 1970).
8. Barnsley, M., J. Math. Phys. **14**, 299 (1973).
9. Caputo, M., F. Mainardi, Riv. Nuovo Cimento (Ser. 2) **1**, 161 (1971).
10. Christensen, R. M., "Theory of Linear Viscoelasticity", Appendix C, Academic Press (New York 1971).

(Received April 22, 1982)

Authors' address:

Prof. Dr. F. Mainardi, Dr. G. Turchetti
Istituto di Fisica "A. Righi"
Università degli Studi di Bologna
Via Irnerio, 46
I-40126 Bologna

Internal state variables and normality in viscoplastic flow

J. D. Weber and A. Eberhardt

CREST-Faculté des Sciences, Université de Metz (France)

Abstract: Isothermal rheological behavior is described by free energy $\phi(\varepsilon, p)$ and potential $\Omega(p, P)$ with $\dot{p} = \partial\Omega/\partial P$, $P = -\partial\phi/\partial p$, $\sigma = \partial\phi/\partial\varepsilon$. A number of usual rheological effects can be explained with only one or two (scalar or tensorial) internal state variables p.

Key words: Viscoplastic flow, internal state variable, normality

1. Introduction

Two functions are needed to describe isothermal rheological behavior with instantaneous elasticity in the formalism of internal state variables [1, 2]:

(i) the free energy $\phi(\varepsilon, p)$ relative to a material element which has unit volume in the reference configuration; ε is a right deformation tensor, such as the Green tensor; p are scalar or tensorial variables;

(ii) the pseudopotential of dissipation $\Omega(p, P)$ in which $P = -\partial\phi/\partial p$ is the thermodynamic action associated to variable p; Ω is a convex function of P for fixed p.

The evolution is called normal when $\dot{p} = \partial\Omega/\partial P$; this generalises Onsager's relations to non-linear behavior.

The dissipation is $P(\partial\Omega/\partial P) \geqslant 0$.

The stress tensor conjugate to ε is $\sigma = \partial\phi/\partial\varepsilon$.

The formalism is invariant when variables $q(p)$ are substituted to variables p. It describes any system of connected non-linear springs and dashpots having instantaneous elasticity.

We shall suppose here that Ω is a function of p and P, but that it is independent of ε. Uniqueness of quasi-static deformation of such media can be proven, when instantaneous elasticity is stable. Linear viscoelasticity is obtained when ϕ is a quadratic function of ε, p and Ω is independant of p and a quadratic function of P. We shall give here some examples of this formalism.

2. One-dimensional rheological model with two internal state variables

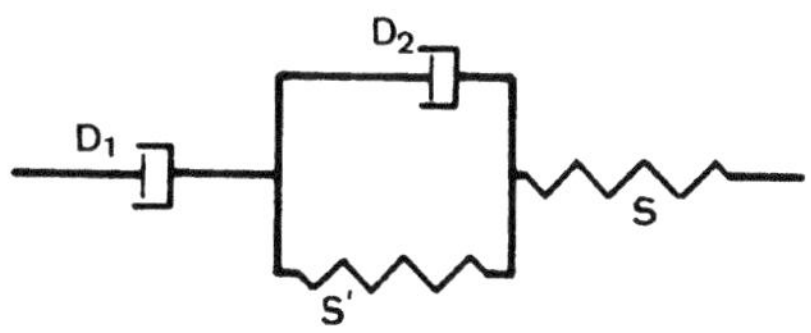

p_1 gives the state of dashpot D_1 i.e. the position of the piston in the cylinder; p_2 gives the state of dashpot D_2. The springs S and S' have stiffness k and k'.

We have

$$\phi = \frac{k}{2}(\varepsilon - p_1 - p_2)^2 + \frac{k'}{2}p_2^2,$$

$$\sigma = k(\varepsilon - p_1 - p_2),$$

$$P_1 = -\frac{\partial\phi}{\partial p_1} = k(\varepsilon - p_1 - p_2),$$

$$P_2 = -\frac{\partial\phi}{\partial p_2} = k(\varepsilon - p_1 - p_2) - k'p_2.$$

A possible expression of Ω will be

$$\Omega = l_1|P_1|^{\alpha_1} + l_2|P_2|^{\alpha_2}.$$

There results

$$\dot{p}_1 = \alpha_1 l_1 \operatorname{sgn} P_1 |P_1|^{\alpha_1 - 1}$$

$$\dot{p}_2 = \alpha_2 l_2 \operatorname{sgn} P_2 |P_2|^{\alpha_2 - 1}.$$

Since

$$\dot{\sigma} = k(\dot{\varepsilon} - \dot{p}_1 - \dot{p}_2),$$

it will be possible to find l_1, α_1, l_2 and α_2: $\sigma\dot{\varepsilon}$ can be measured in tension tests, p_1, p_2 can be measured after sudden unloading of the specimen.

3. Internal state variables and chemical reactions

In a reaction between perfect gases, volume v replaces ε, pressure π replaces σ, the internal state variable ξ characterizes the number of moles which have reacted.

Ω is supposed to be a quadratic function of thermodynamic action $-\partial\phi/\partial\xi$.

Consider for instance the reaction

$$\nu_A M_A + \nu_B M_B \to \nu_C M_C .$$

$n_A^0 n_B^0 n_C^0$ are the number of moles at instant $t = 0$, $n_A^0 - \xi\nu_A$, $n_B^0 - \xi\nu_B$, $n_C^0 + \xi\nu_C$ are the number of moles at instant t. We suppose that there exists a constant K_A such that

$$\phi_A = n_A RT \log \frac{K_A n_A}{v} = (n_A^0 - \xi\nu_A) RT \log \frac{K_A(n_A^0 - \xi\nu_A)}{v},$$

and that similar relation hold for ϕ_B, ϕ_C.

There follows

$$\frac{\partial\phi_A}{\partial\xi} = -\nu_A RT \log \frac{eK_A(n_A^0 - \xi\nu_A)}{v} = -\nu_A RT \log \frac{eK_A\pi_A}{RT},$$

$$\frac{\partial\phi}{\partial\xi} = \frac{\partial\phi_A}{\partial\xi} + \frac{\partial\phi_B}{\partial\xi} + \frac{\partial\phi_C}{\partial\xi} = RT \log \frac{K}{K_0},$$

$$K = \frac{\pi_C^{\nu_C}}{\pi_A^{\nu_A}\pi_B^{\nu_B}},$$

$$K_0 = \left(\frac{e}{RT}\right)^{\nu_A + \nu_B - \nu_C} \cdot K_A^{\nu_A} K_B^{\nu_B} K_C^{-\nu_C}.$$

The formalism of internal state variables gives us

$$\dot{\xi} = -g(n_A^0 n_B^0 n_C^0 \xi) RT \log \frac{K}{K_0}, \quad g \geqslant 0 .$$

At equilibrium ($\dot{\xi} = 0$, $K = K_0$),

$$\frac{\pi_C^{\nu_C}}{\pi_A^{\nu_A}\pi_B^{\nu_B}} = K_0 .$$

4. Maxwell fluid for finite deformations

The fluid is characterized by three constants: the two Lamé constants λ and μ of instantaneous elasticity, and the viscosity η.

Instead of ε we use $F_{ij} = \partial x_i/\partial a_j$. The internal state variable is a second-order symmetric tensor p such that $\det p = 1$.

$$\phi = \frac{\lambda}{2}(\theta - \alpha)^2 + \mu \operatorname{tr}(\varepsilon_e^2),$$

$$\theta = \log\det F,$$

$$\alpha = \frac{\mu}{3\lambda} \log\det p,$$

$$\varepsilon_e = \frac{1}{2}\log(FpF),$$

$$\Omega = \frac{1}{\eta} p_{ih} p_{jk} P_{ij} P_{hk} .$$

p is the left tensor of transformation from unstressed state to reference state.

5. Viscoplastic flow with one tensorial internal state variable

The internal state variable is the plastic deformation p_{ij}; for infinitesimal deformation

$$\phi = \frac{\lambda}{2}(\varepsilon_{ii} - p_{ii})^2 + \mu(\varepsilon_{ij} - p_{ij})(\varepsilon_{ij} - p_{ij}),$$

$$\sigma_{ij} = \frac{\partial \phi}{\partial \varepsilon_{ij}} = \lambda(\varepsilon_{hh} - p_{hh})\delta_{ij} + 2\mu(\varepsilon_{ij} - p_{ij}),$$

$$P_{ij} = -\frac{\partial \phi}{\partial p_{ij}} = \sigma_{ij}.$$

We call s_{ij} the deviator of P_{ij}. When

$$2J = s_{ij}s_{ij},$$

$$\Omega = f(J),$$

we obtain

$$\dot{p}_{ij} = \frac{\partial \Omega}{\partial P_{ij}} = f' \, s_{ij},$$

$$\dot{\sigma}_{ij} = \lambda \dot{\varepsilon}_{kk}\delta_{ij} + 2\mu\dot{\varepsilon}_{ij} - 2\mu f' \, s_{ij}.$$

6. Two tensorial internal state variables

By analogy with the model of sect. 2, we consider here the two tensorial internal state variables $p_{1,ij}$ and $p_{2,ij}$.

$$\phi = \frac{\lambda}{2}(\varepsilon_{ii} - p_{1,ii} - p_{2,ii})^2$$

$$+ \mu(\varepsilon_{ij} - p_{1,ij} - p_{2,ij})(\varepsilon_{ij} - p_{1,ij} - p_{2,ij})$$

$$+ \frac{\lambda'}{2}(p_{2,ii})^2 + \mu' p_{2,ij}p_{2,ij}.$$

For instance, let us call $s_{1,ij}$ the deviator of $P_{1,ij}$ and $s_{2,ij}$ the deviator of $P_{2,ij}$:

$$2J_1 = s_{1,ij}s_{1,ij}, \quad 2J_2 = s_{2,ij}s_{2,ij}.$$

If

$$\Omega = f_1(J_1) + f_2(J_2)$$

we have

$$\dot{p}_{1,ij} = f_1' s_{1,ij} \quad \dot{p}_{2,ij} = f_2' s_{2,ij}.$$

7. Electric systems

The formalism describes as well connected non-linear resistances and capacities. Strain and internal state variables are represented by quantities of electricity.

8. Conclusion

Internal state variables and the normality rule provide a good frame work for describing solids and fluids, with small or large deformation, linear and non-linear behavior.

References

1. Mandel, J., Variables cachées, Puissance dissipée, Dissipation normale. Cah. Groupe Français Rhéol., Numéro spécial Janvier 1980, p. 37.
2. Halphen, B., Nguyen Quoc Son, J. Mécanique **14**, 41 (1975).

(Received April 15, 1982)

Authors' address:

Prof. J. D. Weber, A. Eberhardt
CREST-Faculté des Sciences
Ile du Saulcy
F-57000 Metz

Rheologica Acta Rheol. Acta **21**, 406 – 408 (1982)

Materialgleichungen zur Beschreibung des sekundären und tertiären Kriechverhaltens anisotroper Stoffe

J. Betten

Rheinisch-Westfälische Technische Hochschule Aachen

Zusammenfassung: Zur Beschreibung des sekundären und tertiären Kriechverhaltens anisotroper Stoffe werden Materialgleichungen allgemein formuliert.

Die Anwendbarkeit der Kriechpotentialhypothese auf das Sekundärkriechen wird überprüft. Kriechschäden im tertiären Stadium werden tensoriell berücksichtigt.

Abstract: Considering both the secondary and tertiary creep stage of anisotropic materials constitutive equations are formulated.

The validity of the creep potential hypothesis is investigated. The anisotropic creep damage phenomenon is tensorially described.

Key words: Constitutive equation, secondary creep, tertiary creep, anisotropy, creep potential, anisotropic damage

1. Einleitung

Im Vortrag werden kontinuumstheoretische Gesichtspunkte erörtert, die bei der Aufstellung von Materialgleichungen zur Beschreibung des sekundären und tertiären Kriechverhaltens anisotroper Stoffe unter mehraxialer Beanspruchung zu beachten sind.

Neben der im Werkstoff anfangs vorhandenen (z. B. durch das Herstellungsverfahren bedingten) Anisotropie treten im tertiären Bereich Kriechschädigungen auf, die aufgrund ihres strukturellen Charakters richtungsabhängig sind und somit nur tensoriell erfaßt werden können. Beide Effekte müssen durch geeignete Argumenttensoren in Materialgleichungen berücksichtigt werden.

2. Sekundäres Kriechverhalten

Zur Darstellung des mehraxialen isotropen Kriechens wird von mehreren Autoren ein Kriechpotential benutzt, z. B. in [1, 2]. Die Theorie des Kriechpotentials ergibt sich wie die Theorie des plastischen Potentials [3 – 5] aus dem Prinzip der größten spezifischen Dissipationsleistung, woraus nach der Lagrangeschen Multiplikatorenmethode unter Einbeziehung einer Kriechbedingung $F(\sigma_{ij})$ = const. als Nebenbedingung die Fließregel

$$\dot{\varepsilon}_{ij} = \dot{\Lambda}\,[\partial F(\sigma)/\partial\sigma_{ij}] \tag{1}$$

folgt. Darin sind $\dot{\varepsilon}_{ij}$ die Koordinaten des Verzerrungsgeschwindigkeitstensors $\dot{\boldsymbol{\varepsilon}}$ und σ_{ij} die Koordinaten des Cauchyschen Spannungstensors $\boldsymbol{\sigma}$. Der Lagrangesche Multiplikator $\dot{\Lambda}$ wird in [2] für isotrope und in [6] für anisotrope Körper unter Berücksichtigung des Norton-Baileyschen Kriechgesetzes bestimmt. Die aus Gl. (1) in [2] gewonnenen Materialgleichungen isotroper Körper enthalten die von Odquist und Hult in [7] angegebenen Stoffgleichungen als Sonderfall.

Im isotropen Fall kann das Kriechpotential $F = F(\sigma)$ bekanntlich als Funktion der drei „Grundinvarianten" $S_\nu \equiv \mathrm{tr}\,\boldsymbol{\sigma}^\nu$, $\nu = 1, 2, 3$, des Spannungstensors $\boldsymbol{\sigma}$ dargestellt werden. Dann ist das Kriechpotential eine isotrope Tensorfunktion und erfüllt die Invarianz-Bedingung $F(a_{ip}a_{jq}\sigma_{pq}) \equiv F(\sigma_{ij})$ unter irgendeiner orthogonalen Transformation ($a_{ik}a_{jk} = \delta_{ij}$).

Für anisotrope Stoffe enthält das Kriechpotential neben dem Spannungstensor noch Stofftensoren zur

Charakterisierung der Anisotropie, z. B. durch einen Stofftensor vierter Stufe:

$$F = F(\sigma_{ij}, A_{ijkl}) . \tag{2}$$

Das Hauptproblem bei der Darstellung des Kriechpotentials (2) besteht in dem Auffinden eines irreduziblen Invariantensystems (*Integritätsbasis*) der gegebenen Tensorvariablen σ_{ij}, A_{ijkl}.

In [6, 8] wird auf die Konstruktion einer solchen Integritätsbasis näher eingegangen und das Kriechpotential (2) gemäß

$$F = F[S_1(\sigma), \ldots, S_3(\sigma); \quad S_1(A), \ldots, S_6(A);$$

$$\Pi_2^{[\mu]}, \ldots, \Pi_4^{[\mu]}] \tag{3}$$

dargestellt. Darin sind $S_\nu(\boldsymbol{\sigma})$, $S_\nu(\boldsymbol{A})$ die irreduziblen Invarianten der einzelnen Argumenttensoren $\boldsymbol{\sigma}$, $\boldsymbol{A}$, während

$$\left.\begin{aligned} &\Pi_2^{[\mu]} \equiv \sigma_{ij} A_{ijkl}^{(\mu)} \sigma_{kl}, \quad \Pi_3^{[\mu]} \equiv \sigma_{ij} A_{ijkl}^{[\mu]} \sigma_{kl}^{(\mu)}, \\ &\Pi_4^{[\mu]} \equiv \sigma_{ij}^{(2)} A_{ijkl}^{(\mu)} \sigma_{kl}^{(2)}, \quad \mu = 1, 2, \ldots, 5 \end{aligned}\right\} \tag{4}$$

Simultaninvarianten sind. Durch Differentiation im Sinne der Fließregel (1) erhält man schließlich aus (3) die Materialgleichungen des Sekundärkriechens.

3. Vergleich der Kriechpotentialhypothese mit der Darstellungstheorie

Nach der Darstellungstheorie sind Materialgleichungen vollständig durch die Linearkombination $\dot{\varepsilon} = \sum_\alpha \varphi_\alpha \boldsymbol{G}_\alpha$ darstellbar. Darin sind φ_α skalarwertige Funktionen der Integritätsbasis und $\boldsymbol{G}_\alpha$ symmetrische Tensorgeneratoren zweiter Stufe, die aus den betrachteten Argumenttensoren erzeugt werden. Somit erhält man die Materialgleichungen

$$\dot{\varepsilon}_{ij} = \frac{1}{2} \sum_{\lambda,\mu,\nu} \varphi_{(\lambda,\mu,\nu)} (\sigma_{ik}^{(\lambda)} A_{kjpq}^{(\mu)} \sigma_{pq}^{(\nu)} + \sigma_{jk}^{(\lambda)} A_{kipq}^{(\mu)} \sigma_{pq}^{(\nu)}) . \tag{5}$$

Im Sinne des Hamilton-Cayleyschen Theorems für Tensoren zweiter und vierter Stufe [9] sind in (5) die Potenzen λ, $\nu = 0, 1, 2$ und $\mu = 0, 1, 2, \ldots, 5$ möglich. Für $\mu = 0$ ist jedoch zu beachten, daß die Kombinationen $(\lambda, 0, \nu)$ in λ und ν symmetrisch sind und (wiederum aufgrund des Hamilton-Cayleyschen Theorems) der Einschränkung $\lambda + \nu \leqq 2$ unterliegen. Mithin sind 3 irreduzible Kombinationen $(\lambda, 0, \nu)$ möglich, so daß die Linearkombination (5) insgesamt *48* irreduzible Tensorgeneratoren enthält. Im Vergleich dazu bestehen die nach der Kriechpotentialhypothese, d. h. aus (1) und (3) ermittelten Stoffgleichungen nur aus *23* Termen.

Dieser Vergleich läßt den Schluß zu, daß im anisotropen Fall die Kriechpotentialhypothese mit der klassischen Fließregel (1) keine vollständigen Materialgleichungen liefern kann, auch dann nicht, wenn ein allgemeines Kriechpotential (3) zugrundegelegt wird.

Um eine *formale* Übereinstimmung der Kriechpotentialhypothese mit der Darstellungstheorie (5) zu erzielen, ist demnach eine Modifikation der klassischen Fließregel (1) erforderlich. Mögliche Modifikationen werden in [10] vorgeschlagen.

Zur Vervollständigung der Darstellung (5) ist noch die ihr angepaßte Integritätsbasis aufzustellen, durch welche die skalarwertigen Funktionen $\varphi_{(\lambda,\mu,\nu)}$ auszudrücken sind. Dazu beachte man, daß sie Spur eines jeden Tensorproduktes, das aus den beteiligten Argumenttensoren der darzustellenden Tensorfunktion gebildet werden kann, eine orthogonale Invariante ist. Um mit Hilfe dieser Aussage ein *vollständiges* und *irreduzibles* Invariantensystem zu erzielen, bilde man in Verbindung mit der Darstellung (5) die Spuren $\mathrm{tr}\,\dot{\boldsymbol{\varepsilon}}\boldsymbol{\sigma} \equiv \dot{\varepsilon}_{ij}\sigma_{ji}$ und $\mathrm{tr}\,\dot{\boldsymbol{\varepsilon}}\boldsymbol{A} \cdot \boldsymbol{\sigma} \equiv \dot{\varepsilon}_{ij} A_{jirs} \sigma_{rs}$ und sortiere alle redundanten Elemente aus. Da die spezifische Dissipationsleistung $\dot{D} \equiv \dot{\varepsilon}_{ij}\sigma_{ji}$ durch $\mathrm{tr}\,\dot{\boldsymbol{\varepsilon}}\boldsymbol{\sigma}$ gegeben ist und durch sie der Kriechprozeß charakterisiert werden kann, hat das aus $\mathrm{tr}\,\dot{\boldsymbol{\varepsilon}}\boldsymbol{\sigma}$ gefundene Invariantensystem vorrangige Bedeutung.

4. Darstellung des tertiären Kriechverhaltens

Bei der Formulierung von Materialgleichungen unter Einbeziehung eines anisotropen „Schadenstensors“ ω und eines Anisotropietensors A, der die im schadlosen Zustand vorhandene Anisotropie charakterisiert, ist darauf zu achten, daß man für $\omega \to \mathbf{0}$ zwanglos die Stoffgleichungen des sekundären Kriechens erhält, während der Tensor $\dot{\boldsymbol{\varepsilon}}$ bei vollständiger Schädigung $(\omega \to \boldsymbol{\delta})$ singulär werden muß. Im Hinblick auf eine Polynomdarstellung ist es daher sinnvoll, anstatt ω den Tensor

$$D_{ij} \equiv (\delta_{ij} - \omega_{ij})^{(-1)} \tag{6}$$

als Argumenttensor zu benutzen, so daß die tensorwertige Funktion

$$\dot{\varepsilon}_{ij} = f_{ij}(\sigma_{pq}, D_{pq}, A_{pqrs}) \tag{7}$$

darzustellen ist. Mögliche Formen werden in [11] diskutiert. Die Darstellung vereinfacht sich, wenn man die Tensoren $\boldsymbol{\sigma}$ und $\boldsymbol{D}$ in (7) gemäß der Überschiebung

$$t_{ij} = \tfrac{1}{2}(D_{ip}D_{jq} + D_{iq}D_{jp})\,\sigma_{pq} \tag{8}$$

zu einem symmetrischen Tensor t zweiter Stufe zusammenfaßt.

Dann wird statt (7) die Tensorfunktion

$$\dot{\varepsilon}_{ij} = f_{ij}(t_{pq}, A_{pqrs}) \tag{9}$$

betrachtet, deren Darstellung von Ziffer 3 übernommen werden kann, indem man in (5) den Cauchyschen Spannungstensor $\boldsymbol{\sigma}$ durch den Tensor (8) formal ersetzt. Auf die mechanische Deutung des durch (8) eingeführten Tensors t wird in [11] näher eingegangen.

Literatur

1. Rabotnov, Yu. N., Creep Problems in Structural Members (engl. Übersetzung, herausgegeben von Leckie, F. A.), North-Holland (Amsterdam 1969).
2. Betten, J., Rheol. Acta **14**, 317 – 323 (1975).
3. Mises, R. v., Z. Angew. Math. Mech. **8**, 161 – 185 (1928).
4. Hill, R., The Mathematical Theory of Plasticity, Oxford University Press (Oxford 1950).
5. Betten, J., Materials Science and Engineering **41**, 183 – 192 (1979).
6. Betten, J., Representation of Constitutive Equations in Creep Mechanics of isotropic an anisotropic Materials, in: Ponter, A. R. S. and Hayhurst, D. R. (eds.), Creep in Structures, S. 179 – 201, Springer-Verlag (Berlin 1981).
7. Odquist, F. K. G., J. Hult, Kriechfestigkeit metallischer Werkstoffe (Berlin 1962).
8. Betten, J., Rheol. Acta **20**, 527 – 535 (1981).
9. Betten, J., Z. Angew. Math. Mech. **62**, T 7038 (1982).
10. Betten, J., About the Representation of the Plastic Potential of Anisotropic Solids, vorgetragen auf dem "Colloque International du CNRS n° 319 Comportement platique des solides anisotropes, Grenoble, 16 – 19. Juni 1981, erscheint in den Proceedings.
11. Betten, J., "Damage Tensors in Continuum Mechanics", vorgetragen auf dem Euromech Colloquium 147 in Cachan/Paris am 22. 09. 1981.

(Eingegangen am 14. April 1982)

Anschrift des Verfassers:

Prof. Dr.-Ing. J. Betten
Institut für Werkstoffkunde
RWTH Aachen
Templergraben 55
D-5100 Aachen

Rheologica Acta Rheol. Acta **21**, 409 – 412 (1982)

Ermittlung von Fließkurven in verschiedenen Ausführungsformen des Stauchversuches

W. Nester und K. Pöhlandt

Institut für Umformtechnik der Universität Stuttgart (Prof. Dr.-Ing. K. Lange)

Abstract: Several modifications of the upsetting test at room temperature are discussed under the aspect of suppressing friction and other possible sources of error.

In the upsetting test according to Rastegaev, friction is reduced, and the test-piece remains cylindrical up to high strains, however the specimen's height is determined with a large error. Therefore it is proposed to measure the increase of the specimen's diameter instead of the reduction of height.

Upsetting under sticking friction conditions can formally be treated like upsetting with lubrication for calculating the stress-strain curve. However, the applicability and the limits of this method have not yet been studied sufficiently.

In the plane-strain upsetting test, friction is not suppressed completely, but the error of height measurement propagates only weakly into the calculated stress-strain curve.

Key words: Upsetting test, stress-strain curve, friction suppression

1. Einleitung

Das Umformvermögen metallischer Werkstoffe nimmt im allgemeinen mit wachsender hydrostatischer Druckspannung zu, deshalb gestattet der Stauchversuch die Fließkurvenaufnahme im Prinzip im größten Wertebereich.

Es wird zunächst der herkömmliche Zylinderstauchversuch betrachtet. Für den Umformgrad φ gilt bei Vernachlässigung von Korrekturen*) [1]

$$\varphi(F) \approx \ln\left[\frac{h_0}{h_0 - s(F)}\right]. \quad (1)$$

Hierin ist h_0 die Anfangshöhe und $s(F)$ der Stauchweg bei der Kraft F.

Für die Fließspannung k_f gilt mit Berücksichtigung der Volumenkonstanz

$$k_f(F) \approx \frac{F}{\pi r^2(F)} = \frac{F[h_0 - s(F)]}{\pi r_0^2 h_0}. \quad (2)$$

*) Der Umformgrad im Stauchversuch wird hier als positive Größe definiert.

818

Hierin ist $r(F)$ der Probenradius bei der Kraft F mit $r(0) = r_0$.

2. Fehlerquellen beim Stauchversuch

Beim Stauchversuch stören vor allem solche Fehler, die im Bereich hoher Umformgrade wesentlich sind. Diese Fehler sind:

a) Die Reibung, die sich zweifach auswirkt:
   - Sie erfordert einen zusätzlichen Kraftbetrag zum Erreichen eines gegebenen Umformgrades. Statt Gl. (2) gilt die Beziehung [1]:

$$\frac{F(\varphi)}{\pi r^2(\varphi)} = k_w(\varphi) \approx k_f(\varphi)\left[1 + \frac{2\mu r(\varphi)}{3h(\varphi)}\right]. \quad (3)$$

   Hierin ist $k_w(\varphi)$ der Umformwiderstand und μ die Reibzahl.
   - Die Reibung bewirkt zudem eine Aufwölbung der Probe. Als Folge davon wird das Umformvermögen im Vergleich zu einachsiger Beanspruchung herabgesetzt [2].

b) Der Meßfehler des Stauchweges, der sich bei abnehmender Probenhöhe besonders stark auf die berechnete Fließspannung auswirkt [3].

3. Ausführungsformen des Stauchversuches

3.1. Stauchen mit Haftreibung

Bei Haftreibung aufgenommene Meßkurven weichen für $h/d > 0{,}5$ nur geringfügig von Meßkurven aus Versuchen mit sehr guter Schmierung ab [4, 5]. Erklärt werden kann dies dadurch, daß sich die auftretenden Fehler aufgrund der inhomogenen Formänderungsverteilung gegenseitig ausgleichen. Deshalb kann dieser Versuch formal wie der herkömmliche Zylinderstauchversuch ausgewertet werden. Die Anwendungsgrenzen sind aber noch nicht hinreichend untersucht worden.

3.2. Zylinderstauchversuch nach Rastegaev

Die Stauchproben besitzen an den Stirnflächen flache Ausdrehungen, in die z. B. Paraffin als Schmierstoff eingefüllt wird [6–11], s. Abb. 1.

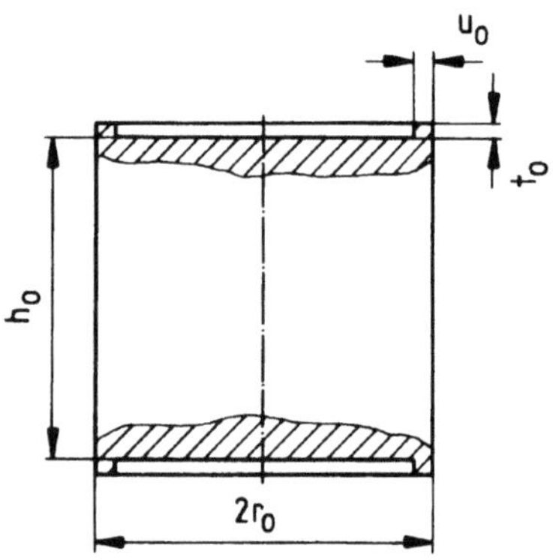

Abb. 1. Stauchprobe nach Rastegaev

Vorteile des Stauchversuches nach Rastegaev sind:

- Erhaltung der Zylinderform bis zu höheren Umformgraden,
- Verminderung der Reibzahl μ.

Den Verfassern ist keine andere Methode bekannt, um einen Körper bis zu derart hohen Umformgraden homogen umzuformen.

Die homogene Umformung kann zu der irrtümlichen Annahme führen, daß die Reibzahl praktisch Null ist. Tatsächlich tritt folgender Effekt auf: der hydrostatische Druck des eingeschlossenen Schmierstoffes wirkt den trotz Schmierung noch vorhandenen Reibschubkräften entgegen [8].

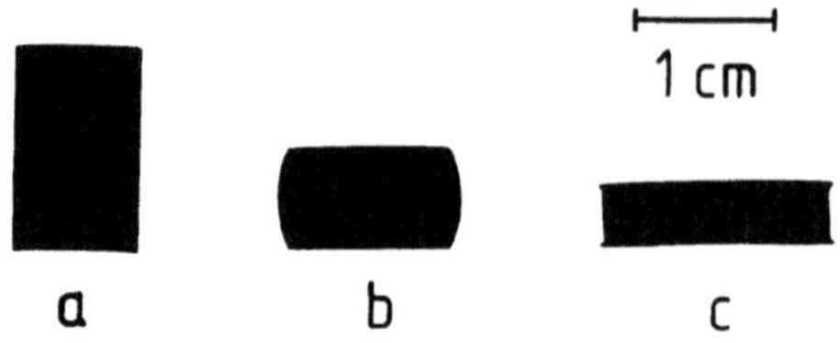

Abb. 2. Zylinderstauchproben ohne und mit Schmiertaschen nach Rastegaev: a) ungestaucht, b) ohne Schmierung ($\varphi \approx 0{,}7$), c) Rastegaev-Probe ($\varphi \approx 1{,}3$)

Die Fließspannung berechnet sich aus dem gemessenen Umformwiderstand wie folgt [9]:

$$k_f(\varphi) \approx k_w(\varphi)\left\{1 - \frac{2}{3h(\varphi)}\left[\chi r(\varphi) + 3\mu u(\varphi)\right]\right\}. \tag{4}$$

Hierin ist χ die Reibzahl im Bereich des Schmierstoffes und μ die Reibzahl für den ungeschmierten Bereich zwischen Wulst und Stauchbahn.

Für die Wulstabmessungen an den Probenstirnflächen (s. Abb. 1) gibt es in der Literatur sich teilweise widersprechende Angaben. In eigenen Versuchen erwiesen sich die Angaben von Krokha [10, 11] als gut geeignet, um die Zylinderform bis zu höheren Umformgraden zu erhalten. Krokha empfiehlt:

$$\frac{t_0}{u_0} \approx 0{,}4 \quad \text{(für Stahlproben)} \tag{5}$$

und

$$u_0 \approx 0{,}45 \ldots 0{,}6\,\text{mm für } r_0 = 3{,}5 \ldots 10\,\text{mm}\,. \tag{6}$$

Die optimalen Werte für t_0 und u_0 hängen vom Probenwerkstoff ab.

Nachteile des Stauchversuches nach Rastegaev sind:

- Die Schmiertaschen führen zu einer ungenaueren Wegmessung.
- Die gute Schmierung verringert den zulässigen Schlankheitsgrad.

3.3. Stauchversuch nach Rastegaev mit Messung der Durchmesserzunahme

Zur Ermittlung der Fließkurve ist im Prinzip auch folgende Vorgehensweise möglich: Es wird nicht die

Höhenabnahme der Probe gemessen, sondern die Zunahme w des Durchmessers.

Abbildung 3 zeigt die zu erwartende Meßkurve. Im Vergleich zur Messung des Stauchweges s ergibt sich ein günstigeres Verhalten bezüglich der Fehlerfortpflanzung. Voraussetzung ist, daß die Probe zylindrisch bleibt. Dies trifft für den Stauchversuch nach Rastegaev zu.

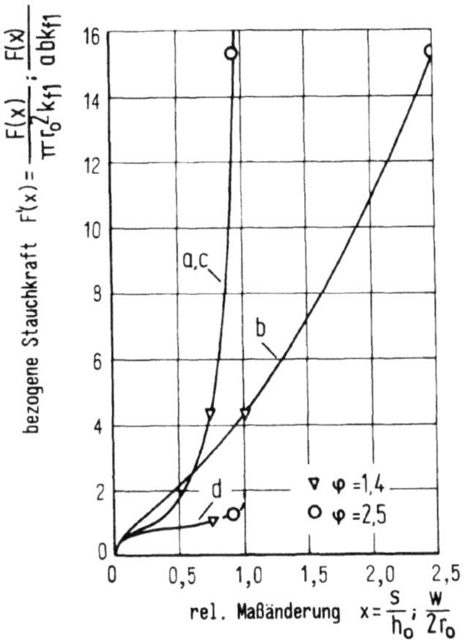

Abb. 3. Zu erwartende Meßkurven bei vier Formen des Stauchversuches (schematisch, Annahmen: $k_f(\varphi) = k_{f1} \cdot \varphi^n$ mit $n = 1/4$, $\partial F/\partial\mu = 0$ bei Haftreibung) [3, 12].
a) Rastegaev-Versuch, Messung der Höhenabnahme, b) Rastegaev-Versuch, Messung der Durchmesserzunahme, c) Zylinderstauchversuch mit Haftreibung, d) Flachstauchversuch

3.4. *Flachstauchversuch*

Dieser Versuch wird in Abbildung 4 schematisch dargestellt. Wichtige Merkmale sind [12–14]:

- Die belastete Fläche bleibt konstant. Es gilt

$$k_f(F) \approx \frac{F}{a\,b}. \tag{7}$$

  Daher geht die Höhenabnahme nicht in die berechnete Fließspannung ein, so daß sich ein Meßfehler der Höhenabnahme nicht so stark auswirkt wie beim Zylinderstauchversuch.
- Die Stauchwerkzeuge müssen sehr genau geführt werden, da ein Seitenversatz die gestauchte Fläche verringert.
- An den Längskanten der belasteten Flächen besteht eine Kerbwirkung. Hierdurch kann es „vorzeitig" zum Bruch kommen.

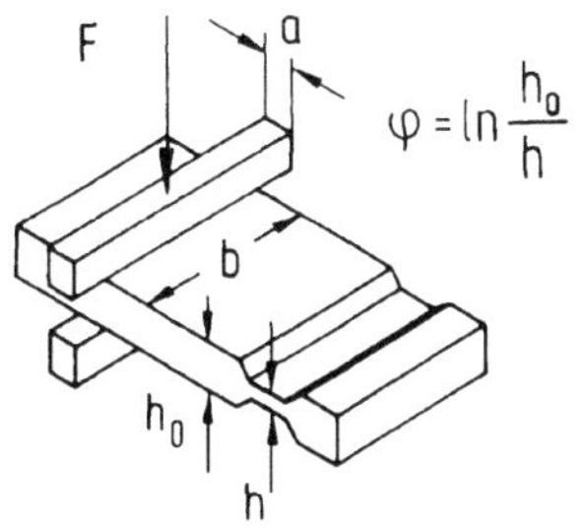

Abb. 4. Schema des Flachstauchversuches

4. Meßfehlerfortpflanzung

Die Fehlergrenze G_{k_f} der Fließspannung ist im allgemeinen Fall

$$|G_{k_f}| \leq \left|\frac{\partial k_f}{\partial F} G_F\right| + \left|\frac{\partial k_f}{\partial x} G_x\right| + \left|\frac{\partial k_f}{\partial \mu} G_\mu\right| + \left|\frac{\partial k_f}{\partial \chi} G_\chi\right|. \tag{8}$$

Hierin ist G_F die Fehlergrenze der Kraft, G_x die Fehlergrenze der relativen Maßänderung (bezogener Stauchweg oder relative Durchmesserzunahme) und G_μ und G_χ sind die Fehlergrenzen der Reibzahlen gemäß Gl. (3) and (4).

Gewöhnlich kann der Fehler der Kraft vernachlässigt werden [3], so daß der erste Term auf der rechten Seite von Gl. (8) entfällt. Mit Hilfe von Gl. (8) lassen sich die in Tab. 1 dargestellten relativen Fehlergrenzen der Fließspannung ableiten [12]. Es handelt sich dabei um eine grundsätzliche Fehlerbetrachtung.

Der Fehlerbeitrag der Maßänderung wirkt sich besonders stark beim Stauchversuch nach Rastegaev mit Messung der Höhenabnahme aus; allerdings ist dieser Fehler auch beim Stauchen mit Haftreibung wesentlich. Außerdem besitzt der Stauchversuch nach Rastegaev unter den beschriebenen Verfahren den geringsten zulässigen Schlankheitsgrad h_0/d_0.

Der Fehlerbeitrag der Reibzahl ist weder beim Stauchversuch nach Rastegaev noch beim Flachstauchversuch zu vernachlässigen, da in diesen Fällen die Fließspannung aus den Meßwerten für den Umformwiderstand k_w berechnet wird.

Tab. 1. Relative Fehlergrenze der Fließspannung $|G_{k_f}/k_f|$

	Höhenabnahme $s \pm G_s$	Durchmesserzunahme $w \pm G_w$	Reibzahl $\mu \pm G_\mu$	Reibzahl $\chi \pm G_\chi$
Zyl.-Stauchversuch (Rastegaev). Messung der Höhenabnahme	$\frac{G_s}{h_0} \cdot e^{\varphi}$		$\frac{2u_0 G_\mu}{h_0} \cdot e^{3\varphi/2}$	$\frac{2r_0 G_\chi}{3h_0} \cdot e^{3\varphi/2}$
Zyl.-Stauchversuch (Rastegaev). Messung der Durchmesserzunahme		$\frac{G_w}{r_0} \cdot e^{-\varphi/2}$	$\frac{2u_0 G_\mu}{h_0} \cdot e^{3\varphi/2}$	$\frac{2r_0 G_\chi}{3h_0} \cdot e^{3\varphi/2}$
Zyl.-Stauchversuch mit Haftreibung	$\frac{G_s}{h_0} \cdot e^{\varphi}$			
Flachstauchversuch			$\frac{aG_\mu}{2h_0} \cdot e^{\varphi}$	

Ein allgemeingültiges „bestes“ Verfahren gibt es nicht. Der Stauchversuch nach Rastegaev ist, soweit bekannt, die einzige Methode, bei der die Probe bis zu höheren Umformgraden zylindrisch bleibt. Das bietet den Vorteil, daß der Probendurchmesser eine genau definierte Größe darstellt, was experimentell genutzt werden kann.

Literatur

1. Lange, K. (Hrsg.), Lehrbuch der Umformtechnik, Bd. 1 und 2, Springer-Verlag (Berlin 1972 und 1974).
2. Stenger, H., Diss. TH Aachen 1965.
3. Pöhlandt, K., W. Nester, Draht **32**, 70–75 und 119–121 (1981).
4. Wiegels, H., R. Herbertz, Stahl und Eisen **99**, 1380–1390 (1979).
5. Schey, J. A. u.a., J. Mech. Working Technol., im Druck.
6. Rastegaev, M. V., Zavod. Lab. **6**, 354 (1940).
7. Turno, A.: Obrobka Plastyczna **11**, 123–127 (1972).
8. Latocha, J., Z. Rafalski, Rudy i Metale **21**, 345–348 (1976).
9. Suyarov, D. I. u.a., Zavod. Lab. **21**, 97–99 (1955).
10. Krokha, V. A., Ind. Lab. **40**, 754–758 (1974).
11. Krokha, V. A., Stamp. Proiz. **10**, 29–47 (1977).
12. Nester, W., K. Pöhlandt, Arch. Eisenhüttenwes. **53** (1982).
13. Lippmann, H., O. Mahrenholtz, Plastomechanik der Umformung metallischer Werkstoffe, Bd. 1, Springer-Verlag (Berlin 1967).
14. Pawelski, O. u.a., Stahl und Eisen **98**, 181–189 (1978).

(Eingegangen am 16. April 1982)

Anschrift der Verfasser:

Dipl.-Ing. W. Nester, Dr.-Ing. K. Pöhlandt
Institut für Umformtechnik
Universität Stuttgart
Postfach 560
D-7000 Stuttgart 1

Rheologica Acta Rheol. Acta **21**, 413–415 (1982)

Strain-rate history effects and microstructural aspects in lithium fluoride and aluminium single crystals after dynamic loading

C. Y. Chiem and J. Duffy

Laboratoire de Physique du Métal, Ecole Nationale Supérieure de Mécanique, Université de Nantes and Division of Engineering, Brown University, Providence (R.I.)

Abstract: Single crystals of LiF and Al are deformed in shear at a number of constant strain-rates in the range 10^{-4} to 1600 s^{-1}. These constant rate tests are supplemented by a series of jump tests in which a sharp increment in strain rate is imposed during the quasi-static straining. Dislocation arrangements are observed by etch-pits technique for LiF crystals and by TEM for Al crystals. It is shown that cell sizes vary inversely with flow stress and strain-rate sensitivity.

Key words: Strain-rate history effect, dynamic torsional loading, single-crystal, dislocation arrangement, Kolsky bar

1. Introduction

An understanding of the plastic behavior of metals during deformation at high strain-rates is of considerable importance in a number of engineering applications. Furthermore, dynamic plasticity often plays a considerable role in other processes; one example is fracture where, depending on the degree of ductility, plastic flow may occur on a large scale or else only near the crack tip. In addition, investigations into dynamic plasticity can lead to knowledge of material behavior on a more fundamental level. The present investigation is intended to fall in this latter category.

2. Results

The apparatus consists of a Kolsky (split-Hopkinson) bar modified for torsional loading so that the specimens are deformed in shear, as described in [1, 2]. Specimen preparations and observations are done as described in [1, 2].

For LiF single-crystals, results show that the flow stress of LiF is strongly sensitive to strain-rate, but that the effects of strain-rate history, although present, are not pronounced (see fig. 1). Results show a considerable difference in the work-hardening rate between the quasi-static and the dynamic strain-rates. During quasi-static deformation at a constant strain-rate the work-hardening rate is constant and rather high. Dynamically, it is also high but only initially, then the flow stress drops to a lower value where it forms a plateau before rising again. The incremental strain-rate tests show the same effect, but the

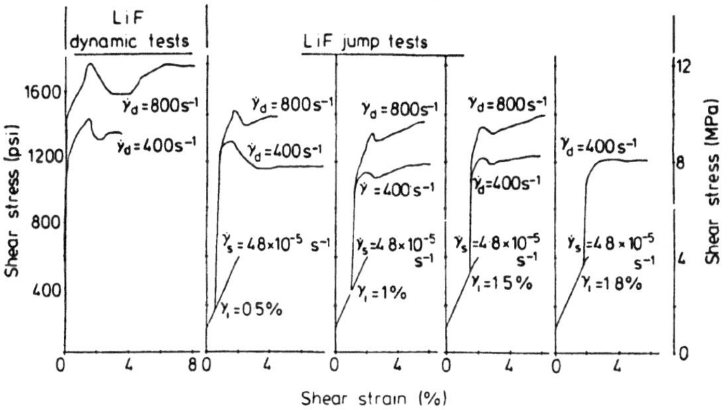

Fig. 1. Results of incremental strain rate experiments in LiF single crystals. γ_i = amount of pre-strain

833

difference in stress between the upper and lower yield points decreases as the amount of pre-strain increases.

The etch-pit patterns reveal a distribution that is uniform when the strain-rate is quasi-static. However, following an increment in strain-rate, the dislocations are seen to form wide bands. At a constant dynamic strain-rate, the dislocation distribution may also form bands or may be homogeneous. Details of those observations are mentioned in [1].

For Al single-crystals, the constant strain-rate tests show an increase in flow stress with strain-rate (fig. 2). A plot of flow stress as a function of the logarithm of strain-rate shows a linear increase up to 500 s^{-1} followed by a much more rapid increase at the higher strain-rates (fig. 3); for strain-rates less than about 500 s^{-1}, the strain-rate sensitivity is approximately equal to that found by other investigations for polycristalline aluminium of high purity and exceeds that of aluminium alloys [3–6]. At strain-rates above about 500 s^{-1}, the flow stress becomes more sensitive to strain-rate and this change in strain-rate sensitivity is greater than in polycrystalline aluminium and pronounced enough to simply a change in the dominant dislocation mechanism of dynamic plastic deformation.

Macroscopic results also show a strain-rate history effect qualitatively similar to that observed with polycristalline aluminium (fig. 4). However, this history effect is not erased by continued straining after a jump in strain-rate, as it is in polycristalline aluminium [7, 8], unless the amount of prestrain is small.

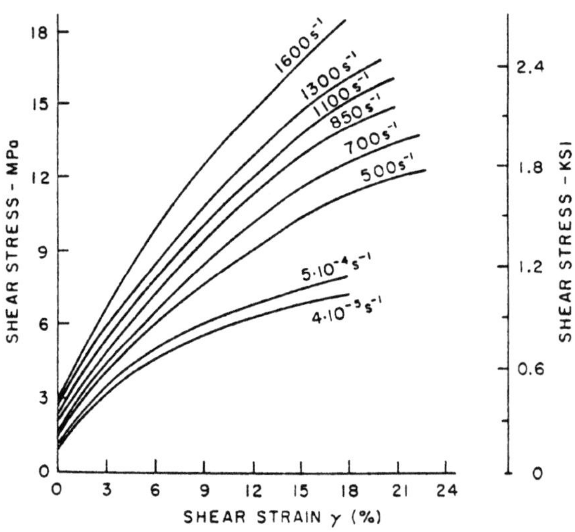

Fig. 2. Shear stress vs. shear strain for aluminium single crystals deformed at the strain rates indicated

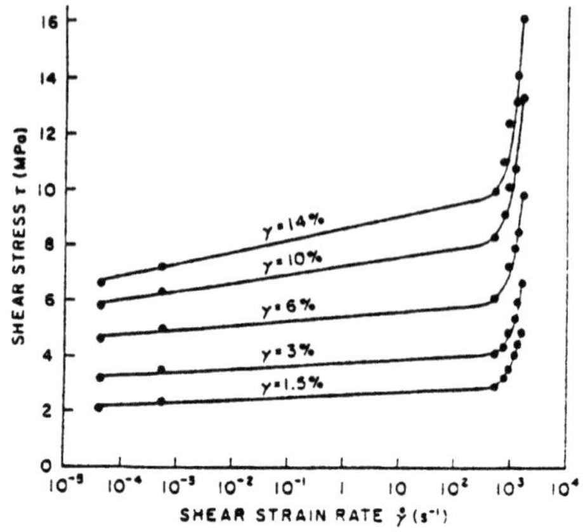

Fig. 3. Flow stress vs. strain rate for aluminium single crystals

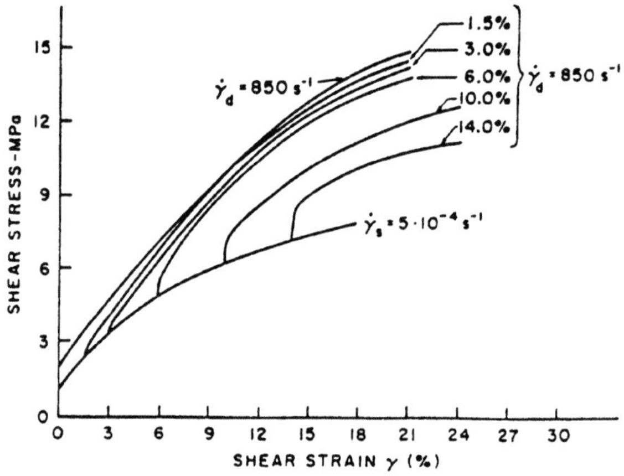

Fig. 4. Shear stress vs. shear strain for aluminium single crystals tested at constant strain rates and subjected to incremental strain rate deformation

After completing of the macroscopic tests, thin foils of representative specimens were examined by transmission electron microscopy to reveal the dislocation substructure and to measure the dislocation density ρ. For a quasi-static deformation rate it was found that ρ increases approximately linearly with the imposed strain. For the dynamic and jump tests, ρ was measured only at the conclusion of the test. However, the results show that dislocation density is greater the higher the strain-rate and that for the same total strain, a quasi-static prestrain decreases the observed dislocation density (fig. 5).

TEM observations after deformation at quasi-static strain-rates reveal a cellular dislocation arrangement that develops into well-outlined subgrains as strain increases. Cells are also visible after dynamic deformation, but these cells are about one quarter the

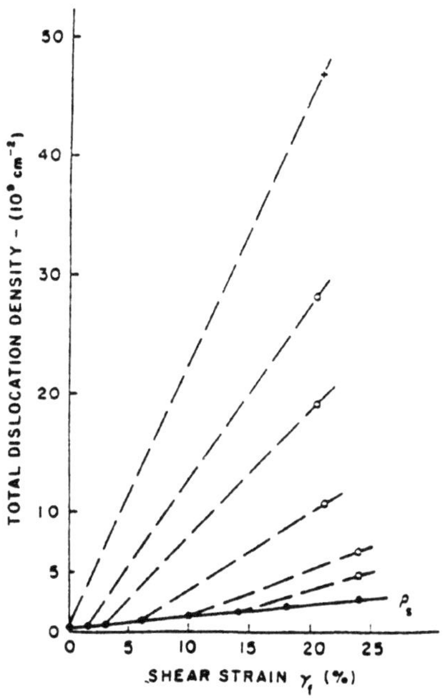

Fig. 5. Total dislocation density vs. shear strain from microscopic examination of specimens. ($\dot{\gamma}_s = 5 \cdot 10^{-4}\ s^{-1}$ for quasi-static test, $\dot{\gamma}_d = 850\ s^{-1}$ for dynamic and jump tests)

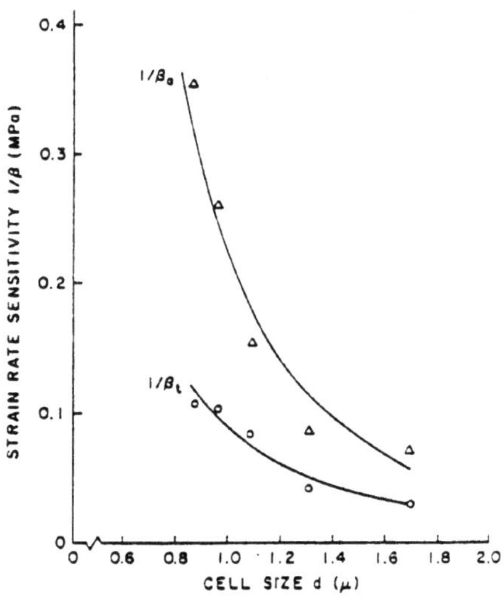

Fig. 6. Strain rate sensitivities, apparent and true, as functions of dislocation cell size

size of those produced by quasi-static deformation and their walls are not as well-defined. Details of these observations can be seen in [2]. For the strain-rates covered by the present tests, the results indicate that as cell size is reduced both the flow stress and the strain-rate sensitivity increases (fig. 6).

TEM observation after the jump tests reveal that the substructure depends on the quasi-static prestrain. With a small prestrain ($\gamma_i < 6\%$) the dislocation distribution generally is uniform and includes numerous dislocation loops. For a greater prestrain a cellular pattern begins to emerge and the number of loops decreases [2]. This is consistent with a continuing dislocation multiplication process at low prestrain and a higher work-hardening rate in this range.

3. Conclusion

Very little work has been performed on the behavior of single-crystals except at strain-rates beneath the dynamic range. Dynamic experiments should be performed, preferably in shear and with different crystal structures; these should include varieties of strain rate histories and microscopic analysis.

In addition, for composite materials, one needs to determine the behavior under various kinds of strain-rate histories as mentioned above. Very few work has been done and much more efforts are required.

References

1. Chiem, C. Y., J. Duffy, Mat. Sci. and Engng. **48**, 207 (1981).
2. Chiem, C. Y., J. Duffy, Brown Univ. Tech. Rep. MEA 79-23742/3 and MRL.E.-137 (October 1981).
3. Hauser, F. E., F. A. Simmons, J. E. Dorn, in: P. G. Shewmon, V. F. Zackay (eds.), "Response of Metals to High Velocity Deformation", p. 93, Interscience (New York 1961).
4. Yoshida, S., N. Nagata, Trans. J. I. M. **7**, 273 (1966).
5. Green, S. J., C. J. Maiden, S. G. Babcock, F. L. Shierloh, in: "Battelle Colloquium on Inelastic Behavior of Solids", p. 521, McGraw-Hill (New York 1969).
6. Dowling, A. R., J. Harding, J. D. Campbell, J. Inst. Metals **98**, 215 (1970).
7. Frantz, R. A., J. Duffy, J. Appl. Mech., Trans. ASME **39**, 939 (1972).
8. Klepaczko, J., J. Duffy, in: J. Harding (ed.), "Mechanical Properties at High Rates of Strain", p. 91, Institute of Physics (London 1974).

(Received April 16, 1982)

Authors' addresses:

Prof. Dr. C. Y. Chiem
E.N.S.M.
Laboratoire de Physique du Mètal
1 rue de la Noë
F-44072 Nantes Cedex

Prof. J. Duffy
Division of Engineering
Brown University
Box D
Providence, RI 02912 (U.S.A)

Rheologica Acta Rheol. Acta 21, 416 – 419 (1982)

Torsional vibrational properties of rubber cylinders

T. E. R. Jones and J. M. Davies

Department of Mathematics, Statistics and Computing, Plymouth Polytechnic, Plymouth, Devon (UK)

Abstract: This paper describes an experimental and theoretical study of the torsional vibrational properties of rubber cylinders. Experiments were carried out on long cylinders using a modified Weissenberg Rheogoniometer. A comparison is made with a theory based on a modified generalized Maxwell model.

Key words: Torsional vibration, rubber cylinder, Maxwell model

1. Introduction

Previous work (Treloar [1], Ferry [2]) on the vibrational properties of rubber cylinders has been mainly concentrated on the longitudinal vibration of a sample under tension. Duperray [3] utilized a parallel-plate viscometer to measure complex modulus data for very short unvulcanised rubber cylinders. In this paper we shall be concerned with a similar method for the measurement of vulcanised rubber. These rubbers have to be bonded by resin to the platens which will introduce end effects for short cylinders. Therefore, our experiments will be carried out with long samples for a range of temperatures and extensions. A comparison will be made with a linear viscoelastic theory for a rubberlike material.

2. Theory

The theory for a cylindrical rubber sample contained between two parallel plates (shown in figure 1) is based upon the Walters [4] analysis for small amplitude oscillatory shear. The dynamic properties of the rubber can be described in terms of the complex modulus

$$G^* = G' + iG'' \tag{1}$$

where G' is the elastic modulus and G'' is the loss modulus. For a particular choice of relaxation modulus given by a modified generalised Maxwell model (Bird [5]), it can be shown that

$$G'' = \sum_{k=1}^{\infty} \frac{\omega \eta_k}{1 + (\lambda_k \omega)^2}, \tag{2}$$

$$G' = G_0 + \sum_{k=1}^{\infty} \frac{\omega^2 \eta_k \lambda_k}{1 + (\lambda_k \omega)^2} \tag{3}$$

where ω is the frequency, η_k the viscosity, λ_k are time constants and G_0 is the elastic modulus at zero-frequency.

By using the following empirical equations (Spriggs [6]),

$$\eta_k = \frac{\eta_0 \lambda_k}{\sum_k \lambda_k} \quad \text{and} \quad \lambda_k = \frac{\lambda}{k^\alpha} \tag{4}$$

where α is a constant and η_0 the zero-shear viscosity, the total number of parameters can be reduced to four and G'' and G' are given as follows:

$$G'' = \frac{\eta_0}{\zeta(\alpha)} \omega \sum_{k=1}^{\infty} \frac{k^\alpha}{k^{2\alpha} + (\lambda \omega)^2}, \tag{5}$$

$$G' = \frac{\eta_0 \lambda \omega^2}{\zeta(\alpha)} \sum_{k=1}^{\infty} \frac{1}{k^{2\alpha} + (\lambda \omega)^2}, \tag{6}$$

where $\zeta(\alpha)$ is the Riemann zeta function. These expressions are not particularly appropriate for computation of the functions G'' and G', since for G'' in particular a large number of terms are required for convergence especially when α is near 1.

Spriggs [6] has proposed a pair of expressions which are valid for large relaxation times or large frequencies (derived from eqs. (5) and (6)):

$$G'' = m\omega^{1/\alpha}, \tag{7}$$

$$G' = G_0 + \frac{m\omega^{1/\alpha}}{\tan(\pi/2\alpha)}. \qquad \text{for} \quad \lambda\omega \gg 1 \tag{8}$$

It should be noted that the constant m is related to the constants in eqs. (5) and (6) by the expression

$$m = \frac{\eta_0 \pi \lambda^{(-1+1/\alpha)}}{\zeta(\alpha) 2\alpha \cos(\pi/2\alpha)}. \tag{9}$$

We also note that the condition for this model to accurately predict vibrational behaviour is that $(G' - G_0)/G''$ must be a constant.

3. Experiments

The experiments were carried out on a rubber-nitrile compound using a Weissenberg Rheogoniometer. The experimental set-up is shown in figure 1. The cylindrical rubber specimen was bonded to

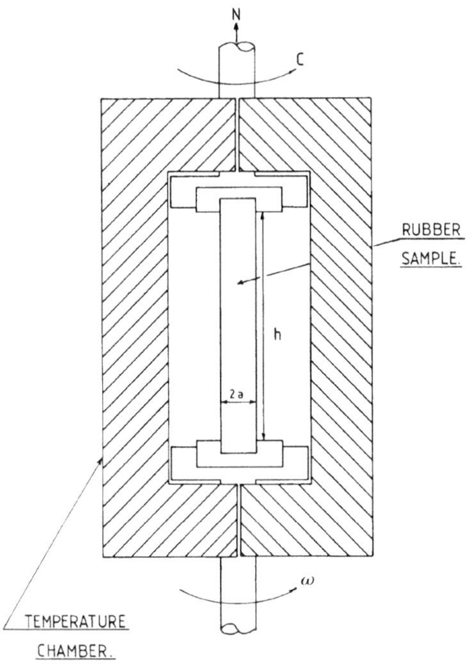

Fig. 1. Schematic diagram of experimental set-up

perspex holders which were secured to the shafts of the instrument in the usual way. An oscillatory shear frequency of small amplitude is applied to one end of the rubber cylinder and the movement of the other end, which is constrained by a torsion wire, is monitored. If ξ is the amplitude ratio and c the phase lag between the upper and lower platens, it can be shown that (Walters [7])

$$G'' = \frac{-S\omega\xi \sin c}{(\xi^2 - 2\xi\cos c + 1)}, \tag{10}$$

$$G' = \frac{\omega S\xi(\cos c - \xi)}{(\xi^2 - 2\xi\cos c + 1)} \tag{11}$$

where

$$S = \frac{2h[K - I\omega^2]}{\pi a^4 \omega} \tag{12}$$

and h is the length of the sample, a the radius, K the restoring constant of the torsion wire and I the moment of inertia of the system.

An amplitude of 0.01 rads was used for our experiments at room temperature which was found to be well within the range of amplitudes for which linear viscoelastic theory would hold. Experiments were carried out with various lengths of the same sample in order to determine the end effect on G^*. Results indicate that this effect for our samples is small for lengths greater than 7 cms.

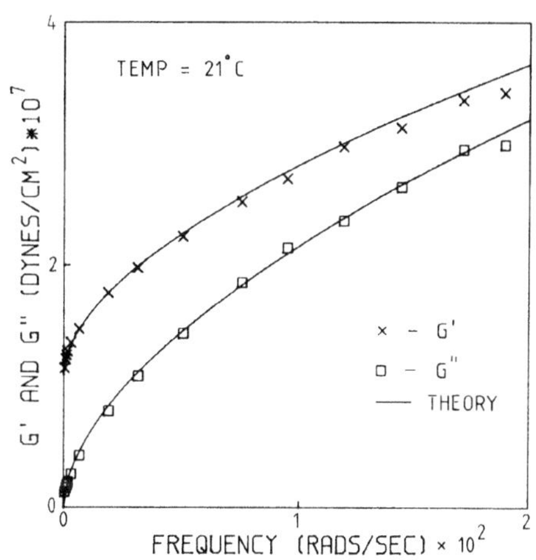

Fig. 2. Experimental G'' and G' vs. frequency for an amplitude of 0.01 rads. Continuous line – theory based on eqs. (7) and (8) with $m = 1.55 \times 10^6$ dynes $\sec^{1/\alpha}/\mathrm{cm}^2$, $\alpha = 1.75$ and $G_0 = 1.1 \times 10^7$ dynes/cm²

Figure 2 shows the complex modulus behaviour with change in frequency using a 10 cm unstretched sample at room temperature. The results are typical of what you would expect from a viscoelastic material with a very large relaxation time except for the solid-like behaviour of G' at low frequencies. Also included in this figure are the theoretical predictions based on eqs. (7) and (8). This model accurately predicts the vibrational properties of this particular rubber sample at room temperature.

An experiment was carried out in order to determine the effect of extension on the experimental data. Results show no change in G'' but there is a slight change in G'. This change is likely to be due to either a non-uniform sample radius near the upper and lower platens or the deformation of the sample is becoming non-linear. The sample radius used for all experiments was 0.765 cms.

One aim of this work was to produce a model to predict torsional vibrational behaviour for a range of temperatures. The experimental variations with temperatures are shown in figures 3 and 4. Examination of the variations of G'' and G' at high frequencies indicates an apparent transition from a "fluid" at room temperature to a "solid" at $-10\,°C$. However, at low frequencies the sample appears to remain a "fluid". At temperatures down to 10 °C there is no appreciable change in complex behaviour with

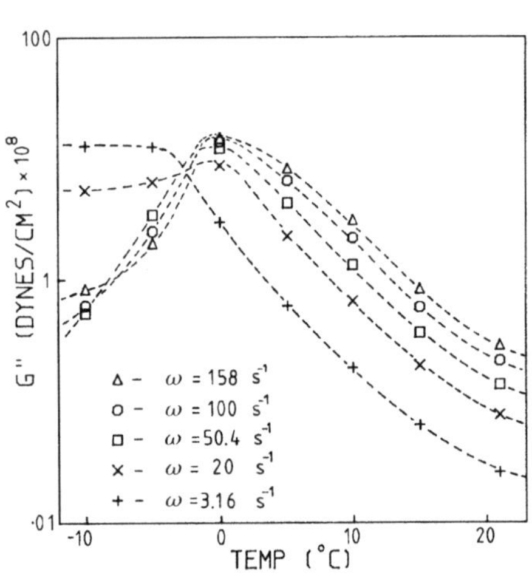

Fig. 3. Experimental G'' vs. temperature (°C) for a range of frequencies from $(+)\omega = 3.16\ \text{sec}^{-1}$ to $(\Delta)\omega = 158\ \text{sec}^{-1}$

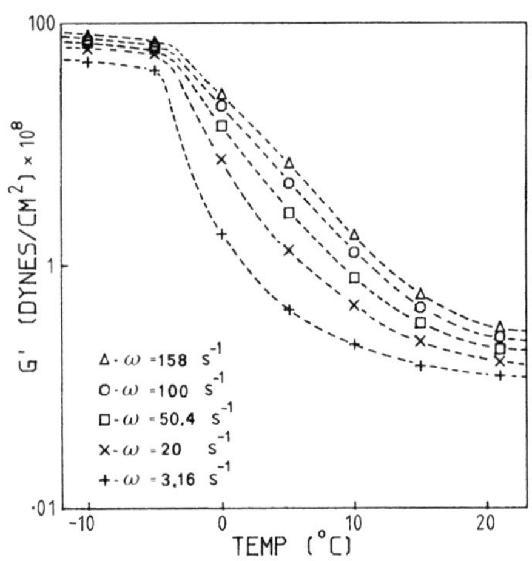

Fig. 4. Experimental G' vs. temperature (°C); frequency range as in figure 3

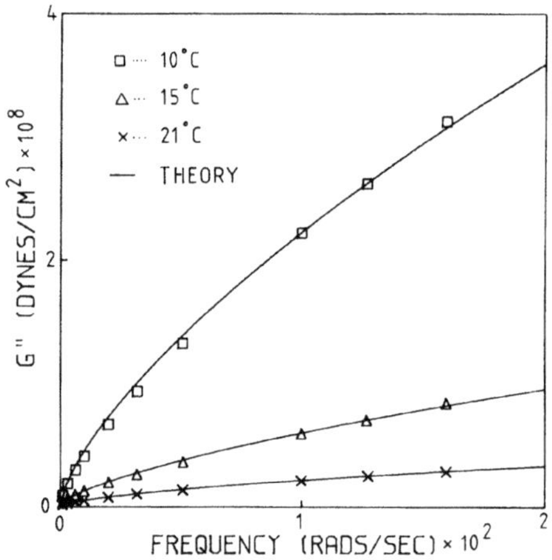

Fig. 5. Experimental G'' vs. frequency for temperatures (i) 21 °C; (ii) 15 °C and (iii) 10 °C. Continuous line – theory based on eqs. (7) and (8) with (i) $m = 1.26 \times 10^6$, $\alpha = 1.61$, $G_0 = 1.1 \times 10^7$; (ii) $m = 2.8 \times 10^6$, $\alpha = 1.5$, $G_0 = 1.2 \times 10^7$; (iii) $m = 9.3 \times 10^6$, $\alpha = 1.45$, $G_0 = 1.15 \times 10^7$.

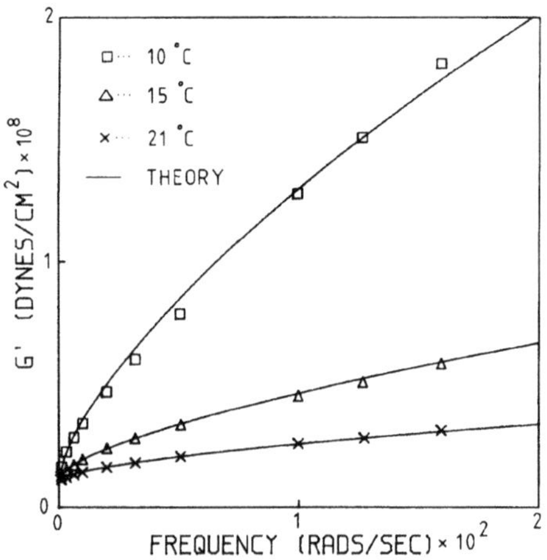

Fig. 6. Experimental G' vs. frequency; experimental and theoretical conditions as for figure 5

amplitude. However, below this temperature there are some changes which imply that we are no longer in the linear viscoelastic region. Therefore, it is not surprising that this model does not predict the experimental behaviour at low temperatures below 10 °C.

The proposed model in expressions (7) and (8) gives very good agreement with experiment for temperatures down to 10 °C. Figures 5 and 6 show this comparison.

4. Conclusions

(i) The experimental system can be used to demonstrate the effect of extension and temperature on the torsional vibrational properties of rubbers.
(ii) Extensions up to 15% do not affect the complex modulus behaviour.
(iii) The proposed large relaxation time model predicts the complex modulus behaviour of a filled rubber compound for a range of frequencies and limited temperatures.
(iv) At low temperatures linear viscoelastic theory is not applicable unless very small amplitudes are considered.
(v) The transition region may be a non-linear viscoelastic effect.

Acknowledgement

We would like to thank the Rubber and Plastics Research Association and Dr. Paul Howgate in particular for all the help and encouragement during the preparation of this work. We would also like to thank Dr. A. Chakrabarti for help with the experimental work.

References

1. Treloar, L. R. G., Physics of Rubber Elasticity (Oxford 1967).
2. Ferry, J. D., Viscoelastic Properties of Polymers, Wiley (New York 1970).
3. Duperray, B., paper presented at BSR-BGR conference "Rheology of Rubbery Materials and Rubber Compounds", Liege, 1981.
4. Walters, K., R. A. Kemp, Rheol. Acta **7**, 1 (1968).
5. Bird, R. B., R. C. Armstrong, O. Hassager, Dynamics of Polymeric Liquids, Vol. I, John Wiley & Sons (New York 1977).
6. Spriggs, T. W., Chem. Eng. Sci. **20**, 931 – 940 (1965).
7. Walters, K., Rheometry, Chapman & Hall (London 1975).

(Received April 15, 1982)

Authors' address:

Dr. T. E. R. Jones, Dr. J. M. Davies
Department of Mathematics, Statistics and Computing
Plymouth Polytechnic
Drake Circus
Plymouth, Devon PL4 8AA (U.K.)

Rheologica Acta Rheol. Acta **21**, 420 – 422 (1982)

Dynamic methods for the determination of the viscoelastic properties of epoxy materials used for refections of civil engineering constructions

F. Boulpaep, A. Cardon, and C. Hiel

Department of Applied Continuum Mechanics, Faculty of Applied Sciences, Free University Brussels (V.U.B.)

Key words: Viscoelasticity, dynamic method, epoxy resin

1. Introduction

The last twenty years epoxy resins were frequently utilised in order to repair cracks and other defects in structural construction elements. If those epoxy resins are very easy to inject, even in narrow cracks and defects, their mechanical response is not elastic but viscoelastic. It is thus important to know the influence of time, frequency, temperature, ageing, humidity and general environmental conditions on the mechanical behaviour of those epoxies and especially on their bonding characteristics.

We present some results of dynamic methods for the viscoelastic characterisation of those epoxy resins for different frequencies and temperatures. With some specific adhesion tests we have tried to obtain some correlation between these viscoelastic properties and the bonding properties.

2. Experimental method

The measuring method was the forced non-resonance method, and the obtained data are the storage modulus E_1, and a measure of the loss modulus $\tan\delta = E_2/E_1$. The instrumentation is commercially available from IMASS and has a frequency range between 0.1 and 100 Hz and a temperature range from $-180\,°C$ to $200\,°C$. In the actual investigation the frequency range is limited to $0.1-10$ Hz and the temperature range to $25-50\,°C$.

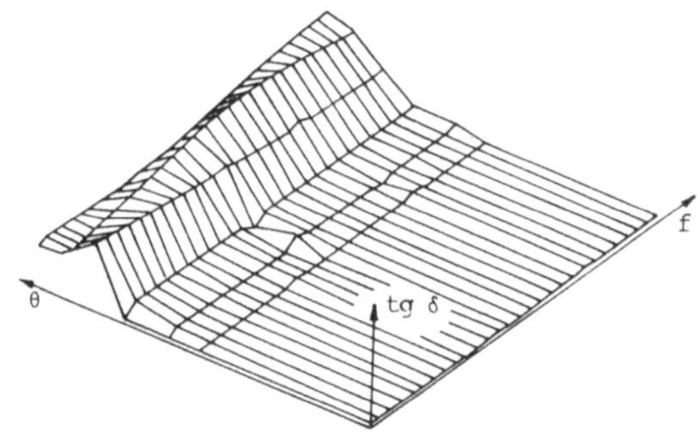

Fig. 1. Three dimensional representation of $\tan\delta$ as a function of frequency (f) and temperature (θ) of an epoxy adhesive

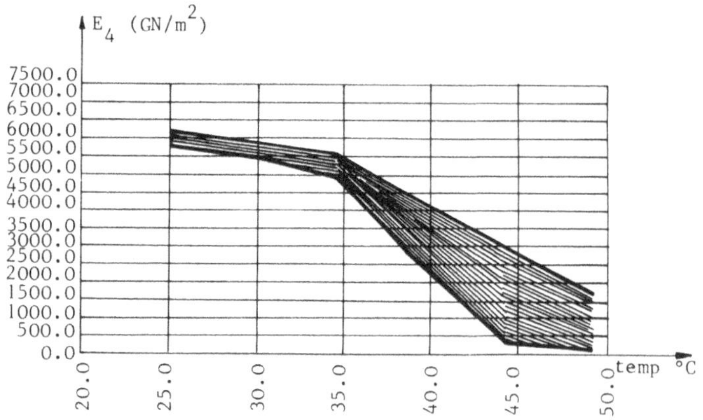

Fig. 2. Storage modulus (E_1) as a function of temperature (θ) for different frequencies of a current epoxy adhesive

A transformation of this standard equipment gives us a direct output connection to an HP-1000 minicomputer and we can obtain a direct plot of E_1 and (or) $\tan\delta$ versus the frequency and the temperature (see fig. 1).

Corrections for the rigidity of the apparatus and for different elementary loading devices are taken into account and are detailed in [1].

3. Viscoelastic properties

Figure 2 gives the variation of E_1 with the temperature for different frequencies of one of the current epoxies utilised for refections in civil engineering.

The transition between the high stiffness region, glassy state, and the low stiffness region, rubberlike state, is situated for all frequencies between 35 and 40 °C. This is a temperature value easily performed at the surface of civil engineering structures, and it is necessary to obtain more detailed informations about the mechanical behaviour and the bonding qualities of those refection materials when the temperature fluctuates between 25 and 50 °C.

4. Experimental results

A first series of experiences was performed in pure tension at different temperatures and constant strain rates. Figure 3 gives the stress-strain curves for temperatures between 24 and 50 °C with a constant strain rate of 0.5 m/min.

As could be expected we find that:

$$(\sigma_{\text{limit}})_{\theta=45\,^\circ\text{C}} = 78\%\,(\sigma_{\text{lim}})_{\theta=24\,^\circ\text{C}}$$

and

$$(\delta_{\max})_{\theta=45\,^\circ\text{C}} = 1.6\,(\delta_{\max})_{\theta=24\,^\circ\text{C}}\,.$$

In order to obtain some correlation between mechanical properties of the pure epoxy and his bonding properties in contact with classical construction materials (steel, concrete, aluminium, etc.), a series of shear tests was performed.

Figure 4 gives the results of such tests at constant strain rate and constant temperature.

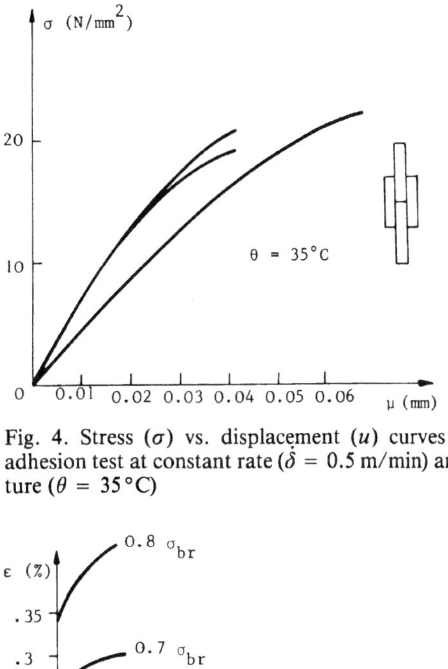

Fig. 4. Stress (σ) vs. displacement (u) curves of a shear adhesion test at constant rate ($\dot{\delta}$ = 0.5 m/min) and temperature (θ = 35 °C)

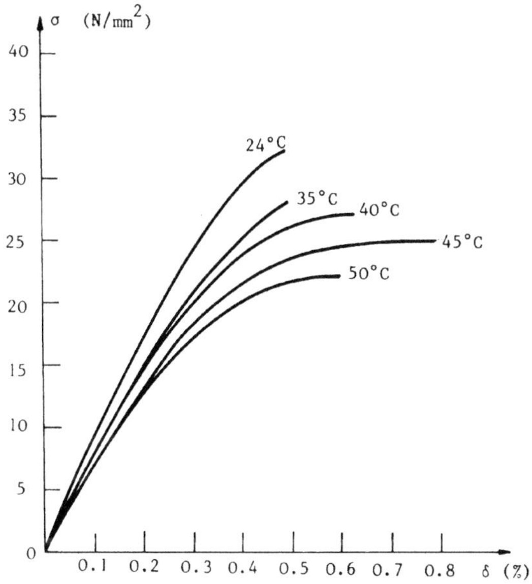

Fig. 3. Stress (σ) vs. strain (δ) curves in pure tension at different temperatures and constant strain rate $\dot{\delta}$ = 0.5 m/min

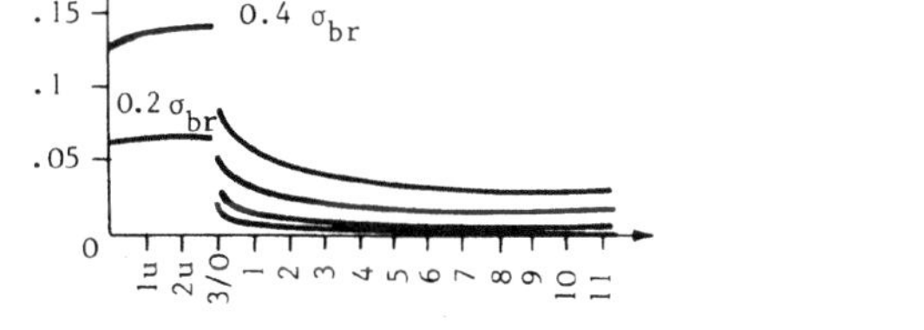

Fig. 5. Creep and recovery curves at constant temperature (θ = 35 °C) and different stress levels ($\alpha\sigma_{\text{limit}}$; $0 < \alpha < 1$)

Actually a comparison is made with adhesion tests at different temperatures between 25 and 50 °C.

5. Nonlinear viscoelastic effects

Those epoxies could be considered as linear viscoelastic materials only when $\sigma \leqslant 0.6\ \sigma_{\mathrm{lim}}$.

Figure 5 gives the creep curves at different stress levels and $\theta = 35\,^{\circ}\mathrm{C}$.

For $\sigma \geqslant 0.6\ \sigma_{\mathrm{lim}}$ a nonlinear viscoelastic representation must be introduced.

Acknowledgement

This research was sponsored by the Belgian National Science Foundation – N.F.W.O. – F.K.F.O. 2.9003.79.

References

1. Cardon, A., Cl. Hiel, "Forced Non-Resonance Method for the Determination of Viscoelastic Response of Composite Materials", in: A. P. S. Selvadurai (ed.), "Mechanics of Structured Media", part B – pp. 47 – 55, Elsevier Scientific Publishing Company (Amsterdam 1981).

(Received April 15, 1982)

Authors' address:

F. Boulpaep, Prof. Dr. A. Cardon, C. Hiel
Faculteit der Toegepaste Wetenschappen
Vrije Universiteit Brussel
Pleinlaan, 2
B-1050 Brussel

Rheologica Acta Rheol. Acta 21, 423–426 (1982)

Viscoelastic properties of silicone polymers with liquid crystalline behaviour

W. Oppermann, K. Braatz, H. Finkelmann, W. Gleim, H. J. Kock, and G. Rehage

Institute of Physical Chemistry, Technical University Clausthal, Clausthal-Zellerfeld

Key words: Liquid crystalline elastomer, polysiloxane, mesogenic side group, viscoelasticity, thermoelastic properties, photoelastic properties

1. Introduction

If small mesogenic molecules which are able to form a liquid crystalline phase are linked as side groups to a polymer chain, one obtains a substance that combines liquid crystalline and polymer specific properties [1, 2]. Above the glass transition temperature, the polymer exhibits an almost free motion of chain segments, the so-called micro-Brownian motion. Due to this liquid-like state of the chain backbone, the mesogenic side groups can build up ordered structures and thus form a liquid crystalline phase [3].

If the polymer molecules are crosslinked to form a network, the micro-Brownian motion of chain segments is not essentially influenced [4]. The mesogenic groups possess the same freedom as they do in a linear polymer. However, the macro-Brownian motion, that is the motion of the entire molecules, is prevented by crosslinking. This changes the viscous melt into an elastomer. That means we have now a combination of liquid crystalline properties with rubber elasticity. A first insight into the rheological behaviour of such liquid-crystalline elastomers will be discussed in this paper.

2. Brief sample characterization

The structure of the samples studied is seen from figure 1 which shows the reaction scheme for the synthesis. We used a siloxane copolymer that consists of equal parts of dimethylsiloxane units and such units that have a methyl group and a mesogenic group attached to the silicone atom. The length of the flexible spacer between the mesogenic groups and the polymer backbone, a requirement to obtain a liquid-crystalline phase, varies from 3 to 6 methylene units. The crosslinking was achieved by short chains consisting of 10 dimethylsiloxane units. The molar ratio of crosslinker and mesogenic groups was 1 : 8. The linkage of the mesogenic groups to the polymer as well as the crosslinking was done in a one step reaction in a 30% toluene solution. The crosslinked polymers were carefully extracted with toluene, then deswollen and dried [4].

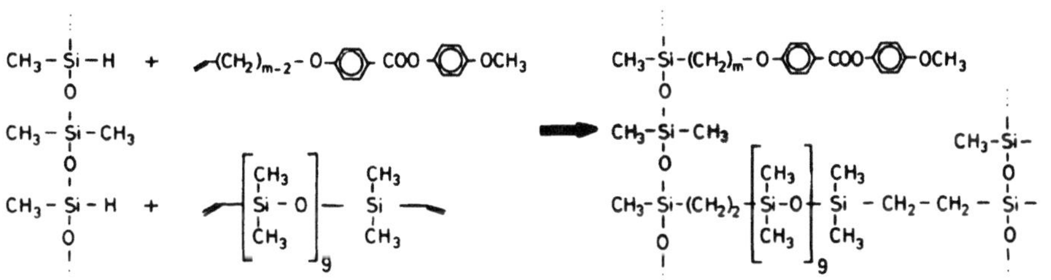

Fig. 1. Reaction scheme for the synthesis of liquid crystalline elastomers

828

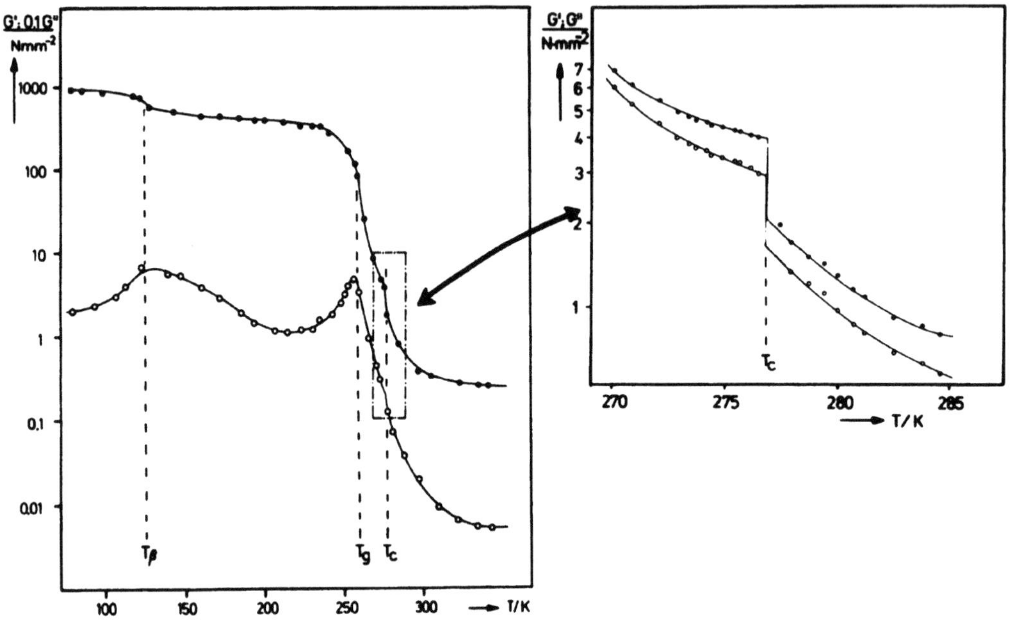

Fig. 2. a) Temperature dependence of the storage modulus G' (filled circles) and loss modulus G'' (open circles), measured at 0.5 Hz, for a liquid crystalline elastomer with spacer length m = 3. b) Detailed section of figure 2a

3. Results and discussion

3.1. Dynamic mechanical properties

The idea to study the viscoelastic properties of these elastomers resulted from the understanding that at the transition liquid crystalline to isotropic the mobility of the mesogenic groups changes considerably. This behaviour should be reflected somehow in the temperature dependences of G' and G''.

Figure 2 shows the results for an elastomer with spacer length m = 3, obtained by torsional vibration experiments. There is a β-relaxation process at about 120 – 130 K. The glass transition occurs around 260 K. The transition nematic to isotropic is indicated by a step at 277 K in the temperature dependence of G' and G''. It occurs in the tail of the glass transition. The detailed investigation in the temperature range close to T_c (fig. 2b) however, shows evidently that there is a clear jump in G' and G'' as well, indicating the first-order transition liquid crystalline to isotropic.

The samples having spacer lengths 4 and 6 give similar results. Table 1 summarizes the data. T_g decreases slightly with increasing spacer length, the clearing temperature increases strongly and hence the temperature range where the liquid crystalline phase exists, broadens considerably from 20 to 80 K*). As the spacer length increases, the motions of the polymer backbone and of the mesogenic groups are more and more decoupled. This is the reason for the decreasing of T_g. If they were totally independent, T_g should be much lower; e.g. polydimethylsiloxane has the glass transition at 140 K. On the other hand, increasing decoupling enables a better arrangement of the mesogenic groups. So the clearing temperature T_c increases.

Table 1. Transition temperatures T_g and T_c and step values $\Delta G'/G'$ and $\Delta G''/G''$ at T_c obtained from torsional vibration experiments at 0.5 Hz for samples with different spacer lengths

Spacer length (m)	Transition temperatures		Changes at T_c	
	T_g/K	T_c/K	$\Delta G'/G'$	$\Delta G''/G''$
3	258	277	0.55	0.50
4	255	302	0.10	0.11
6	253	335	0.08	0.09

*) The transition temperature obtained by differential scanning calorimetry differs sometimes up to ± 3 K, reflecting the difficulty to synthesize exactly identical samples.

The relative changes $\Delta G'/G'$ and $\Delta G''/G''$ which occur at the clearing temperature, decrease considerably with increasing spacer length. The polymer having spacer length $m = 3$, exhibits a nematic phase whereas the other ones are smectic. As the smectic phase is of higher order, a greater step of G' and G'' at T_c was expected with spacer lengths 4 and 6. The experimental results show the opposite behaviour.

What we have to conclude from this fact is that the rheological properties are largely dominated by the properties of the polymer backbone. The mesogenic side groups with their specific transition can only act as a small disturbance, and their influence is the smaller the weaker the coupling between the mesogenic groups and the polymer main chain is.

This view is supported by the fact that G' and G'' are in the usual order of magnitude found with common polymers, in the glassy state and in the rubbery region as well.

Frequency dependent measurement of G' and G'' were also carried out in the frequency range from 10^{-3} to 1 Hz. A specific influence of the phase transition on the data could not be detected. The curves below and above T_c appeared to be similar in shape.

3.2. Thermoelastic and photoelastic behaviour

Figure 3 shows the results of an experiment which is quite usual in rubber elasticity. The sample is stretched to a certain extent and then kept at this length. The retractive force divided by the crossrectional area of the undeformed sample, σ^0, is measured as function of temperature. In the upper temperature range, $T > T_c$, we find the typical behaviour of rubbers. At low elongation, σ^0 decreases with increasing temperature. This is because of the thermal extension of the sample. At a certain elongation, in our case at about 12%, σ^0 is independent of T. This is the so-called thermoelastic inversion point. At still higher elongation – we could not go so far because the sample broke – one should find an increase of σ^0 with rising temperature.

The clearing temperature of this sample is 283 K. At this point, the retractive force decreases substantially, even going to some negative values. Although the specimen had been stretched at higher temperature, now the tensile force changes into a compressive force. What we observe in the experiment is that the sample, a thin rubber strip, buckles when we go from higher to lower temperatures. This is an absolutely unusual and surprising behaviour in the rubbery region and has to be due to the ordering of the mesogenic side groups.

In the undeformed state above T_c, the network is isotropic, the statistical segments being randomly distributed. When the sample is stretched, the statistical segments and hence the mesogenic groups have an anisotropic distribution, but there is not yet a correlation among the mesogenic groups. When cooling down below T_c at this fixed length, the mesogenic groups become anisotropically ordered in the liquid crystalline state. As the network is already deformed uniaxially, the mesogenic groups adapt this macroscopic orientation, yielding a macroscopically ordered liquid crystal.

If now the force is removed (still below T_c), the sample will not move all the way back to its initial

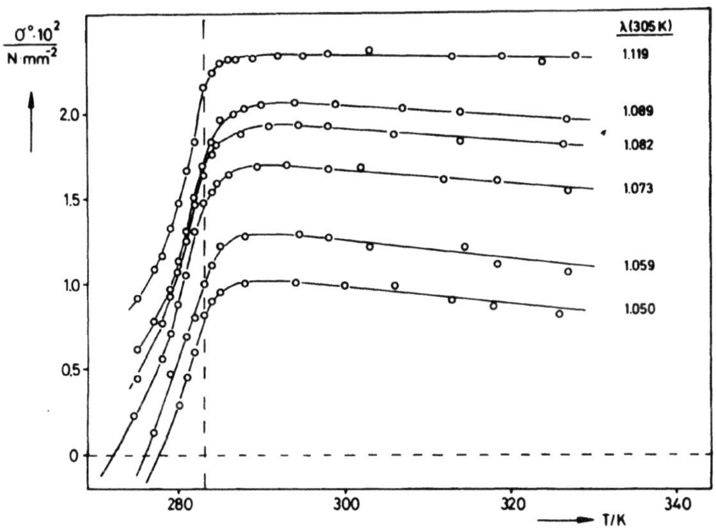

Fig. 3. Reduced stress σ^0 at constant sample length as function of temperature for a liquid crystalline elastomer with spacer length $m = 3$. The deformation was applied at the highest temperature. Different stretching ratios are indicated

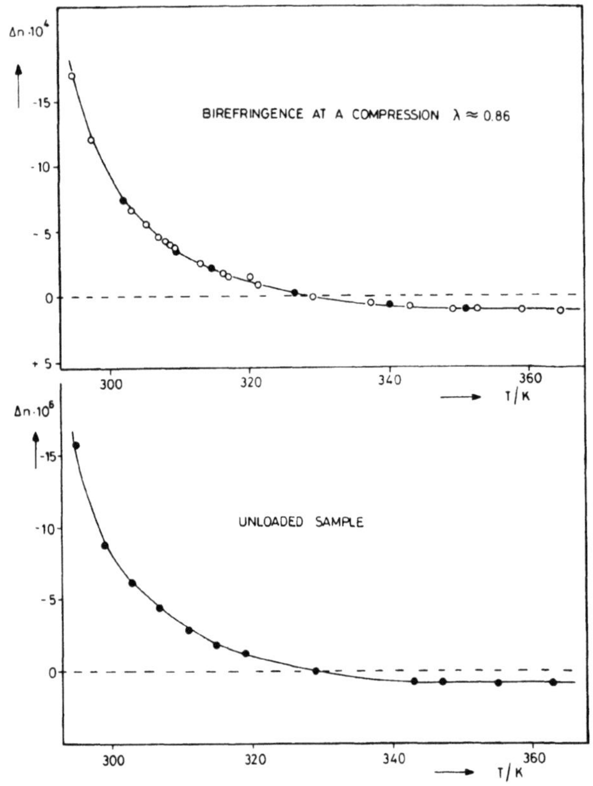

Fig. 4. a) Birefringence, Δn, as function of temperature for a liquid crystalline elastomer with spacer length $m = 3$; open circles: from constant stress data ($0.84 < \lambda < 0.87$); closed circles: from constant strain data ($\lambda = 0.86$). b) Birefringence of the unloaded sample after annealing for 3 days at 360 K

length but remain at some intermediate length. The reason for this behaviour which was also detected by direct measurement of the length of the unloaded sample, is that the anisotropic order of mesogenic groups causes the network to remain also to some extent anisotropic. The increase of the length of the unloaded sample below T_c is on the other hand the reason for the strong drop of the retractive force if the sample length is kept constant.

An important point is that after a deformation has been applied once, it seems to be impossible to bring the sample back into the totally isotropic state. Even after annealing at high temperatures, it will always remember some preferential direction, maybe because of some unrecovered strain, and then show the behaviour described.

Stress-optical measurements easily prove that this consideration is correct. Such measurements are only possible above T_c because of insufficient transparency of thick samples below T_c. Figure 4a shows the birefringence, Δn, plotted versus temperature. There is a considerable temperature dependence untypical for common rubbers. It is due to some pretransitional effects of the mesogenic groups which are well known for conventional liquid crystals [5].

If the sample is unloaded, stored at higher temperatures for some days, and thereafter the birefringence is measured as function of temperature, instead of zero birefringence the same shape of the curve is found, the absolute values just two orders of magnitude smaller (fig. 4b). This indicates that the sample has not regained its isotropy.

4. Conclusions

The experiments show that the viscoelastic properties of elastomers having mesogenic side groups are quite different from those of common polymers. At temperatures above the transition liquid crystalline to isotropic, the elastic behaviour is the usual one found for common rubbers. The photoelastic properties, however, are extraordinary, indication of a strong interrelation between the deformation of the network and the ordering of the mesogenic groups even in the pretransformational regime. Below T_c, this interrelation gives rise to exceptional effects of the elastic behaviour. The anisotropy of the network due to uniaxial strain causes a macroscopic orientation of the mesogenic groups and vice versa. These unusual effects are of great scientific and technological interest. To come to a more profound understanding, further and more detailed studies of such effects are in progress.

References

1. Shibaev, V. P., N. A. Plate, Pol. Sci. USSR **19**, 1065 (1978).
2. Finkelmann, H., H. Ringsdorf, J. H. Wendorff, Makromol. Chem. **179**, 273 (1978).
3. Finkelmann, H., G. Rehage, Makromol. Chem. Rapid Comm. **1**, 31 (1980).
4. Finkelmann, H., H.-J. Kock, G. Rehage, Makromol. Chem. Rapid Comm. **2**, 317 (1981).
5. Kelker, H., R. Hatz, Handbook of Liquid Crystals, p. 206 (Weinheim 1980).

(Received June 28, 1982)

Authors' address:

Dr. W. Oppermann et al.
Institut für Physikalische Chemie
Technische Universität Clausthal
Adolf-Römer-Str. 2a
D-3392 Clausthal-Zellerfeld

Influence of molar mass and of nucleating agents on the orientation of polyethylene terephthalate

H. G. Zachmann and B. Günther

Abteilung Angewandte Chemie, Universität Hamburg

Key words: Polymer, orientation, crystallization, molar mass

If amorphous polyethylene terephthalate is drawn at temperatures of about 90 °C, this is slightly above the glass transition temperature, the material is deformed completely homogeneously without necking; the molecules become oriented. Which parameters influence the orientation obtained? In crystalline polymers the orientation is determined mainly by the drawing ratio. In amorphous polymers also the drawing temperature and the drawing rate are of great interest. This has been shown already some years ago for polystyrene by Berdelew et al. [1]. Measurements with polyethylene terephthalate have been performed in our laboratory [2]. Figure 1 shows some results. As a measure of orientation the birefringence Δn_0 is used. Complete orientation corresponds to a value of Δn_0 equal to $265 \cdot 10^{-3}$. Similar results were obtained by De Vries et al. [4].

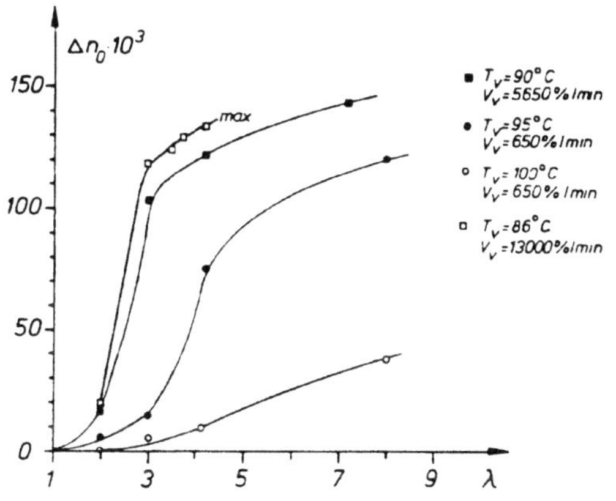

Fig. 1. Birefringence Δn_0 of amorphous polyethylene terephthalate drawn at the temperature T_v with the rate V_v as a function of the drawing ratio λ

775

There arises the question of how the orientation obtained is influenced by the molar mass. Some results obtained on a material which was synthesized in our laboratory by using calcium acetate as a catalyst are shown in figure 2. Two different drawing rates have been applied namely 400%/min and 5000%/min. One can see the following: If one uses the smaller drawing rate, the birefringence obtained at a given drawing ratio is the larger the higher the molar mass is. For the larger drawing rate there is also an influence of the molar mass; this influence, however, is much smaller than in the case of the lower drawing rate.

What is to be learned from these results? The influence of the molar mass on the orientation obtained can be caused in principle by two effects:

(i) by an increase of the entanglement density with molar mass,
(ii) by a decrease of slipping of the chains with increasing molar mass.

As the influence of the molar mass on the orientation obtained becomes smaller with increasing rate of drawing the first explanation cannot be the correct one. This means that the increase of orientation obtained with molar mass is due rather to chain slipping than to an increase of entanglement density.

In addition to samples polycondensated with calcium acetate also samples which contained manganese acetate as a catalyst were investigated. No significant difference was obtained.

It is interesting to compare the orientation behavior of amorphous samples with that of crystalline samples. Samples with different molar masses and

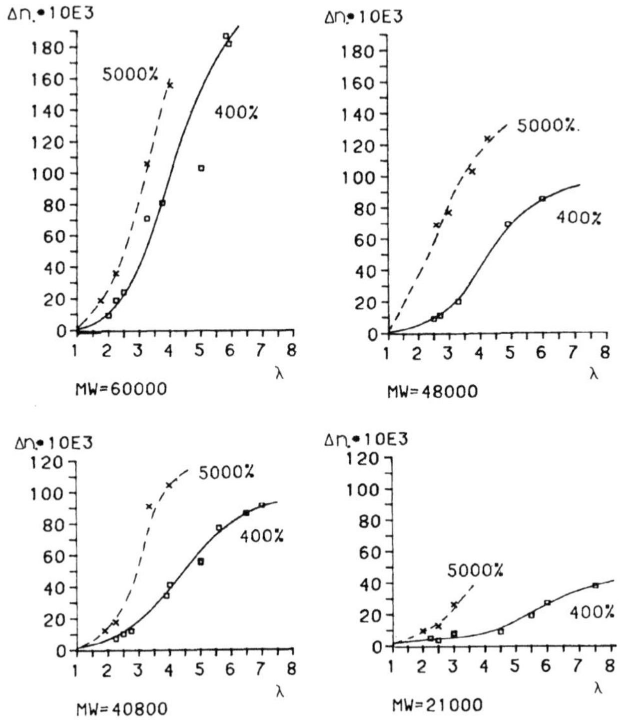

Fig. 2. Birefringence of amorphous polyethylene terephthalate drawn at 92 °C with the drawing rates 5000%/min and 400%/min respectively as a function of the drawing ratio λ. Four different molecular weights MW have been investigated

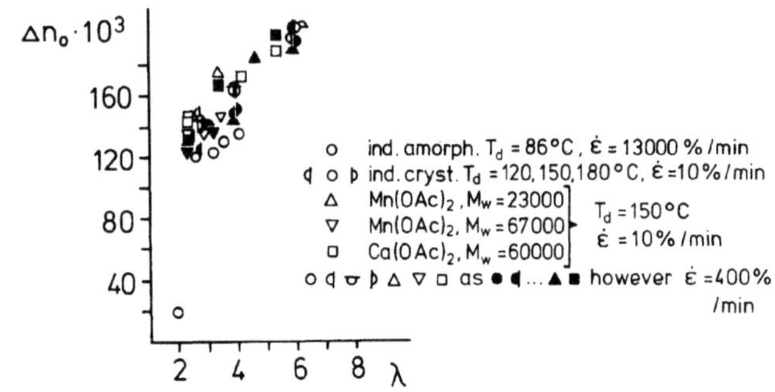

Fig. 3. Birefringence Δn_0 of amorphous and crystalline polyethylene terephthalate as a function of the drawing ratio λ for samples with different molar masses, different drawing temperatures and different drawing rates, as indicated above

different catalysts were crystallized at 190 °C and drawn at 150 °C. In the case of the crystallized samples it was not possible to perform the drawing homogeneously; always necking occurred. In figure 3 the birefringence is plotted as a function of drawing ratio λ obtained within the neck. One sees that the drawing rate, the molar mass and the catalysts do not influence the birefringence obtained. In addition also experiments with a film from the company Kalle were performed. The drawing rate was 10%/min and 400%/min respectively. The results obtained agree completely with those obtained on the samples synthesized in our laboratory. In addition, also the results obtained for amorphous samples with drawing conditions giving the maximum birefringence from figure 1 are introduced into this figure 3 (open circles). These results coincide with those obtained on the crystalline samples. From this it has to be concluded that the netpoint density in the crystalline material is the same as that in the amorphous material if one is drawing with a large drawing rate at a low temperature. As in the amorphous material the chain entanglements are the netpoints one has to conclude that also in the crystalline material the netpoints are formed by chain entanglements. The crystals have the effect to prevent chain slipping so that the maximum density of chain entanglements is always effective. Obviously the crystals can do this although they are partly distroyed during drawing.

References

1. Rettig, W., Colloid & Polymer Sci. **257**, 689 (1979); see also G. M. Bartenev and Ju. V. Zelenev: "Die Physik der Polymeren", VEB Deutscher Verlag für Grundstoffindustrie (Leipzig 1979).
2. Biangardi, H. J., H. G. Zachmann, Progr. Colloid & Polymer Sci. **62**, 71 (1977).
3. Biangardi, H. J., Habilitationsschrift, Berlin 1980.
4. De Vries, A. J., C. Bonnebeat and J. Beautemps, J. Polymer Sci.: Polymer Symposium **58**, 109 (1977).

(Received May 11, 1982)

Authors' address:

Prof. Dr. H. G. Zachmann, Dipl. Chem. B. Günther
Institut für Anorganische und Angewandte Chemie
Universität Hamburg
Martin-Luther-King-Platz 6
D-2000 Hamburg 13

Rheologica Acta Rheol. Acta **21**, 429 – 430 (1982)

The deformation behaviour of annealed polypropylene film tapes

J. Lenz

Chemiefaser Lenzing AG

Key words: Polypropylene, film, annealing, deformation, conformation

As a striking difference to other polymers the stress-strain-curves of monoaxially stretched polypropylene do not show any stress-hardening region or yield point. The curves look parabolic without any sharp bend, as known about polyamide, polyethyleneterephthalate or polyacrylonitrile. It seems, that the deformation mechanism of sufficiently stretched polypropylene is relative simple.

In our experiments isotactic polypropylene was extruded to a flat film and cooled on a chill roll. It was slit into strips, which were stretched in a ratio of 1 to 8 on a hot plate, giving the maximum tensile strength. Immediately after passing the stretching field the tapes were annealed on a number of heated stretching rollers. As the annealing was done within 20 seconds under the stretching tension, without shrinkage, this arrangement allowed the application of temperatures far above the melting point. The stress prevents melting owing to the decrease of the entropy change at the fusion point about 1 J/K mol. After annealing the stress-strain-curves were plotted.

We could describe the different curves of all annealing experiments with an empirical formula: $\sigma^0 = E^* \cdot \varepsilon^n$ where δ^0 means the conventional stress, E^* a constant parameter, ε the relative elongation and n a constant exponent < 1.

The differentiation of the stress to the elongation results in the Young's modulus. The integration of the stress to the elongation at break results in the breaking work.

E^* is inversely proportional to the annealing temperature, similar to the relationship between tenacity and annealing temperature. In contrast n remains constant up to an annealing temperature of 196 °C and decreases abruptly above this temperature. This feature corresponds to the tear strength in transverse direction. It rises suddenly between 190 and 200 °C. Above this temperature the tapes, which normally manifest a brittle behaviour, become ductile, flexible and nonfibrillating. The change of the deformation properties was so significant, that we suggested a transformation of the microstructure:

As expected, on annealing the crystallinity rises with increasing temperature. Above 196 °C however it falls off. This effect could be reduced to the changes of the long identity period. It increases up to 189 °C. Above this temperature the long spacing falls off, too. A similar effect was found regarding the lateral identity period calculated from the equatorial scattering.

The crystalline orientation function shows a maximum at 195 °C, the average orientation function a curve similar to the variation of exponent n in relation to the annealing temperature.

In order to get a better understanding of this disorientation during annealing under tension, we determined the dichroitic ratio of the π-polarized IR-bands ascribed to the regular helix-conformation as a measure of the molecular orientation.

The dichroitic ratio of these bands, versus the annealing temperature, differs as well from the crystalline as from the amorphous orientation function and correlates well with the parameter E^*. E^* is found to be the higher, the better the orientation of the molecules in regular helical conformation is.

The physical meaning of the relationship between this deformation parameter and the molecular orientation is revealed by the correlation existing between the parameter E^* and the small angle X-ray scattering intensity. The density of the amorphous phase increases with increasing amorphous orientation [1]. Consequently a disorientation of the helical molecules in the noncrystalline domains with rising annealing temperature must result in a less dense package of molecules and hence in a greater difference of electron density between the crystallised and amorphous regions. The relationship of E^* as well with the dichroitic ratio as with the scattering intensity leads to the conclusion, that this parameter is a function of the orientation of the so called tie molecules.

The assumption, that all tie molecules have the regular helix conformation, is not generally true: According to Takanayagi [2] the conformation of the so called taut tie molecules, formed during stretching, may not be a regular helix but an irregular one. He suggests a high potential barrier, when transforming the latter into the normal helix conformation. When annealing is done with free ends, a transformation temperature of 155 °C has been found.

We examined, whether such a transformation also occurs, if annealing is performed with clamped ends under high tension. For this reason we observed the variation of the absorbances in non-polarized infrared spectra, as function of the annealing temperature. We compared the absorbance of the band at 998 cm^{-1}, which is doubtless assigned to a regular helix, with the one at 900 cm^{-1}, which is characteristic for crystallized polypropylene but not typical for a regular helix [3].

At an annealing temperature of 196 °C the absorbance of the 900 cm^{-1}-band decreases, whereas that of the 998 cm^{-1}-band increases. Thus we find a transformation temperature, at which the irregular helical conformation is transformed into a regular one. This temperature lies 40 °C above the one found by Takayanagi, because annealing was done under tension.

These taut tie molecules must also be oriented. We therefore investigated the dependence of the dichroitic ratio of the band at 900 cm^{-1} upon the annealing temperature: The dichroitic ratio abruptly increases between 190 and 200 °C, which means in this case a decrease of the orientation, because this band is perpendicularly polarized.

If this band is characteristic for taut tie molecules a relationship with the deformation behaviour must exist. Indeed a correlation was found between the exponent n in our formula and the dichroitic ratio of the band at 900 cm^{-1}. n was found to be the higher, the better the orientation of the taut tie molecules. In contrast to the tie molecules they retain their orientation until high annealing temperatures on a constant level. At a certain transformation temperature however they disorient abruptly and regain their regular helical conformation. It is the same point, where the crystalline orientation function falls off and apparently the internal tension of the molecular network collapses under the influence of the thermal energy, which breaks the intermolecular bonds.

These results make our formula suggestive: If n becomes 1, crystalline domains and taut tie molecules are completely oriented, the formula is identical with Hooke's law and E^* means the Young's modulus. If n is zero, no taut tie molecules are existing at all, crystalline blocks and tie molecules are randomly oriented, and E^* gets the significance of the yield stress in a steady elongational flow.

Proceeding from this formula we would like to propose a deformation mechanism as follows:

If a filmtape is strained, at first it yields by orientation of the tie molecules between the crystalline domains. Stress and strain increase proportionally to the orientation.

The slope is the steeper, the higher the orientation of the tie molecules in the undeformed tape. So far as the orientation of the crystalline domains remains constant, in this process stress and strain are direct proportional to each other. If these areas are flexible, however, the crystallites will rotate into the deformation-direction and act as additional yield points.

The orientation of the crystalline regions is restrained, because bundles of taut tie molecules connect the crystalline domains to a physical network, and impede the rotation of the crystallites. At annealing above a certain temperature however the taut tie molecules are transformed into molecules with a regular helical conformation. Now the stress-strain-curve shows the course of a flat parabola.

A detailed description of the experiments and conclusions outlined above has recently been given in [4].

References

1. Farrow, G., J. M. Ward, Polymer **1**, 330 (1960).
2. Yamada, K., M. Kamezawa, M. Takayanagi, Journ. Appl. Polymer Sci. **26**, 49 – 60 (1981).
3. Raff, R. A. V., K. W. Doak, Crystalline Olefin Polymers, part I, in: H. Mark, P. S. Flory, C. S. Marvel, H. W. Melville (eds.), High Polymers, Vol. XX, p. 804, J. Wiley & Sons (New York 1965).
4. Lenz, J., Rheol. Acta **21**, 255 – 269 (1982).

(Received April 15, 1982)

Author's address:

Dr. J. Lenz
Chemiefaser Lenzing AG
Postfach
A-4860 Lenzing

Rheological properties of high polymers at high rates of strain

J.-L. Lataillade and J. Pouyet

Laboratoire de Mécanique Physique, E.R.A.CNRS 769, Université de Bordeaux I, Talence (France)

Key words: Impact rate loading, dynamic viscoplasticity, non-linear viscoelasticity

1. Introduction

Rheological characterization of solids at high strain rates is of great importance because of the ultimate severe conditions encountered in their use or treatment: shocks in the former case, transformation processes in the latter one.

The problems involved can be split into two : On the one hand it is necessary to establish a mechanical relationship including strain rate effects – such as the increase of the instantaneous Young's modulus, and of the yield limit. On the other hand it is necessary to interpret such results in relation to the structural properties in so far as the strain rate has an effect on the interaction mechanisms between strain links, groups or phases, etc. Thus we present an experimental way of characterization by means of the Split Hopkinson bar technique and some methods of rheological analysis.

2. Experimental set-up and governing equation for data analysis

The apparatus used consists of aligned pressure bars, sandwiching a sample to be tested. Its quasi-static loading is achieved owing to a stress or torque impulse travelling down the input bar.

Because of the impedance mismatch, a reflection occurs governing the rate of the tensile or shear deformation; the wave transmitted through the specimen into the output bar is related to the shear (or compressive) stress, supposed uniformly in the active length of the tubular specimen, and during a time at most equal to $130 \cdot 10^{-6}$ s for compression, to $2 \cdot 10^{-3}$ s for torsion; for example, the torsion device is schematically described in figure 1.

The active part of the input bar is previously twisted to a desired value of torque T between the drive unit and the clamp. Fracture of the notched bolt implies the sudden release of the stored energy: incident (I), reflected (R) and transmitted (T) waves are registered through semi-conductor strain gauges during the time t.

The governing equations for the compression and for the torsion are respectively

$$\dot{\varepsilon}(t) = (2C_b^1/e)\varepsilon_R(t)\,, \tag{1a}$$

$$\dot{\gamma}(t) = (2r_m/eJ\rho_b)C_b^2(t)\,, \tag{1b}$$

where C_b^1 is the longitudinal wave velocity in the bar, C_b^2 the shear wave velocity, ρ_b the specific mass of the bar, J the polar initial moment of the bar; r_m and e the characteristics of the specimen, i. e. the radius and the active length.

Assuming a quasi-static stress state in the specimen one gets

$$\sigma(t) = \rho_b(C_b^1)^2\varepsilon_T(t), \tag{2a}$$

$$\tau(t) = \frac{\mathrm{T_T}(t)}{2\pi r_m^2 a} \tag{2b}$$

where a is the thickness of the tubular specimen and

$$\sigma(t) + K\dot{\varepsilon}(t) = \sigma_I(t), \quad 2K = \rho_b C_b^1 e, \tag{3a}$$

$$J\rho C_b^2 e\dot{\gamma}(t) + 4\tau(t)\pi e r_m^3 = T_I(t)\,. \tag{3b}$$

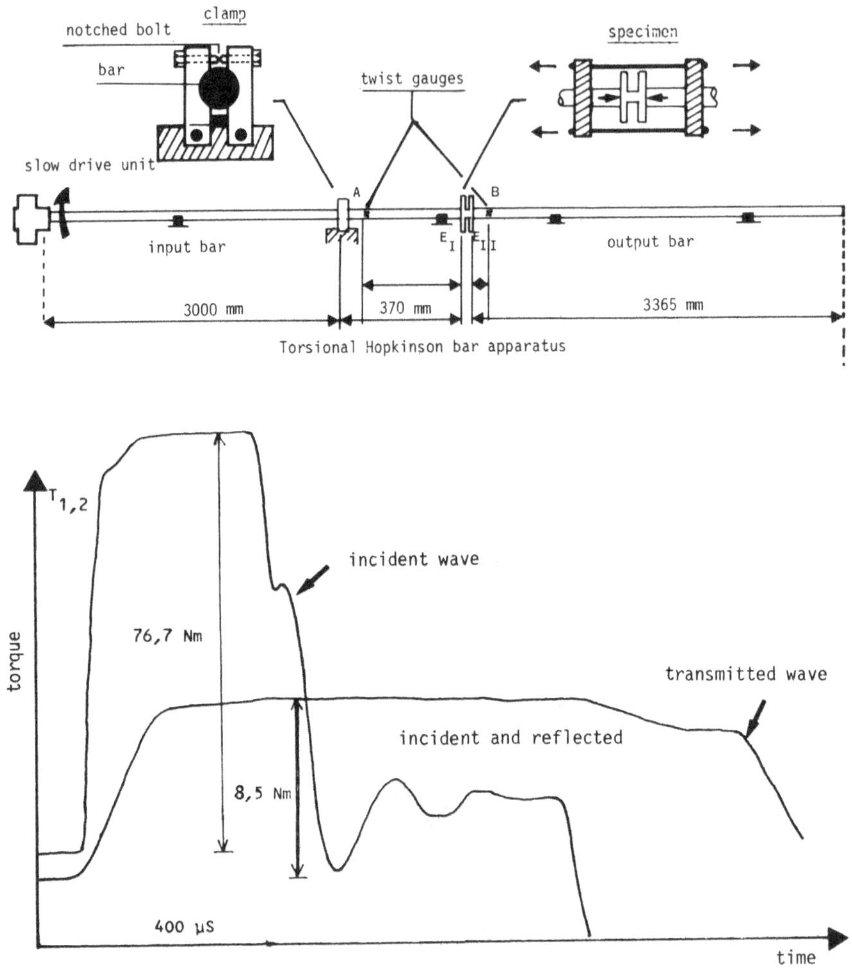

Fig. 1. Torsional Hopkinson bar apparatus (a), and propagation of incident, reflected and transmitted waves (b)

3. Modelization: examples of compression results

3.1. Analog way

Considering the total deformation ε as a sum of an elastic, a viscoelastic and a plastic part, and supposing the Von Mises' criterion verified, one finds (with σ_{ii} = Trace σ_{ij}, T the temperature and α the thermal coefficient):

$$\dot{\varepsilon}_{ij} = \frac{1+\nu}{E}\sigma_{ij} - \frac{\nu}{E}\delta_{ij}\sigma_{ii} + \left[\frac{1}{2\eta} + (1-f)\right]\left[\sigma_{ij} - \frac{\sigma_{ii}}{3}\delta_{ij})\right] + \alpha\dot{T}\delta_{ij}\,; \tag{4}$$

ν and E are respectively the Poisson's ratio and the glassy Young's modulus of the material, η a generalized viscosity and δ_{ij} the Kronecker's symbol.

In the uniaxial case:

$$f = 1 \Rightarrow \dot{\sigma}(t) = E\left(\dot{\varepsilon}(t) - \frac{\sigma(t)}{3\eta} - \alpha\dot{T}\right)$$

$$\text{if} \quad s_{ij}\dot{s}_{ij} - 2kT < 0\,,$$

$$f = 0 \Rightarrow \sigma(t) = k\sqrt{3} = Y(t)$$

$$\text{if} \quad s_{ij}\dot{s}_{ij} - 2k\dot{T} > 0\,,$$

which leads to:

$$\dot{\sigma}(t) = fE\left(\dot{\varepsilon}(t) - \frac{\sigma(t)}{3\eta} - \alpha\dot{T}\right) + (1-f)\,\dot{Y}(t)\,. \tag{5}$$

Regarding to the apparatus equation (3 b), and for very weak variations of temperature, one obtains:

(i) for $\sigma(t) < Y(t)$, or $\sigma(t) < \sigma_I(t)\dfrac{\tau}{\tau_0}$:

$$\dot\sigma(t) + \sigma(t)\left(\frac{1}{3\tau_1} + \frac{1}{\tau_0}\right) = \frac{\sigma_I(t)}{\tau_0}, \qquad (6)$$

(ii) for $\sigma(t) = Y(t)$:

$$\kappa\dot\varepsilon(t) = \sigma_I(t) - Y(t).$$

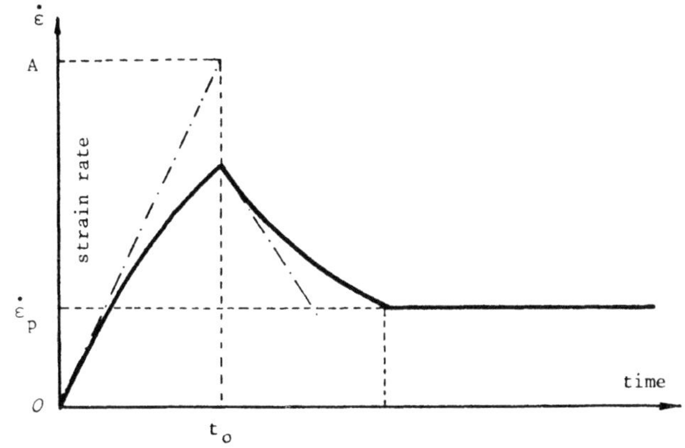

Fig. 2. Viscoelastoplastic material ($t_0 \simeq 15$ μs is the rise time of incident wave)

Figure 2 represents some particular interesting values;

$$OA = \frac{\sigma_0}{K}\left[1 - \frac{\tau}{\tau_0} + \left(\frac{\tau}{\tau_0}\right)^2\right],$$

$$\tau_1 = K/E, \quad \tau_0 = \eta/E,$$

$$\frac{\tau}{\tau_0} = \frac{K}{K+\eta}, \quad \tau^{-1} = (3\tau_1)^{-1} + (\tau_0)^{-1}.$$

With such an approach it is then possible to compute a viscosity η from the value of τ, the value of E being determined from the curve (σ, ε).

Reciprocally the validity of the uniaxial relations in compression and in shear, leads to a generalization in the three-dimensional case, provided that no coupling between shear mechanisms and volumic effects occurs, which can be supposed for small deformation.

For example, for a homopolymer of polypropylene one gets:

$E = 270 \cdot 10^7$ Pa, $Y = 11.25 \cdot 10^7$ Pa, $\eta = 5.625$ Pa s in the range $10^{-2}\,\mathrm{s}^{-1} < \dot\varepsilon < 10^3\,\mathrm{s}^{-1}$.

In some cases the behaviour exhibits more than one relaxation time, than a second method based on a graphical construction in the plane $(\sigma, \dot\varepsilon, t, \varepsilon)$ and fully described in [1] can be employed. For an interpenetrated network system (polyepoxy-polyurethane) it has been measured in the same ranges of strain rates as previously quoted, the respective values are reported in table 1.

Table 1

	100/0	90/10	80/20	70/30
σ_{S1}	59 MPa	70 MPa	80 MPa	97 MPa
σ_{S2}	125 MPa	110 MPa	100 MPa	68 MPa
η_1(P1)	18.61	21.94	27.08	37.1
η_2(P1)	8	8.57	6.25	5.2

The dependence of η_2 on the composition seems to be consistent with other thermophysical results published in [2] and showing a synergistic effect because of the particularly sharp interpenetration for

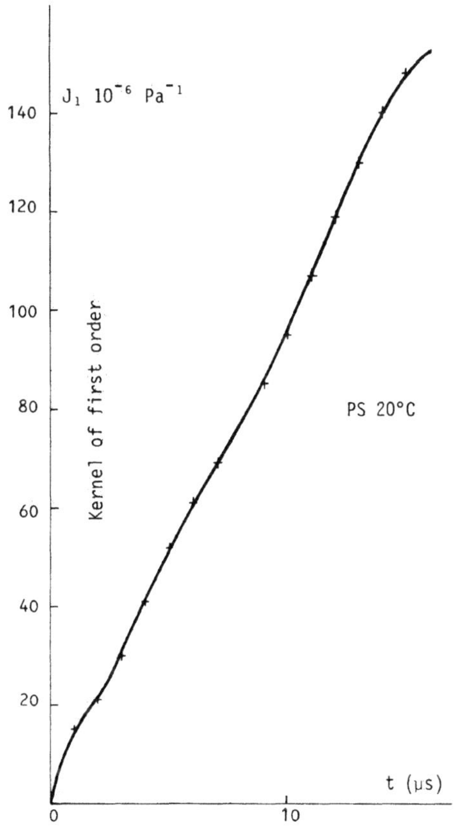

Fig. 3. Variation of the compliance with time for a polystyrene

the blends with a compositions around 90/10. This phenomenon could be due to the mechanical entanglements at the boundaries between two phases, implying an extent of the relaxation processes along the time axis.

3.2. Functional modelization

Another way of our studies concerns the illustration of nonlinearities, due to specific non-linear mechanisms, or to the amplitude of sollicitations which can be involved in a viscoelastic material.

For uniaxial configurations and among various analysis methods (see Lockett [4]) our modelization is based on a Frechet expansion by means of Volterra integrals restricted to the third rank:

$$\varepsilon(t) = \sum_1^n \int \ldots \int J_n(\tau_1 \ldots \tau_n) \prod_1^n \dot{\sigma}(t - \tau_n) d\tau_n , \quad n \leqslant 3 . \tag{7}$$

The experimental method consists in maintaining the stress rate in the material constant. It allows us to treat eq. (7) in a rather simple way by means of an identification method for the kernels J_n: the numerical method is the systems' one. Corresponding developments are fully described in [3].

Figure 3 shows the variation of the compliance J_1 during the time of the sollicitation. The two first kernels can be determined with reliability by this method.

4. Conclusion

Hopkinson pressure bars devices seem to be a good tool for the investigation of solid polymers at high strain rates. Because of the principle of the loading it is impossible to obtain a constant strain rate but nevertheless the data can be treated for deducing a rheological relationship. In a second step, it is intended to continue this work by taking into account the coupling between thermal and mechanical processes, referring to the thermodynamics of irreversible processes. As a third step we are interested in treating the problem of biaxiality.

References

1. Lataillade, J. L., J. Pouyet, A. Hamdy, Comptes-Rendus à l'Académie des Sciences, Série B, **291**, 59–62 (1980).
2. Poussin, D., J. C. Pommier, R. Daviaud, J. L. Lataillade, J. Pouyet, J. P. Robert-Arnouilh, Interpenetrating networks based on a polyepoxyde-polyurethane system. Thermomechanical investigation, 6th European Plastics Conference, Paris, June 7–9, 1982.
3. Pouyet, J., J. L. Lataillade, C. Signoret, Cahiers Groupe Français Rhéologie, **V**, n° 6, février 1982.
4. Lockett, F. J., Non-linear viscoelastic solids, Academic Press (New York 1972).

(Received April 15, 1982)

Authors' address:

Prof.-Ass. J.-L. Lataillade, Prof. J. Pouyet
Laboratoire de Mécanique Physique
Université de Bordeaux I
351 cours de la Libération
F-33405 Talence Cedex

Extrusion of thermoplastic elastomers in the solid state

M. Dröscher and U. Bandara

Institut für Makromolekulare Chemie der Albert-Ludwigs-Universität, Freiburg i. Br.

Key words: Elastomer, solid state extrusion, block copolyester, structure

1. Introduction

The deformation in the solid state is a well known orientation process for polymers. An up-to-date account of developments in the production, structure, and properties of solid state extruded homopolymers can be found in [1]. Very recently, solid state extrusion has been applied for copolymers as well [2, 3]. Copolymers are extruded under lower forming pressures than homopolymers but the extrudates show lower stiffness together with higher die swell and poorer creep properties.

Here we report on the solid state extrusion of segmented copoly(ether ester)s [4, 5]. These materials combine the elastic recovery and energy absorption behavior of elastomers with the processing convenience of thermoplastics. The properties are based on the phase separation behavior of the materials. A certain fraction of the ester segments crystallizes and forms the hard phase, whereas the remaining ester segments mix with the ether sequences and build up the soft matrix (see fig. 1):

$$\{-O-(CH_2)_4-\}_n O-CO-C_6H_4-CO-$$

$$\{O-(CH_2)_4-O-CO-C_6H_4-CO-\}_m$$

Material **1A**: $\bar{n} = 14$, $\bar{m} = 5$, **1B**: $\bar{n} = 14$, $\bar{m} = 6$, **1C**: $\bar{n} = 14$, $\bar{m} = 12$.

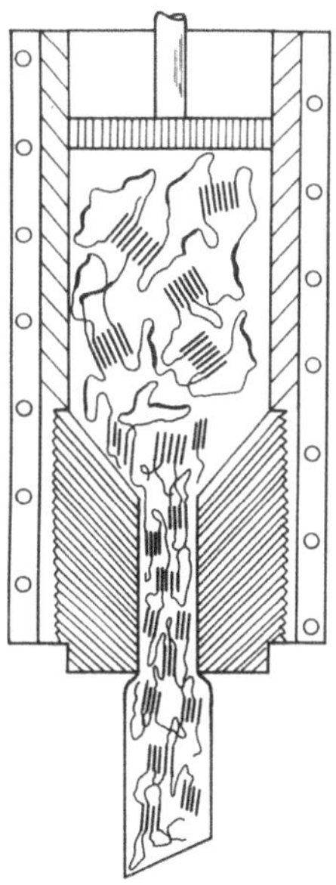

Fig. 1. Schematic representation of the solid state extrusion experiment under ram conditions

2. Results and Discussion

Figure 1 is a schematic representation of the extrusion process in the solid state. The experiments were carried out with a commercial rheometer (Göttfert Rheograph 2000). An injection molded billet was extruded under ram conditions through a circular die with a conical inlet. The optimum cone semi angle was found to be 41°. The nominal draw ratio R_N, which is the ratio of the cross sections of the die inlet and outlet, was varied between 2 and 7. Because of the die swell which occurred in all cases, the actual draw ratio was always smaller than R_N. The largest die swell was observed for material **1A** with 34 – 40% for a R_N of 4 and a rate of extrusion of 0.4 mm^3/s. The smallest effect was found for material **1C** with 12% under similar extrusion conditions. The die swell increased with the rate of extrusion to a value of 24% at 384 mm^3/s and $R_N = 4$ for sample **1C**.

Typical pressure time curves for a set of extrusion experiments are presented in figure 2. At 190 °C the extrusion pressure, as determined perpendicular to the extrusion direction, did not depend on the extrusion rate. At 138 °C the pressure increased with the rate of extrusion significantly.

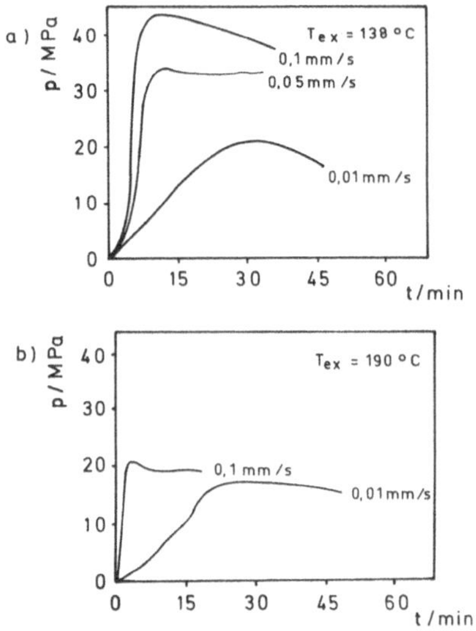

Fig. 2. Extrusion pressure as determined perpendicular to the extrusion direction versus extrusion time. Parameter ram speed. Diameter of the die inlet 7 mm, die outlet 3.5 mm. Material **1C**

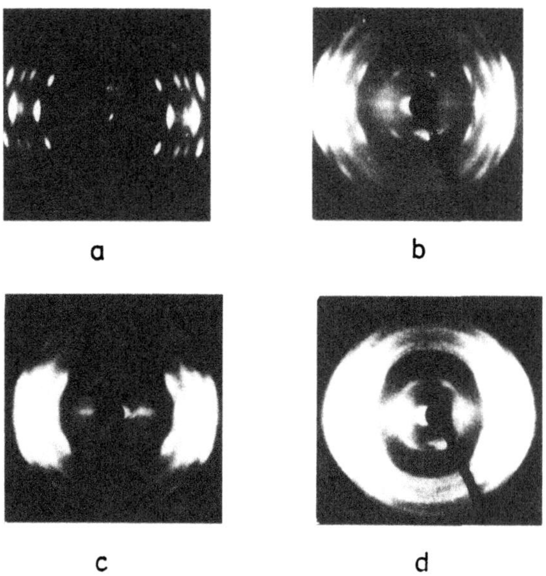

Fig. 3. Fiber diagrams of material **1C** extruded at (a) 180 °C, (b) 143 °C, (c) 93 °C, (d) 65 °C. Rate of extrusion 0.4 mm³/s, nominal draw ratio 4

As indicated in figure 1, the lamellar crystallites which built up the hard phase of the elastomer were oriented upon the deformation process. The degree of orientation of the crystallites depended strongly on the temperature of extrusion. This is demonstrated in figure 3, where a set of wide angle X-ray fiber diagrams is shown. The second moment orientation function f_c [6], which becomes 1 for full orientation and -0.5 for orientation in the perpendicular direction, decreased from 0.97 to 0.86 as the extrusion temperature was lowered from 180 °C to 65 °C.

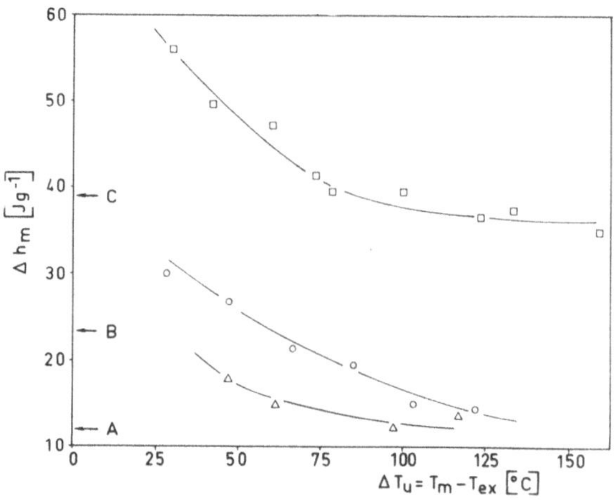

Fig. 4. Heat of melting versus undercooling of extrusion. (△) **1A**, (○) **1B**, (□) **1C**, arrows indicate heats of melting of billets. Rate of extrusion 0.4 mm³/s, nominal draw ratio 4

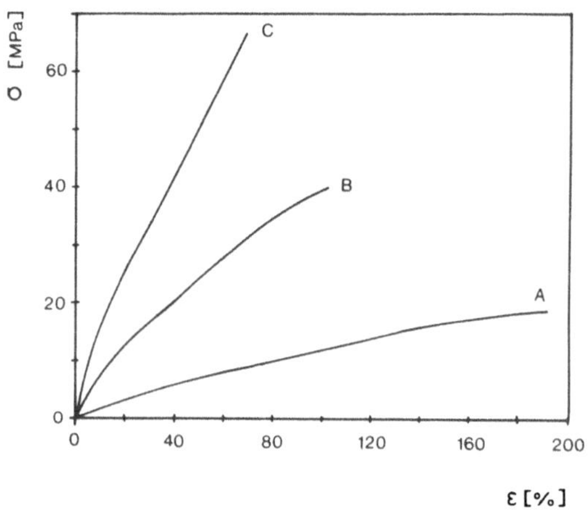

Fig. 5. Stress strain curves of extrudates **1A**, **1B**, and **1C**. Initial sample length 25 mm, strain rate 1 mm/s, 25 °C. Rate of extrusion 0.4 mm³/s, nominal draw ratio 4, undercooling 60 °C

The crystallinity of the extrudates varied with the extrusion temperature as shown in figure 4, where the heat of melting is plotted as a function of the undercooling of extrusion. The arrows indicate the heats of melting of the isotropic billets. It is obvious that the annealing effect which occurred at high extrusion temperatures was overcome by deformation effects at low temperatures.

In figure 5 typical tensile properties of the extrudates are presented. The stress-strain experiments were performed on microtomed ribbons of 0.1 mm thickness and 25 mm length at a strain rate which corresponded to a cross-head speed of 1 mm/s. The initial tensile modulus increased from 14 MPa for sample **1A** to 208 MPa for **1C**. The graph does not show the ultimate strains of the three materials which was in the order of 200 to 400%. Up to a strain of 40% most of the deformation is reversible, as was shown with hysteresis experiments [5].

References

1. Ciferri, A., I. M. Ward (ed.), Ultra-High Modulus Polymers, Applied Science (London 1979).
2. Hope, P. S., B. Parsons, Polym. Eng. Sci. **20**, 589 (1980).
3. Dröscher, M., Adv. Polym. Sci., in press.
4. Dröscher, M., W. Regel, Polym. Bull. **1**, 551 (1979).
5. Bandara, U., M. Dröscher, Angew. Makromol. Chem., in press.
6. Stein, R. S., F. H. Norris, J. Polym. Sci. **21**, 381 (1956).

(Received April 16, 1982)

Authors' address:

Dr. M. Dröscher, Dr. U. Bandara
Institut für Makromolekulare Chemie
Albert-Ludwigs-Universität
Hermann-Staudinger-Haus
Stefan-Meier-Straße 31
D-7800 Freiburg

 Rheol. Acta 21, 438 – 440 (1982)

Some applications of thermo-mechanical rheology for testing of plastics

J. Hetenyi

Laboratory for Polymeric Materials, National Testing Institute, Borås (Sweden)

Abstract: Short-time and long-term loading stress capacity are the fundamental mechanical engineering properties of plastics. The proper mechanical measurement of these properties is complicated and expensive, since it involves long-term testing. the aim of this paper is to present the possibilities of the DTA (Differential Thermal Analysis)- method to basic calculation of determination these loading capacity of plastics.

Key words: Creep property, loading capacity, differential thermal analysis, plastic

Creep properties are the fundamental mechanical engineering properties of plastics. The proper mechanical measurement of creep properties is expensive, since it involves long-term testing. The aim is to present the possibilities of the DTA (differential thermal analysis) method to simplify basic calculation of determination of short-time stress-strain properties and long-term behavior of plastics at several constant temperatures.

At ordinary temperatures the molecules of polymer are in constant thermal agitation. When a force is imposed on the plastics an influence will be set up in their motion and cold flow will occur. The greater the force, and the higher the temperature, the greater the cold flow will be.

Tensile short-time test are an important indication of strength in plastics. The force necessary to pull the specimen apart is determined until it causes flow failure limits (rupture or yield). This is illustrated in figure 1.

The area under the curves represents the mechanical energy density which is absorbed when a given stress-strain is applied. This makes it possible to determine the mechanical energy density which is absorbed until the flow failure limit. It can be calculated from the integral as illustrated in the figure. The flow limit is conventionally represented by the failure.

$$w = \int_0^{\varepsilon} \sigma(\varepsilon) \cdot d\varepsilon \qquad (1)$$

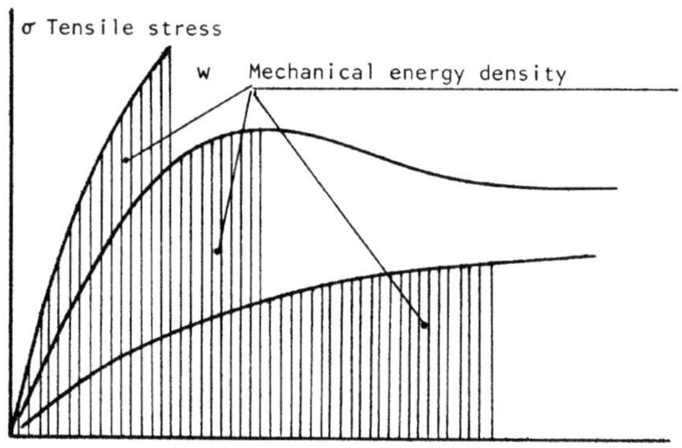

Fig. 1. Vide text

where w [kJ/m^3] is the mechanical energy density, $\sigma(\varepsilon)$ [MPa] the mechanical tensile stress and ε [m/m] the mechanical tensile strain.

It is known that the properties of plastics as defined by laboratory short-time loading may not be the same as the properties of long-time loading. From a mechanical deformation point of view, the two most fundamental types of long-term loading are represented by creep external deformation at constant stress and relaxation internal deformation at constant strain curves.

Furthermore, it has been found that in thermoplastics the initial damage is normally microscopic or submicroscopic and that it occurs under static as well as under dynamic loading conditions. The line of damage runs below the line of failure and is referred

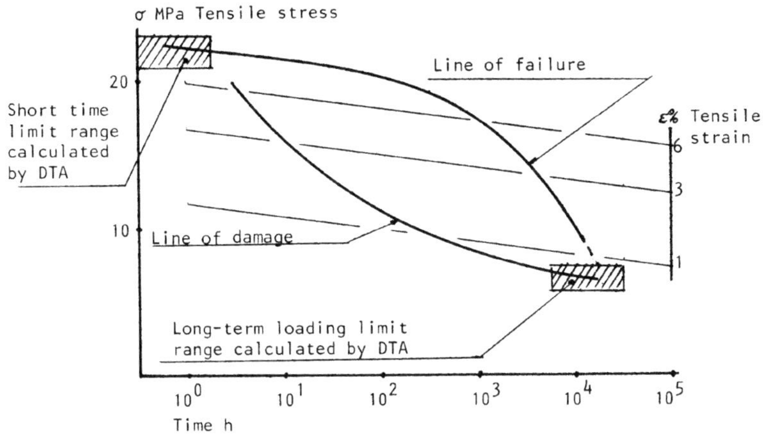

Fig. 2. Vide text

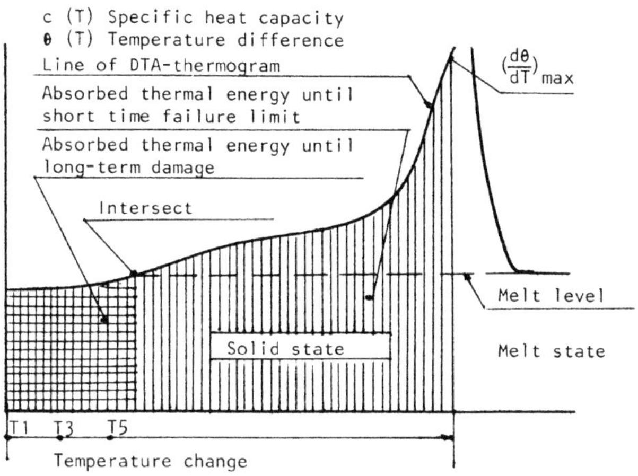

Fig. 3. Vide text

to as critical stress-strain limit on the time scale. The initial crack formations are called crazes. Figure 2 shows models of creep rupture and crazes crack formation lines. The tensile rupture stress and the tensile crazes stress are plotted against time for varying levels of tensile strain.

The data are also particularly important in engineering design because of their values in determining failure limits in terms of critical stress-strain conditions.

DTA makes it possible to determine thermal effects which occur in a given temperature range. The area under the curve in figure 3 represents the heat of energy density which is absorbed when a given heat flow is applied.

The energy density can be calculated, as figure 3 illustrates, from the integral

$$w_q = \bar{\rho} \int_{T_0}^{T} c(T) \cdot dT, \tag{2}$$

where w_q [kJ/m^3] is the thermo-energy density, $\bar{\rho}$ [kg/m^3] the mean value of density, T [°C] the furnace-program testing temperature, and $c(T)$ [kJ/kg · K] the specific heat capacity.

In the following we shall consider flow processes during mechanical loading as well as during DTA thermal analysis. The time for carrying out both tests, mechanical and DTA, must be the same, for instance: the speed of mechanical testing between 0.1/60 – 0.05/60 mm/s and the heating rate 3/60 – 7/60 K/s.

The intersection of the horizontal line from the melt state thermo-line with the solid state thermo-line will give the limit of the heat energy density of damage, that is the long-term effect of crazes.

The thermogram, figure 3, illustrates in a simplified form the basis of the calculation of the stress-strain properties of thermoplastics.

The strain relevance ε_r may be determined from the formula

$$\varepsilon_r = \sum_{i=1}^{N} \Delta\varepsilon_{ri} = \sum_{i=1}^{N} \left(\frac{\bar{\theta}(T)}{\Delta T \cdot N}\right)_i, \tag{3}$$

where $\theta(T)$ [K] is the DTA-temperature difference (sample-reference), ΔT [K] the temperature change, $T_n - T_0$, N the Simpsson's integral number, that is the number of integral sections, and i the index denoting the sections number.

The strain ratio k_ε may be determined from the relationship between the mechanical and relevance strain

$$\varepsilon = k_\varepsilon \varepsilon_r. \tag{4}$$

The stress relevance $\sigma_{ri}(\varepsilon_{ri})$ may be determined from the formula

$$\bar{\sigma}_{ri}(\varepsilon_{ri}) = \frac{\Delta w_{qi}}{\Delta \varepsilon_{ri}} \tag{5}$$

where w_{qi} [kJ/m^3] is the section thermal energy density, that is

$$w_q = \bar{\rho} \int_{T_0}^{T_n} c(T)\,dT \approx \sum_{i=1}^{N} \Delta w_{qi} = \bar{\rho} \sum_{i=1}^{N} \bar{c}_i(T) \cdot \Delta T \tag{6}$$

where $\bar{\rho}$ [kg/m^3] is the mean value of density, T [°C] the program testing temperature, and $c(T)$ [kJ/kg · K] the specific heat capacity.

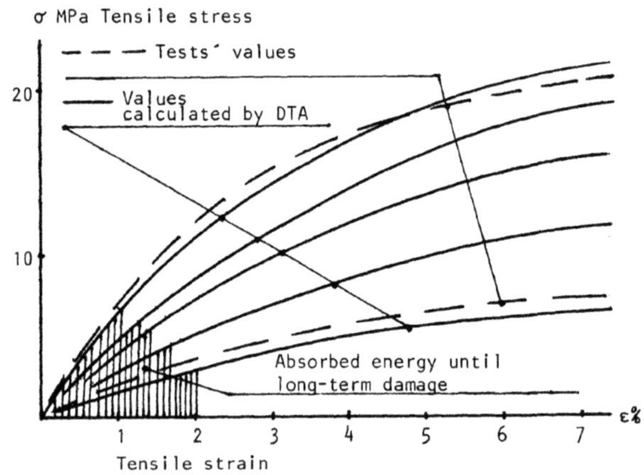

Fig. 4. Vide text

The energy relevance w_r may be determined from the formula

$$w_r = \int_0^{\varepsilon_r} \sigma_r(\varepsilon_r)\,d\varepsilon_r \approx \sum_{i=1}^{N} \bar{\sigma}_{ri}(\varepsilon_r)\Delta \varepsilon_{ri}\,. \tag{7}$$

The energy ratio k_w may be determined from the relationship between the mechanical and the relevance thermal energy density

$$w = k_w w_r\,. \tag{8}$$

The thermal mechanical section (step) stress $\sigma_{Mi}(\varepsilon_M)$ and ε_{Mi} values may be determined from the formulae

$$\sigma_{Mi}(\varepsilon_{Mi}) = \frac{k_w}{k_\varepsilon} \cdot \sigma_{ri}(\varepsilon_r) \tag{9}$$

and

$$\varepsilon_{Mi} = k_\varepsilon \sum_{i=1}^{m \le N} \Delta \varepsilon_{ri}\,. \tag{10}$$

The practical points of view are the energy ratio k_w and the strain ratio k_ε constant in the range of temperature change.

Based on these calculations, figure 4 illustrates (i) the stress-strain properties at several constant temperatures, like short time tests, (ii) stress temperature properties at several constant strain, like short time failure limit, and long term damage limit (crazes), and (iii) strain temperature properties at several constant stress, like short time failure limit, and long term damage limit (crazes).

(Received June 21, 1982)

Author's address:

J. Hetenyi
Laboratory for Polymeric Materials
National Testing Institute
Box 857
S-50115 Borås

Energy considerations of the time-dependent failure of polymers

O. S. Brüller

Department of Material Science, University of Duisburg

Key words: Viscoelastic linearity, crazing, yielding, fracture

1. Introduction

Failure of polymers does not mean only fracture or rupture. Phenomena such as transition from linear to nonlinear viscoelasticity, crazing and yielding are also failure phenomena. While crazing and fracture can be directly observed and yielding becomes visible from the strain-stress representation of the loading history, the transition from linear to nonlinear viscoelasticity can be detected only by using more sophisticated procedures. On the other hand, the observation of first crazes is relatively simple in so-called long-term tests (such as creep and stress relaxation) but very difficult in short-term tests, whereas yielding and fracture can be studied better in short-term tests.

If, in addition, one remembers the strong dependence of the behaviour of mechanically loaded polymers upon temperature, environmental conditions and primarily upon time, one can understand that prediction of failure of polymers is a very difficult problem. The present contribution is an attempt to explain the different forms of failure of polymers from an energetical point of view.

2. Energy considerations of failure phenomena

During loading, the work done by external forces on a viscoelastic material is converted into a stored part (potential energy) and a dissipated part, each of which may be divided into two other parts: the isotropic one, connected with volume changes and the deviatoric one, associated with shape changes. According to the Reiner-Weissenberg theory of strength [1], failure occurs when the stored deviatoric strain energy reaches a certain value which is assumed to be a material constant.

Consequently, failure depends upon the loading history and one must asses the effect of history on the nature of a deformation process. Thus, one must consider the relative participation of the two primary mechanisms, viscous flow and elastic deformation [2]. The participation of these two processes varies with rate of deformation and since the viscous processes require appreciable time whereas the elastic ones are quasi-instantaneous, the higher the rate, the greater the elastic part. For that reason, attempts made for predicting failure must take into consideration the whole loading history of the material.

With polymers, the application of an energy-related theory of strength renders possible the establishment of the limit of various forms of failure, allowing also predictions. It is obvious that the obtained limits are no more "strain limits" or "stress limits", but energetical limits, depending upon both, strain and stress.

2.1. *Linear and nonlinear viscoelasticity*

The transition from linear to nonlinear viscoelasticity can be considered as a first sign of failure. The experimental determination of the limit of linear viscoelasticity is rather difficult and contains a certain amount of uncertainty. Nevertheless, some useful results can be found for the same material, polymethylmethacrylate (PMMA), in [3] under creep loading and in [4] under constant strain rate tests. In [3] the creep compliance of PMMA is depicted vs. stress at temperatures of 22 and 80 °C after loading periods of 10 and 1000 s. The transition points from linearity to nonlinearity can be established with a reasonable accuracy. In [4] the results obtained on PMMA in constant strain rate experiments, are depicted in form of log $(\sigma/\dot{\varepsilon})$ vs. log t. The deviation of an individual curve, obtained with a certain strain rate, from the common "linear" curve marks the transition to nonlinearity.

Computations of the stored deviatoric energies at the transition points have confirmed the Reiner-Weissenberg theory of strength. The experimental points, as well as some computed points, belong obviously to a common curve, established on the basis of the constancy of the stored energy, as depicted in figure 1 (for more details see [5]).

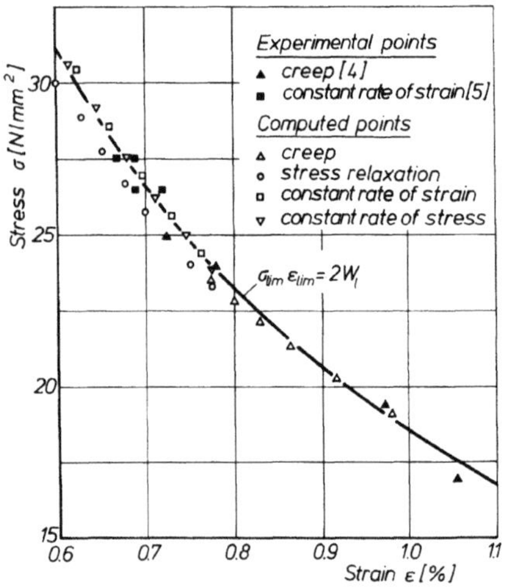

Fig. 1. The limit of linear viscoelasticity of PMMA

From the physical point of view it seems that at the linear viscoelastic limit first irreversible submicroscopic damages are produced in the material [6]. This leads to the conclusion that the relations governing the behaviour of the material in the linear range will no more be valid beyond this limit.

2.2. *Crazing*

The phenomena of crazing of transparent and translucent polymers can easily be observed, especially under long-term loading such as creep and stress relaxation. The energy related study of crazing under creep loading has shown that crazes become visible when the time-dependent part of the conserved energy reaches a certain constant value. This means that the elastic energy related to the quasi-instantaneous loading step has not to be considered [7]. Later investigations have demonstrated that at crazing, the dissipated energy is also a constant. This allows the use of a very simple relation for predicting crazing [8]. As an example let us consider the same material, PMMA. Figure 2 shows the fair agreement between experimental results [9] and computational ones based on the constancy of the dissipated energy ($W_{diss} = 0.06$ N/mm^2). This was confirmed with other materials too, as for example with PC [9].

In stress relaxation, after the quasi-instantaneous loading phase no more energy is supplied to the material. Internal time-dependent redistribution of energy takes place and a part of the energy is dis-

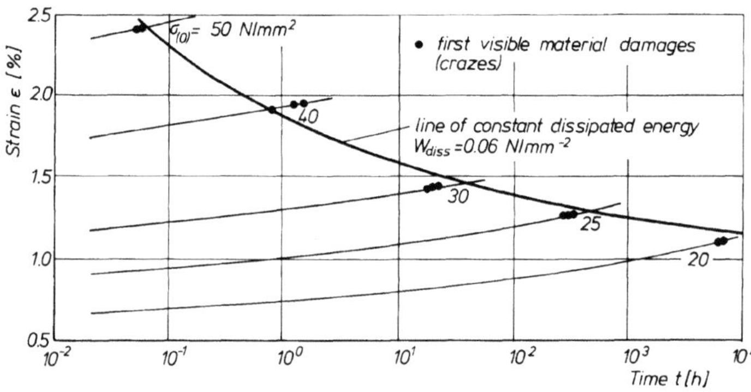

Fig. 2. Crazing of PMMA in creep

sipated. Starting again from the same value of the dissipated energy of PMMA at crazing, a line of constant dissipated energy can be established. This is depicted in figure 3. It is seen in this figure that the experimental points of first visible crazes are in the vicinity of the line of constant dissipated energy $W_{diss} = 0.06$ N/mm^2 which is the same value as under creep loading.

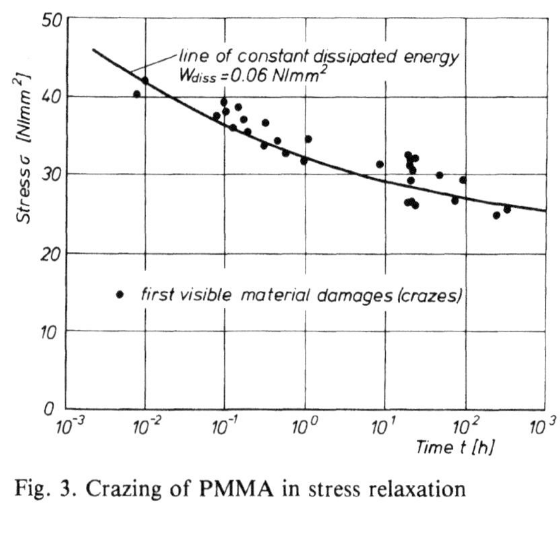

Fig. 3. Crazing of PMMA in stress relaxation

The ease with which crazes can be discovered depends on the nature of the tested polymer. With the translucent acrylo-butadiene-styrene (ABS) their observation is easier. The results, presented in [10] show a good agreement between theory and experiment.

2.3. *Fracture*) and yielding*

As shown in the previous section, an energy related strength theory enables a suitable prediction of crazing in long-term experiments such as creep or stress relaxation. In order to avoid laborious computations, we try to apply an approximated criterion based on the constancy of the viscous flow energy.

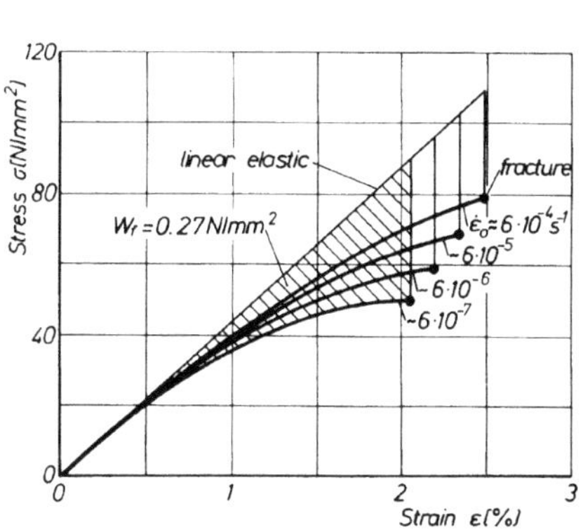

Fig. 4. Fracture of PMMA in constant-rate-of-strain experiments

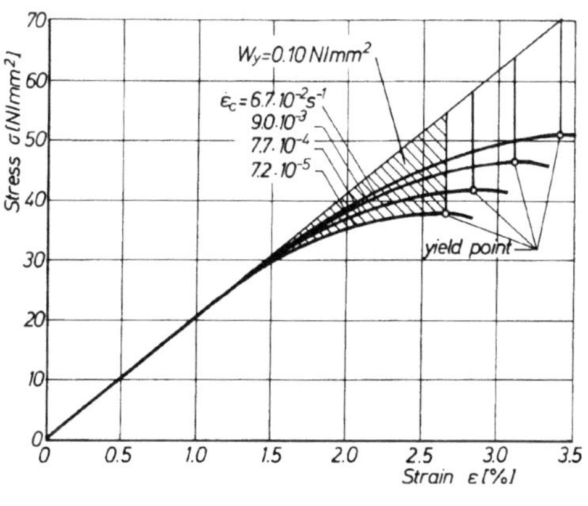

Fig. 5. Yielding of ABS in constant-rate-of-strain experiments

*) Fracture is defined in this section as the breaking of a specimen which occurs *before* zero slope in the stress-strain curve is reached.

For constant strain rate experiments for instance, the specific viscous flow energy is the difference between the specific elastic energy of an ideal (non-existent) elastic body having a modulus of elasticity equal to the initial modulus of the tested polymer and the specific energy at the considered time.

This is depicted in figure 4 on the example of PMMA. W_f represents the specific viscous flow energy at fracture. Planimetric measurements conducted in figure 4, confirm the assumption: only small differences between the surfaces delimited by the curves obtained at different strain rates have been noted. (Mean value: $W_f = 0.27$ Nmm/mm^3).

For materials which reach a yield point (zero-slope in the σ-ε curve), the situation seems to be the same. An example is presented in figure 5. Here, a slight increase of the viscous flow energy denoted by W_y can be observed with increasing rates of strain [11].

Similar results were obtained also under constant rate-of-stress conditions. The tested material was a cured styrene-acrylo-nitrile (SAN). The results are shown in figure 6.

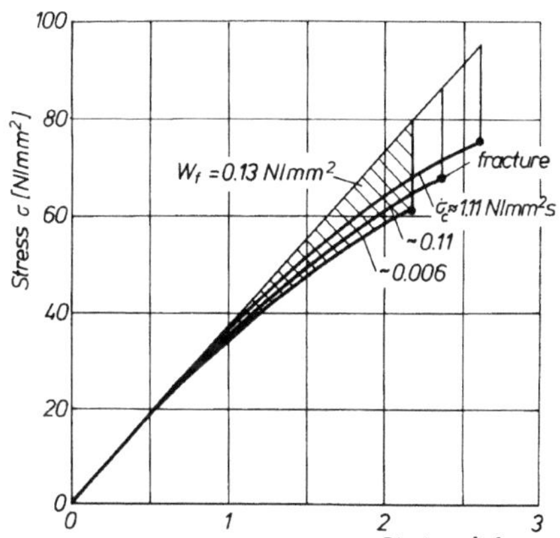

Fig. 6. Fracture of SAN in constant-rate-of-stress-experiments

References

1. Reiner, M., K. Weissenberg, Rheol. Leafl. **10**, 12 – 20 (1939).
2. Eirich, F. R., High Speed Testing **5**, 271 – 300 (1966).
3. Bertilsson, H., J. F. Jansson, J. Appl. Polym. Sci. **19**, 1971 1978 (1975).
4. Foux, A., O. S. Brüller, Proc. 1971 Intern. Conf. Mech. Behav. Mater. Japan **1**, 237 – 244 (1972).

5. Brüller, O. S., H. H. Schmidt, Polym. Eng. Sci. **19**, 883 – 887 (1979).
6. Menges, G., Kunststoffe **63**, 95 – 100, 173 – 177 (1973).
7. Brüller, O. S., Polym. Eng. Sci. **18**, 42 – 44 (1978).
8. Brüller, O. S., Polymer **19**, 1195 – 1198 (1978).
9. Schmidt, H., Kunststoff-Rundschau **19**, 1 – 7, 56 – 65, 145 – 150, 196 – 199, 320 – 326 (1972).
10. Brüller, O. S., Crazing of Polymers under Stress Relaxation, Proc. 40th Ann. Techn. Conf. Soc. Plast. Eng., San Francisco, Ca., 121 – 122, 1982.
11. Brüller, O. S., Predicting Yield and Fracture of Polymers, Paper Nr. R 150, presented at 1980 SESA Spring Meeting, Boston, Ma., May 1980.

(Received July 1, 1982)

Author's address:

Dr. O. S. Brüller
Universität (GH) Duisburg
FB 7 – Werkstofftechnik
Lotharstraße 1 – 21
D-4100 Duisburg 1

Rheology and stircasting of (organic) alloys

J. C. van Dam and F. H. Mischgofsky

Laboratory of Inorganic and Physical Chemistry, Delft University of Technology, and Delft Soil Mechanics Laboratory

Abstract: Partially solidified metal alloys behave thixotropic during stircasting. Some transparent, low melting organic alloys such as NPA behave similarly. We studied the rheological behaviour of NPA during and after stirred nucleation. The shear stress reaches a maximum τ_{max} and subsequently decreases. This decrease is explained by the coarsening of dendrites into spherical particles. The maximum shear stress increases with increasing cooling rate and also with decreasing final temperature. The latter increase is more pronounced if the shear rate during solidification is lower. Both effects can be explained in terms of coarsening and particle sizes.

Key words: Stircasting, solidification, thixotropy, coarsening, alloy

1. Introduction

Stircasting or rheocasting is an innovative casting technique which was the topic of interest in a number of studies [1 – 6] in the last ten years, due to the well promising properties of the castings. The difference with other casting techniques is that the alloy is not fully liquid but is partially solidified under vigorous stirring before the alloy is cast. The partially solidified alloys behave thixotropic, so that the fluidity of the slurries can be influenced by shear rate and by time. Joly and Mehrabian [3] studied the thixotropic behaviour for a Sn-15% Pb alloy in detail. The microstructure of stircast alloys is not dendritic which it would be after conventional casting, but it consists of spherical or rosette shaped particles. The fundamental understanding of the origin of the spherical particles and the thixotropic behaviour of the partially solidified slurries is still poor.

In order to get more information on the influence of the shear rate and cooling rate on the particle formation and the thixotropic behaviour one should do simultaneously microscopic observations and viscosimetric measurements. Metal alloys do not allow this. For the case of *unidirectional* solidification, however, it is well known that some transparent organic compounds solidify at low temperatures in the same way as metals do and these are therefore widely used as model substances [8 – 12].

For two model substances we proved [7, 13, 15] that these also during *stircasting* behave similarly as metals do.

We started our investigations with succinodinitrile [7] but due to its toxity we are now using neopentylalcohol (NPA) [13, 15] as a model substance for stircasting. The microscopic observations were performed in a transparent stircast apparatus and were recorded on micrographs and microholograms using a pulse laser. These observations will be published separately [13]. In this paper we investigate the rheological behaviour of NPA during and after stirred nucleation.

2. Experimental procedure

The rheological measurements were performed with a Searle-type viscometer, in which the cup is fixed and the rotor stirs the melt and measures the torque. The rotor is a hollow cylinder which fits in the cylindrical gap of the cup (system NV from Haake). The slit at both sides of the cylinder is 0.35 mm. Figure 1 shows the experimental procedure to yield partial solidification. The cup was filled with 99% pure neopentylalcohol, which has a freezing range of more than six degrees. The impurities mainly consist of water. The temperature in the gap of the viscometer is measured with a thermocouple and recorded

805

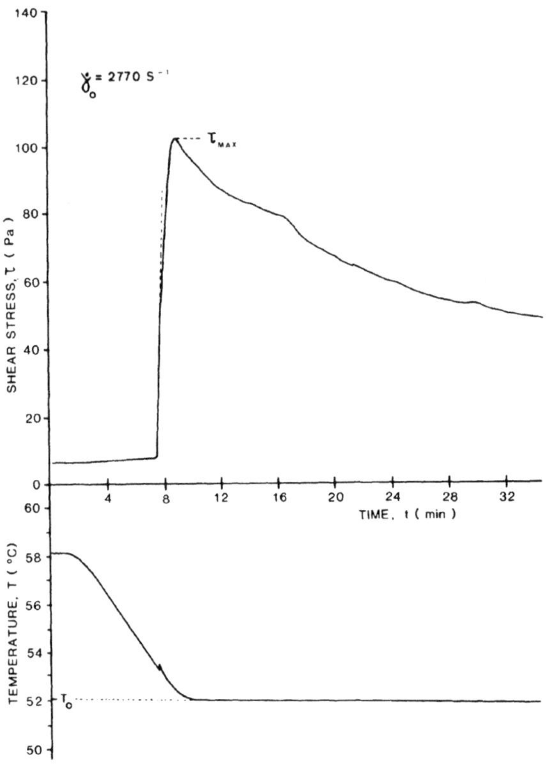

Fig. 1. Shear stress τ during stirred nucleation, growth and coarsening

together with the shear stress (fig. 1). The temperature of the melt is lowered with a constant cooling rate ε from above liquidus to a final temperature T_0 between liquidus and solidus, while the shear rate $\dot{\gamma}_0$ remains constant.

3. Results

During the cooling nucleation occurs suddenly and causes a rapid increase of the shear stress. The nucleation also raises the temperature in the gap as can be seen in the lower part of figure 1, but is also visible as a change in the slope of the increasing shear stress, as is indicated by the dotted line in figure 1.

In the greater part of the experiments the shear stress reaches a clearly visible maximum value (which we will denote as τ_{max}) after which the shear stress immediately decreases and finally reaches an equilibrium value. In some cases, however, the shear stress exhibits a "plateau" at τ_{max}. Till now no explanation could be given to the different curves. Several experiments were done at various cooling rates, shear rates and final temperatures. The experiments clearly showed that the same samples gave different results under the same conditions. However, some systematic trends can be found as shown in figures 2 and 3.

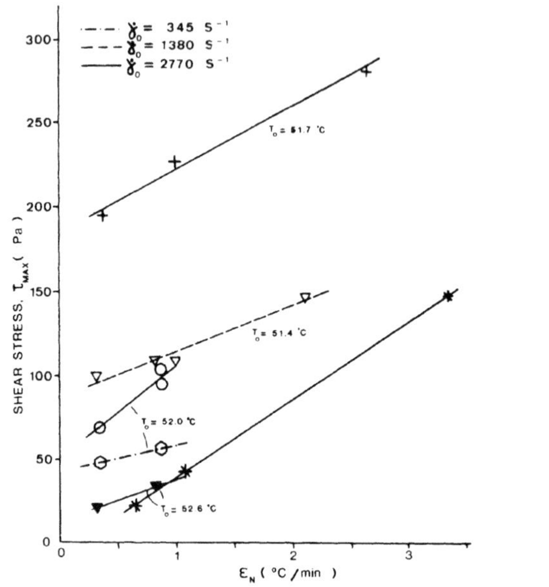

Fig. 2. The influence of the cooling rate at the nucleation temperature ε_N, on the maximum shear stress τ_{max}. T_0 is the final temperature

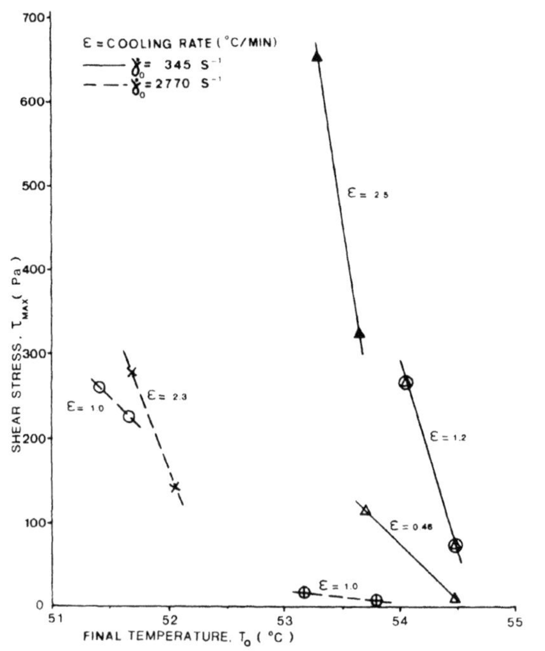

Fig. 3. The influence of the final temperature T_0 on the maximum shear stress τ_{max} for two different shear rates

Each of the symbols in these figures represents measurements on a single sample done at a same day. Figure 2 shows the influence of the cooling rate ε at the nucleation temperature T_N (we will denote this cooling rate as ε_N) on the maximum shear stress τ_{max} for measurements with the same final temperature T_0 and the same shear rate $\dot{\gamma}_0$. From these measurements it is clear that the maximum shear stress increases with increasing cooling rate.

The influence of the final temperature T_0 on the maximum shear stress is shown in figure 3 for experiments with the same cooling rate ε and the same shear rate $\dot{\gamma}_0$. A lowering of the final temperature, which causes a higher fraction of solid, clearly raises τ_{max}. The increase of τ_{max} is stronger for lower shear rates $\dot{\gamma}_0$ even at much lower fractions of solid. Experiments with the same shear rate and at the same temperatures but at a higher cooling rate ε also clearly show a higher maximum shear stress which is in accordance with the results of figure 2.

4. Discussion

4.1. Influence of coarsening

Preliminary microscopic observations with the pulse-laser in the transparent stircast apparatus showed that the primary particles, which form during the first stage of the nucleation of NPA exhibit a dendritic structure and are not spherical at all [13]. When the melt (slurry) reaches its final temperature the particles become more and more spherical [13]. This means that the dendritic shape disappears and that the surface of the particles becomes more or less smooth (this is sometimes called "coarsening of dendrites"), so that can be expected that simultaneously the shear stress will decrease and will reach a steady state. Indeed, both the coarsening [13] and the shear stress reach a steady state after about one hour. The mechanism for this surface transformation at constant temperature is probably an analogon to "Ostwald Ripening" in which case the surface tension causes the disappearance of regions with a small radius of curvature and so leads to a reduction of the total surface area. Huang and Glickmann [14] recently published data in which they described the isothermal coarsening of dendrites. This also is explained by reduction of the interfacial energy.

4.2. Influence of cooling rate

It is known that higher cooling rates yield smaller particles [3] at the same initial shear rate. So one might expect smaller maximum shear rates for increasing cooling rates which is in contradiction with the results of figure 2. On the other hand, due to the higher cooling rate the coarsening of the dendrites will be less *during* the cooling, which will increase τ_{max} as shown in figure 2. Apparently the latter effect is dominant.

4.3. Influence of shear rate and fraction of solid

From the experiments of Joly et al. [3], it is known that for a constant cooling rate and a constant temperature lower shear rates $\dot{\gamma}_0$ give a larger average particle size (which is also confirmed for NPA [13]) and that larger particles give a higher apparent viscosity. This explains the higher shear stress τ_{max} for lower shear rates in figure 3. It can be seen for the curves with $\varepsilon \simeq 1$ that τ_{max} is several orders of magnitude larger for $\dot{\gamma}_0 = 345\ s^{-1}$ than for $\dot{\gamma}_0 = 2770\ s^{-1}$. This could be observed directly only for $T_0 > 53.5\,°C$ since for lower T_0 (and so for a larger fraction of solid) τ_{max} exceeds the measuring range of the viscosimeter for $\dot{\gamma}_0 = 345\ s^{-1}$.

5. Conclusions

The partial solidification of a stircast organic alloy clearly shows the influence of the cooling rate on the maximum shear stress. Higher cooling rates give higher shear stresses, which means higher casting pressures on stircaster and plunger. Decreasing final temperature gives increasing fraction of solid and will give also higher shear stresses.

An increasing shear rate during solidification gives smaller particles which give easier castable slurries.

The variations in τ_{max} can be explained in terms of average particle sizes and of the coarsening of primary dendrites into spherical particles.

Acknowledgement

This study is part of the research program of the "Stichting voor Fundamenteel Onderzoek der Materie" (Foundation for Fundamental Research on Matter).

References

1. Spencer, D. B., R. Mehrabian, M. C. Flemings, Met. Trans. **3**, 1925 (1972).
2. Flemings, M. C., R. G. Riek, K. P. Young, Mater. Sci. Eng. **25**, 103 (1976).
3. Joly, R. A., R. Mehrabian, J. Mater. Sci. **11**, 1393 (1976).
4. Vogel, A., R. D. Doherty, B. Cantor, in: J. Hunt, I. G. Davies, R. Jordan, A. Hellawell, J. O. Ward, A. Nichol-

son, P. Beeley, A. G. Fuller, H. A. Davies (eds.), Solidification and casting of metals, The Metals Society 1979, p. 479.
5. Lee, H. I., R. D. Doherty, E. A. Feest, J. M. Titchmarsh, in: "Solidification Technology in the Foundry and Casthouse", The Metals Society (Warwick, September 15–17, 1980), paper 21.
6. Apaydin, N., K. V. Prabhakar, R. D. Doherty, Mater. Sci. Eng. **46**, 145 (1980).
7. van Dam, J. C., F. H. Mischgofsky, J. Mat. Sci. **17**, 989 (1982).
8. Jackson, K. A., J. D. Hunt, Acta Met. **13**, 1212 (1965).
9. Hunt, J. D., K. A. Jackson, Trans. Met. Soc. AIME **236**, 843 (1966).
10. Jackson, K. A., J. D. Hunt, D. R. Uhlmann, T. P. Seward III, Trans. Met. Soc. AIME **236**, 149 (1966).
11. McDonald, R. J., J. D. Hunt, Trans. Met. Soc. AIME **245**, 1993 (1969).
12. Glickmann, M. E., R. J. Schaefer, J. D. Ayers, Met. Trans. **7A**, 1747 (1976).
13. Smeulders, R. J., F. H. Mischgofsky, H. J. Frankena, to be published.
14. Huang, S. C., M. E. Glickmann, Acta Met. **29**, 717 (1981).
15. van Dam, J. C., F. H. Mischgofsky, submitted to J. Mat. Sci.

(Received May 24, 1982)

Authors' addresses:

Ir. J. C. van Dam
Laboratory of Inorganic and Physical Chemistry
Delft University of Technology
P.O. Box 5045
NL-2600 GA Delft

Dr. F. H. Mischgofsky
Laboratory of Inorganic and Physical Chemistry
Delft University of Technology,
and Delft Soil Mechanics Laboratory
P.O. Box 69
NL-2600 AB Delft

Viscoelastic properties of concentrated polymer solutions and melts: a reduced form of linear viscoelasticity

G. Marin, J. P. Montfort, and Ph. Monge

Université de Pau et des Pays de l'Adour, I.U.R.S. – Laboratoire de Thermodynamique, Pau (France)

Key words: Linear viscoelasticity, polymer solution, bulk polymer

Introduction

From the very large number of papers issued on polymer rheology during the last 15 years, one of the main features is that the behavior of different polymer melts at low frequencies (or long times) is very similar and characterizes the "polymeric" nature of the material, whatever the chain length, polymer concentration, temperature or even the chemical structure of the material.

From this qualitative remark, it seemed to us important to show that it is possible to quantify this picture, i. e. express the linear viscoelastic behavior of polymers independently of chain length, concentration, temperature and chemical structure. In this paper, we will deal only with narrow distribution (quasi-monodisperse) linear polymers in the entangled and concentrated regime.

It is possible to use various functions to characterize the linear viscoelastic properties of polymers: these functions are essentially the creep function $J(t)$, the relaxation function $G(t)$, the complex viscosity $\eta^*(\omega)$, the complex modulus $G^*(\omega)$, and the complex compliance $J^*(\omega)$. These functions are interrelated by either algebraic or integral relations, given by the mathematical theory of linear viscoelasticity [1 – 3]. As a consequence, only one of these functions is necessary to characterize completely the linear viscoelastic behavior.

The various interrelations between the functions of linear viscoelasticity can be written in a rather simple form when one introduces a distribution of relaxation times. In polymer rheology, authors are using the relaxation spectrum $H(\tau)$ defined from complex modulus as [2, 3]:

$$G^*(\omega) = i\omega \int_{-\infty}^{+\infty} \frac{\tau H(\tau)}{1 + i\omega\tau} d\ln\tau . \tag{1}$$

We will look in this paper for a "master behavior" of polymers in bulk or solutions.

1. Time-temperature superposition

It has been widely shown in polymer rheology that the curves characterizing one of the above functions obtained at various temperatures have similar shapes and can be superposed by translating the curves at different temperatures onto that at a reference temperature [2]. This time-temperature superposition can be expressed for the relaxation function as:

$$G_1(t, T_1) = b_T G_2(t/a_T, T_2) , \tag{2}$$

G_1 being the modulus at temperature T_1 and G_2 at temperature T_2.

We can write for the complex moduli $G^*(\omega)$:

$$G_1^*(\omega, T_1) = b_T G_2^*(\omega a_T, T_2) \tag{3}$$

and for the relaxation spectrum:

$$H_1(\tau, T_1) = b_T H_2(\tau/a_T, T_2) . \tag{4}$$

The time (or frequency) shift factor a_T is a function of T_1 and T_2.

The modulus shift factor b_T is given by [2]:

$$b_T = \frac{\rho_1 T_1}{\rho_2 T_2} ,$$

ρ_i being the polymer density at temperature T_i.

773

2. Time-chain length superposition

In the terminal region, the shapes of the $G^*(\omega)$ or $H(\tau)$ curves obtained for different molecular weights (same polymer) are the same, and it is possible to get a master curve at an arbitrary reference molecular weight (playing the same role as the reference temperature in the previous paragraph), using a time shift factor a_M.

When working at the same concentration (bulk polymers, for example) and same temperature, we get

$$G_1^*(\omega, M_1) = G^*(\omega a_M, M_2)\,, \tag{5}$$

$$H_1(\tau, M_1) = H_2(\tau/a_M, M_2) \tag{6}$$

where G_i^* is the complex modulus of the polymer having molecular weight M_i. For entangled systems, $a_M = (M_2/M_1)^{3.4}$.

3. Time-concentration superposition

In the same way as done in the last two sections, it is possible to superpose the $G^*(\omega)$ or $H(\tau)$ curves obtained for the same polymer, at different concentrations in a low-viscosity solvent, using a vertical (modulus) shift factor b_c and a frequency shift factor a_c:

$$G_1^*(\omega, \phi_1) = b_c G_2^*(\omega a_c, \phi_2)\,, \tag{7}$$

G_i^* being the complex modulus at polymer volume fraction ϕ_i in the same solvent. We have recently shown [4, 5] that in the concentrated-entangled regime the a_c and b_c factors could be expressed as a function of the respective polymer volume fractions ϕ_1 and ϕ_2:

$$b_c = (\phi_1/\phi_2)^{2.25},$$

and a_c varies as $(\phi_2/\phi_1)^{-2.25}\,(\phi_2/\phi_1)^{4.0}$ at same chain mobility (i. e. same free volume) and independently of molecular weight (good solvent case [4]).

4. Reduced form of viscoelasticity

We can now derive a reduced form for $G^*(\omega)$ for a narrow molecular weight distribution polymer, independently of temperature, concentration, and chain length.

It can be easily shown [5] that G_N^0 and the product $(\eta_0 J_N^0)$ have the same variations (vs. chain length, temperature, and concentration) as the b and a coefficients defined here above.

So a plot of $(J_N^0 G^*)$ vs. $(\omega \eta_0 J_N^0)$ will be a master plot for one polymer. This reduced form is in fact much more general, because it is also independent of the chemical structure of the polymer: i. e. it will give the same curves for polyethylene and polystyrene, for example.

In this way we got a function characterizing the polymeric nature of our materials, independently of temperature, molecular weight, polymer concentration and polymer nature.

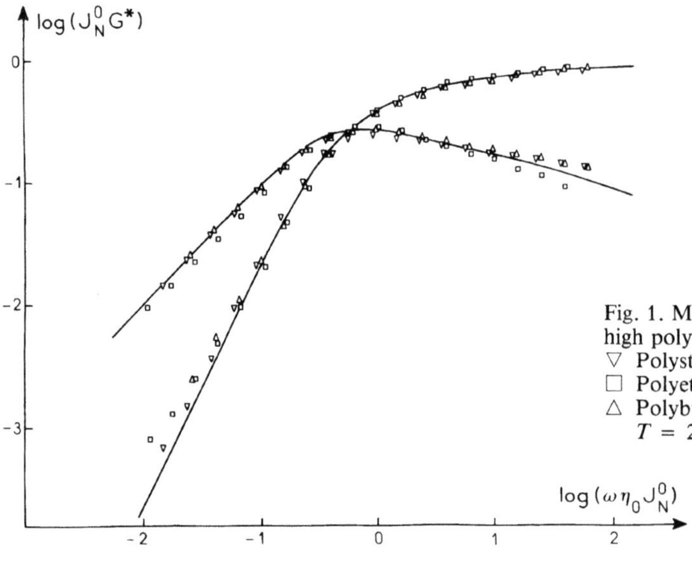

Fig. 1. Master curve for the linear viscoelastic behavior of high polymers in the terminal region:
▽ Polystyrene, bulk (M = 860000, T = 190 °C)
□ Polyethylene, bulk (M = 340000, T = 130 °C)
△ Polybutadiene solution (M = 350000, Φ = 0.43, T = 25 °C)

In figure 1 we show a plot of $(J_N^0 G^*)$ vs. $(\omega \eta_0 J_N^0)$ for one bulk polyethylene (M = 340000, T = 130 °C), one bulk polystyrene (M = 860000, T = 190 °C), and one polybutadiene solution (M = 350000, T = 25 °C, ϕ = 0.43).

The solid lines on that figure have been calculated from the equation for the complex compliance:

$$\frac{J^*(\omega)}{J_N^0} = \frac{1}{i\omega \eta_0 J_N^0} + \frac{1.4}{1 + (i\omega \eta_0 J_N^0 \times 0.6)^{0.5}} + 1 \tag{8}$$

derived from our previous work on the linear viscoelasticity of polymer melts [6].

This demonstrates that it is possible to forecast the linear viscoelastic behavior of one polymer in the terminal region at any concentration, temperature, or chain length, from the behavior of another polymer, if we know only: (a) the plateau moduli of both polymers, (b) the activation energy (or W. L. F. equation) of both polymers, and (c) the zero-shear viscosity at one temperature and one molecular weight for each polymer.

References

1. Gross, B., Mathematical structure of the theories of viscoelasticity, Herman (Paris, 1953).
2. Ferry, J. D., Viscoelastic Properties of Polymers, J. Wiley (New York 1970).
3. Marin, G., J. P. Montfort, J. Arman, Ph. Monge, Rheol. Acta **18**, 629 (1979).
4. Marin, G., E. Menezes, V. R. Raju, W. W. Graessley, Rheol. Acta **19**, 462 (1980).
5. Raju, V. R., E. V. Menezes, G. Marin, W. W. Graessley, L. J. Fetters, Macromolecules **14**, 1668 (1981).
6. Marin, G., W. W. Graessley, Rheol. Acta **16**, 527 (1977).

(Received April 14, 1982)

Authors' address:

Dr. G. Marin, J. P. Montfort, Ph. Monge
Université de Pau et des Pays de l'Adour
I.U.R.S. – Laboratoire de Thermodynamique
Avenue Philippon
F-64000 Pau

Rheologica Acta Rheol. Acta **21**, 452 – 455 (1982)

Extensional flow of HDPE/LDPE blends

F. P. La Mantia, D. Acierno, and D. Curto

Istituto di Ingegneria Chimica, Università di Palermo

Abstract: Elongational viscosity data, obtained through the converging flow analysis by Cogswell, are presented for two types of HDPE/LDPE blends at various compositions and different temperatures. The results relative to the homopolymer parents compare favourably with literature data obtained also with different and more sophisticated techniques. Those relative to blends show peculiar features for the two cases: when the newtonian viscosity of the LDPE is higher all the blends show a behaviour typical of the LDPE with a maximum in η_{el}/η_0 enhanced at small percentage of HDPE; when the newtonian viscosity of the LDPE is similar to that of the HDPE there is only a gradual change in the properties.

Key words: Elongational viscosity, converging flow, polymer blend

1. Introduction

Extensional flow of polymer melts has recently received more and more attention [1 – 3] due to its implications in many technological processes such as fiber spinning, film blowing, etc. However the techniques adopted to study this flow are often very lengthy and difficult; therefore the data, especially as steady-state values are concerned, are sparse and sometimes contradictory [1, 2, 4 – 6]. The situation is particularly difficult in the field of polymer blends for which data are available only in [7], in spite of the increasing relevance that such materials are gaining [8].

In this work the converging flow analysis by Cogswell [9] has been employed to evaluate the extensional viscosity of HDPE/LDPE blends, which may be considered of interest in the production of though blown films.

2. Experimental

The materials employed in this work were commercially available high density and low density polyethylenes manufactured and kindly supplied by Solvay (HDPE) and Montedison (LDPE). The MFI of the samples are reported in table 1.

Table 1

Sample code	M.F.I.
HDPE-1	4.90
HDPE-2	0.98
LDPE-2	0.74
LDPE-3	0.07

Two types of blends (HDPE-1/LDPE-3 and HDPE-2/LDPE-2) with various compositions of 20, 40, 60, 80 weight percentage of LDPE were prepared by melt mixing the homopolymers parents in a laboratory mixer (Rheocord, Haake) at 210 °C at about 20 rpm. The mixing time was approximately 20 min, sufficient to obtain in all cases steady-state temperature and torque. The homopolymers were also subjected to the same procedure.

3. Rheological characterization

A constant-force viscometer and a constant-rate viscometer have been employed for determining the rheological characteristics of both the blends and the polymer parents: more precisely the former has been used in the low shear-rate region and the latter in the high shear-rate region.

Three different capillaries have been adopted in all cases: the diameter was always 1 mm and the length to diameter ratios were 40, 10 and 5; the entrance angle was 90°.

The test temperatures have been 160, 180 and 200 °C.

4. Analysis of results and discussion

Following the analysis of Cogswell [9], the elongation viscosity can be found through the relation:

$$\eta_{el} = \frac{9}{32}(n+1)^2 \frac{\Delta P_{ent}^2}{\dot{\gamma}^2 \eta} \tag{1}$$

where ΔP_{ent} is the loss pressure due to the entrance effect, $\dot{\gamma}$ is the apparent shear rate $32Q/D^3$, η the shear viscosity and n the slope in the log-log shear stress-shear rate plot.

The stretching rate is

$$\dot{\varepsilon} = \frac{4}{3(n+1)} \eta \frac{\dot{\gamma}^2}{\Delta P_{ent}}. \tag{2}$$

It is generally considered that ΔP_{ent} can be found by extrapolating a straight line to $L/D = 0$ in a ΔP vs. L/D plot. It is however true [10] that the so found value, is the sum of both entrance and exit pressure losses:

$$\Delta P_{ends} = \Delta P_{ent} + \Delta P_{exit}. \tag{3}$$

It has been further shown that ΔP_{ends} is proportional to the principal normal stress difference [11], and ΔP_{exit} is the sum of the first and second normal stress difference [12], so that

$$\Delta P_{ends} = \alpha \sigma_{11}, \tag{4}$$

$$\Delta P_{exit} = \sigma_{11} + \sigma_{22}. \tag{5}$$

The value of α has been found $\simeq 2$ for many polymers [11], and σ_{22} is a negative fraction of σ_{11}

$$\sigma_{22} = -\varepsilon \sigma_{11} \tag{6}$$

where ε is between 0 and 0.6 [12, 13].

Shortly from eqs. (3)–(6):

$$\Delta P_{ent} = \frac{\alpha - 1 + \varepsilon}{\alpha} \Delta P_{ends}. \tag{7}$$

The value of ε is very uncertain, and however in accordance with Han [14] we have taken the values of 0.4 and 0.5 for LDPE and HDPE respectively. For want of experimental data relative to the blends we have considered for ε a weighted average of the values of the polymer parents.

With the above assumption, from the ΔP_{ends} values taken in the ΔP vs. L/D plots, ΔP_{ent} have been evaluated through eq. (7). With these values and the rheological characteristics of the same blends already reported in [14], η_{el} and $\dot{\varepsilon}$ have been evaluated. Figures 1 and 2 show such data for both types of blend at

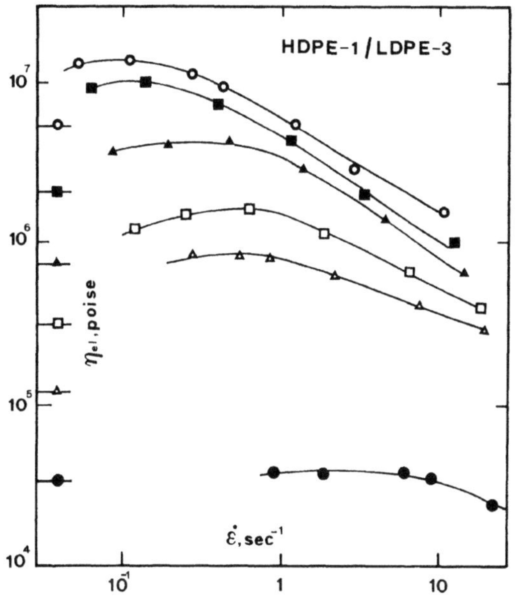

Fig. 1. Elongational viscosity vs. stretching rate for HDPE-1/LDPE-3 at 180 °C. ○ LDPE, ■ 80% LDPE, ▲ 60% LDPE, □ 40% LDPE, △ 20% LDPE, ● HDPE

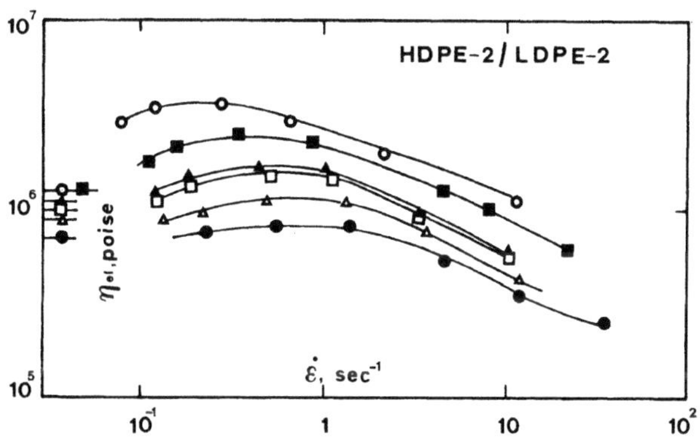

Fig. 2. Elongational viscosity vs. stretching rate for HDPE-2/LDPE-2 at 180 °C. Key for symbols as in figure 1

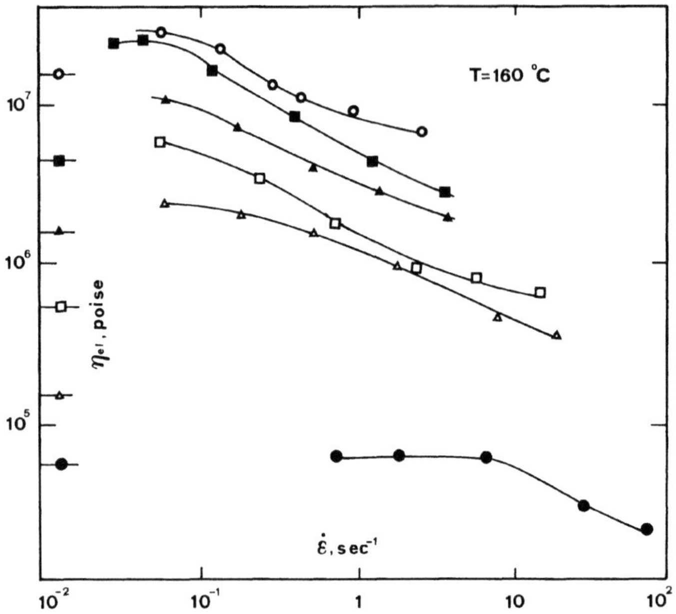

Fig. 3. Elongational viscosity vs. stretching rate for HDPE-1/LDPE-3 at 160 °C. Key for symbols as in figure 1

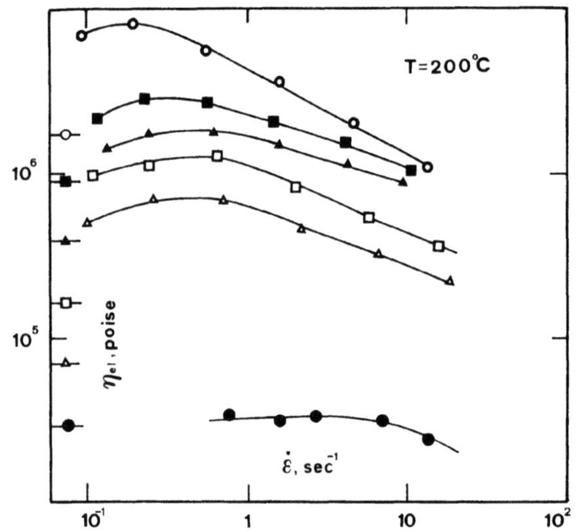

Fig. 4. Elongational viscosity vs. stretching rate for HDPE-1/LDPE-3 at 200 °C. Key for symbols as in figure 1

T = 180 °C. Figures 3 and 4 further show data for HDPE-1/LDPE-3 at 160 and 200 °C respectively. In all cases the Trouton value $3\eta_0$ has been reported in the figures for comparison.

The curves relative to the pure LDPE show in all cases a significant maximum whose value, relative to the Trouton viscosity, increases when η_0 decreases. Also a shift of the maximum toward higher stretching rates is evident decreasing the Trouton value.

The data of the HDPE show only a very slight maximum in the case of HDPE-2, which has a higher viscosity. All these features are in good agreement with literature results [4].

The viscosities of the blends lie for all the cases between those of the pure polymers. However there are typical features peculiar of each type of mixture. When the shear viscosity of the LDPE is much higher than that of the HDPE, all blends show a behaviour very similar to that of the low density polyethylene and the ratio $\eta_{el}/3\eta_0$ becomes much larger increasing the HDPE content. This unexpected result is probably due to the fact that the blend behaves like a LDPE with a lower molecular weight. The shift of the maximum toward higher stretching rates seems to confirm this hypothesis.

The system HDPE-2/LDPE-2 with similar newtonian viscosities, shows a gradual change from one type of behaviour to another, and indeed, when the concentration of the high density polyethylene is 80%, the maximum almost disappears.

Such a behaviour can be explained with the hypothesis that in this case, where the molecular weights of the two homopolymers are not too different, the primary effect is on the average a decrease in the degree of branching with increasing HDPE content.

Qualitatively figures 3 and 4 confirm the above indication with the obvious quantitative differences due to the temperature (and therefore newtonian viscosity) variations. The same features are present also in the data at different temperatures for the system HDPE-2/LDPE-2 not reported here.

This work has been carried out with the financial support of "Progetto Finalizzato del C.N.R. Chimica Fine e Secondaria".

References

1. Cogswell, F. N., Appl. Polym. Symp. **27**, 1 (1975).
2. White, J. L., Appl. Polym. Symp. **33**, 31 (1978).
3. Petrie, C. J. S., Elongational Flow, Pitman (London 1979).
4. Münstedt, H., Joint Meeting of British, Netherlands and Italian Societies of Rheology (Pisa 1977).
5. Laun, H. M., H. Münstedt, Rheol. Acta **17**, 415 (1978).
6. Raible, T., A. Demarmels, J. Meissner, Polymer Bulletin **1**, 397 (1979).
7. Cogswell, F. N., Trans. Soc. Rheol. **16**, 383 (1972).
8. Paul, D. R., S. Newmann (eds.), Polymer Blends, Academic Press (London 1978).
9. Cogswell, F. N., Polym. Eng. Sci. **12**, 64 (1972).
10. Han, C. D., Rheology in Polymer Processing, Academic Press (London 1976).
11. White, J. L., A. Kondo, J. Appl. Polym. Sci. **21**, 2289 (1977).
12. Han, C. D., Trans. Soc. Rheol. **18**, 163 (1974).
13. Lee, B. L., J. L. White, Trans. Soc. Rheol. **18**, 467 (1974).
14. Curto, D., D. Acierno, F. P. La Mantia, submitted for publication.

(Received April 15, 1982)

Authors' address:

Dr. F. P. La Mantia, Prof. D. Acierno, Dr. D. Curto
Istituto di Ingegneria Chimica
Università di Palermo
Viale delle Scienze
I-90100 Palermo

Rheologica Acta Rheol. Acta 21, 456–459 (1982)

Dehnviskosität von Kautschukmischungen

T. S. Ng

Deutsches Kunststoff-Institut, Darmstadt

Abstract: It is shown that the extensional viscosity measured by uniaxial stretch may give useful information as to the processability of rubber compounds. A simple apparatus was constructed to measure this rheological property at constant rate of strain. The influence of rubber type, filler, aging, mastication and degree of crosslinking on the reduced stress is represented by diagrams. No steady-state extensional viscosity was obtained for highly filled rubber compounds. It was found that the flow behaviour of rubber compounds can be better characterized with an extensional rheometer than with a high-pressure capillary rheometer.

Key words: Extensional viscosity, uniaxial stretch, processability, rubber compound, aging

1. Einleitung

Im Zusammenhang mit der Untersuchung von Materialien wie z.B. Pech durch einaxiale Verstreckung hat Trouton [1] für das Verhältnis von Zugspannung und Dehngeschwindigkeit den Begriff „Dehnviskosität" eingeführt. Mit dem Fließverhalten von Elastomeren bei Verstreckung haben sich eingehend Vinogradov [2, 3] und Stevenson [4] befaßt. Vor allem Meissner [5] hat mit seinen profilierten Probenabzugsrollen viel zur Entwicklung der Meßapparatur auf dem Gebiet der Dehnviskosität beigetragen. 1978 haben Cotton und Thiele [6] angefangen, die Dehnviskosität von rußgefüllten Kautschukmischungen zu messen, wobei sie eine einfache Apparatur auf Basis des Meissnerschen Prinzips verwendeten. Zur Erfassung der nachfolgend diskutierten Meßergebnisse sind die Zustandsgleichungen gut anwendbar, die von Smith [7, 8] für viskoelastische Materialien unter einaxialer Verstreckung bei konstanten Dehngeschwindigkeiten entwickelt worden sind.

2. Ziel der Untersuchung

Das Hauptziel der Untersuchungen ist, zu überprüfen, inwieweit die Dehnviskosität der Kautschukmischungen sich bei Zimmertemperatur messen läßt, und inwieweit diese im Hinblick auf die Verarbeitbarkeit als charakteristische Eigenschaften gedeutet werden kann. Insbesondere wird versucht, den Einfluß der Lagerung auf die Rohmischungen festzustellen.

3. Experimenteller Teil

Für die vorliegende Untersuchung ist ein einfaches Meßgerät nach dem schon erwähnten Konzept von Cotton und Thiele [6] konstruiert und gebaut worden. Abbildung 1 gibt eine schematische Darstellung der Meßapparatur.

Ein strangförmiger Probekörper von 2–5 mm Durchmesser und ungefähr 100 cm Länge wird über zwei Rollen der Probenhalterung geführt. Zur Vermeidung von Einschnürungen sind diese Rollen in einem Abstand von 8 cm befestigt. Die Probenhalterung ist mit einer Kraftmeßdose verbunden. Ein Meßverstärker und ein Schreiber sorgen für die Registrierung der auftretenden Kraft. Als Zugvorrichtung dienen ein Paar profilierte Rollen (Zahnradpaar), die mit konstanter Drehzahl und entgegengesetztem Drehsinn von einem regelbaren Motor angetrieben werden. Außer an den Rollen treten bei dieser Prüfmethode keine Wandeffekte auf.

Die strangförmigen Probekörper wurden mit Hilfe des Göttfert-Hochdruck-Kapillarviskosimeters herge-

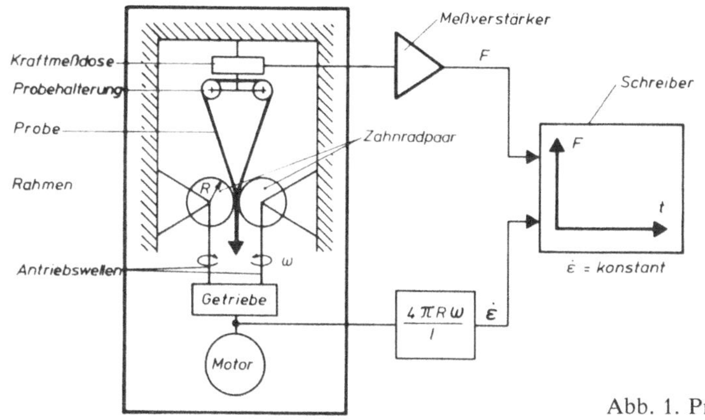

Abb. 1. Prinzipskizze des Dehnrheometers

stellt. Hierbei besteht die Beschränkung, daß nur stark gefüllte Mischungen zu Extrudaten mit guter Oberflächenbeschaffenheit verarbeitet werden können. Die jetzigen Untersuchungen beschränken sich deshalb auf Mischungen, die vergleichsweise viel Füllstoff enthalten.

4. Meßergebnisse

4.1. Einfluß von Kautschuk und Füllstoff

Verschiedene rußgefüllte Kautschukmischungen, die die wichtigen Allzweck-Kautschuke Natur-, Styrol-Butadien- und Butadienkautschuk (NR, SBR und BR) enthielten, wurden auf ihr Dehnverhalten untersucht, wobei verschiedene Dehngeschwindigkeiten berücksichtigt wurden.

Abbildung 2 zeigt die normierte Spannung für die Kautschukmischung BR/65 Teile Ruß, aufgetragen gegen die Zeit.

Ähnliche Ergebnisse, jedoch mit unterschiedlicher Höhe der $\sigma/\dot{\varepsilon}$-Kurven, erhält man für NR/35 Teile Ruß und SBR/50 Teile Ruß. Die Kurven zeigen einen stetig steigenden Verlauf. Ein stationärer Zustand wurde bei allen drei Kautschukmischungen nicht erreicht. Weitere Dehnung würde zum Bruch führen. Bei kleinen Zeiten überlagern sich die Kurven. Bei größeren Dehngeschwindigkeiten beobachtet man eine Abweichung vom linearen viskoelastischen Verhalten. Die Kurven biegen bei größeren Zeiten nach oben ab, und zwar bei den größeren Dehngeschwindigkeiten merklich früher.

4.2. Einfluß von Lagerung und Mastikation

Durch die Lagerung finden in den Rohmischungen, je nach den Lagerungsbedingungen, thermochemisch oder photochemisch ausgelöste oxidative Reaktionen statt, die zum einen zu Abbau und zum anderen zu Vorvernetzung führen können. Man kann vorwegnehmen, daß durch Lagerung bei höherer Temperatur

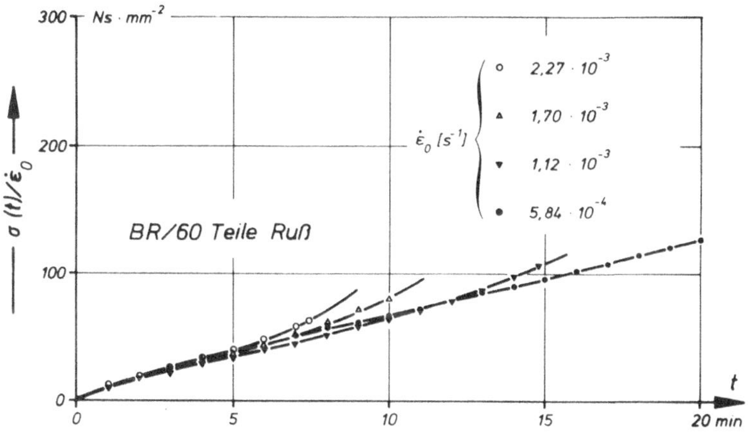

Abb. 2. Normierte Zugspannung einer rußgefüllten BR-Mischung als Funktion der Zeit t, gemessen bei verschiedenen Dehngeschwindigkeiten $\dot{\varepsilon}$. $\sigma(t)$ = wahre Zugspannung zur Zeit t

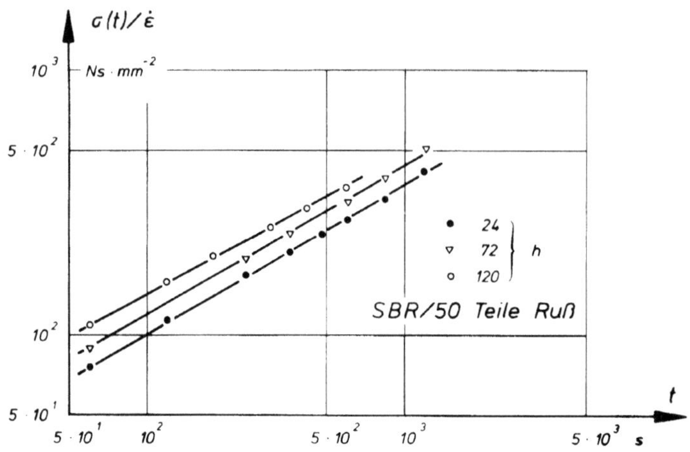

Abb. 3. Einfluß der thermischen Alterung auf die normierte Zugspannung einer rußgefüllten SBR-Mischung

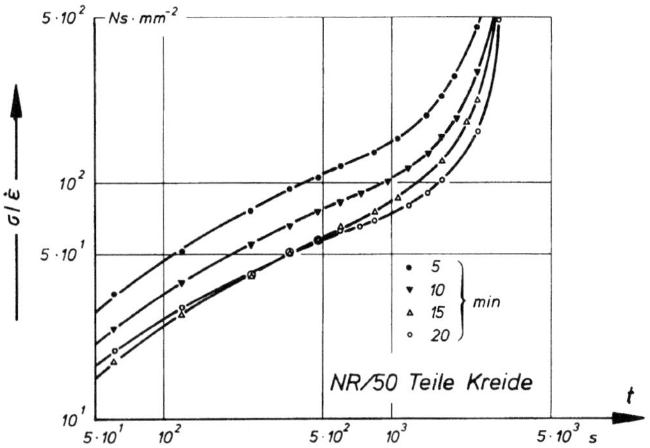

Abb. 4. Einfluß der Mastikation auf die normierte Zugspannung einer kreidegefüllten NR-Mischung

oder Bestrahlung mit UV-Licht die Veränderungen in der Rohmischung beschleunigt werden.

Abbildung 3 zeigt die Ergebnisse einer Alterung bei 70 °C in Luft bis zu fünf Tagen für SBR/50 Teile Ruß. Durch die Alterung fand eine deutliche Erhöhung der Steifigkeit der Rohmischung statt. Auch die Bestrahlung mit UV-Licht hat die gleiche Wirkung, d. h. die Dehnviskositätswerte erhöhen sich.

Mechanischer Abbau der Mischungen durch gesteigerte Mastikation hat eine Erniedrigung der entsprechenden normierten Zugspannungskurven zur Folge, wie Abbildung 4 das für NR/50 Teile Kreide zeigt.

5. Diskussion

Es können die folgenden Schlußfolgerungen gezogen werden: Die Dehnviskosität erreicht bei den rußgefüllten NR-, SBR- und BR-Mischungen keinen konstanten Wert. Nur bei sehr kleinen Dehngeschwindigkeiten deutet der Bereich längerer Belastungszeiten der Überlagerungskurve, deren Verlauf flach ist, auf einen stationären Zustand hin, wobei die Dehnviskosität einen endlichen Wert annimmt. Diese Befunde sind in Übereinstimmung mit den theoretischen Überlegungen von Denn und Marucci [9], Lodge [10] und Petrie [11].

Lagerungsprozesse in den Rohmischungen, in den Versuchen beschleunigt durch Alterung bei höherer Temperatur und UV-Bestrahlung, sowie Abbau durch Mastikation, können durch das Dehnrheometer besser als mit dem Hochdruck-Kapillarviskosimeter charakterisiert werden.

Die beschriebenen Messungen liefern ohne Zweifel wichtige Informationen für die Praxis der Gummiindustrie. So wie die Dehnviskositätsbestimmung an

Hochpolymeren durch einaxiale Verstreckung in den letzten Jahren einen großen Aufschwung erfahren hat, was die zahlreichen Publikationen auf dem Gebiet beweisen, dürfte sie sicherlich auch für Rohkautschukmischungen eine Lücke in der Reihe der möglichen und praktikablen Meßtechniken schließen.

Danksagung

Für die Zurverfügungstellung der Kautschukmischungen sei Herrn Dr. G. Angerer von der Firma Dunlop Hanau herzlich gedankt, ebenso der Arbeitsgemeinschaft Industrieller Forschungsvereinigungen e.V. für finanzielle Unterstützung.

Literatur

1. Trouton, F. T., Proc. Roy. Soc. **A77**, 426 (1906).
2. Vinogradov, G. V., B. V. Radushkevich, V. D. Fikhman, J. Polymer Sci. **A-2**, 1 (1970).
3. Vinogradov, G. V., Rheol. Acta **14**, 942 (1975).
4. Stevenson, J. F., AIChE J. **18**, 540 (1972).
5. Meissner, K., Trans. Soc. Rheol. **16**, 405 (1972).
6. Cotton, G. R., J. L. Thiele, Rubber Chem. Technol. **51**, 749 (1978).
7. Smith, T. L., J. Polymer Sci. **20**, 89 (1956).
8. Smith, T. L., Trans. Soc. Rheol. **6**, 61 (1962).
9. Denn, M. M., G. Marrucci, AIChE J. **17**, 101 (1971).
10. Lodge, A. S., Elastic Liquids, S. 114, Academic Press (London 1964).
11. Petrie, C. J. S., J. Non-Newtonian Fluid Mech. **2**, 228 (1977).

(Eingegangen am 14. April 1982)

Anschrift des Verfassers:

Dr. T. S. Ng
Deutsches Kunststoff-Institut
Schloßgartenstraße 6 R
D-6100 Darmstadt

Rheologica Acta Rheol. Acta **21**, 460 – 463 (1982)

The description of the profile balance and the elastic properties at die-swell

J. Vlček

VÚGPT Gottwaldov (ČSSR)

Key words: Die-swell, velocity profile, elastic properties, polyethylene

The die-swell phenomenon is found, for example, in the extrusion of plastics. It manifests itself as a change in the form of the material after it leaves the die, thus influencing the final dimensions of the product. The aim of this paper is to suggest a way of obtaining a model for the description of die-swell. The amount of the die-swell consists of two contributions. The first one is caused by a redistribution of the velocity profile, the other one arises from the elastic properties of the material.

We shall deal first with the velocity profile rearrangement. For non-Newtonian liquids that are not elastic the well-known relation for the change of radius holds (R_∞ – final radius, R_0 – die radius)

$$\xi^2 = \left(\frac{R_\infty}{R_0}\right)^2 = \frac{2n+1}{3n+1} \tag{1}$$

where n is the exponent of the power law. For $0 \leqslant n \leqslant 1$ this gives $\xi \leqslant 1$. This value does not agree with experimental values for polymer melts, which show $\xi > 1$. Relation (1) was derived from the laws of mass and momentum conservation. The derivation will be modified so as to also include the influence of the elasticity. The velocity profile in the die is given by the following relation

$$v(r) = v_0\left[1 - \left(\frac{r}{R_0}\right)^{(1/n)+1}\right] \tag{2}$$

where v_0 is the maximum velocity. After leaving the die, the velocity profile is changed to a constant value of velocity in the whole cross-section. This value is denoted by v_∞. The continuity equation has the form

$$\rho\pi R_\infty^2 v_\infty = 2\rho\pi \int_0^{R_0} v_0\left[1 - \left(\frac{r}{R_0}\right)^{(1/n)+1}\right] r\,dr$$

$$= \rho\pi v_0 R_0^2 \frac{n+1}{3n+1} \tag{3}$$

where ρ is the density.

The momentum conservation equation has the form

$$\frac{dp_\infty}{dt} = \frac{dp_0}{dt} \tag{4}$$

whith

$$\frac{dp_\infty}{dt} = \pi\rho R_\infty^2 v_\infty^2,$$

$$\frac{dp_0}{dt} = 2\pi\rho \int_0^{R_2} v^2(r)\, r\,dr,$$

where p_0 is the momentum at the die exit and p_∞ its final value. From relation (4) we can obtain relation (1). If the melt is under the influence of a force F, eq. (4) takes the more general form

$$\frac{dp_\infty}{dt} = \frac{dp_0}{dt} - F. \tag{5}$$

For $F = 0$, we get eq. (1), which is the relation for nonelastic materials. We now suppose that the force F is related to the normal stresses. We suggest that the intensity of the force F is equal to intensity of the normal-stress difference. The work W_F done by the force F is proportional to F, the kinematic viscosity ν_w

at the wall and to the wall surface. It is inversely proportional to the volume V because the bigger the volume the relatively smaller the surface influence. Therefore

$$W_F = F v_w \frac{s}{V}, \tag{6}$$

and for a circular cross-section

$$W_F = F v_w \frac{2}{R_0}.$$

Now we evaluate the normal-stress difference. We suppose that points closer to the surface have a greater influence on the die-swell. This we express with the factor r/R_0. For the normal stresses we have $|T_{xx}| = |T_{rr}| = \sigma(\dot\gamma)/2$ where $\sigma(\dot\gamma)$ is the first normal-stress difference and $\sigma(r) = \sigma_w(r/R_0)$, with σ_w its value at the wall. Then we get

$$W(r) = 2\pi \int_0^r \sigma_w \left(\frac{x}{R_0}\right)^2 v_0 \left[1 - \left(\frac{x}{R_0}\right)^{(1/n)+1}\right] x\,dx. \tag{7}$$

The mean value can be obtained by integration over the circular cross-section. With the aid of relations

$$\dot\gamma_w = \frac{1+n}{n}\frac{v_0}{R_0}, \quad \mu_w = \rho v_w, \quad \tau_w = \mu_w \dot\gamma_w$$

and if $W_F = \bar W_\sigma$, we find that

$$F = \frac{dp_0}{dt}\frac{(3n+1)(2n+1)(11n+1)}{(5n+1)(7n+1)n}\frac{\sigma_w}{24\tau_w}. \tag{8}$$

After substitution into relation (5) we obtain

$$\xi^2 = \left(\frac{R_\infty}{R_0}\right)^2 = \frac{2n+1}{3n+1}\frac{1}{1 - \dfrac{(3n+1)(2n+1)(11n+1)}{(5n+1)(7n+1)n}\dfrac{\sigma_w}{24\tau_w}}. \tag{9}$$

This is the resulting relation for the influence of the rearrangement of the velocity profile. When the shear rate is small enough $\sigma_w/\tau_w \to 0$, and eq. (9) changes into eq. (1). However, in the case of high shear rates, ξ may be larger or smaller than one according to the size of the ratio σ_w/τ_w. The typical dependence for PE is shown in figure 1.

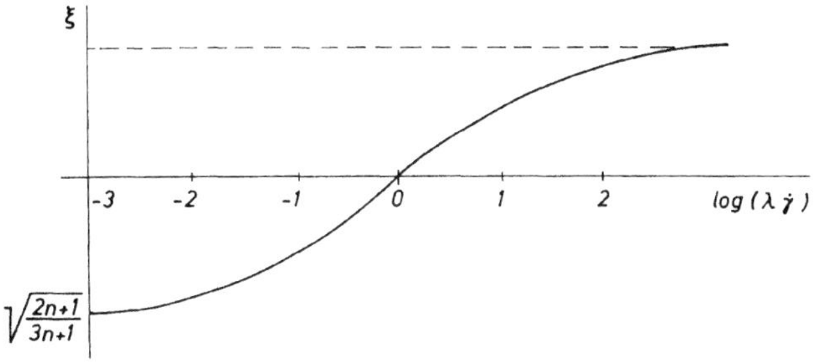

Fig. 1

Now we turn our attention to the elastic component of die-swell. This problem has been studied by several authors [1–3]. Their treatments resulted in the following equations

$$\xi^6 = 1 + \frac{1}{2}\left(\frac{\sigma_w}{G} - \frac{\tau_w^2}{G^2}\right) \quad \text{(Tanner [1])},$$

$$\xi^4 - \xi^{-2} = \frac{1}{2}\left(\frac{\sigma_w}{G} - \frac{\tau_w^2}{G^2}\right) \quad \text{(Vlachopoulos et al. [2])},$$

$$\xi^4 - \xi^{-2} = \frac{1}{2}\left(\frac{\sigma_w}{G} - \xi^2\frac{\tau_w^2}{G^2}\right) \quad \text{(Vlček [3])}, \tag{10}$$

where G is the shear modulus. The above equations were used with the aid of the relation

$$G = \frac{\tau_w^2}{\sigma_w}. \tag{11}$$

It was further supposed that G depended on $\dot\gamma$ resulting in the dependence of ξ on $\dot\gamma$.

Later it was found [4] that for plastic melts the dependences of σ and τ on $\dot\gamma$ have almost the same exponent for high shear rates (that is $\sigma(\dot\gamma) = K\tau(\dot\gamma)$). Writing

$$\tau_w = K_1\dot\gamma_w^n, \quad \sigma_w = K_2\dot\gamma_w^n, \quad G = K_3\dot\gamma_w^m$$

(m can be different from n) we get

$$\frac{\sigma_w}{G}\left(1 - \frac{\tau_w^2}{\sigma_w G}\right) = \frac{K_2}{K_3}\dot\gamma^{n-m}\left(1 - \frac{K_1^2}{K_2K_3}\dot\gamma^{n-m}\right).$$

It is evident that with increasing $\dot\gamma$ we cannot obtain a monotonically increasing function of ξ for any m and n. The eqs. (10) result from the assumption that the melt behaves as an elastic material at the moment of die-swell. We can anticipate, however, that the behaviour of melt during die-swell is also viscoelastic. The melt is liquid and the elastic component of deformation is not changed by the relative motion of the moving layers i. e. the melt elasticity cannot recover in this shear-motion. Therefore relations (10) can be rewritten as

$$\xi^2 = 1 + \frac{1}{2}\frac{\sigma_w}{G}, \tag{12a}$$

$$\xi^4 - \xi^{-2} = \frac{1}{2}\frac{\sigma_w}{G}. \tag{12b}$$

If eq. (11) is valid and above dependences are true relations (12) give only constant values of ξ. However, for $G = \eta_0/\lambda_0$ (η_0 – limit viscosity, λ_0 – time constant of Carreau's model) eqs. (12a) and (9) provide values of ξ that are in good agreement with experiment (cf. fig. 2).

The first normal-stress difference can only be measured with difficulty. Therefore, we tried to determine its value from the co-rotational model. In this model

$$\tau_{xr} = \dot\gamma \int_{-\infty}^{t} f(t,t')\cos(\dot\gamma(t-t'))\,dt' \tag{13}$$

$$\tau_{xx} = -\tau_{rr} = \dot\gamma \int_{-\infty}^{t} f(t,t')\sin(\dot\gamma(t-t'))\,dt' \tag{14}$$

where τ_{xr}, τ_{rr}, τ_{xx} are stresses and $f(t,t')$ the memory function.

For the function $f(t',t)$ we used the relation

$$f(t,t') = \frac{2\eta_0}{\lambda_0 \Gamma\left(\frac{1-n}{2}\right)\Gamma\left(\frac{1+n}{2}\right)} K_n\left(\frac{t-t'}{\lambda_0}\right) \tag{15}$$

where K_n is the generalized Bessel function. By substituting this into eqs. (13) and (14) we obtain

$$\eta(\dot\gamma) = \frac{\eta_0}{(1+\lambda_0^2\dot\gamma^2)^{(1-n)/2}} \cdot \frac{\left(1+\frac{\lambda_0\dot\gamma}{(1+\lambda_0^2\dot\gamma^2)^{1/2}}\right)^n + \left(1-\frac{\lambda_0\dot\gamma}{(1+\lambda_0^2\dot\gamma^2)^{1/2}}\right)^n}{2}, \tag{16a}$$

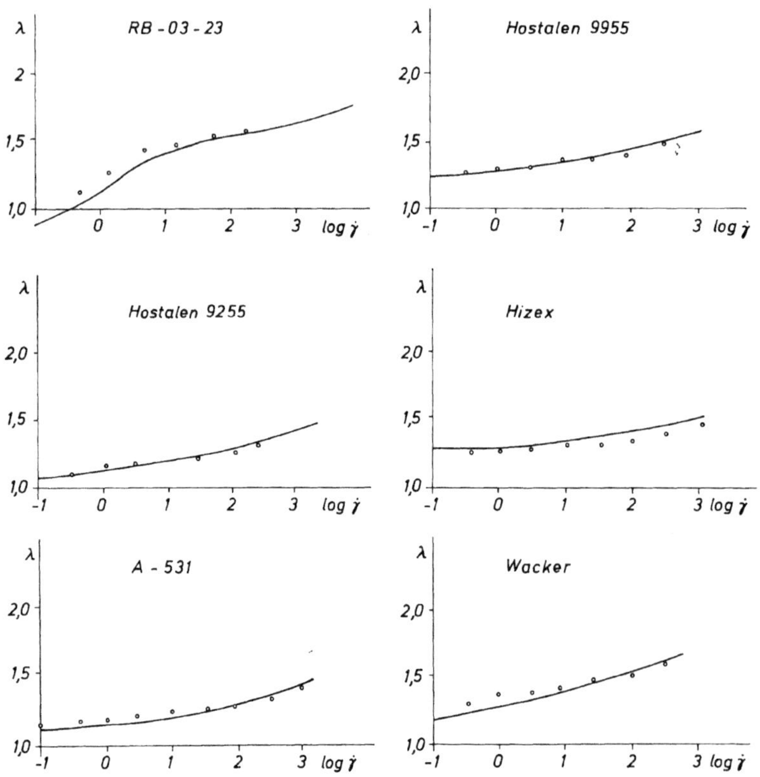

Fig. 2

$$\sigma(\dot{\gamma}) = \frac{2\eta_0\dot{\gamma}\cot\left(\frac{\pi}{2}n\right)}{(1+\lambda_0^2\dot{\gamma}^2)^{(1-n)/2}}$$

$$\cdot \frac{\left(1+\frac{\lambda_0\dot{\gamma}}{(1+\lambda_0^2\dot{\gamma}^2)^{1/2}}\right)^n - \left(1-\frac{\lambda_0\dot{\gamma}}{(1+\lambda_0^2\dot{\gamma}^2)^{1/2}}\right)^n}{2} \qquad (16b)$$

where η_0, λ_0, n describe the given material. For large $\dot{\gamma}$ it gives

$$\frac{\sigma(\dot{\gamma})}{\tau(\dot{\gamma})} = 2\cot\left(\frac{\pi}{2}n\right). \qquad (17)$$

In this paper a model has been suggested for the description of die-swell. We make use of the resulting eqs. (9) and (12) for comparing theory and experiment. The measurements were performed on fifteen LDPE and HDPE specimens, some of the results being shown in figure 2. For the other melts we also obtained similar agreement.

References

1. Tanner, R. I., J. Polym. Sci. **A2, 8**, 2051 (1970).
2. Vlachopoulos, J., H. Michiniko, L. Stathis, Trans. Soc. Rheol. **16**, 669 (1972).
3. Vlček, J., Polym. Mech. **14**, 273 (1978).
4. Han, C. D., K. N. Kim, N. Siskovic, C. R. Huang, Rheol. Acta **14**, 533 (1975).

(Received April 16, 1982)

Author's address:

Dr. rer. nat. J. Vlček
VÚGPT
ČSSR-76422 Gottwaldov-Louky

Rheologica Acta Rheol. Acta **21**, 464 – 469 (1982)

Elastic properties of polyethylene melts at high shear rates with respect to extrusion

H. M. Laun

Kunststofflaboratorium, BASF Aktiengesellschaft, Ludwigshafen/Rhein

Abstract: At high shear rates a steady state of shear flow with constant shear rate, constant shear stress, and constant recoverable shear strain is observed in the short-time sandwich rheometer after some few shear units already. The melt exhibits rather high elastic shear deformations and the recovery occurs at much higher speed than it is observed in the newtonian range. The ratio of first normal stress difference and twice the shear stress, being equal to the recoverable strain in the second-order fluid limit, significantly underestimates the true elastic shear strains at high shear rates. The observed shear rate dependence of shear stress and first normal stress difference as well as of the (constrained) elastic shear strain is correctly described on the basis of a discrete relaxation time spectrum. In simple shear a stick-slip transition at the metal walls is found. Necessary for the onset of slip is a critical value of shear stress and a certain amount of elastic shear deformation or orientation of the melt.

Key words: Polyethylene melt, recoverable shear strain, normal stress coefficient, extrudate swell, stick-slip transition

1. Experimental

A short-time sandwich rheometer (SSR) has been developed which makes it possible to measure time-dependent rheological properties like shear creep, stress relaxation, and elastic recovery in a homogeneous flow field up to high shear rates relevant to extrusion and injection moulding processes.

Two compression moulded sample films of equal thickness $h = 0.2$ to 2 mm are sheared in the gap between a central plate and a pneumatically driven cage (fig. 1). The whole sandwich is surrounded by a

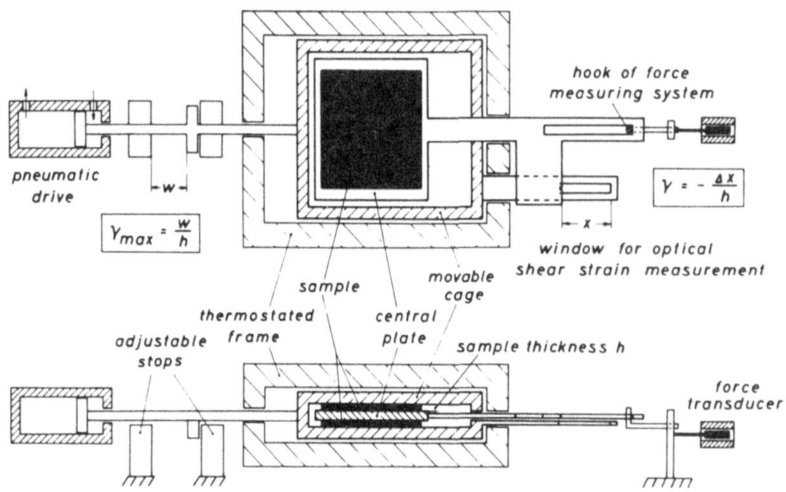

Fig. 1. Schematic drawing of the short time sandwich rheometer SSR

fix thermostated frame. In the shear creep test the cage is pulled to the left at constant force while the central plate is clicked into the hook of the force-measuring system. The shear strain γ is obtained from the relative displacement of central plate and cage. The force acting on the central plate gives the shear stress p_{21}. Evaluation of tests has been described in [1] for a low speed sandwich rheometer.

Parts of the central plate and of the cage are constructed such that they form a window which is illuminated from one side and the width x measured optically (Digital Dimension Meter 1000 C, Zimmer, Darmstadt). The time-dependent stress and strain are both recorded by a transient recorder of 0.1 ms resolution. Maximum force of the pneumatic is 500 N, maximum speed of the piston 1 m/s, maximum film area 0.02 m^2. The adjustable stop at the left makes it possible to suddenly stop deformation after a total strain γ_{max} to measure the relaxation of the shear stress. For recovery experiments, however, after a strain of $\gamma < \gamma_{max}$ the cage is suddenly moved in reverse sense by the pneumatic till the right stop. Due to the slit in the central plate at the hook the central plate is released within much less than 10 ms. It can now move relative to the cage according to the elastic recoil of the sample. The optics measures the true strain independent of the position of the cage. A deformation of the sample during displacement of the whole sandwich due to inertia forces can be neglected.

The SSR enables high shear rate measurements of short duration. Hence, adiabatic heating can be shown to be less than 1 °C for the PE melts investigated here, even at the highest shear rates of 1000 s^{-1}.

Figure 2 shows a creep test followed by a recovery test. The short oscillation of p_{21} at the beginning of the test is caused by a short vibration of the whole

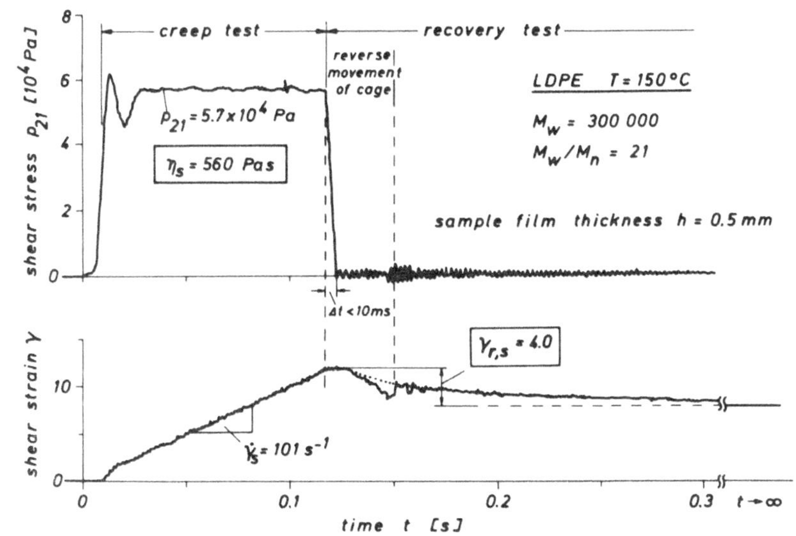

Fig. 2. Measurement of viscosity and elastic recovery

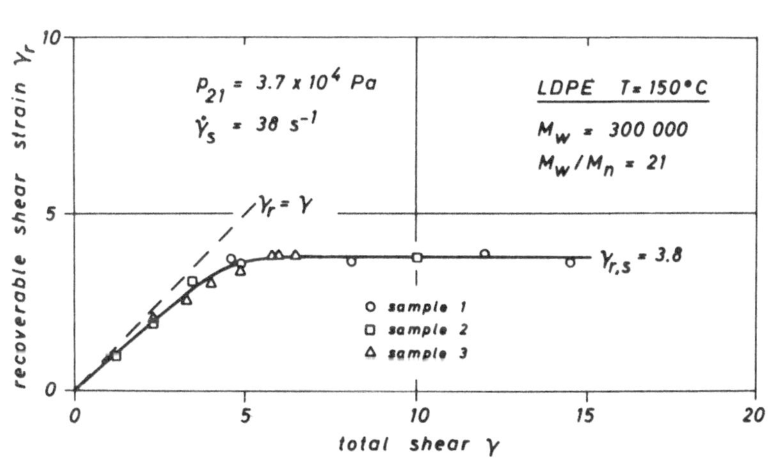

Fig. 3. Dependence of recoverable shear strain on total strain in a shear creep test

apparatus. After a small amount of shear the shear rate becomes constant and a steady shear flow is achieved. Having reached a total strain of about 12 the sample was unloaded by the reverse movement of the cage. The sample shows a rather fast elastic recovery. A recoverable strain of 4 is obtained within 1 s only, whereas in the newtonian range the total recovery would take several minutes!

During the fast reverse movement of the cage the optically measured strain is less than the true strain (dotted line) due to the time necessary for one scan over the window (1 ms). As the displacement of the cage is recorded, too, this effect can be corrected for.

During a creep test the recoverable strain increases with growing total strain in the first instance (fig. 3) and most of the strain is recoverable as indicated by the broken line of a completely elastic deformation. For a total strain bigger than five the recoverable strain becomes constant. This strain-independent value gives the steady-state recoverable strain $\gamma_{r,s}$. It should be noted that the total strain necessary to achieve a steady shear flow is strongly depending on the magnitude of $\gamma_{r,s}$.

2. Material functions in shear

In figures 4 and 5 experimental results obtained on two LDPE melts by three different apparatuses are compared. The material functions plotted versus shear rate are the steady-state values of viscosity, recoverable shear strain, and normal stress coefficient. Low shear rate viscosities were determined with a Weissenberg Rheogoniometer (WRG) modified by Meissner [2] (unfilled circles). In the high shear rate range a capillary viscometer was used (circles with tic). The viscosities obtained from the high speed sandwich SSR (full squares) are in very good agreement with the cone and plate and capillary data thus demonstrating the reliability of viscosity measurements with the SSR.

The second material function for which results from the SSR and the WRG can be compared is the steady-state recoverable shear strain. Recoverable strains at low shear rates can very accurately be measured by the servo-controlled WRG [2] (unfilled triangles). In the newtonian range $\gamma_{r,s}$ increases proportional to the shear rate [1, 3]. In the shear thinning range the increase becomes less pronounced. Unfortunately the speed of the servo-controlled WRG is not fast enough to give the true elastic strains after high shear rate tests which caused erroneous results earlier.

The recoverable strains from the SSR are represented by the full triangles. The WRG data agree with those in the overlapping range but only up to a shear rate of 0.1 s^{-1}. At still higher shear rates the WRG would give too small values. These results are not

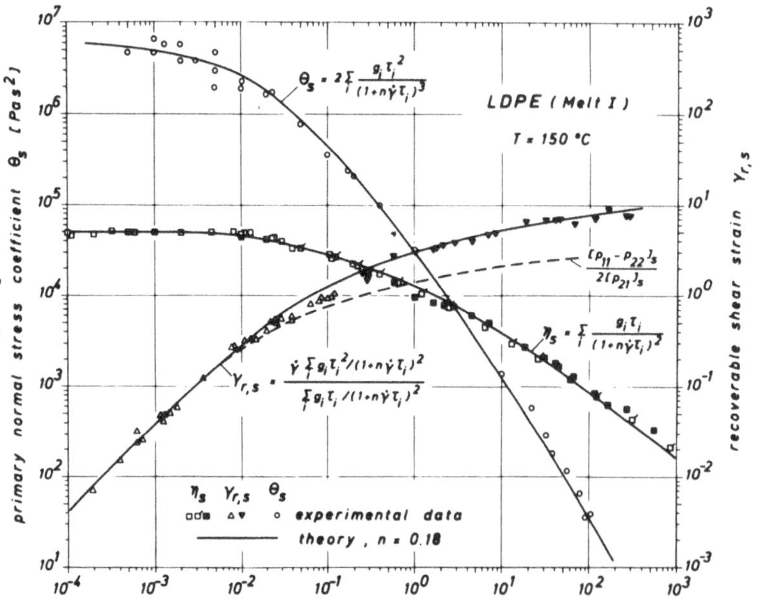

Fig. 4. Comparison of measured and predicted shear rate dependence of viscosity, normal stress coefficient, and recoverable strain is the steady state (LDPE melt, M_w = 300000, M_w/M_n = 21)

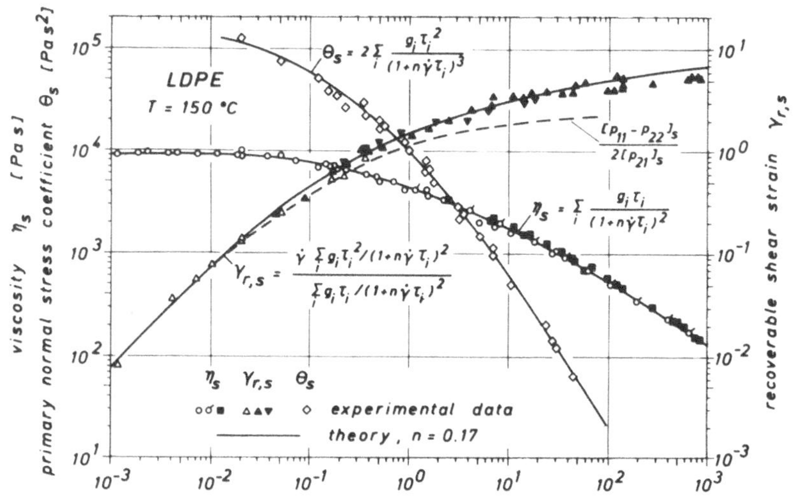

Fig. 5. Comparison of calculated and measured material functions is shear (LDPE Melt I)

contained in the figures for clarity. The recoverable shear strain measured on LDPE Melt I at 300 s^{-1} is 9. This magnitude is rather surprising.

The third material function, the normal stress coefficient, could be obtained from the WRG only (unfilled rhombs). Very often the ratio of first normal stress difference $[p_{11}-p_{22}]_s$ and twice the shear stress $[p_{21}]_s$ in the steady state is taken equal to the recoverable shear strain also in the non-newtonian range. The relation holds in the linear viscoelastic flow range of the melt [4]. From the measured normal stress coefficients and viscosities the stress ratio was computed. It is represented by the broken line. One can clearly see that the stress ratio significantly underestimates the true elastic strains up to more than a factor of two at high shear rates!

The full lines in figures 4 and 5 represent the theoretical predictions by means of a discrete relaxation time spectrum and an integral constitutive equation with strain-dependent and factorized memory function. The τ_i being the relaxation times and g_i the corresponding relaxation strengths. The nonlinearity of the flow behaviour is introduced by the so called damping constant n following Wagner [5]. Whereas the analytical expressions for the viscosity and the normal stress coefficient have been published earlier [6], the theory for the recoverable shear strain was recently worked out in collaboration with Fischer [7]. The agreement between experiment and theory over more than 6 decades of shear rate (Melt I) is very encouraging.

3. Stick-slip transition of HDPE

Linear polyethylene melts of high molecular weight exhibit a drastic increase of flow rate if the extrusion pressure reaches a critical value. It is assumed that this spurt effect is caused by a stick-slip transition at the die wall [8]. The low molecular weight PE in figure 6 does not show a flow instability up to 1000 s^{-1}. Compared with the long chain branched polyethylenes of similar viscosity (fig. 5) the linear melt shows significantly smaller recoverable strains. Contrary to the low viscosity HDPE the high molecular weight HDPE exhibits a pronounced flow instability for wall shear rates $\dot{\gamma} \gtrsim 250\ s^{-1}$. Up to that shear rate agreeing results from the cone and plate, capillary, and sandwich rheometers are obtained.

The recoverable shear strain of the high molecular weight HDPE remains nearly constant over the shear rate range investigated, the highest recoverable strains being about 3. The data points in the vicinity of the instability region are represented by stars as in fact they are no true steady-state values.

Figure 7 shows the shear stress and the apparent shear strain – as calculated from the relative displacement of cage and central plate – versus time. This quantity only gives the true shear strain as long as the melt sticks at the wall. A black coloured mark in the sample films was used to make the true shear deformation visible. The sample was sheared up to different amounts and then was cooled down under shape constrain in order to take photographs of the flow profiles. The arrows indicate the corresponding parts of the test where the sample was frozen in.

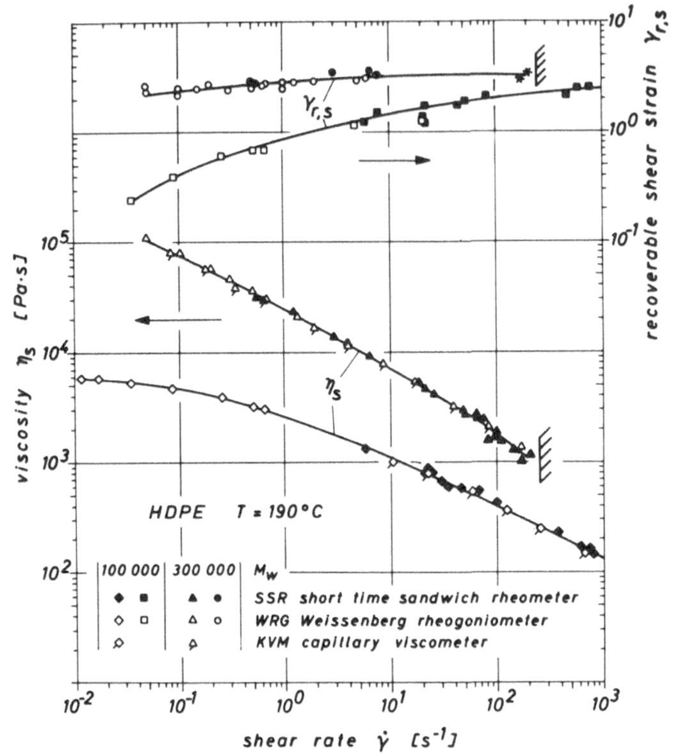

Fig. 6. Viscosity and recoverable shear strain of HDPE melts

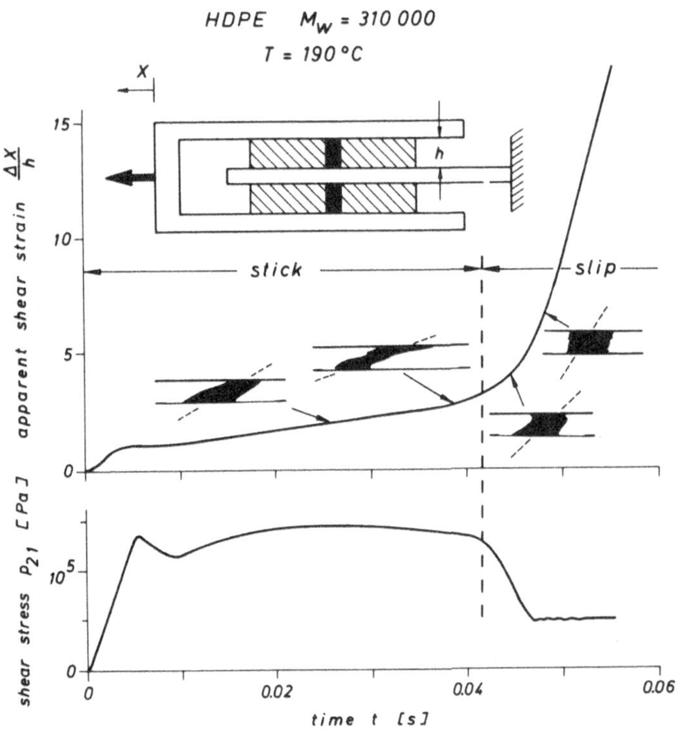

Fig. 7. Stick-slip transition of a HDPE melt in the sandwich rheometer

After approximately 10 ms a nearly constant shear stress acting at the melt and a nearly constant apparent shear rate is observed. The sample sticks at the metal plates and is deformed as expected. This is evident from the inclination of the flow profiles.

After a certain shear deformation, however, the stress transmitted to the central plate decreases suddenly and the apparent shear rate increases very rapidly. The flow profiles show that now slip occurs at the metal plates. During this period the shear deformation imposed in the stick phase recovers nearly totally such that the sample finally appears nearly undeformed! That means that the preceding shear deformation has been essentially elastic. It can be concluded that for the stick-slip transition not only a critical shear stress but also a certain amount of (elastic) shear deformation or orientation of the melt must be achieved.

4. Extrudate swell and elastic shear strain

Extrudate swell of a long die is unquestionably attributed to the elastic shear strains built up within the die [4]. We can measure extrudate swell B of long dies, calculate the corresponding true wall shear rate $\dot{\gamma}_w$ and compare extrudate swell with the recoverable shear strains measured in the sandwich at the same shear rate. This is a recommendable procedure to examine expressions interrelating extrudate swell and recoverable shear strain. The table gives a compilation of data obtained on Melt I at 150 °C.

Extrudate swell for dies with a length over radius ratio of $L/R = 60$ was measured on annealed samples. The influence of interfacial tension has been corrected for [9]. The values are only slightly higher than those of a die of $L/R = \infty$. At a wall shear rate of 144 s^{-1} where the measured recoverable strain was 8.1 the stress ratio only had a value of 2.7. Therefore, if the elastic strain is replaced by the stress ratio the magnitude of predicted extrudate swell would be significantly smaller.

Table

$\dot{\gamma}_w$ [s^{-1}]	B ($L/R = 60$)	$\gamma_{r,s}$	$\frac{[p_{11} - p_{22}]_s}{2[p_{21}]_s}$
0.107	1.28	1.3	0.76
1.19	1.50	3.2	1.5
13.0	1.67	5.5	2.2
144	1.73	8.1	2.7

References

1. Laun, H. M., J. Meißner, Rheol. Acta **20**, 211 (1981).
2. Meißner, J., Rheol. Acta **14**, 201 (1975).
3. Laun, H. M., Colloid & Polymer Sci. **259**, 97 (1981).
4. Lodge, A. S., Elastic Liquids (London-New York 1964).
5. Wagner, M. H., Rheol. Acta **15**, 136 (1976).
6. Laun, H. M., Rheol. Acta **17**, 1 (1978).
7. Laun, H. M., E. Fischer, to be published elsewere.
8. Uhland, E., Rheol. Acta **18**, 1 (1979).
9. Laun, H. M., H. Münstedt, Rheol. Acta **17**, 415 (1978).

(Received April 28, 1982)

Author's address:

Dr. H. M. Laun
Polymerphysik, Festkörperphysik – G 201
Kunststofflaboratorium
BASF Aktiengesellschaft
D-6700 Ludwigshafen/Rhein

Rheologica Acta Rheol. Acta 21, 470–474 (1982)

The time, temperature and shear dependence of the viscosity of polypropylene and its influence upon the extrusion process

E. Závadský, J. Karniš, and V. Pechoč

Research Institute of Man-Made Fibres, Svit (ČSSR)

Abstract: The polypropylene melt viscosity dependence upon temperature, shear and molecular weight, and the molecular weight dependence upon temperature and time were measured and mathematically described. For the dependence of the melt viscosity upon temperature, shear and molecular weight (melt index), a master curve has been found. The influence of the progressive degradation of polypropylene during the extrusion upon the temperature, pressure and dissipated energy down-channel profiles is given.

Key words: Polypropylene, viscosity, degradation, master curve, extrusion

1. Introduction

The basic condition for the design of polymer processing equipment as well as for the modeling of physical processes taking place in it is the knowledge of the viscosity of the polymer melt and of its dependence upon the shear, temperature and molecular weight, as well as the mathematical description of the functional relationship of these quantities.

This knowledge is the more important in processes where the polymer melt undergoes high shear and heat treatment and where degradation of the polymer occurs. In some cases, the degradation of polymer is intentional to obtain the desired characteristics of the final polymer melt.

The aim of this work has been the mathematical description of flow curves of a degradable type of polypropylene TE 430. The relationships obtained were used for the design of plasticating screw extruders and of their operating conditions assuring a polymer melt with properties required for the production of polypropylene fibres, where the polypropylene has to be degraded from the initial melt index of 5 g/10 min to final value of 30 to 50 g/10 min.

2. The measurement and evaluation of flow curves

The flow curves were measured on a Göttfert laboratory measuring extruder with diameter 20 mm and with a die 3 mm in diameter and 30 mm long in the temperature range 190 to 310 °C and shear rates 1 to 300 s^{-1}. For each measurement, the final melt index was determined on a Takara plastometer. The effective viscosity η was calculated as the ratio of the wall shear stress $\tau = \Delta p d/4L$ to the nominal shear rate D at the wall, calculated without any correction as $D = 8u/d$ with d and L as diameter and length of the die and u as the mean flow velocity. To get a master curve for all temperatures, the method of Münstedt [1] with a temperature shift factor a_T was used. The shift factor is defined as

$$a_T = \exp\left(\frac{E}{R}\,\frac{T_0 - T}{T_0 T}\right) \tag{1}$$

with E as the activation energy of flow, R the universal gas constant, T_0 the absolute reference temperature, T the absolute temperature of measurement, and the general form of the master curve is

$$\eta/a_T = f(a_T D)\,. \tag{2}$$

The measured data of the flow velocity u and pressure difference Δp converted to η and D and reduced to the master curve by the temperature shift factor a_T with reference temperature T_0 = 503 K are plotted in logarithmic coordinates in figure 1. It is

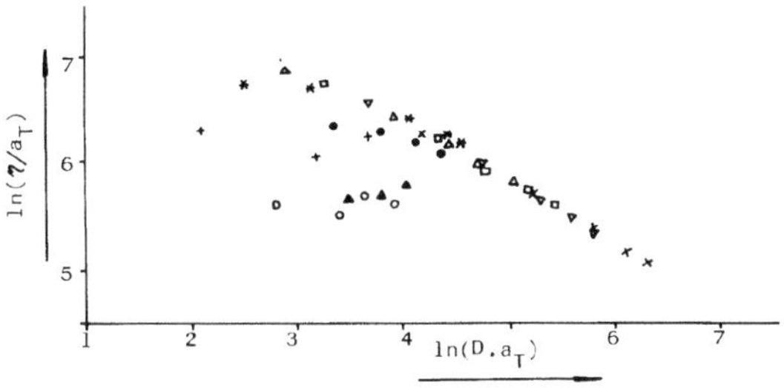

Fig. 1. Experimental data plotted as $\ln(\eta/a_T) = f \cdot [\ln(D \cdot a_T)]$; × 190, ▽ 210, □ 230, △ 250, * 270, ● 280, + 290, ▲ 300 and ○ 310 °C

evident that the data for temperatures lower than 280 °C lie really on a single curve, whereas the data for temperatures 280 °C and higher do not. This can be easily explained by the degradation of polypropylene at these higher temperatures. To get a master curve even with the degradation taken into account, a modified shift factor a_s was used, i. e. the ratio of the temperature shift factor a_T and the melt index MI, $a_s = a_T/MI$. The data reduced by this shift factor are plotted in figure 2, where all the data lie very nearly on a single curve.

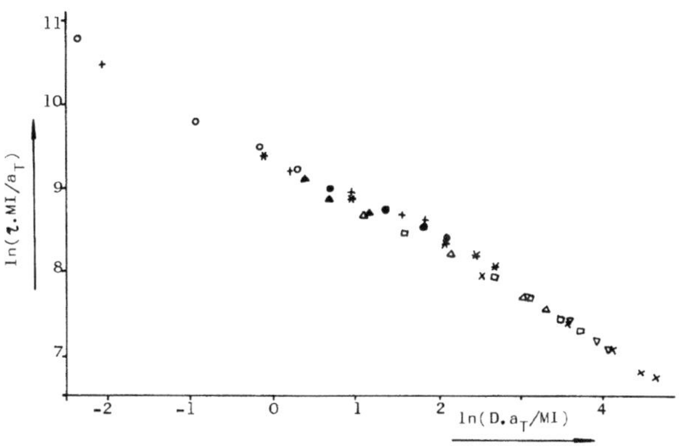

Fig. 2. Experimental data plotted as $\ln(\eta \cdot MI/a_T) = f \cdot [\ln(D \cdot a_T/MI)]$; × 190, ▽ 210, □ 230, △ 250, * 270, ● 280, + 290, ▲ 300 and ○ 310 °C

The master curve was defined generally by a power series

$$\ln \eta = \ln a_s + \sum_{i=1}^{5} a_i (\ln a_s D)^{i-1} \tag{3}$$

and the shift factor a_s as

$$a_s = (MI)^{a_8} \exp(a_6 + a_7/T) \,. \tag{4}$$

By the nonlinear regression with the least-square-error criterion [2], the following values of the coefficients a_1 to a_8 were determined:

$a_1 = 10.306, \quad a_2 = -0.52327, \quad a_3 = -0.019027,$

$a_4 = a_5 = 0,$

$a_6 = -10.718, \quad a_7 = 4263, \quad a_8 = -0.87968.$

The coefficient of determination was 0.998. Eq. (3) is valid in the range of temperatures $190 \leqslant t \leqslant 310\,°C$, melt indexes $5.1 \leqslant MI \leqslant 100$ g/10 min and shear rates $1 \leqslant D \leqslant 300\,s^{-1}$, with a maximum error 11% in η.

The melt index depends upon the temperature and the residence time in the melt. The residence time of the polymer melt in the experimental device was determined from the volumetric flow rate and the volume of the device from the die to the place of the supposed origine of the melt. The dependence of the melt index upon the temperature t and the residence time t_c was determined in the form

$$MI = 5.1 + t_c \cdot 1.0725^{(t-321.5)} + t_c^2 \cdot 1.2082^{(t-338.1)} \,. \tag{5}$$

This equation, determined with a coefficient of determination 0.997, is valid for temperatures 190 to 310 °C and residence times 0 to 420 s, with maximum error 20% at the melt index of 10 g/10 min. The relationship (5) is presented together with the experimental data in figures 3 and 4.

Eq. (3) is not very convenient for further calculations. The flow curves are mostly expressed in the form of the power law

$$\eta = m_0 \exp(-kt) D^{n-1} \tag{6}$$

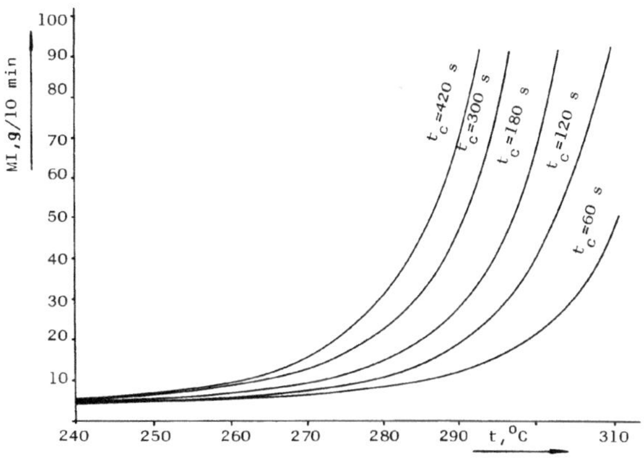

Fig. 3. The effect of temperature and residence time on melt index

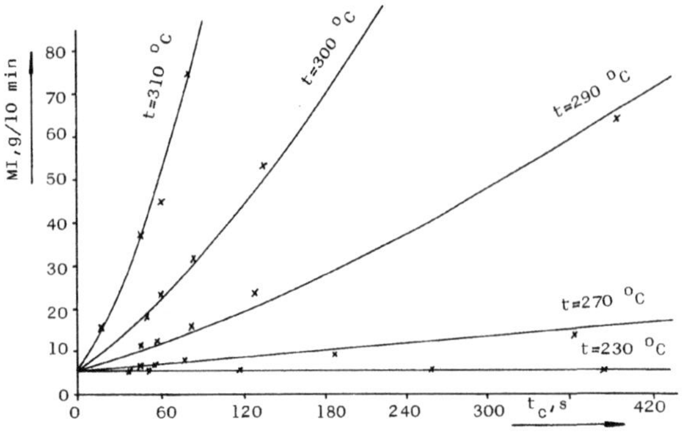

Fig. 4. The effect of temperature and residence time on melt index; × – experimental data

which is simple and very suitable for calculations of transport problems of polymer melts in pipes and extruders. The parameters k and n of the power law can be determined by the comparison of the partial derivatives of eqs. (6) and (3):

$$\frac{\partial \ln \eta}{\partial \ln D} \equiv n - 1 = a_2 + 2a_3 \ln(a_s D)\,, \tag{7}$$

from which

$$n = 1 + a_2 + 2a_3 \ln(a_s D)\,, \tag{8}$$

$$-\frac{\partial \ln \eta}{\partial t} \equiv k = (1 + a_2 + 2a_3 \ln(a_s D))\, a_7/T^2 \tag{9}$$

or

$$k = n a_7/T^2\,. \tag{10}$$

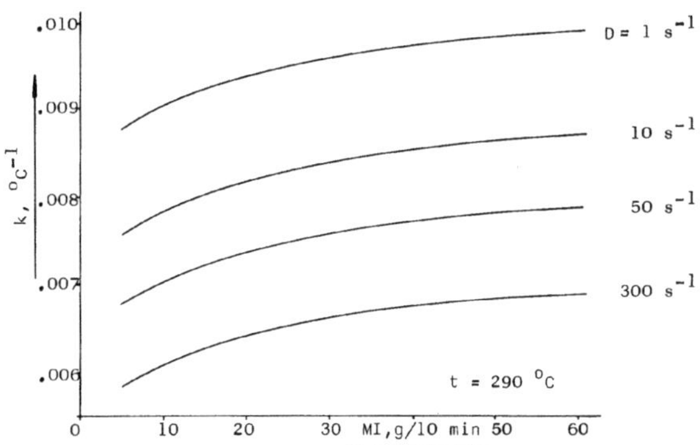

Fig. 5. The effect of melt index and shear rate on parameter k of power-law model

The flow consistency index m_0 can be calculated from eq. (6) written for m_0

$$m_0 = \eta \exp(kt) D^{1-n} \tag{11}$$

by substitution of η, n and k calculated from eqs. (3), (8) and (10).

It can be seen from eqs. (8), (10) and (11) that the parameters of the power law change due to the degradation of the polymer even for constant shear rate and temperature. The influence of the melt index upon the parameters of the power law and upon the effective viscosity is demonstrated in figures 5 to 8.

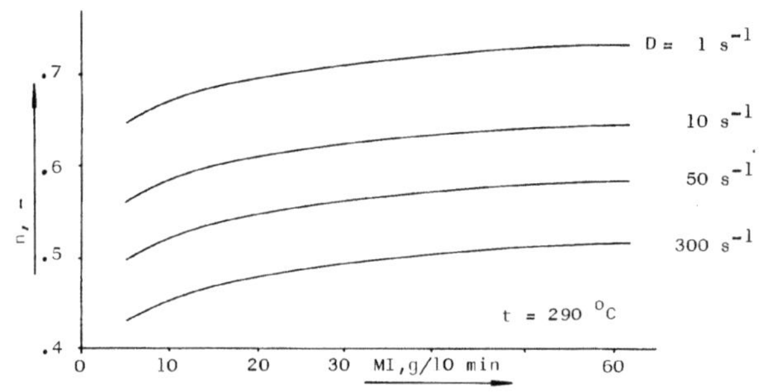

Fig. 6. The effect of melt index and shear rate on parameter n of power-law model

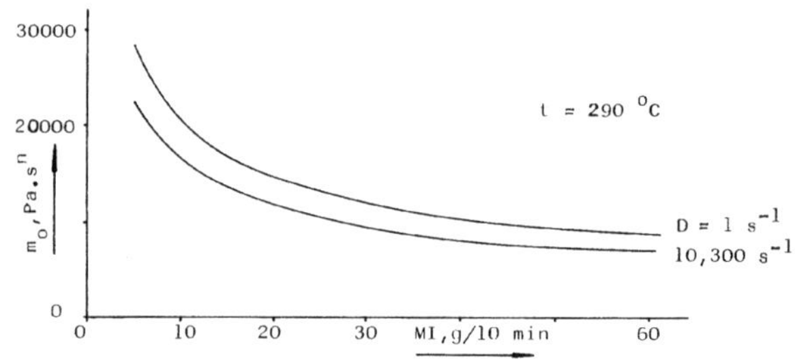

Fig. 7. The effect of melt index and shear rate on parameter m_0 of power-law model

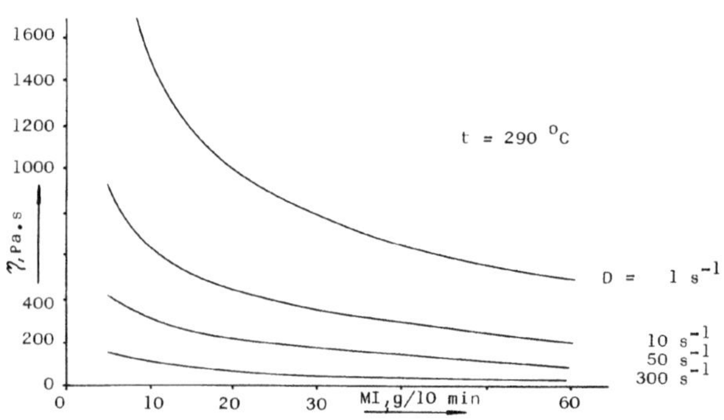

Fig. 8. The effect of melt index and shear rate on apparent viscosity

3. Degradation and the extrusion process

The degradation of the polymer influences the extrusion process and the properties of the melt produced. The influence of the polypropylene degradation upon the energy dissipation, temperature and pressure down channel profiles is presented in figures 9 to 16. The solid lines in these figures represent the profiles with eq. (5) – i. e., with progressive degradation – taken into account, the dashed and the dotted lines the results of calculations with the assumption of constant initial and final melt indexes, respectively. The parameters of the power law were calculated from eqs. (8), (10) and (11), the profiles predicted by

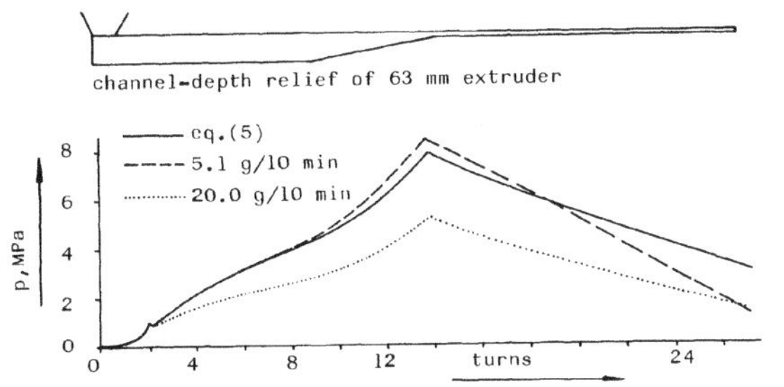

Fig. 9. Down-channel pressure profiles

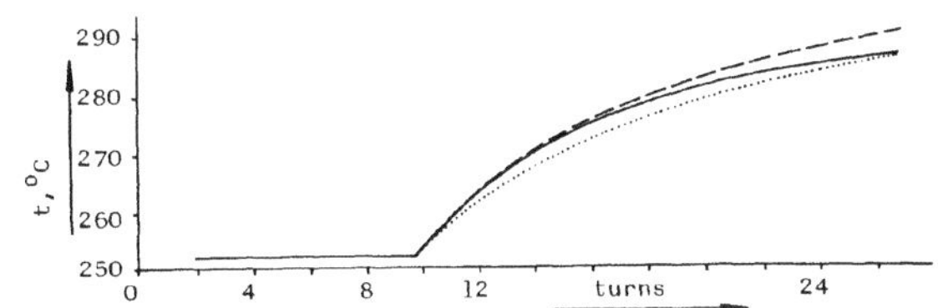

Fig. 10. Down-channel melt-temperature profiles

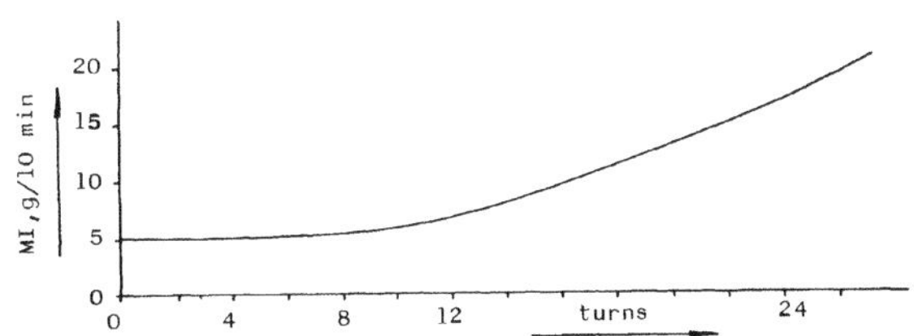

Fig. 11. Down-channel melt index profile

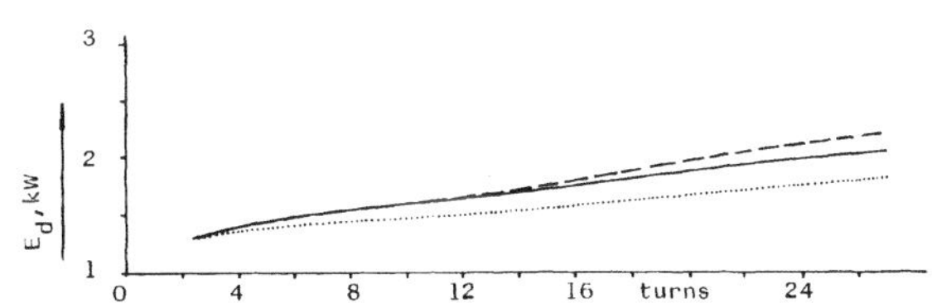

Fig. 12. Down-channel dissipated-energy profiles

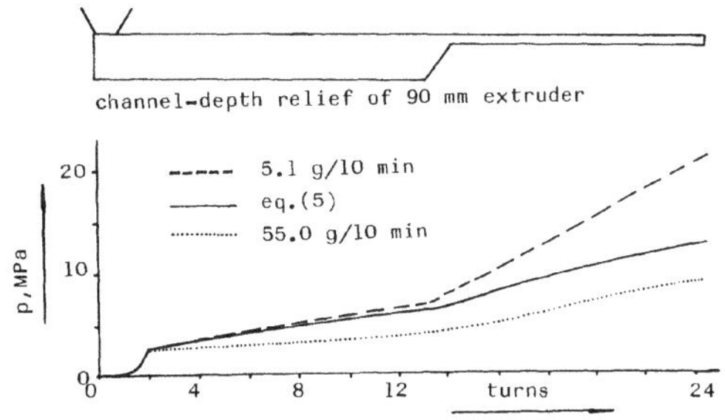

Fig. 13. Down-channel pressure profiles

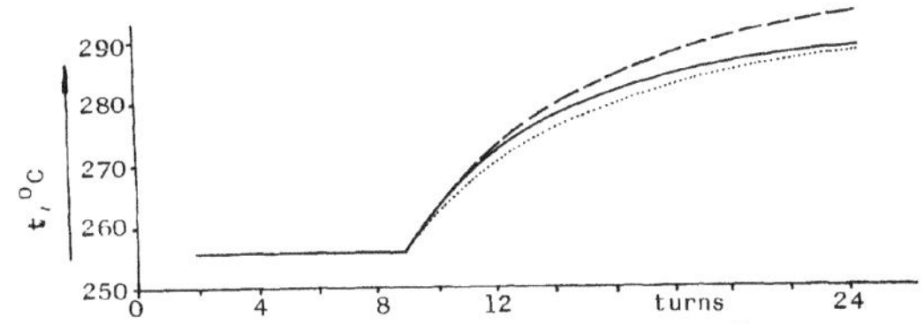

Fig. 14. Down-channel melt-temperature profiles

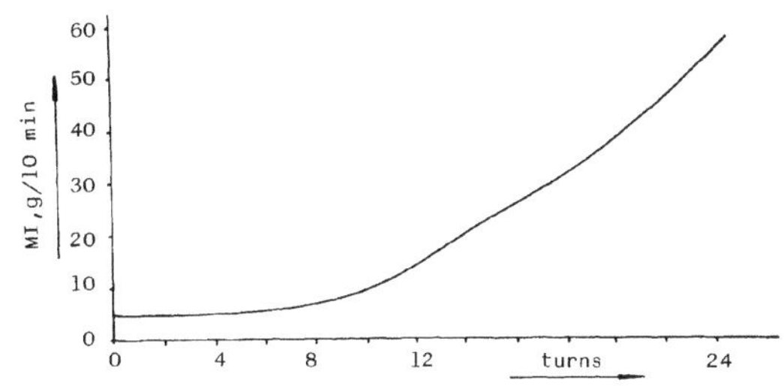

Fig. 15. Down-channel melt index profile

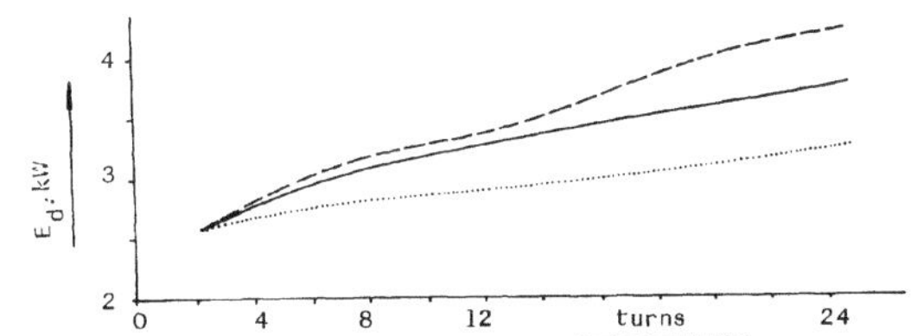

Fig. 16. Down-channel dissipated-energy profiles

an extruder program EXTR 2, developed for the simulation of extrusion process and for extruder design [3]. The geometry of extruders, operating conditions of extrusion and polymer properties are summarized in the appendix.

4. Discussion

From figures 1 and 2 it follows that it is possible to reduce the relationship between effective viscosity, temperature, shear rate and molecular weight to one master curve. This master curve can be defined either by eq. (3) or by the power law (6).

The coefficient a_7 in eq. (4) is in fact the ratio of the activation energy of flow and the gas constant, reported for polypropylene as 4500 – 5500 K in the literature [1, 4 – 7]. The value of 4263 K found from our experimental data agree satisfactorily with the reported values.

The influence of the shear rate and melt index upon the parameters k and n of the power law is distinct but diminishes with increasing values of both quantities. The influence of the shear rate upon the parameter m_0 is negligible but this parameter is strongly dependent upon the melt index (fig. 7).

To show the influence of the polymer degradation upon the extrusion process, two extruders with different down-channel pressure profiles have been chosen. Figs. 9 to 12 represent the melt index, temperature and energy dissipation profiles for an extruder 63 mm in diameter with a pronounced maximum in pressure profile, figures 13 to 16 the same profiles for a 90 mm extruder with progressively increasing pressure. The higher melt index in figure 15 in comparison with figure 11 has been caused by a 3 °C higher exit melt temperature and by a 7 min residence time against 3 min in 63 mm extruder.

The extrusion process calculated with the constant initial melt index 5.1 g/10 min shows the maximum dissipated energy (figs. 12 and 16) which causes the maximum exit temperature of the melt (figs. 10 and 14).

The differences between the values calculated for constant initial and final melt indexes, respectively, are 20 – 30% in dissipated energy, 1 – 3% in temperature and 60 – 120% in pressure. It is evident that the degradation of the polymer has to be taken into account to get correct computer predictions.

Appendix

The geometry of extruders, their operating conditions and the properties of polypropylene used for the computer predictions presented in figures 9 – 16.

Inner diameter of the barrel, mm	63.0	90.0
Screw lead, mm	63.0	90.0
Width of the flight parallel with the screw axis, mm	6.5	6.5
Flight clearance, mm	0.1	0.1
Inlet depth of the channel, mm	11.8	19.0
Outlet depth of the channel, mm	2.8	4.0
Mass flow of the polymer, kg/h	22.3	35.3
Revolutions of the screw, min^{-1}	40.0	31.2
Inlet polymer pressure, MPa	0.1	0.1
Inlet polymer temperature, °C	25.0	25.0
Barrel temperature, °C		
– for first two turns	70.0	70.0
– for further turns	285.0	285.0

Bulk density of the polymer
kg/m³, (Pa) $\rho_b = \rho - (\rho - 531)\exp(-1.7 \cdot 10^{-8} p)$
Polymer density, kg/m³ $\rho = 900$
Specific heat of the polymer, J/kg · K, (°C) $c_p = 1563.6 - 8.89\,t$
Specific heat of the melt, J/kg · K, (°C) $c_p = 2334 + 2.3\,t$
Thermal conductivity of the polymer, W/m · K 0.235
Thermal conductivity of the melt, W/m · K 0.216
Melt density, kg/m³, (°C) $\rho_m = 853.6 - 0.51\,t$
Melting point, °C 165
Heat of fusion, J/kg 80000
Heat transfer coefficient, W/m² · K
at the barrel 300
at the screw 50
Friction coefficient polymer-metal, (°C)

$$f = 0.0019 \cdot t^{1.7586} \exp(-0.0295\,t)$$

References

1. Münstedt, H., Kunststoffe **68**, 92 – 98 (1978).
2. Marquardt, D. W., J. Soc. Indust. Appl. Math. **11**, 431 – 441 (1963).
3. Závadský, E., J. Karniš, V. Pechoč, Technical Report No. 25/80, Research Institute of Man-Made Fibres, Svit 1980.
4. Van der Vegt, A. K., Plast. Inst. Trans. J. **32**, 165 – 169 (1964).
5. Ui, J., et al., SPE Trans. **4**, 295 – 305 (1964).
6. Pechoč, V., Research Report No. 260. Research Institute of Man-Made Fibres, Svit 1968.
7. Vinogradov, G. V., N. N. Prozorovskaja, Rheol. Acta **3**, 156 – 163 (1964).

(Received April 16, 1982)

Authors' address:

Dipl.-Ing. E. Závadský, Dipl.-Ing. J. Karniš,
Dipl.-Ing. V. Pechoč, C.Sc.
Research Institute for Man-Made Fibres
ČSSR-05921 Svit

Rheologica Acta Rheol. Acta **21**, 475 – 477 (1982)

A shear stress transducer for molten polymers

J. M. Dealy

Department of Chemical Engineering, McGill University, Montreal, Quebec (Canada)

Abstract: Present rheometrical techniques are inadequate for the measurement of viscoelastic properties associated with shearing at high rates. A possible solution to this problem is to use a sliding plate rheometer together with a device for measuring the local wall shear stress away from the ends and edges of the plates. Such a device has been constructed, and the results of preliminary tests are encouraging.

Key words: Wall shear stress, transducer, molten polymer

1. Introduction

Present methods for determining the shear stress exerted on a solid wall by a fluid flowing past it are based on the measurement of either the total force or torque on the wall or the pressure drop for flow through a channel. In both cases, the measurement is subject to severe restrictions, and this has limited our ability to study the nonlinear aspects of the viscoelasticity of polymeric liquids [1]. In particular, it has not been possible to make quantitative measurements of well-defined viscoelastic properties involving shearing at high rates. In addition it is not possible to make measurements of local wall shear rates in plastics processing equipment.

Measurements of viscoelastic properties, such as the complex viscosity and the relaxation modulus, are carried out using rotational rheometers with cone-plate, parallel-plate, or concentric-cylinder fixtures. Edge and end effects limit the use of these flow geometries to rather low shear rates, often below $1\ s^{-1}$.

Capillary and slit rheometers can be used at much higher shear rates, but well-defined viscoelastic properties cannot be determined, because the shear rate is nonuniform and its distribution depends on the fluid properties. Although it has been suggested that such quantities as entrance pressure drop, exit pressure and hole pressure error, which can be measured in capillaries and/or slits, can be used to infer rheological properties besides viscosity, the interpretation of such data is based on semi-empirical relationships, and the additional information available is limited to viscometric functions.

The limitations of traditional melt rheometry are of increasing importance in the light of recent reports of remarkable long-term strain history effects. The studies of this phenomenon reported to date [2, 3] have made use of devices such as extruders and torque rheometer mixing heads, in which the flow pattern is neither uniform nor well-controlled. A study making use of a uniform shearing deformation [4], on the other hand, was limited to very low shear rates.

To carry out tests involving uniform transient deformations, it appears that a rectilinear drag flow is required. In this way, the instabilities associated with rotational motions are avoided, and the nonuniform shear rate associated with pressure flows is eliminated. The sliding plate rheometer is a device for generating a uniform rectilinear shear flow. The shear stress at the wall is inferred from the total force required to move the sliding plate [5, 6], and any friction stemming from the structures supporting the plate will cause an error in the calculated shear stress. Because of this, sliding plate rheometers for use with melts are often built in the form of a sandwich, so that the melt itself is used to maintain the plate spacing [7 – 9]. Laun [10] has developed a very sophisticated sandwich creepmeter for use at high shear stress.

Another problem that arises in the use of parallel plate rheometers is the variable end effect that causes the effective shearing area to decrease as the experiment procedes. This effect has limited the use of parallel plate and sandwich rheometers to rather small total strains. However, the various problems associated with this type of device could be avoided entirely if it were possible to measure the local wall shear stress directly rather than inferring it from the total force.

Local wall shear stress measurements would also be useful in the study of flow in extruders, dies and molds. Revesz [11], has described a screw tip torque sensor for use in extruders, but it measures only the average shear stress on a surface over which the flow is quite nonuniform.

2. Direct measurement of wall shear stress

In order to measure the shear force on a small test section imbeded in a wall, the test section must be free of contact with the wall area around it. Otherwise the force on it would simply be transmitted directly to the surrounding wall rather than to the test section's force-sensitive mounting. This, however, makes it possible for the fluid to flow into the unavoidable gap around the force-sensing wall element. Whether or not this will pose an insurmountable obstacle to local shear stress measurement can best be determined by experience, and this paper reports the results of preliminary tests of a simple prototype transducer.

In designing the first prototype, a number of possible arrangements were considered. One approach is to use a section of quartz crystal cut in such a way that its piezoelectric output responds to shearing stresses [12]. However, the mounting of such a crystal is not straightforward. It must be insulated electrically from the wall and its outer surface protected from abrasion. In addition, the crystal cannot be held in place by compression.

Another approach is to mount the sensing element in such a way that there is negligible friction or elastic force impeding its lateral motion in the direction of flow. A sensitive displacement transducer senses any motion and the signal is used as the error signal for a linear servo-motor that supplies a restoring force sufficient to keep the plate from moving. The voltage required by the linear motor is a measure of the restoring force and thus of the shear stress exerted by the fluid on the sensing element. This method has been used successfully by Dickinson [13] in the design of a skin friction balance for use in aerodynamic studies.

The simplest approach is to make the sensitive element part of an elastically deformable structure and to measure its displacement under load. This technique has been used to measure shear stresses in soils [14 – 17]. Because it permitted the testing of the basic concept with a minimum of apparatus complexity, this was the approach taken in the present work.

3. A Prototype shear stress transducer

The active element of the shear stress transducer is a cantilever beam, 4 cm in length, with a flat plate, 1 cm square, at its free end. The beam is mounted inside a shell that makes the transducer self-contained. Strain gauges are bonded to the beam to permit the measurement of its deflection.

The disadvantage of this type of assembly is that a significant displacement of the end plate will occur at high values of stress. This will result in a slight tilting of the plate and any melt that flows into the gap around it will damp its response to transient stresses associated, for example, with stress growth and relaxation tests. We designed our prototype unit to have a gap of 0.5 mm around the end plate.

To test the transducer, a parallel-plate shearing apparatus was designed and constructed. The lower plate is mounted on linear bushings that move along rods and is driven by a linear actuator coupled to a servo-motor with a constant speed control system. The transducer is mounted in the upper, stationary plate. The spacing between the two plates was one mm in our preliminary experiments, but it can be easily changed. The fluid used was VISTANEX LM-MS polyisobutylene supplied by Exxon Chemicals. Measurements were made at 22.0 °C.

Measurement of the steady state shear stress at various constant speeds permitted the determination of the viscosity as a function of shear rate, and these data agreed well with results from a Rheometrics Mechanical Spectrometer. Measurements of the stress relaxation following cessation of steady shear, however, revealed a very poor frequency response of the transducer. The transducer was then modified by thinning the edges of the end plate to reduce the squeezing surface acting on the melt. Also, a thicker cantilever beam was used to reduce its deflection. The resulting stress relaxation data fell close to results obtained using a Mechanical Spectrometer.

4. Conclusions

It is felt that there is a strong possibility of developing a wall shear stress transducer for use with molten plastics. For improved precision, range and frequency response, a null-type mechanism involving a displacement transducer and a linear servo-motor is probably the best basis for such an instrument. Design for use at melt temperatures, however, will pose many challenging problems.

References

1. Dealy, J. M., Rheometers for Molten Plastics: A Practical Guide to Testing and Property Measurement, (an SPE book), Van Nostrand Reinhold Co. (New York 1982).
2. Maxwell, B., E. J. Dormier, F. P. Smith, P. P. Tong, Polym. Eng. Sci. **22**, 280 (1982).
3. Rokudai, M. et al., J. Appl. Polym. Sci. **23**, 463; **23**, 3289; **23**, 3295; **26**, 1343; **26**, 1427 (1979–81).
4. Dealy, J. M., W. Tsang, J. Appl. Polym. Sci. **26**, 1149 (1981).
5. Fruh, S. M., F. Rodriguez, AIChEJ **16**, 907 (1970).
6. Fujiyama, M., M. Takayanagi, Kogyo Kagaku Zasshi **72**, 1163 (1969).
7. Middleman, S., Trans. Soc. Reol. **13**, 123 (1969).
8. LaMantia, F. P., B. de Cindio, E. Sorta, D. Acierno, Rheol. Acta **18**, 369 (1979).
9. Laun, H. M., J. Meissner, Rheol. Acta **19**, 60 (1980).
10. Laun, H. M., Proc. 2nd World Congr. Chem. Eng. **6**, 411 (1981).
11. Revesz, H., Kunststoffe **64**, 35 (1974).
12. Kistler Instrumente AG, "Measure Cutting Forces?", Winterthur, Switzerland.
13. Dickinson, J., N. D. Vinh, Proc. 5th Canadian Congr. Appl. Mech., p. 515 (1975).
14. Agarwal, S. L., S. Venkatesan, Instruments and Apparatus for Soil and Rock Mechanics, ASTM STP 392, 152 (1965).
15. Broers, G., Proc. THTR Symp., p. 356 (1968).
16. Smid, J., J. Novosad, Power Technology **4**, 322 (1970/71).
17. Tcheng, Y., J. Iseux, Proc. 5th European Conf. Soil Mech. **1**, 207 (1972).

(Received June 25, 1982)

Author's address:

Prof. J. M. Dealy
Department of Chemical Engineering
McGill University
3480 University Street
Montreal, PQ, Canada H3A 2A7

Rheologica Acta Rheol. Acta **21**, 478 – 480 (1982)

Rheological measurements with technical polymers on the injection moulding machine

W. Knappe

Institut für Kunststoffverarbeitung, Montanuniversität Leoben (Österreich)

Key words: Injection moulding, melt flow index, moulding compound

In injection moulding, process variables like temperatures, pressures and the intervals of the cycle will be controlled to get mouldings of constant quality. But if the rheological properties of the moulding compound will fluctuate, the quality of the mouldings will also change and often no longer fulfil the quality requirements. Therefore a control of the rheological properties of the moulding compound during the processing cycle is of importance.

For this purpose we measured the energy for injection into the mould (injection energy W_E) and the plasticizing time during the backward movement of the screw. Figure 1 shows a sketch of the injection moulding machine (screw diameter 30 mm) and the equipment for our experiments.

Plasticizing time is measured, when the screw moves backward and the two switches (4) will start and stop the time counter UZ (7). The interval of measurement was limited to the middle range of the way of the screw, thus avoiding difficulties from start and stop. Measurements carried out over 30 cycles showed little scatter in the results (coefficients of variation ≈ 1%). However, it should be pointed out that the results are not only sensitive to the rheological properties of the melt but also to the tribological behavior of the granules in the feeding section of the screw, see our recently published results [1].

In most cases the speed of the screw during injection v_E will be held constant by control. Therefore the

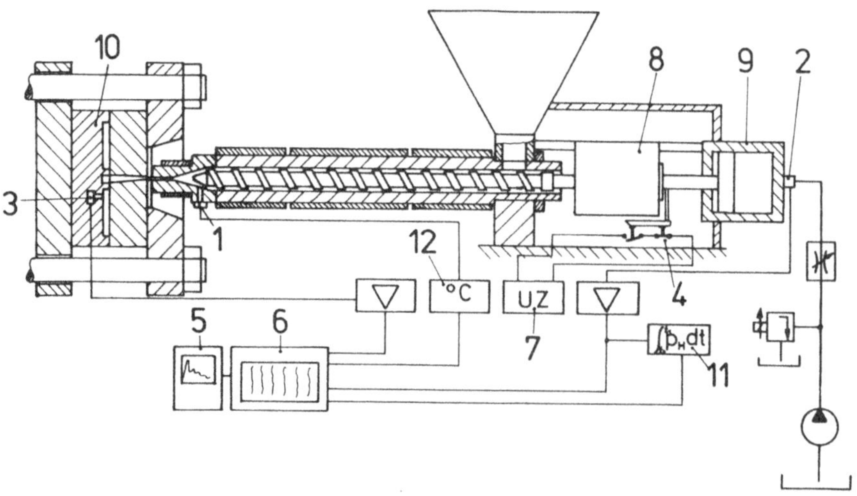

Fig. 1. Injection moulding machine with measuring equipment. 1 – temperature probe, 2 and 3 – pressure transducers, 4 – micro switches, 5 – oscilloscope, 6 – recorder, 7 – time counter, 8 – hydraulic drive for screw rotation, 9 – hydraulic cylinder for transversing the screw, 10 – tool, 11 – integration unit, 12 – amplifier for temperature measurements.

injection energy will be given by the following relation:

$$W_E = \int_{s_1}^{s_2} p_H dV = A_K \int_{t_1}^{t_2} p_H v_E dt = A_K v_E \int_{t_1}^{t_2} p_H dt . \quad (1)$$

p_H is the hydraulic pressure measured with a pressure transducer at (2), A_K is the area of the piston, which moves the screw forward (9). The integral $\int_{t_1}^{t_2} p_H dt$ can be easily evaluated by an electronic device (11). Further on the pressure in one cavity of the tool (10) was measured by a piezoelectric pressure transducer (3). The integration may be performed over the same interval over which the plasticizing time is measured.

Our first measurements were carried out with polypropylenes with different values of the melt flow index (MFI), see figure 2. The processing variables of the machine were held constant. The end of the injection time t_F will be given, when the pressure in the mould cavity p_i (dashed line) reaches 50 bar. Values of W_E were got by measuring the areas under the curves of p_H during t_F. In later experiments with the electronic integrator we used only the interval of 30 mm fixed by the switches, see figure 1. One sees, that with a higher MFI the injection energy decreases and the pressure p_i in the filled mould increases since the pressure drop in nozzle and gates becomes lower. By this t_F becomes something smaller with growing MFI.

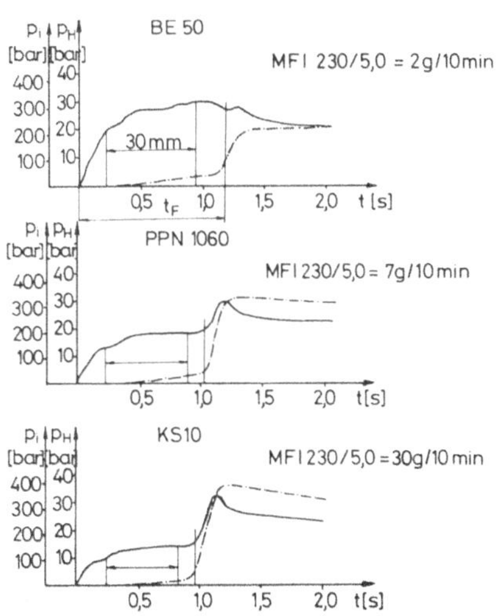

Fig. 2. Hydraulic pressure p_H (full line) and pressure in the cavity p_i (dashed line) under constant processing variables for PP with different MFI (230 °C/5 kg). BE 50 and KS 10 from Chemie-Linz, PPN 1060 from Hoechst

If we plot W_E over the reciprocal value of the MFI we get a linear correlation with some deviation for the lowest MFI. Extrapolation leads to the value W_{ER} which describes that part of energy which is consumed to overcome the friction between the cylinder and the moving screw. This is more than a third of the highest value of W_E.

It seems surprising that we get a rather good linear correlation between W_E and MFI^{-1} although the shear gradient in the mould is much greater ($>10^3\ s^{-1}$ in the gates) than in the case of MFI-measurement. With a slit rheometer and measurements in the range

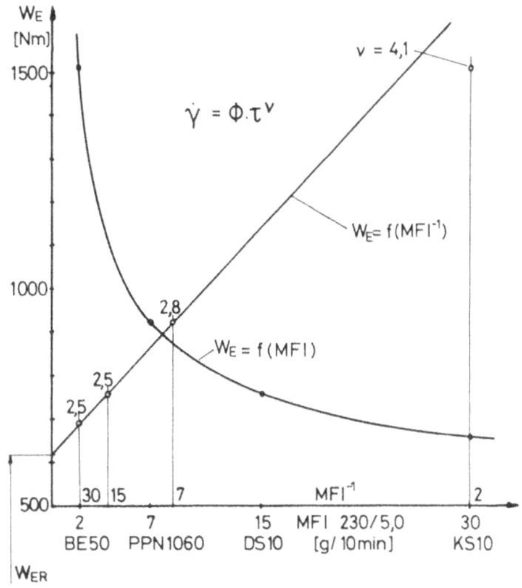

Fig. 3. Injection energy as function of MFI for PP (DS 10 from Chemie-Linz, see also figure 2)

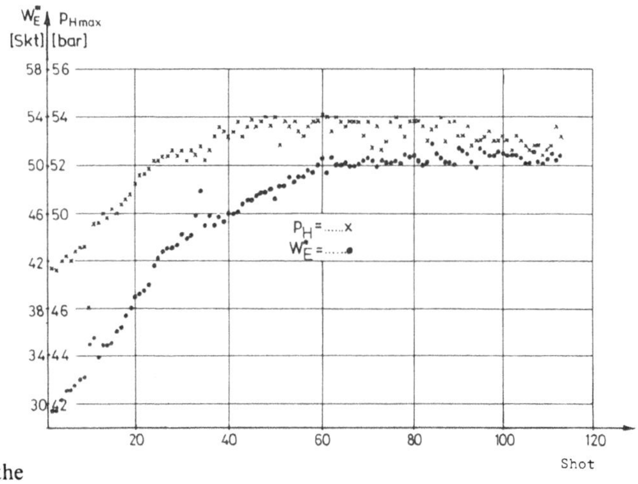

Fig. 4. Maximum hydraulic pressure $p_{H\,max}$ and injection energy W_E^*, electronically integrated, in arbitrary units, during production run

of 10^2 to 10^3 s^{-1} we found a considerable pseudoplasticity of our compounds, characterized by the exponents ν of the power law, see figure 3. In contrast to MFI, which is measured with constant shear stress τ, W_E is caused by a viscosity at constant shear gradient $\dot{\gamma}$. Therefore, the differences in viscosity under constant τ and $\dot{\gamma}$ for two different compounds should be given by the correlation [2]

$$\Delta(\mathrm{MFI}^{-1}) \sim \Delta\eta_\tau = \nu \cdot \Delta\eta_{\dot{\gamma}}, \quad (2)$$

if the power law will hold. Eq. (2) can qualitatively explain the deviation from linearity for $\nu = 4.1$.

In figure 4 injection energy W_E^* and the maximum hydraulic pressure were plotted versus the number of shots during the begin of a production run. Here W_E^* is measured by electronic means in the middle range (30 mm) of the injection stroke. It may be seen that the scatter in W_E is rather low (coefficient of variation $\approx$ 1%). The results indicate that thermal equilibrium was reached after 60 shots and that the cylinder without any throughput had a higher temperature than in equilibrium, thus causing a higher melt temperature and a lower viscosity at the beginning of the run.

References

1. Gissing, K., W. Knappe, Kunststoffe **72**, 78 – 82 (1982).
2. Semjonow, V. Rheol. Acta **3**, 98 – 100 (1963).

(Received April 30, 1982)

Author's address:

Prof. Dr. W. Knappe
Institut für Kunststoffverarbeitung
Montanuniversität Leoben
A-8700 Leoben

Rheologica Acta Rheol. Acta **21**, 481 – 483 (1982)

A computer model of plasticating extrusion

E. E. Agur and J. Vlachopoulos

Department of Chemical Engineering, McMaster University, Hamilton, Ontario (Canada)

Key words: Plastic, extrusion, mathematical model

1. Introduction

The development of mathematical models for single screw plasticating extruders has been covered extensively in the literature [1 – 7]. By far, the majority of these studies have concentrated on melt extruders or the melt conveying section of plasticating extruders. The main reason for this is that the analysis and modelling of melt flow is far simpler than that of solids conveying and melting of the polymer. Applications for melt fed extruders can be found in the polymer processing industry, for example, in the homogenization and pelletization of the polymer immediately after polymerization. Most extruders used in the plastics industry, however, are plasticating extruders, that is, they are fed by polymer in solid granular form. The solid polymer granules flow by gravity in the hopper and into the screw channel where they are compressed, conveyed down the screw channel, and then melted.

This paper gives a brief description of a fully predictive computer model of a single screw plasticating extruder developed at the Department of Chemical Engineering, McMaster University. More details can be found in a full version of the present paper [8] and in a thesis [9].

Given the material and rheological properties of the polymer, the screw geometry and dimensions, and the extruder operating conditions (screw speed and barrel temperature profile), the model is used to predict:

(i) flow rate of the polymer,
(ii) pressure and temperature profiles along the extruder and in the die,
(iii) extrudate swell at the die exit.

2. Description of the extruder model

The computer model consists of six interdependent sections, as shown in figure 1: the feed hopper, the solids conveying, melting and melt conveying zones, the die and the swelling of the extrudate at the die exit.

The base pressure in the *feed hopper* is determined by analysing the pressure distribution in a pellet bed

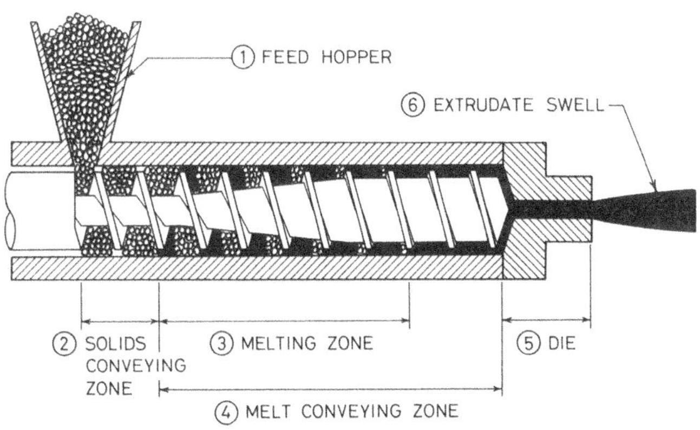

Fig. 1. Schematic diagram of a plasticating extruder showing the components of the present computer model

along Walker's [10] lines. In the *solids conveying zone* it is assumed that the solid bed is isothermal and travels as a solid plug with constant velocity. By applying a force and torque balance on a differential element of the solid bed in the down channel direction, the pressure in the solids conveying zone is expressed in the form [5]:

$$p = p_0 e^{-\lambda z_b}. \quad (1)$$

In the *melting zone* the solid and melt phases coexist. Tadmor's melting model [3, 4] is used as a basis for modelling this section of the extruder. Mass and energy balances on the melt film and solid bed are carried out to obtain the solid bed profile and the rate of melting [9]. In the *melt conveying zone* the equations of conservation of mass, momentum and energy are solved simultaneously by a finite difference method [9]. The rheological behavior of the melt is described in terms of a temperature and shear rate dependent constitutive relation in the form:

$$\tau_{yx} = \eta \frac{\partial v_x}{\partial y}, \quad \tau_{yz} = \eta \frac{\partial v_z}{\partial y}, \quad (2)$$

$$\eta = f\left(T, \frac{\partial v_x}{\partial y}, \frac{\partial v_z}{\partial y}\right). \quad (3)$$

A finite difference method is also used for the determination of pressure, velocity and temperature distributions in *the die*. The *extrudate swell* [11] is determined by using a two-constant model, based on Tanner's [12] ideas, in the form:

$$\frac{d}{D} = \left[\frac{A^2}{4} \frac{(4-b)}{(2+b)} \tau_w^{2b-2} + 1\right]^{1/6} + 0.12 \quad (4)$$

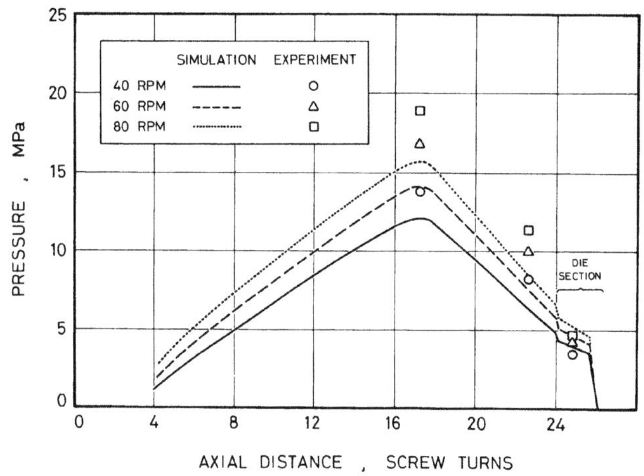

Fig. 2. Predicted pressure profiles in the axial (or down channel) direction for a low-density polyethylene resin in the extruder and die section. Data points represent pressure measurements in the extruder screw channel and adapter

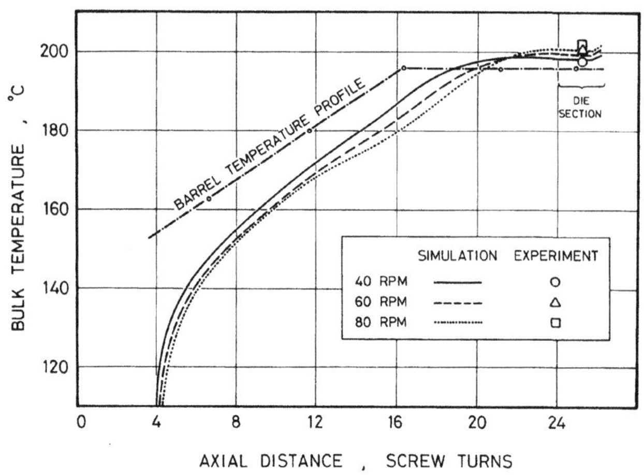

Fig. 3. Predicted bulk temperature profiles in the axial (or down channel) direction for a low-density polyethylene resin in the extruder and die section. Data points represent melt temperature measurements in the adapter

where A and b are obtained by curve-fitting Instron rheometer data [9].

3. Results of simulations and experiments

The present computer model was used to simulate a $1\frac{1}{2}$ inch (38 mm) diameter, 24:1 L/D single screw plasticating extruder. Various screw speeds and barrel temperature profiles were used for both high- and low-density polyethylene resins. The simulations were confirmed with experimental runs on the extruder.

A sampling of the simulation and experimental results are presented in tables 1 and 2 and figures 2 and 3 for a low-density polyethylene resin. The discrepancies are due in part to the uncertainty in some of the polymer properties.

Table 1. Predicted versus measured mass flow rates

Screw speed [rpm]	Mass flow rate [kg/s]	
	Predicted	Measured
40	$1.965 \cdot 10^{-3}$	$1.911 \cdot 10^{-3}$
60	$2.929 \cdot 10^{-3}$	$2.948 \cdot 10^{-3}$
80	$3.886 \cdot 10^{-3}$	$3.968 \cdot 10^{-3}$

Table 2. Predicted versus measured extrudate swell at the die exit

Screw speed [rpm]	Extrudate swell (d/D)	
	Predicted	Measured
40	1.692	1.646
60	1.721	1.674
80	1.742	1.713

References

1. Bernhardt, E. C. (ed.), "Processing of Thermoplastic Materials", Reinhold (New York 1959).
2. McKelvey, J. M., "Polymer Processing", Wiley (New York 1962).
3. Klein, I., D. I. Marshall (eds.), "Computer Programs for Plastics Engineers", Reinhold (New York 1968).
4. Tadmor, Z., I. Klein, "Engineering Principles of Plasticating Extrusion", Reinhold (New York 1970).
5. Tadmor, Z., C. G. Gogos, "Principles of Polymer Processing", Wiley-Interscience (New York 1979).
6. Fenner, R. T., "Extruder Screw Design", Iliffe (London 1970).
7. Fenner, R. T., Polymer **18**, 617 (1977).
8. Agur, E. E., J. Vlachopoulos, to appear in Polym. Eng. Sci.
9. Agur, E. E., Ph. D. Thesis, Dept. of Chemical Engineering, McMaster University, Hamilton, Canada (1982).
10. Walker, D. M., Chem. Eng. Sci. **21**, 975 (1966).
11. Vlachopoulos, J., Rev. Def. Beh. Mat. **3**, 219 (1981).
12. Tanner, R. I., J. Polym. Sci. **A2**, **8**, 2067 (1970).

(Received April 16, 1982)

Authors' addresses:

Dr. E. E. Agur
Xerox Research Centre of Canada
2660 Speakman Drive
Mississauga, Ontario L5K 2L1 (Canada)

Prof. J. Vlachopoulos
Department of Chemical Engineering
McMaster University
Hamilton, Ontario L8S 4L7 (Canada)

Rheologica Acta Rheol. Acta **21**, 484 – 487 (1982)

Stresses in polymer melts at the beginning of flow instabilities (melt fracture) in cylindrical capillaries

W. Gleißle

Institut für Mechanische Verfahrenstechnik und Mechanik, Universität Karlsruhe

Key words: Melt fracture, critical stress ratio, normal-stress measurement, normal-stress calculation, constitutive equation

The flow of fluid polymers is often accompanied by flow instabilities. This phenomenon can be observed at the outlet of capillaries and is a characteristic phenomenon of polymer melts [1, 2]. In plastic processing this effect is called melt fracture. Because of the low Reynolds numbers at which melt fracture starts, it is assumed that these instabilities are caused by the viscoelastic properties of those melts. Up to now these assumptions could not be proved, because in most cases the elastic properties of the polymers are unknown at shear rates κ where melt fracture begins.

To investigate these assumptions, the shear stress τ and the first normal-stress difference $\sigma_1 - \sigma_2$ were measured with a special cone and plate rheometer [3 – 5] up to $\kappa > 10^3\ s^{-1}$. The instabilities were investigated with a capillary rheometer (Rheograph 2000 Göttfert). The capillaries had various lengths, diameters and inlet cones. In the absence of proved measuring methods the beginning of the flow instabilities was defined by visual criteria alone. The criteria were the observation of rough surfaces, small deformations and unsteady flow of the extrudate.

For a great number of the fluids the instabilities began at shear rates higher than $10^3\ s^{-1}$. In these cases the normal stress differences were calculated directly from the viscosity functions using the "mirror relations" [5 – 7]. The first mirror relation combines the transient zero-shear viscosity $\eta_0(t)$ with the viscosity function $\eta(\kappa)$:

$$\eta(\kappa) = \eta_0(t) \quad \text{if} \quad \kappa t = 1\,, \tag{1}$$

and the second one the transient zero-normal-stress coefficient $\theta_0(t)$ with the shear rate dependent coefficient $\theta(\kappa)$:

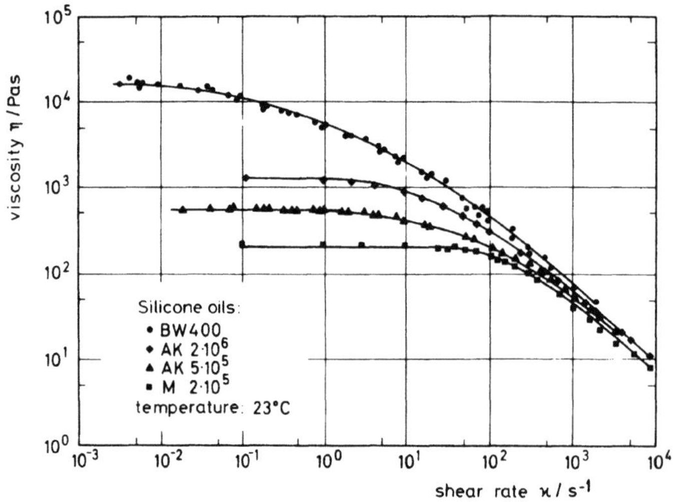

Fig. 1. Viscosity functions for various silicone oils

$$\theta(\kappa) = \theta_0(t) \quad \text{if} \quad \kappa t = k\,; \quad 2 \leq k \leq 3 \qquad (2)$$

where

$$\theta(\kappa) = [\sigma_1(\kappa) - \sigma_2(\kappa)]/\kappa^2 . \qquad (3)$$

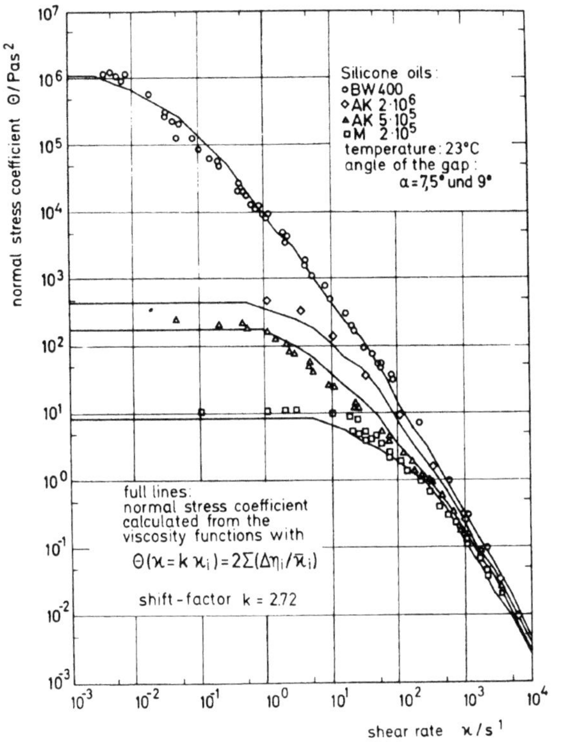

Fig. 2. First normal stress coefficient measured (points) and calculated (curves) for different silicone oils

With the help of eqs. (1) and (2) and the equation of the linear theory of rubber-like liquids [9] one can derive a simple constitutive equation combining viscosity η and normal-stress coefficient θ, as follows:

$$\theta(\kappa^* = k\kappa) = -2\int \frac{d\eta}{\kappa}, \qquad (4)$$

$$\theta(\kappa^* = k\kappa_i) = -2\sum_i (\Delta\eta_i/\bar{\kappa}_i). \qquad (5)$$

Using eq. (5) it is possible to calculate the normal stresses from the viscosity function measured with a capillary rheometer. $\Delta\eta_i$ is the viscosity difference between two shear rates and $\bar{\kappa}_i$ is the mean value of the shear rate in this interval. Fig. 1 shows the viscosity functions $\eta(\kappa)$ for various silicone oils.

In figure 2 the normal-stress coefficients measured and calculated with eq. (5) from viscosity data, are ploted against the shear rate for the same silicone oils. Measurement and calculation agree well. The same good agreement is obtained for polymer melts. An example, namely LDPE-melt (Lupolen 1800S), is given in figure 3. As is shown by figures 1 – 3, the calculation of normal stress data from viscosity measurements for higher shear rates appears to be reasonable.

The double logarithmic plot of figure 4 shows the shear stress function $\tau(\kappa)$ and the normal stress functions $[\sigma_1(\kappa) - \sigma_2(\kappa)]$ for polyisobutylene at different temperatures. The beginning of instabilities is marked with MF. As other investigators observed [1, 2] melt fracture begins at the same critical shear stress τ_{crit} independent of the temperature. From the normal-stress functions it can be seen that the critical

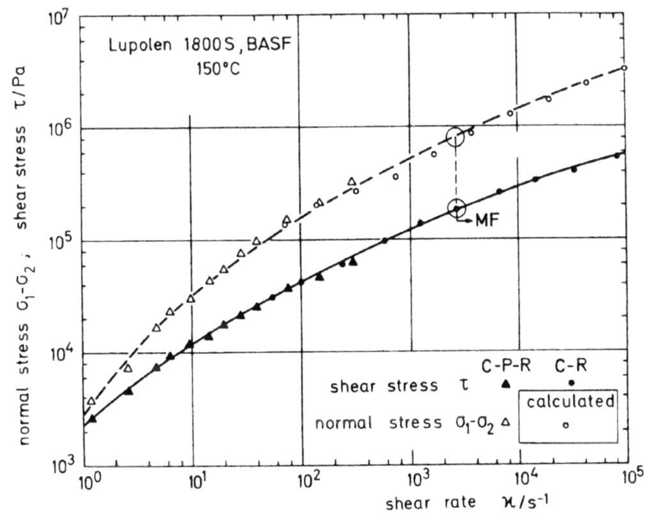

Fig. 3. Shear stress τ and first normal stress coefficient measured and calculated of a low density polyethylene melt (CPR: cone and plate – CR: capillary rheometer)

normal stress differences $(\sigma_1 - \sigma_2)_{crit}$ are also independent of the temperature, and therefore the ratio $(\sigma_1 - \sigma_2)_{crit}/\tau_{crit}$ is constant.

Table 1 shows the critical values for the beginning of flow instabilities of very different polymer fluids.

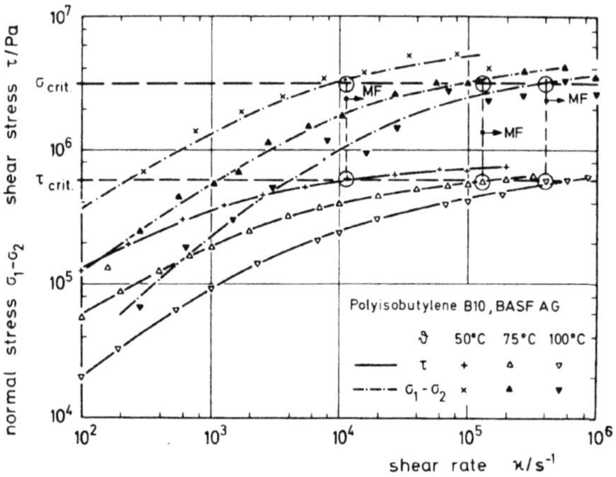

Fig. 4. Shear stress τ and first normal-stress difference $\sigma_1 - \sigma_2$ of polyisobutylene at different temperatures. MF: melt fracture

The critical values in this table show that neither the viscous data η and τ nor the elastic data θ and $\sigma_1 - \sigma_2$ are criteria for flow instabilities, because these critical data vary in the range of many decades. But comparing the ratio $R = (\sigma_1 - \sigma_2)/\tau$ (ratio of the critical stresses) one finds that R is nearly constant and its value is independent of the fluid, approximately 4.5 (The mean value R of this arbitrary selection of polymers is 4.63 and the standard deviation $S = \pm 0.26$). This result proves the assumption that the beginning of melt fracture is correlated with a critical viscoelastic stress ratio. In addition, the experiments demonstrate that the critical stress Ratio R is independent of the nature of the fluid, a result which corresponds to perturbation calculations [10] predicting flow instabilities for $R > 4$.

In figure 5 the critical shear stresses τ_{crit} are plotted against the critical normal stresses $(\sigma_1 - \sigma_2)_{crit}$. The curve which describes the measuring points to a very good approximation is a straight line through the origin, an obvious demonstration that the critical stress ratio is constant. The Mohr-circles in this plot (boundary condition: $\sigma_2 = 0$) show that the critical state of stress is nearly a pure tensile stress in the direction of the flow lines.

Table 1. Critical values for the beginning of flow instabilities for different polymers

Material	Temperature	Shear rate		Viscosity		Shear stress		Normal-stress coefficient		Normal-stress		Stress ratio
	$\vartheta/°C$	κ/s^{-1}	exp	η/Pa s	exp	τ/Pa	exp	θ/Pa s^2	exp	$\sigma_1 - \sigma_2$/ Pa	exp	$R = (\sigma_1 - \sigma_2)/\tau$
Silicone oils												
BW 400	23	6.25	2	1.21	2	7.59	4	9.22	−1	3.60	5	4.74
AK $2 \cdot 10^6$	23	4.08	3	2.02	1	8.25	4	2.15	−2	3.60	5	4.36
AK $5 \cdot 10^5$	23	1.50	4	6.23	0	9.34	4	2.00	−3	4.49	5	4.79
M $2 \cdot 10^5$	23	1.65	4	5.25	0	8.76	4	1.47	−3	4.00	5	4.57
LDPE1800S	150	2.45	3	6.94	1	1.70	5	1.25	−1	7.50	5	4.41
PMMA	200	1.24	3	3.50	2	4.33	5	1.30	0	2.00	6	4.62
PE	200	2.46	3	1.46	2	3.58	5	2.94	−1	1.78	6	4.97
SB	200	4.92	2	3.27	2	1.61	5	3.01	0	7.29	5	4.53
ABS 90	200	1.10	3	1.92	2	2.11	5	8.26	−1	1.00	6	4.74
PS 143 E	200	7.50	3	3.00	1	2.25	5	1.75	−2	9.84	5	4.38
HDPE5261Z	190	1.50	2	2.07	3	3.10	5	6.44	1	1.45	6	4.68
PIB10	75	1.44	5	4.10	0	5.90	5	1.11	−4	3.10	6	5.24
PA N12	220	2.64	4	1.75	1	4.62	5	2.90	−3	2.02	6	4.37
Epoxydharz	100	1.33	4	5.91	1	7.86	5	1.95	−2	3.45	6	4.39
												$\bar{R}$ = 4.63 S = ±0.26

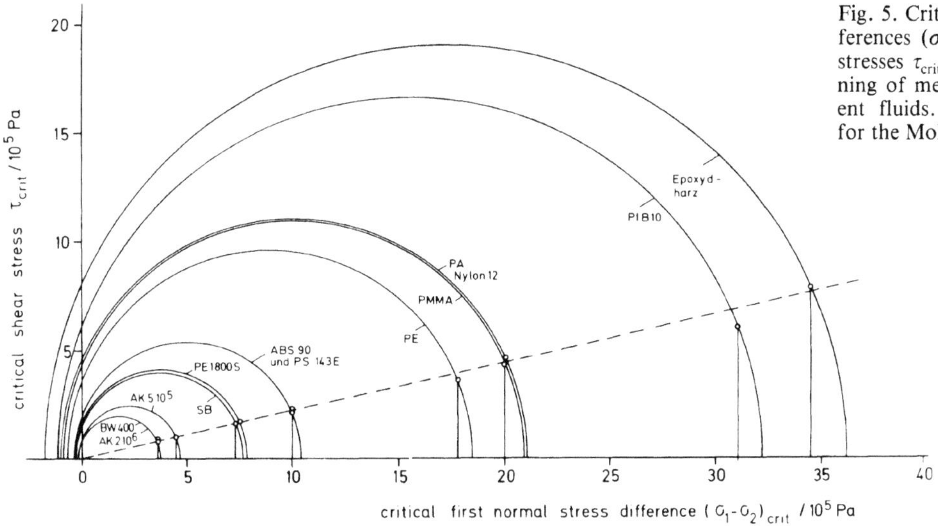

Fig. 5. Critical normal stress differences $(\sigma_1 - \sigma_2)_{crit}$ and shear stresses τ_{crit} indicating the beginning of melt fracture for different fluids. Boundary condition for the Mohr-circles $\sigma_2 = 0$

References

1. Ramsteiner, F., Kunststoffe **61**, 943 (1971).
2. Uhland, E., Diss. Universität Stuttgart 1978.
3. Gleißle, W., H. Reichert, Rheol. Acta **12**, 77–81 (1973).
4. Gleißle, W., H. Reichert, Rheol. Acta **12**, 572–577 (1973).
5. Gleißle, W., Diss. Universität Karlsruhe 1978.
6. Gleißle, W., Rheology (Proc. of the VIII Int. Congr. Rheology, Naples 1980), Vol. II, Plenum Press (New York 1980).
7. Gleißle, W., Jahrestreffen der Deutschen Rheologischen Gesellschaft, Berlin 1981. AIChE-Annual Meeting, New Orleans (1981).
8. Gleißle, W., Preprints CHISA '81, Prag 1981.
9. Lodge, A. S., Elastic Liquids, Academic Press (New York 1964).
10. Akbay, U., St. Sponagel, First Conf. European Rheologists, Graz 1982 (unpublished).

(Received July 9, 1982)

Author's address:

Dr. W. Gleißle
Institut für Mechanische Verfahrenstechnik und Mechanik
Universität Karlsruhe (TH)
Postfach 6380
D-7500 Karlsruhe 1

Rheologica Acta Rheol. Acta **21**, 488 – 490 (1982)

The effect of "shear thinning" on the interfacial stability of two stratified liquids

N. D. Waters

Department of Applied Mathematics and Theoretical Physics, University of Liverpool

Key words: Stability, plane Couette flow, stratified liquid, power-law liquid

1. Introduction

In polymer processing the combination of two melt streams (co-extrusion through a die) has become a very economical method of producing materials with unique properties. One of the main problems that arises in this process is that the interface between the two liquids can become irregular at high enough rates of flow leading to what is known as "melt fracture".

Using hydrodynamic stability analysis Yih [1] has found that for simple flows, even for low Reynolds numbers, viscosity stratification alone is sufficient to cause instability. Li [2] and Khan and Han [3] have shown that the presence of elasticity can not only destabilize simple flows but stabilize them for certain values of the parameters involved.

None of these analyses take into account the readily observed characteristic exhibited by many liquids, including polymer melts, often called shear thinning, which is usually a decrease in apparent viscosity with shear rate.

In this paper we investigate the interfacial stability of two "power-law" liquids flowing together between two parallel planes when one plane is stationary and the other moves with constant speed U in a fixed direction in its own plane.

2. The primary flow

We use rectangular cartesian coordinates (x_1, x_2, x_3) so that the interface between the liquids is $x_2 = 0$, the stationary plane is $x_2 = -d_{II}$ and the moving plane is $x_2 = d_I$ ($d_I > 0$). We assume that the velocity distribution in both layers has the form $(\bar{u}_j(x_2), 0, 0)$ which satisfies continuity immediately; $j = \text{I}$ for the upper layer and $j = \text{II}$ for the lower layer.

The liquids are characterized by equations of state of the form

$$\tau_{ik} = -\bar{p} g_{ik}^{(1)} + \bar{p}_{ik}, \quad \bar{p}_{ik} = 2\mu(S)\bar{e}_{ik}^{(1)} \tag{1}$$

where τ_{ik} is the stress tensor, $\bar{p}_{ik}$ is the extra stress tensor, $\bar{e}_{ik}^{(1)}$ is the rate of strain tensor, $\bar{p}$ is the pressure and the apparent viscosity $\mu(S)$ is a function of the second rate-of-strain invariant $S = 2e_{ik}^{(1)}e_{ki}^{(1)}$ given by

$$\mu(S) = \kappa(S)^{(n-1)/2} \tag{2}$$

where n is called the power-law index and κ is the consistency constant. We shall append the suffix denoting which layer when necessary and drop the overbars when we non-dimensionalize using a typical speed U, typical length d_I and the density ρ_I as follows:

$$x = x_1/d_I, \quad y = x_2/d_I, \quad u = \bar{u}/U,$$

$$p = \bar{p}/U^2\rho_I, \quad p_{ik} = \bar{p}_{ik}/U^2\rho_I . \tag{3}$$

The only non-vanishing stress component is

$$p_{12} = h\left(\frac{du}{dy}\right)^n \quad \text{where } h_j = \left(\frac{U}{d_I}\right)^{n_j - n_I} \frac{\kappa_j}{\kappa_I} . \tag{4}$$

With no pressure gradient the equations of motion for both layers reduce to

$$\frac{d^2u}{dy^2} = 0 \tag{5}$$

subject to the no-slip conditions $u_I(1) = 1$, $u_{II}(-\delta) = 0$, where $\delta = d_{II}/d_I$ and the continuity of velocity and stress conditions

$$u_I = u_{II} \tag{6}$$

and

$$\left(\frac{du_I}{dy}\right)^{n_I} = h_{II}\left(\frac{du_{II}}{dy}\right)^{n_{II}} \quad \text{at} \quad y = 0\,. \tag{7}$$

The solution is

$$u = ay + b \tag{8}$$

where

$$b(= b_I = b_{II}) = a_{II}\delta = 1 - a_I \tag{9}$$

and

$$h_{II}(a_{II})^{n_{II}} = (a_I)^{n_I}\,. \tag{10}$$

a_{II} or a_I can be readily obtained by solving, for example

$$h_{II}(a_{II})^{n_{II}} = (1 - a_{II}\delta)^{n_I} \tag{11}$$

using a Newton-Raphson technique.

3. Stability analysis

We follow Yih [1] and use the result due to Squire [4] that it is sufficient to consider two-dimensional disturbances only; we assume a perturbed velocity distribution $(u + u^*, v^*)$ where u^* and v^* can be expressed in terms of the stream function ψ by

$$u^* = \frac{\partial\psi}{\partial y}, \quad v^* = -\frac{\partial\psi}{\partial x}\,. \tag{12}$$

As is customary in stability analysis, we assume that all the non-dimensional perturbation variables are small and contain an exponential time factor (real part understood) of the form

$$\psi = \phi(y)\,e^{i\alpha(x-ct)} \tag{13}$$

where $t = U\bar{t}/d_I$ is a non-dimensional time, α is a real positive number proportional to the reciprocal of wavelength and c is complex.

On linearizing in the perturbation amplitudes and making use of the equations for the primary flow the equations of motion reduce to an Orr-Sommerfeld type equation for each layer. The boundary and interfacial conditions being essentially the same as those given by Yih [1].

Following Yih we seek solutions of the form

$$\phi = \phi_0 + \alpha\phi_1 + \alpha^2\phi_2 +$$
$$c = c_0 + \alpha c_1 + \alpha^2 c_2 + \,. \tag{14}$$

After working to zero order in α and obtaining c_0 and ϕ_0, c_1 can be obtained by working to first order. Its form is similar to that obtained by Yih [1] and can be written (Waters [5])

$$\mathrm{Re}\, c_1 = 0, \quad \mathrm{Im}\, c_1 = \frac{R}{n_I(a_I)^{n_I - 1}} J(\zeta, N, \delta, \gamma) \tag{15}$$

where $N = n_{II}/n_I$, $\zeta = h^{1/n_I} = \left(\dfrac{U}{d_I}\right)^{N-1}\left(\dfrac{\kappa_{II}}{\kappa_I}\right)^{1/n_I}$, $\gamma = \rho_{II}/\rho_I$ and the Reynolds number $R = \dfrac{U^2\rho_I}{\kappa_I}\left(\dfrac{d_I}{U}\right)^{n_I}$.

4. Discussion

Following Yih [1] we only consider here the case $\gamma = 1$ although when $\gamma \neq 1$ the effect of viscosity stratification on the stability of the interface can be

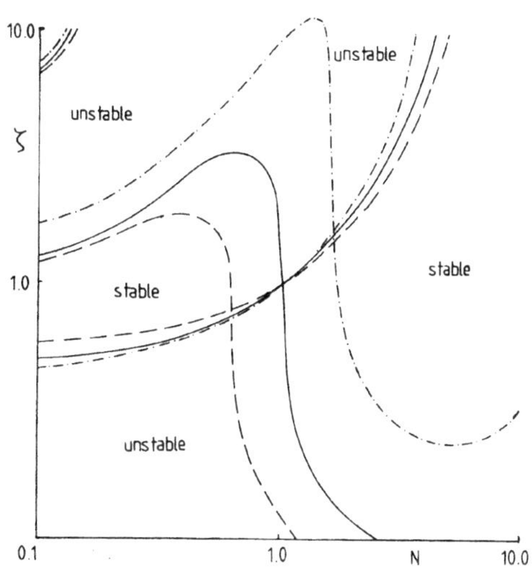

Fig. 1. The $J(\zeta, N) = 0$ curves, showing the regions of interfacial stability in the $\zeta - N$ plane, for $\delta = 1.25$ — · — · —; $\delta = 1.00$ ————; $\delta = 0.80$ – – – –

completely overshadowed by density stratification. We also find that instability can occur for all Reynolds numbers. Figure 1 shows the curves $J(\zeta, N) = 0$ with $\gamma = 1$ and $\delta = 1.25$, 1.00 and 0.80, a log-log scale is used so that the effect of reciprocating the ratios ζ and N can be easily observed. These curves separate the regions of stability and instability of the interface ($J < 0$ for stability). For each value of δ shown $J(\zeta, N) = 0$ on two main curves, one of these is a concave upwards increasing curve which passes through the point (1,1), the slope of which increases with δ, and the other an oscillatory curve which moves from left to right with increasing δ. For smaller values of δ there are only two regions separated by a curve through (1,1); the upper region being unstable. As δ increases the oscillatory curve appears and moves from left to right changing the stable regions to unstable and vice versa, before disappearing for large δ.

We conclude that shear thinning can have a dramatic effect on the stability of the interface and the parameters N and ζ can be chosen to stabilize or destabilize the flow.

References

1. Yih, C.-S., J. Fluid Mech. **27**, 337 (1967).
2. Li, C.-H., Phys. Fluids **12**, 531 (1969).
3. Khan, A. A., C. D. Han, Trans. Soc. Rheol. **20**, 595 (1976); **21**, 101 (1977).
4. Squire, H. B., Proc. Roy. Soc. **A 142**, 621 (1933).
5. Waters, N. D., J. Non-Newtonian Fluid Mech. (to be published).

(Received April 16, 1982)

Author's address:

Dr. N. D. Waters
Department of Applied Mathematics
and Theoretical Physics
The University of Liverpool
P. O. Box 147
Liverpool L69 3BX (U.K.)

Zur analytischen Beschreibung des Füllstoffeinflusses auf das Fließverhalten von Kunststoff-Schmelzen

B. Hinkelmann

Deutsches Kunststoff-Institut, Darmstadt

Abstract: From literature some representative equations have been compiled describing the influence of filler on the viscosity of polymer melts. By application of these on the experimental results obtained from GF-SAN it was found that the relative viscosity η_R, i.e. the ratio of the viscosities of the filled and unfilled melt, shows a pronounced dependence on the shear rate $\dot{\gamma}$ but not on the shear stress τ. Defining η_R with constant τ and not with constant $\dot{\gamma}$ (as it is usually done), an analytical approach is possible independent of $\dot{\gamma}$. Further the influence of pressure, temperature and filler content on the zero-shear viscosity of filled polymer melts may be expressed by a modified Arrhenius equation.

Key words: Polymer melt, glass fiber, relative viscosity, analytical approach

Das Eigenschaftsbild von Kunststoffen läßt sich durch Beigabe von Zusätzen in weiten Grenzen variieren. Dabei werden aber nicht nur die Eigenschaften der Endprodukte modifiziert, sondern oft wird auch die Verarbeitbarkeit erheblich beeinflußt. Durch den Zusatz von Füllstoffen wird sowohl das viskose als auch das elastische Verhalten der Polymerschmelzen geändert [1].

Eine analytische Beschreibung des rheologischen Verhaltens ist in der Literatur z. Zt. vorrangig auf das viskose Stoffverhalten beschränkt und wird häufig als Abhängigkeit der relativen Viskosität η_R vom Füllstoffgehalt dargestellt. η_R wird dabei als Quotient der Viskositäten des gefüllten Systems und der Matrix bei gleicher Schergeschwindigkeit definiert. Einige typische Beispiele aus einer Vielzahl von Gleichungen, deren Anwendbarkeit auf eigene Meßergebnisse an GF-SAN [2, 3] überprüft wurde, sind in Tabelle 1 zusammengestellt [4 – 8]. Die bekannte Formel nach Einstein [4] wurde darin mit aufgenommen, weil eine Vielzahl von Gleichungen auf diese zurückgeführt werden kann. Wegen des linearen Zusammenhangs der Füllgradabhängigkeit von η_R erscheint jedoch eine Anwendung dieser Beziehung nicht sinnvoll.

Aus Abbildung 1 ist zu ersehen, daß nur die Gleichung nach Eilers [7] in der Lage ist, die Meßergebnisse im Füllgradbereich > 13 Vol.-% zu beschreiben.

Tab. 1. Typische Gleichungen aus der Literatur [4 – 8] zur Beschreibung des Füllstoffeinflusses auf die relative Viskosität η_R

Autor	Gleichung	Füllstoff
Einstein [4]	$\eta_R = 1 + k \cdot \Phi$	Kugeln
Mooney [5]	$\eta_R = \exp \frac{k \cdot \Phi}{1 - s \cdot \Phi}$	Kugeln
Brodnyan [6]	$\eta_R = \exp \frac{2{,}5\,\Phi + 0{,}39\,(rp-1)^{1{,}48} \cdot \Phi}{1 - k\,\Phi}$	Fasern
Eilers [7]	$\eta_R = \left(1 + \frac{k \cdot \Phi}{1 - s \cdot \Phi}\right)^2$	Kugeln
Hashin [8]	$\eta_R = 1 + \frac{2 \cdot \Phi}{1 - \Phi}$	Glasfasern

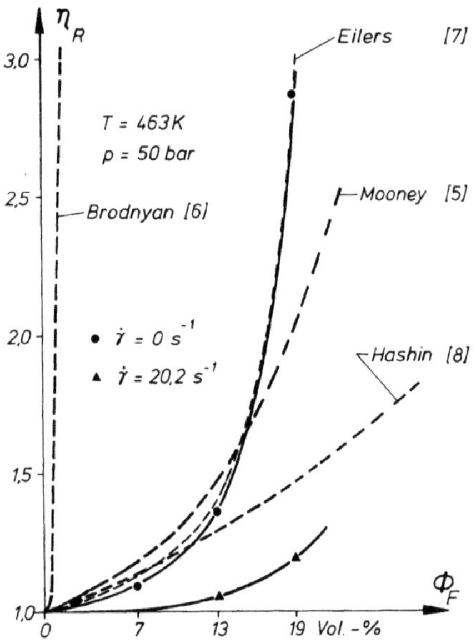

Abb. 1. Relative Viskosität η_R als Funktion des Faservolumenanteils ϕ_F nach [5–8], verglichen mit Meßwerten an GF-SAN

Dies gilt aber nur für die Darstellung der relativen Nullviskositäten und nicht für Viskositäten bei höheren Schergeschwindigkeiten. Bei den übrigen Beispielen ist eine Anpassung an die Meßwerte noch weniger möglich. Eine einfache analytische Beschreibung des Füllgradeinflusses auf die relative Viskosität wird vor allem dadurch erschwert, daß eine starke Schergeschwindigkeits-Abhängigkeit dieser Größe vorliegt [3].

Andererseits ist bekannt, daß sich die Viskositätsfunktion ähnlich wie für Temperatur und Druck [9] auch invariant gegenüber dem Füllstoffgehalt [2] darstellen läßt, wobei dies einer Verschiebung der Viskositätskurven entlang einer Linie konstanter Schubspannung entspricht. Diese Erkenntnis legt den Versuch nahe, die relative Viskosität ebenfalls bei konstanter Schubspannung und nicht, wie üblich, bei konstanter Schergeschwindigkeit zu ermitteln. In [10] wurde, jedoch offenbar auf anderen Überlegungen basierend, ebenfalls diese Möglichkeit der η_R-Definition vorgeschlagen. Aus Abbildung 2 geht hervor, hier am Beispiel von Meßergebnissen an einem 35 Gew.-% GF-SAN, daß eine derart bestimmte relative Viskosität vom Betrag der anliegenden Schubspannung nahezu unabhängig ist.

Eine Anwendung dieser η_R-Ermittlung ist in Abbildung 3 für ein GF-SAN dargestellt. Hierbei entsprechen die Meßpunkte annähernd dem nach einer modifizierten Gleichung von Maron und Pierce [11] errechneten Kurvenverlauf. Die Größe A, ein Maß für das L/D-Verhältnis der Fasern, wurde entsprechend den Angaben in [10] aus einer linearen Approximation der Form $A = 0{,}54 - 0{,}0125\ \bar{p}$ errechnet. Für $A = 0{,}45$ (entsprechend einem L/D von 7) läßt sich die gerechnete Kurve den Meßpunkten sehr gut anpassen. Die Kurve mit $A = 0{,}37$ entspricht dagegen der an GF-SAN bestimmten wirklichen mittleren Faserlänge. Diese Abweichung deutet darauf hin, daß eventuell auch Fragen der Verteilungsart berücksichtigt werden müssen, um eine verbesserte analytische Beschreibung der Meßwerte zu erreichen.

Während nun eine analytische Beschreibung der Füllgradabhängigkeit der relativen Viskosität im Bereich des strukturviskosen Fließverhaltens offenbar noch mit Schwierigkeiten verbunden ist, besteht für den linearen Bereich eine Möglichkeit der eindeutigen Charakterisierung, denn die Füllgradabhängigkeit der Nullviskosität läßt sich in halblogarithmischer Auf-

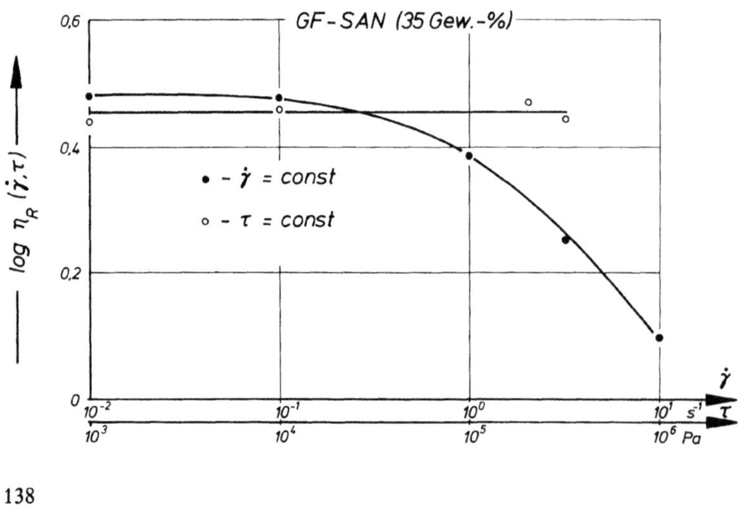

Abb. 2. Relative Viskosität η_R als Funktion der Schergeschwindigkeit $\dot\gamma$ bzw. der Schubspannung τ

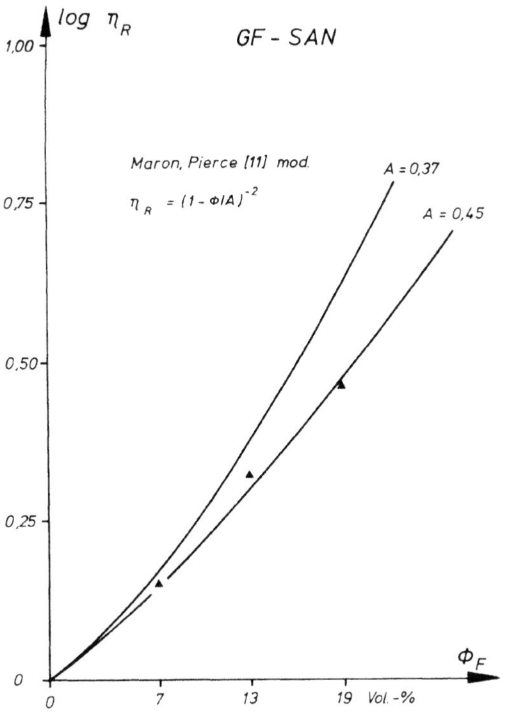

Abb. 3. Relative Viskosität η_R als Funktion des Faservolumenanteils ϕ_F nach Maron und Pierce [11]

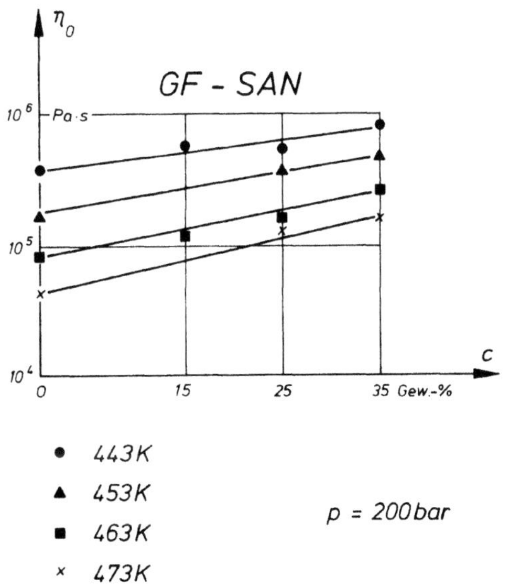

Abb. 4. Nullviskosität η_0 als Funktion der Faserkonzentration c für GF-SAN

tragung als Gerade darstellen (Abb. 4). Dies bedeutet, daß aus dem Zusammenhang

$$\eta_0(\phi_1) = \eta_0(\phi_2)\exp[\gamma_0(\phi_1 - \phi_2)] \tag{1}$$

ein Füllgradkoeffizient

$$\gamma_0 = \frac{\ln[\eta_0(\phi_1)/\eta_0(\phi_2)]}{\phi_1 - \phi_2} \tag{2}$$

formuliert werden kann. Die Abhängigkeit der Nullviskosität von Druck, Temperatur und Füllstoffgehalt läßt sich damit in verallgemeinerter Form in einer erweiterten Arrheniusbeziehung folgendermaßen beschreiben:

$$\eta_0(T,p,\phi) = A\exp(\beta_0/T + \alpha_0 p + \gamma_0\phi)\,. \tag{3}$$

Die jeweiligen Temperatur-, Druck- und Füllgradkoeffizienten können mit den zugehörigen Absolutgliedern für das untersuchte GF-SAN errechnet werden. Mit diesen Werten ist es möglich, Nullviskositäten für beliebige Temperaturen, Drücke und Füllstoffgehalte entsprechend Gl. (3) zu ermitteln.

Danksagung

Der Arbeitsgemeinschaft Industrieller Forschungsvereinigungen e. V. wird für die Förderung der Untersuchungen gedankt.

Literatur

1. Vinogradov, G. V., A. Ya. Malkin, Rheology of Polymers, Mir Publishers Moscow, Springer-Verlag (Berlin, Heidelberg, New York 1980).
2. Mennig, G., B. Hinkelmann, Angew. Makromol. Chem. **74**, 249 (1978).
3. Hinkelmann, B., Rheol. Acta **20**, 561 (1981).
4. Einstein, A., Ann. Phys. **19**, 289 (1906).
5. Mooney, M., J. Colloid Interface Sci. **6**, 162 (1951).
6. Brodnyan, J., Trans. Soc. Rheol. **3**, 61 (1959).
7. Eilers, H., Kolloid-Z. **102**, 154 (1943).
8. Hashin, Z., ref. in Maschmeyer, R. O., C. T. Hill, Adv. Chem. Ser. **134**, 95 (1974).
9. Semjonow, V., Adv. Polym. Sci. **5**, 387 (1967).
10. Kitano, T., T. Kataoka, T. Shirota, Rheol. Acta **20**, 207 (1981).
11. Maron, S. H., P. E. Pierce, J. Colloid Sci. **11**, 80 (1956).

(Eingegangen am 15. April 1982)

Anschrift des Verfassers:

Dipl.-Ing. B. Hinkelmann
Deutsches Kunststoff-Institut
Schloßgartenstraße 6 R
D-6100 Darmstadt

Rheologica Acta Rheol. Acta **21**, 494 – 498 (1982)

Viskositätsmessung mit dem longitudinal schwingenden Band

T. Gast und K. U. Kramm

Institut für Meß- und Regelungstechnik der Technischen Universität Berlin

Abstract: A method for the detection of small masses with the aid of a longitudinally vibrating band is described and further applications, e.g. for viscosity and humidity measurement are discussed. The complex Young's modulus of the band can also be determined.

Key words: Longitudinal vibration, mass detection, viscosity measurement, complex modulus measurement

Am Institut für Meß- und Regelungstechnik der Technischen Universität Berlin sind Arbeiten über Meßwerterfassung mit schwingenden Saiten und Bändern durchgeführt worden, z. B. über das transversal schwingende Band zur Präzisionsmessung von Längen [1], Massen [2] und Kräften sowie zur Bestimmung der Konzentration disperser Stoffe in Flüssigkeiten [3, 4] und Gasen, z. B. der Staubkonzentration am Arbeitsplatz [5] und in Abgasen.

Wegen der zu erwartenden geringeren systematischen Meßfehler wurde auch die Massenbestimmung mit einem longitudinal schwingenden Band bearbeitet [6].

Erwartungsgemäß hat sich gezeigt, daß sich die Eigenfrequenz eines longitudinal schwingenden Bandes ändert, wenn dessen träge Masse variiert. Das schwingende Band dient dabei als Meßfühler für die Staubkonzentration im Rauchgaskanal von Kraftwerken. Hierbei stellte sich die Frage, welchen Einfluß viskose Medien auf das Schwingverhalten des Bandes ausüben [7].

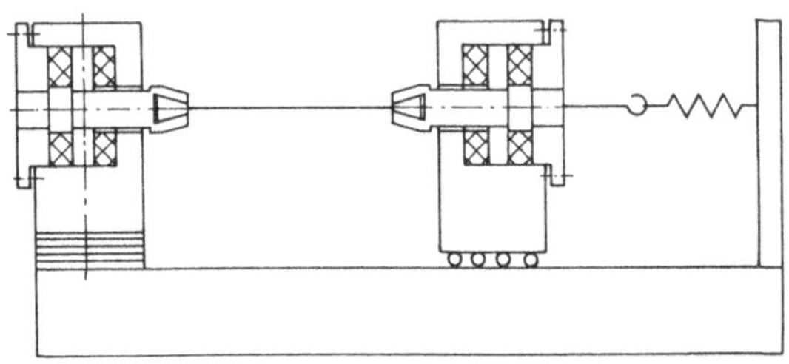

Abb. 1. Schema der Versuchsanordnung zur *E*-Modul-Bestimmung

Abbildung 1 zeigt schematisch die benutzte Versuchsanordnung. Als Sender und Empfänger dienen piezoelektrische Schwinger. Sie stehen sich im Abstand von etwa 10 bis 20 cm gegenüber. Das Band wird von zwei Spannzangen gehalten, die jeweils mit einer scheibenförmigen Verdickung zwischen Piezo-Dickenschwingern eingespannt sind. Die Schwingungen des Senders übertragen sich über Band und zweite Zange auf den Empfänger und erzeugen dort ein Signal. Die direkte Übertragung von Schwingungen über die Halterung wird durch einen eingeschalteten Dämpfer unterdrückt.

Die Frequenz des Senders wird derart gesteuert, daß eine Phasenverschiebung von 90° zwischen der angeregten und empfangenen Wechselspannung besteht und somit eine stehende Welle mit Knoten in den Einspannungen entsteht. Das Verhalten des longitudinal schwingenden Bandes läßt sich mit Hilfe des Schwerpunktsatzes berechnen.

Bei der Herleitung des gesetzmäßigen Zusammenhanges zwischen Frequenzänderung und Elastizitätsmodul wurden folgende Annahmen gemacht:

- Es gilt das Hookesche Gesetz.
- Die äußere Reibung kann zunächst unberücksichtigt bleiben.
- Einschnürungen des Bandes sind vernachlässigbar.

Aus den Dehnungen

$$\varepsilon(x) = \frac{\partial u}{\partial x}, \tag{1}$$

wobei die Verschiebung des Bandelements und x die Längskoordinate bedeuten, ergibt sich die Längs- oder

Normalkraft. Es wird eine innere Dämpfung berücksichtigt. Die Trägheitskraft wird mit $\rho q dx\, \partial^2 u/\partial x^2$ in Ansatz gebracht. Dabei bedeuten $\rho q dx = dm$ die Masse, q den Querschnitt und ρ die Dichte des Bandes.

Die Bewegungsgleichung für ein schwingendes Kontinuum ergibt sich zu

$$\frac{\partial^2 u}{\partial t^2} = \frac{E}{\rho}\,\frac{\partial^2 u}{\partial x^2} - r\frac{\partial u}{\partial t}. \tag{2}$$

Die Dämpfung wird durch äußere Energiezufuhr aufgehoben. Die Gleichung für die Eigenfrequenz lautet:

$$f = \frac{1}{2l}\sqrt{\frac{E}{\rho}}. \tag{3}$$

Daraus ergibt sich der Elastizitätsmodul E zu

$$E = f^2 \cdot \rho \cdot 4\hat{l}^2 \tag{4}$$

mit $\hat{l}$ als der Länge des Bandes zwischen den Einspannungen. Somit ist es möglich, bei gemessenen Frequenzen und bekannten Längen den Elastizitätsmodul zu bestimmen. Unter Berücksichtigung der Dämpfung kann der E-Modul komplex angesetzt werden. Er steht zur Resonanzfrequenz $\hat{f}_{1/2}$ in der Beziehung

$$\hat{f}_{1/2} = -\frac{r}{2m} \pm \sqrt{\frac{r}{2m} - \frac{\hat{E}x^2}{\rho l^2}}. \tag{5}$$

Abbildung 1 zeigt die Anordnung zur Bestimmung des E-Moduls für verschiedene Materialien in Form dünner Folien.

Abbildung 2 stellt die Eigenfrequenz von Makrofol E, 0,4 mm stark, aufgetragen über die Temperatur dar. Hierzu wird das Band eingespannt, von 0 °C stufenweise bis auf 90 °C aufgeheizt und bei dieser Temperatur die Längenmessung wiederholt. Anschließend wird auf 0 °C abgekühlt und der Wiederholwert der Eigenfrequenz bestimmt.

Abbildung 3 schließlich zeigt den errechneten E-Modul als Funktion der Temperatur. Die thermische Längenausdehnung ist berücksichtigt. Es wurde auch der E-Modul von Papier als Funktion der Temperatur gemessen. Hier machte sich der Feuchteeinfluß durch ein Minimum knapp unterhalb der Raumtemperatur

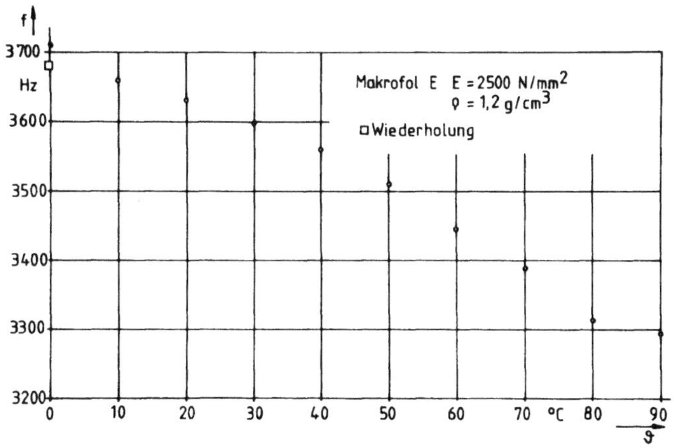

Abb. 2. Eigenfrequenz eines Polycarbonatbandes als Funktion der Temperatur

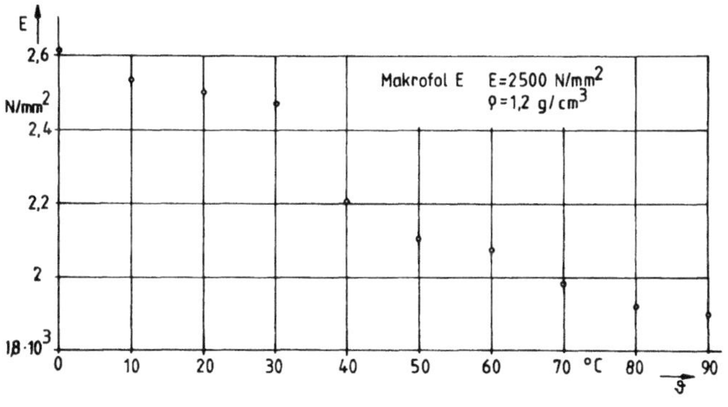

Abb. 3. Realteil des komplexen E-Moduls für ein Polycarbonatband in Abhängigkeit von der Temperatur

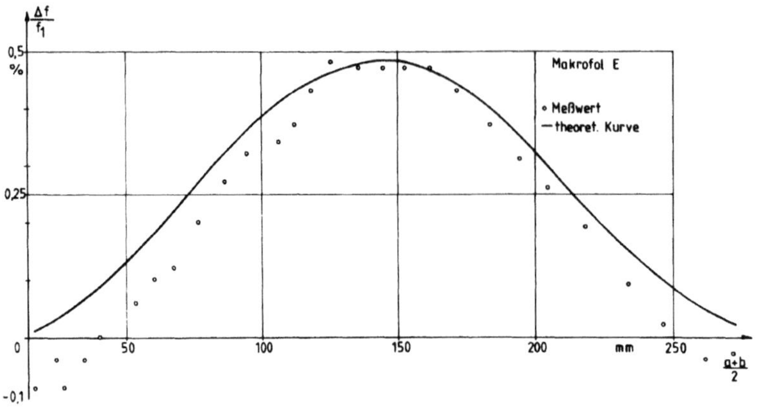

Abb. 4. Einfluß des Ortes einer zusätzlichen Masse auf die Eigenfrequenz eines Bandes

bemerkbar. Durch das aufgenommene Wasser werden gleichzeitig Masse und E-Modul verändert.

Bei der Herleitung der Bewegungsgleichung für eine mitschwingende zusätzliche Masse werden folgende Voraussetzungen gemacht: Für den E-Modul der aufgebrachten Masse gilt $E = 0$, d. h. es handelt sich um disperse Stoffe, die weder eine elastische Rückstellkraft noch eine Dämpfung verursachen. Es ändert sich lediglich die Trägheitskraft. Bei der Massenbestimmung mit dem longitudinal schwingenden Band wird nur der mittlere Bandabschnitt genutzt, wo sich die größte Empfindlichkeit ergibt. Zur Berechnung wird hierbei das Band in drei Abschnitte unterteilt: Band ohne Massenbelegung, Band mit Linien-Massenbelegung und Band ohne Massenbelegung.

Die ausgezogene Kurve in Abbildung 4 wurde für eine längs des Bandes verschobene Linienmasse berechnet, die Meßpunkte sind dagegen durch Versuche gewonnen worden. Hierzu wurde transparentes Zellulose-Klebeband (Tesafilm der Fa. Beiersdorf) an die verschiedenen Stellen eines etwa 30 cm langen Polycarbonat-Bandes aufgeklebt. Die Abweichungen an den Bandrändern lassen sich durch die Steifigkeit des aufgeklebten Materials erklären, die dort besonders wirksam ist, wo die Dehnung maximale Werte annimmt.

Bei mittig aufgebrachter Masse (auch hier Tesafilm) ergeben sich die in Abbildung 5 dargestellten Meßwerte. Der Vergleich mit der berechneten Kurve ergibt eine sehr gute Übereinstimmung. Für ein Band aus Glasfaservlies ergibt sich eine Kurvensteigung, die einer absoluten Empfindlichkeit von 35 Hz/mg entspricht.

Mit zunehmender aufgeklebter Masse nimmt die Amplitude ab, die mitschwingende Materie nimmt diese Energie auf. Die beobachteten Amplitudenänderungen lassen darauf schließen, daß mit der beschriebenen Anordnung auch die Dämpfungen meßbar sind.

Das zweite Stokessche Problem betrifft die Strömung eines viskosen Mediums in der Nähe einer oszillierenden ebenen Wand. Die Geschwindigkeitsverteilung in Bandnähe $v = v(y, t)$ ergibt sich als Lösung einer partiellen Differentialgleichung

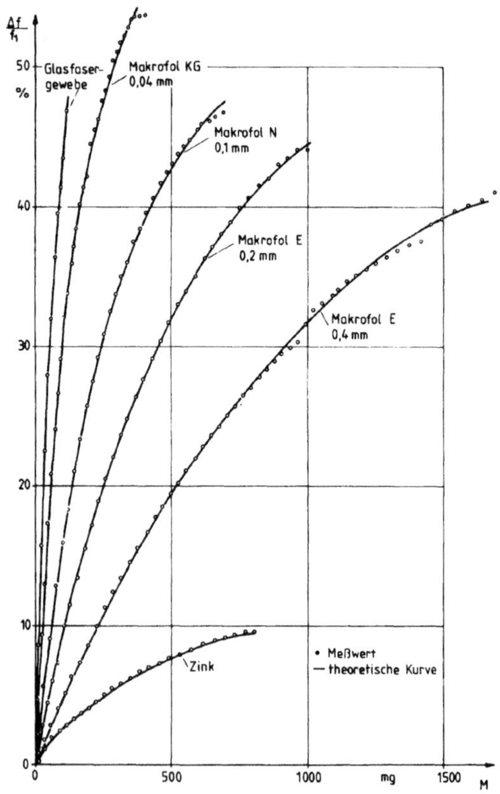

Abb. 5. Darstellung der relativen Frequenzänderung unterschiedlicher Bänder als Funktion einer mittig aufgebrachten Masse

$$\frac{\partial v}{\partial t} = \nu \frac{\partial^2 u}{\partial y^2}. \tag{6}$$

Die Lösung ist aus der Behandlung der Wärmeübertragung bekannt [8]. An der Stelle $y = 0$ ergibt sich die Geschwindigkeit zu

$$v(0, t) = U \cos \omega t. \tag{7}$$

Die dynamische Viskosität ist bekanntlich definiert durch $\tau = \eta \, \partial v / \partial y$. Somit ergibt sich für die Dämpfungskraft

$$F_D = b \nu \rho_F \left(k \omega U - k \frac{\partial u}{\partial t} \right), \tag{8}$$

wo b die Bandbreite, ν die kinematische Zähigkeit, ρ_F die Dichte des Fluids, ω die Kreisfrequenz, u die Verschiebung des Bandelementes, v die Geschwindigkeit und q den Querschnitt des Bandes bedeuten.

Mit der vom Band mitbewegten Flüssigkeitsmasse m_F ergibt sich nach Integration über die Bandlänge die Bewegungsgleichung zu

$$\frac{\partial^2 u}{\partial t^2} = \frac{Eql}{m_F + m_B} \frac{\partial^2 u}{\partial x^2} + \frac{\nu \rho_F k l b r}{m_F + m_B} \frac{\partial u}{\partial t} - \frac{k \omega \nu \rho_F l b}{m_F + m_B} u. \tag{9}$$

Abbildung 6 zeigt den vorgesehenen Versuchsaufbau für das longitudinal schwingende Band zur Messung der viskoelastischen Eigenschaften. Eine Führung erlaubt, die Spaltbreite y zwischen einem piezoelektrischen Scherschwinger und dem Band einzustellen. Mit diesem System werden sich die Schubspannung und der Phasengang der Schwingung als Funktion des Schergefälles beobachten lassen. Die Stirnseite des Abstandstellers läßt sich der Wegfunktion $u(x) = \sin x/l \cdot \pi$ anpassen, so daß sich ein örtlich konstantes Schergefälle ergibt (Abb. 7). Mittelbar über die Anregungsspannung wird das Schergefälle eingestellt.

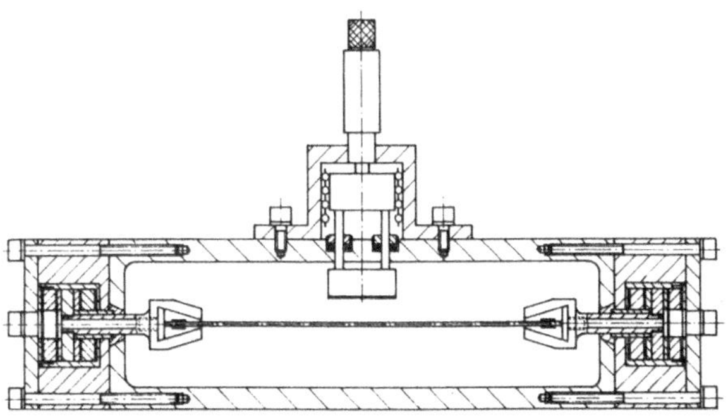

Abb. 6. Schema der Versuchsanordnung zur Viskositätsmessung

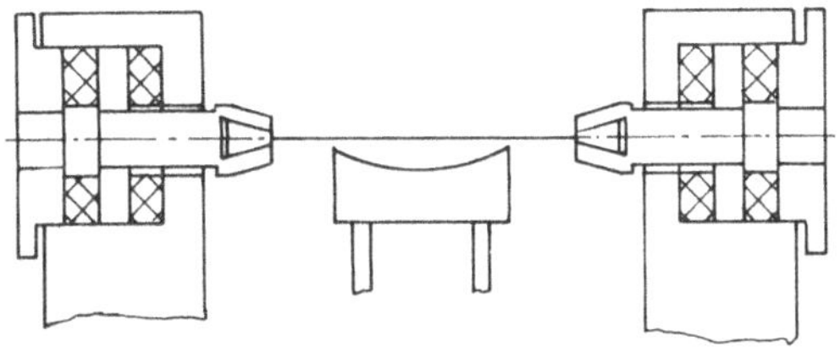

Abb. 7. Versuchsanordnung für ein konstantes Schergefälle in schematischer Darstellung

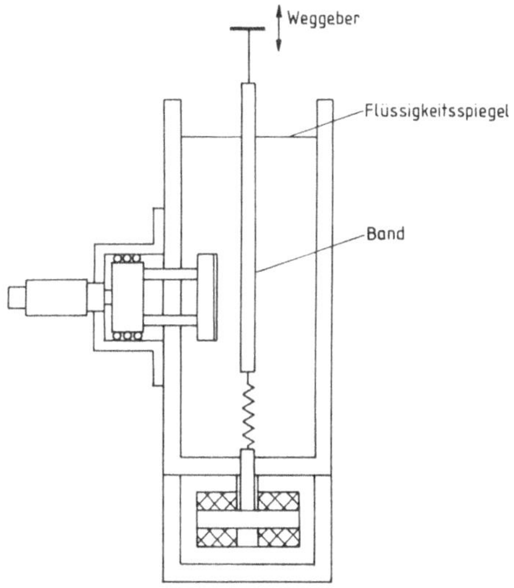

Abb. 8. Viskosimeter mit örtlich konstanter Bandgeschwindigkeit, schematisch

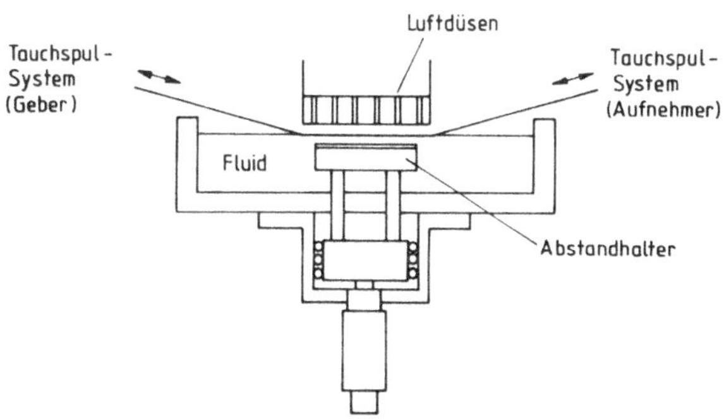

Abb. 9. Prinzipdarstellung eines Viskosimeters mit luftgelagertem Band

Für die frequenzabhängige Bestimmung der Eindringtiefe der Schwingung bzw. des Exponenten des Schergefälles wird die in Abbildung 8 gezeigte Anordnung vorgeschlagen. Das Band taucht in die Flüssigkeit ein. Eine Feder wandelt den vom Band zurückgelegten Weg in eine Kraft um, die piezoelektrisch in die Referenzspannung umgesetzt wird.

Abbildung 9 zeigt ein Viskosimeter mit luftgelagertem Band. Das Luftlager hält das bewegte Band in einer sehr gut definierten Lage zur verstellbaren Gegenplatte.

Die Antriebskraft und Amplitude kann z. B. mit einem Tauchspulsystem erzeugt, der Schwingweg induktiv erfaßt werden. Wegen der geringen Viskosität der Luft ist die Reibung im Luftlager sehr klein. Das Bandmaterial ist bezüglich Dicke, E-Modul und Dichte in weiten Grenzen variabel. Seine Temperaturbeständigkeit und chemische Beständigkeit läßt sich schwierigen Versuchsbedingungen anpassen. Die träge Masse kann sehr klein sein.

Zusammenfassend kann man feststellen, daß das beschriebene Meßprinzip mit longitudinal schwingendem Band folgende Vorteile verspricht:

- weiter Frequenzbereich (Eigenfrequenz bei Kunststoffen 2 – 3 kHz, bei Metallbändern 8 – 20 kHz).
- Einfachheit der Anordnung.
- Die Empfindlichkeit der Anordnung wird durch die nötigenfalls sehr kleine Bandmasse bestimmt.
- Die stehende Welle im Band ist leicht aufrechtzuerhalten. Selbst Papierbahnen und Filterbänder, die äußerlich kaum schwingungsfähig erscheinen, lassen sich zu longitudinalen Eigenschwingungen definierter Frequenz anregen.
- Meßgrößen sind die leicht weiter zu verarbeitenden Frequenzen und Amplituden der Schwingungen.
- Es ergeben sich einfache Möglichkeiten zur Erzeugung eines gewünschten Schergefälles.

Literatur

1. Gast, Th., D. Bouts, Technisches Messen atm **44**, 125 – 130 (1977).
2. Gast, Th., Präzisionsmessungen von Kräften und Massen mit Hilfe schwingender Saiten und Bänder, Dechema-Monographien, Band **67**, Verlag Chemie (Weinheim 1971).
3. Gast, Th., H. Bahner, Staub-Reinhaltung Luft **39**, 109 – 111 (1979).
4. Bahner, H., Ein Beitrag zur Massenbestimmung disperser Stoffe mit Hilfe von Schwingungen, Dissertation TU Berlin 1979.
5. Gast, Th., Ein Gerät zur Registrierung der Staubkonzentration mit Hilfe eines schwingenden Bandes. Measurement and Instrumentation, Acta Imeko 1973, Vol. III, Verlag der ungarischen Akademie der Wissenschaften.
6. Gast, Th., K. U. Kramm, Kraftwerksemissionsmessungen, Vortragsmanuskript zum Statusseminar „Umweltfreundliche Kraftwerkstechnologie" 6. – 9. Okt. 1980, Herausgegeben von Kernforschungsanlage Jülich GmbH.
7. Kalähne, A., Annalen Physik, 4. Folge, **46**, 1 – 39 (1915).
8. Schlichting, H., Grenzschicht-Theorie, S. 77, Verlag G. Braun (Karlsruhe 1965).

(Eingegangen am 12. Mai 1982)

Anschrift der Verfasser:

Prof. Dr.-Ing. habil. Th. Gast,
Dipl.-Ing. K. U. Kramm
Institut für Meß- und Regelungstechnik
Fachbereich 10 – Verfahrenstechnik
Technische Universität Berlin
Budapester Straße 46 – 50
D-1000 Berlin 30

 Rheol. Acta 21, 499 – 501 (1982)

Newtonian reference materials with $\nu < 1$ mm²/s and their application in extending the range of capillary viscometers

H. Bauer and G. Meerlender

Physikalisch-Technische Bundesanstalt, Braunschweig

Abstract: Several newtonian liquids with kinematic viscosities below about 1 mm²/s are investigated in order to find out if the commonly used purity specifications are sufficient to qualify the liquids as viscosity standards which can be used to determine the kinetic energy correction of glas capillary viscometers.

Key words: Viscosity standard, capillary viscometer, kinetic energy correction

1. Introduction

Measurements of kinematic viscosity using standard-type glass capillary viscometers with suspended levels [1] usually yield an accuracy of 0.4 to 0.5%, provided the efflux time is not too short. For viscosities above about 1 mm²/s, this can always be achieved by selecting a suitable capillary diameter. In the range of lower viscosities down to 0.3 mm²/s, the kinetic energy correction [1, 2], which is usually expressed as a correction Δt_H to the measured efflux time t, increases rapidly with decreasing efflux time. For viscometers with trumpet-shaped capillary ends, the following equation is commonly used to calculate this correction [1, 2]:

$$\Delta t_H = \frac{E}{Kt^2}; \quad E = 1.66 \cdot V^{1.5}/L \cdot (2RK)^{0.5}. \qquad (1)$$

V is the volume of liquid which flows during the efflux time t through the capillary with radius R and length L, and K is the viscometer constant. When applied to standard-type viscometers [1], eq. (1) does not give reliable results in the viscosity range considered here. Reducing the capillary diameter would, in principle, solve the problem. In normal laboratory use, however, a value of 0.47 mm (size 0c according to [1]) is considered to be the smallest capillary diameter that can be handled without too many problems. These viscometers are therefore often used to measure viscosities down to 0.3 mm²/s. The application of eq. (1) may cause errors of several per cent. One way to increase the accuracy of viscosity determination in this range is the use of long capillary viscometers. With these instruments (capillary length 400 mm, capillary diameter 0.24 mm and 0.36 mm) the viscosity can be measured with an uncertainty between $\pm 1.5 \cdot 10^{-3}$ and $\pm 3.0 \cdot 10^{-3}$, depending on surface tension and vapour pressure of the liquid [3]. Since long capillary viscometers are not suitable for routine laboratory use, it is necessary to determine the kinetic energy correction of standard-type viscometers more precisely in order to make low viscosity measurements possible.

2. Measurements and results

Experiments with more than 40 standard-type viscometers [1] with capillary diameters of various sizes showed that it is not possible to find a simple relation to describe the kinetic energy correction at low viscosities. At least three or sometimes four individual constants would be necessary for each viscometer. Each viscometer has therefore to be calibrated individually by direct comparison with long capillary viscometers or by measurements with several certified standard liquids. Since both methods would be relatively expensive (compared to the price of a glass viscometer), investigations were made on liquids of high purity which are commercially available for various purposes (spectrophotometry, gas chromatography, etc.) in order to find out if these purity specifications would be sufficient to qualify the liquids as viscosity standards.

Measurements were carried out on samples from two different suppliers [4]. The results of viscosity determinations at 20 °C are given in table 1. In most cases, test portions were taken from different bottles from the same supplier and no significant difference in viscosity was found. Methylene chloride, trichloroethylene and tetrachloroethylene contained small amounts of stabilizer. Methylene chloride was also measured after distillation, but the same viscosity was obtained as for the stabilized sample. Chloroform usually contains 0.5% to 1.0% ethanol as a stabilizer. The observed difference in viscosity of 0.11% must be attributed to an unknown difference in the ethanol content. Sample 2 was labelled as containing 0.8% ethanol, whereas the actual concentration of ethanol for sample 1 was unknown. For chloroform without a stabilizer, a viscosity of 0.3799 mm^2/s was found. The increase in viscosity up to an ethanol concentration of 1.5% was investigated [3]. For methyl ethyl ketone, the observed difference of 0.21% is caused by 0.2% ethyl acetate and 0.1% 4-methylpentanone-2 which were found in sample 2 but not in sample 1. This must be related to the different purposes for which the materials were specified (table 1). With acetone, there have been problems which are likely to arise as a result of unsuitable bottle-sealing material.

Table 1. Liquids proposed for use as viscosity standards; 1 and 2 correspond to supplier 1 and supplier 2 [4]; $\bar{\nu}_{20\,°C}$ is the average value for the kinematic viscosity at 20 °C, $(\nu_1 - \nu_2)/\nu$ the relative difference in viscosity between samples from supplier 1 and supplier 2

Material	minimum purity (%) 1	2	$\bar{\nu}_{20\,°C}$ (mm^2/s)	$\frac{\nu_1 - \nu_2}{\nu}$ 10^{-3}
Methylene chloride	99.5 a)	99 c)	0.3273	0.4
Chloroform	99 d)	99 d)	0.3808	−1.1
Trichloroethylene	99.5 a)	99.6 i)	0.3850	0.03
Acetone	99.5 a)	–	0.4007	–
Methyl ethyl ketone	99.7 c)	99.7 e)	0.4938	−2.1
Tetrachloroethylene	99.7 f)	99.5 a)	0.5524	−0.2
Tetrachloromethane	99.8 a)	99.5 d)	0.6109	0.3
Toluene	99.5 a)	*) h)	0.6778	−0.2
Chlorobenzene	99 g)	99 a)	0.7251	0.1
Benzene	99.7 a)	99 c)	0.7394	0.4
Cyclohexane	99.7 b)	99 c)	1.2581	0.6
n-Undecane	99.5 a)	99*)	1.5861	0.1

*) not specified, a) for analysis, b) for chromatography, c) for gas chromatography, d) for spectrometry, e) for UV spectrometry, f) for IR, NMR spectrometry, g) for synthesis, h) for organic residue analysis, i) MOS quality.

These results show (at least with respect to the suppliers [4]) that for several liquids with viscosities between 0.3 mm^2/s and 1.5 mm^2/s, the common purity classification is sufficient to characterize the viscosity within 0.3%.

With the use of these materials, individual determinations of the kinetic energy correction can be made in order to make measurements down to 0.3 mm^2/s possible with standard-type viscometers, by aid of the following equation:

$$\Delta t_H = t - \nu/K\,, \tag{2}$$

where ν is the kinematic viscosity.

Figure 1 gives some examples of those calibrations for viscometers with capillary diameters of various sizes. The usual correction according to eq. (1) is shown for comparison.

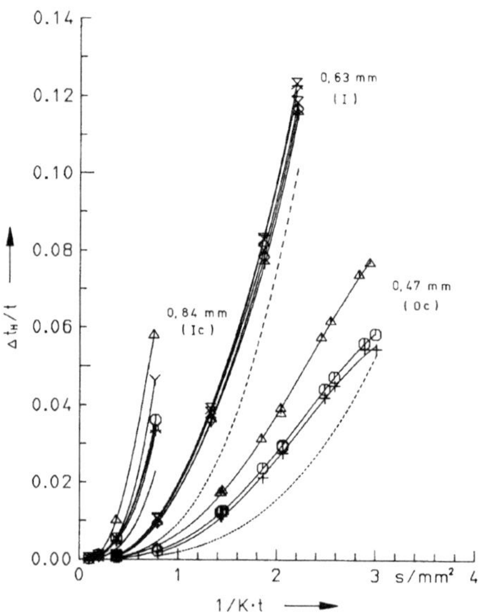

Fig. 1. Individual calibration curves for viscometers [1] with different capillary diameters. Dashed lines correspond to the standard correction according to eq. (1)

This method has been applied to the determination of the "viscosity number" (Staudinger function) J of polycarbonate in methylene chloride using 26 viscometers (size 0c according to [1]):

$$J = \frac{1}{c}\left(\frac{t - t_0}{t_0}\right). \tag{3}$$

c is the polycarbonate concentration, t and t_0 are the corrected efflux times of solution and solvent, respectively. The J values obtained were in the range between 67.1 and 67.9 (average value 67.4, standard deviation 0.3%). Measurements with long capillaries [3] also gave an average value of 67.4. If, however, eq. (1) is used to apply the kinetic energy correction, the results are in the range between 66.5 and 72.3 (average value 70.4, standard deviation 2.1%). Similar results were obtained on PMMA (polymethacrylmethacrylate) in chloroform.

References

1. DIN 51562 (1978): Messung der kinematischen Viskosität mit dem Ubbelohde-Viskosimeter.
2. Cannon, M. R., R. E. Manning, J. D. Bell, Anal. Chem. **32**, 255 (1960).
3. Meerlender, G., H. Bauer, Rheol. Acta (to be published).
4. Supplier 1: E. Merck, Labor- und Industriechemikalien, Darmstadt (FRG). Supplier 2: J. T. Baker, Chemicals B. V., Deventer (The Netherlands).

(Received April 16, 1982)

Authors' address:

Dr. H. Bauer, Dr. G. Meerlender
Physikalisch-Technische Bundesanstalt
Postfach 3345
D-3300 Braunschweig

Rheologica Acta Rheol. Acta **21**, 502 – 504 (1982)

Viskosimetrie bei hohem Schergefälle und hoher Temperatur – eine instrumentelle Entwicklung

J. Klein, H. G. Müller und G. Leidigkeit

Institut für Chemische Technologie, TU Braunschweig

Abstract: A newly designed high shear capillary viscometer will be presented which allows to perform measurements up to 150 °C and shear rates up to $3 \cdot 10^6$ s^{-1}. A description of the instrument will be combined with the discussion of the measurements accuracy and applicational examples related to lubrication oil characterization.

Key words: Capillary viscometer, high shear, lubrication oil

1. Allgemeines

Unter Berücksichtigung verschiedener Anwendungsbereiche wird die Notwendigkeit deutlich, über ein Viskosimeter verfügen zu können, in dem die Viskosität von Fluiden unter sehr hohen Schergefällen gemessen wird.

Betrachtet man z. B. die Problematik der Motorenöle, die im Motor Geschwindigkeitsgefällen bis zu $1 \cdot 10^7\ s^{-1}$ [1] ausgesetzt sind, so erweitert sich die erste Forderung durch den Anspruch, ein Viskosimeter zu besitzen, das außerdem bei hohen Temperaturen zu messen imstande ist [2].

Generell sind zwei Problemlösungen möglich: Das Rotations- oder das Kapillar-Viskosimeter. Rotationssysteme – dabei liegt besonders das Couette-System im Interesse der Entwickler – liefern zwar direkt aus der Rotation die Schergeschwindigkeit und sind mit geringsten Probenmengen zu betreiben, bereiten aber besonders bei der Temperierung große Schwierigkeiten. Aufgrund der geringen Spaltdimensionen ist eine konstruktive Auslegung äußerst aufwendig, hohe Schergeschwindigkeiten können nur schwer erreicht werden.

Demgegenüber ist ein Kapillarviskosimeter konstruktiv wesentlich einfacher aufgebaut und auch in seiner Anwendungstechnik wesentlich einfacher und flexibler. Dies gilt sowohl für Messungen im hohen Scherbereich (bis $3 \cdot 10^6\ s^{-1}$) als auch bei hohen Temperaturen (bis 160 °C).

2. Aufbau des Viskosimeters und Meßprinzip

Das Gerät besteht im wesentlichen aus einer kontinuierlich fördernden Hochdruckpumpe, einem Hochdruckreservoir zur Temperierung der Meßflüssigkeit und einer Kapillare. Zwischen Pumpe und Hochdruckreservoir befindet sich eine Druckmeßeinheit, der Kapillare nachgeschaltet ist ein Volumenmeßsystem (Abb. 1). Mit Kenntnis der Kapillardaten läßt sich aus Druckdifferenz Δp und Volumenfluß Q die dynamische Viskosität η bestimmen:

$$\eta = \frac{\Delta p \cdot \pi \cdot R^4}{8 \cdot Q \cdot L}, \tag{1}$$

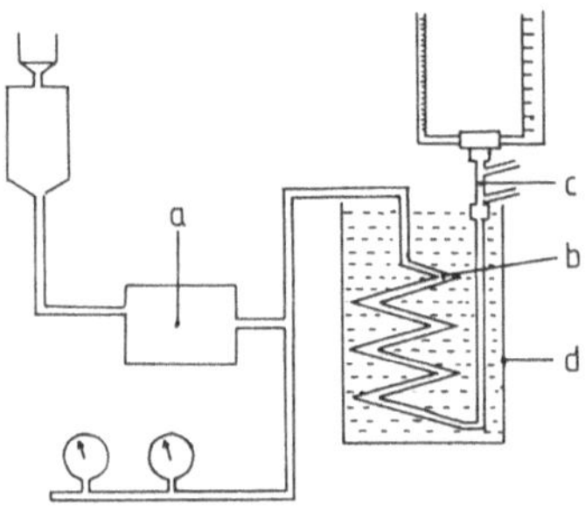

Abb. 1. Schematische Darstellung des Hochdruck-Hochtemperatur-Kapillarviskosimeters.
a) Hochdruckpumpe, b) Hochdruckreservoir, c) Kapillare, d) Thermostat

wobei L die Länge und R den Radius der Meßkapillare bedeuten. Δp ist dabei noch um die Hagenbach-Korrektion vermindert.

Die Kapillare selber ist aus Stahl gefertigt, sie ist in der Regel 10 cm lang und kann im Durchmesser bis zu 0,01 cm eng sein. Sie wird gesondert thermostatisiert, indem Temperierflüssigkeit aus dem Hauptbad turbulent in einen Mantel gepumpt wird, der die Kapillare umgibt. Da die Kapillareinheit leicht austauschbar ist, wird es möglich, größere Viskositäts- oder Scherbereiche problemlos zu erreichen.

3. Meßkriterien

3.1. Reynolds-Zahl und Scherabbau

Eine ganz entscheidende Voraussetzung für ein Kapillarviskosimeter ist die Einhaltung des laminaren Strömungszustandes innerhalb der Kapillare. Dieser Vorgang wird durch die Reynolds-Zahl beschrieben. Durch die Wahl der Kapillardimensionen (Länge/Radius $\geqq$ 1000) ist sichergestellt, daß auch bei höchsten Schergefällen ein Wert von 1000 für die Reynolds-Zahl nicht überschritten wird. Damit ist eine laminare Strömung nach den üblichen Kriterien der Strömungsmechanik gewährleistet.

Eine weitere Voraussetzung ist die Eliminierung oder Minimierung irreversibler Viskositätsänderungen, die durch einen Scherabbau insbesondere bei polymeren Substanzen verursacht werden können. Zur Erfassung dieses Effektes wurde ein polymer-legiertes Öl bei 150 °C unter einem Schergefälle von $2{,}3 \cdot 10^6\ s^{-1}$ durch die Kapillare gepreßt. Die Viskosität dieser so behandelten Probe lag lediglich um 2,2% unter dem Wert des frischen Öls. (Diese Differenz wurde im Ubbelohde-Viskosimeter ermittelt.) Die Abweichung ist gegenüber einem Scherstabilitätstest nach DIN 51382 (30% Viskositätsabfall mit der gleichen Probe) wesentlich geringer und liegt innerhalb der Fehlergrenzen einer Messung. Damit sollte sichergestellt sein, daß bei einmaligem Kapillardurchtritt der Lösung irreversible Viskositätsveränderungen zu vernachlässigen sind.

3.2. Temperatur- und Druckeinflüsse

Einen ganz entscheidenden Einfluß übt die Temperatur auf die Viskosität aus. Bei hohen treibenden Drücken kann die Temperaturdifferenz in der Kapillare zwischen Ein- und Auslauf bis zu 7 °C betragen. Dies bedingt eine Viskositätserniedrigung um bis zu 0,5 mPas, da bei Motorölen allgemein ein Temperatur/Viskositätskoeffizient von 0,16 mPas/°C gefunden wird. Um diesen Temperatureffekt zu mindern, wurde die Kapillarlänge relativ groß gehalten und der Kapillarradius klein gewählt, da der Temperatureffekt mit zunehmender Länge und abnehmendem Radius verringert wird [3]. Weiterhin wirkt die separate Thermostatisierung der Meßstrecke einem größeren Temperaturanstieg in der Kapillare entgegen.

Der Druck übt einen entgegengesetzten Effekt auf die Viskosität aus [4], und zwar gilt:

$$\eta = \eta_0 \cdot e^{(\alpha \cdot p + \beta \cdot p^2)}, \tag{2}$$

wobei η die Viskosität beim Druck p,
η_0 die Viskosität bei 760 Torr,
p den Druck,
α den Druckkoeffizienten und
β den Kompressibilitätsfaktor

bedeuten. Messungen von E. Kuss ergaben, daß bei Motorölen α-Werte zwischen $1{,}3 \cdot 10^{-3}$ und $1{,}8 \cdot 10^{-3}$ und β-Werte in einer Größenordnung um $1{,}5 \cdot 10^{-7}$ erhalten werden. Die Kompressibilitätsfaktoren wirken sich damit auf die Viskositätsänderung nur unwesentlich aus.

Zur Bestimmung der druckabhängigen Viskosität benötigt man die Kenntnis des effektiven Druckverlaufs entlang der Kapillare [5]. Für den Fehler F_m, der durch den Druckeffekt entsteht, ergibt sich an der Stelle m:

$$F_m = \frac{\sum_{0}^{m-1} (1 - \beta \cdot p) e^{\alpha(p-1)}}{m-1} - 1\,. \tag{3}$$

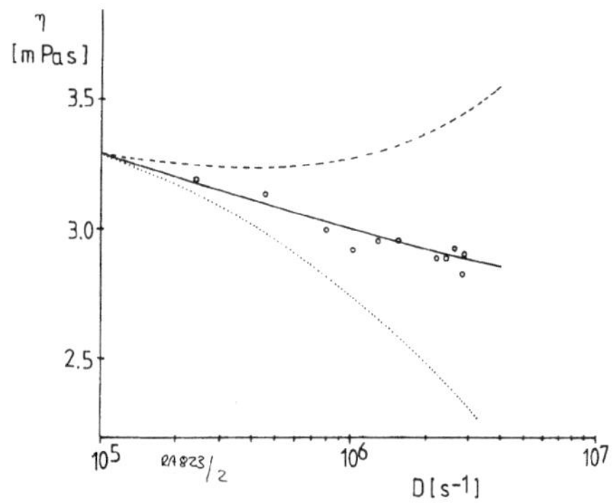

Abb. 2. Meßfehler durch Temperatur- und Druckeffekte.
– – – Viskositäserhöhung durch Druck,
. . . . . . Viskositätserniedrigung durch Temperatur,
______ berechnete Meßkurve,
o Meßpunkte

F_m nimmt einen Wert von 35% an, wenn von einem Druck von 350 bar und $\alpha \approx 2$ ausgegangen wird. Im betrachteten Fall liegt der Fehler um 20%. Abbildung 2 zeigt den Einfluß von Temperatur und Druck auf die Viskosität. Eine entsprechende Korrektur, die beide Effekte zusammenfaßt, zeigt deutlich die Kompensation beider Fehler: Die experimentelle Fließkurve ist der berechneten praktisch deckungsgleich.

Dies beweisen auch Messungen mit newtonschen Flüssigkeiten, die bis in höchste Scherbereiche konstante Viskositäten besitzen (Abb. 3). Wäre irgendein Meßfehler dominierend, so könnte hier keine konstante Viskosität über den gesamten Scherbereich erhalten werden.

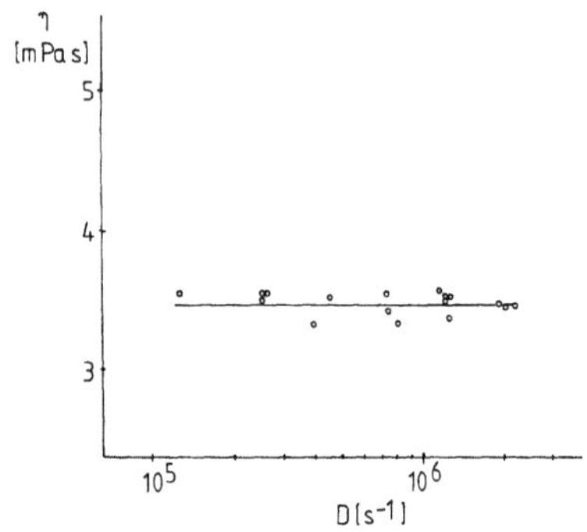

Abb. 3. Fließkurve einer newtonschen Flüssigkeit (Einbereichsöl ohne Polymeradditive)

4. Zusammenfassung

Mit dem beschriebenen Kapillarviskosimeter können Viskositätsmessungen sowie die Bestimmung von Fließkurven bei Temperaturen bis zu 160 °C durchgeführt werden, wobei Schergefälle bis $3 \cdot 10^6\ s^{-1}$ erreicht werden. Die obere Meßgrenze kann mit 10000 mPas angesetzt werden. Dies ist gleichzeitig die maximale pumpbare Viskosität, die von der Hochdruckpumpe bewältigt wird.

Die Kapillare ist in der Weise ausgelegt, daß korrekturfreie Messungen erhalten werden, in denen lediglich die Hagenbach-Korrektion eingeht. Temperatur- und Druckeffekte beeinflussen die Ergebnisse nicht. Irreversible Viskositätsänderungen, die durch Scherung verursacht werden, liegen innerhalb der Fehlergrenzen der Messung. Die Kapillare selber ist leicht austauschbar, so daß große Meßbereiche der Viskosität oder der Schergeschwindigkeit schnell erfaßt werden können.

Das Hochdruckreservoir, das mit minimalem Meßvolumen gefüllt ist, gewährleistet auch bei höchsten Pumpleistungen eine optimale Thermostatisierung der Probe.

Ein weiterer Vorteil ist die kontinuierliche Arbeitsweise des Gerätes, wodurch sich ein zeitaufwendiges Befüllen des Hochdruckreservoirs erübrigt.

Anwendungsmöglichkeiten in der Prozeßkontrolle sind gegeben.

Literatur

1. Bell, J. C., M. A. Voisey, SAE Bericht 770378.
2. Stambaugh, R. L., R. J. Kopko, SAE Bericht 730487.
3. Klein, J., H. Fußer, Rheol. Acta **6**, 76 (1967).
4. Kuss, E., Z. Angew. Phys. **10**, 12 (1858).
5. Brennschede, E., J. Klein, Rheol. Acta **8**, 1 (1969).

(Eingegangen am 30. Juni 1982)

Anschrift der Verfasser:

Prof. Dr. J. Klein, Dipl. Chem. G. Leidigkeit
Institut für Chemische Technologie
Technische Universität Braunschweig
Hans-Sommer-Straße 10
D-3300 Braunschweig

Dr. H. G. Müller
Bayer AG
Fe-DPP-Pa
D-5090 Leverkusen

Pressure influences on the viscosity of polymer solutions

B. A. Wolf, H. Geerissen, R. Jend, and J. R. Schmidt

Institut für Physikalische Chemie der Universität Mainz, Sonderforschungsbereich 41

Key words: Polystyrene solution, viscosity, pressure influence

1. Introduction

Of all physico-chemical quantities characterizing the behavior of liquids, the viscosity coefficient η is normally most sensitive to pressure. Early measurements of $\eta(p)$ date back almost 100 years [1] but they were carried out with very different objectives [2]. Currently, one is particularly interested in polymer solutions, for instance because of the additives to motor oils and of oil recovery. The present paper reports on measurements with polystyrenes of high molecular uniformity in eight different solvents. It confines itself to zero-shear viscosities and discusses the possibilities of discovering the influence of pressure via the molecular structure of the solvent and via its thermodynamic interactions with the polymer.

Two different concentration regimes must be distinguished by means of the reduced polymer concentration [3] $\tilde{c} = c[\eta]$; (c = polymer concentration (mass/vol), $[\eta]$ = intrinsic viscosity (vol/mass)). In dilute solutions ($\tilde{c} < 0.5$) the flow mechanism is non-draining, i.e. the characteristics of the individual polymer coils (as measured by $[\eta]$) play a decisive role. In the case of moderately concentrated solutions ($\tilde{c} \geqslant 2$), the system becomes free draining and the intersegmental interactions (as measured by the Huggins-constant k_H) begin to outweigh the importance of $[\eta]$.

2. Results

2.1. Dilute solutions

Measurements were performed up to 4000 bar with a self-made rolling-ball viscometer that allows the angle of inclination [4] to be varied. Tert.-butyl acetate was chosen as the solvent for the investigation of the thermodynamic influences, since it covers the entire spectrum from highly endo- to highly exothermal conditions. The endothermal theta temperature lies at 15.5, the exothermal at 109.5 °C; for M = 110000, the system becomes athermal at 45 °C [5].

Figure 1 shows the experimental results. A one-to-one correspondence is found for $[\eta]$ and the second osmotic virial coefficient A_2, this is not the case for k_H and A_2 however. The highest pressure influences on $[\eta]$ can be observed at 130 °C where it increases by 75% going from 1 to 3500 bar. The isobaric maximum in $[\eta](T)$ (corresponding to approximately athermal conditions) changes by +7K/100 bar. The influence of the shape of the solvent molecules on $[\eta]$ for given thermodynamic conditions is negligible.

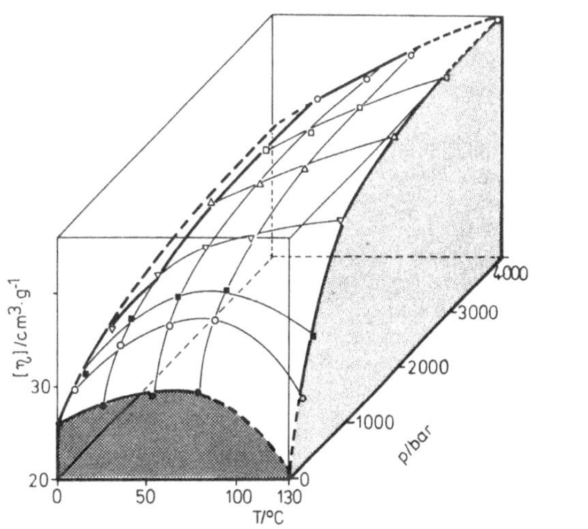

Fig. 1. Temperature and pressure dependence of the intrinsic viscosity of polystyrene (M = 110000) in tert.-butyl acetate

2.2. *Moderately concentrated solutions*

A Searle-type apparatus [6] was used for the more viscous systems up to pressures of 1000 bar. The influence of the thermodynamic conditions on $\eta(p)$ was again studied by means of the solvent tert.-butyl acetate. The effects can be described in terms of the volume of activation $V^{\neq} = RT(\partial \ln \eta / \partial p)$ or $f_{1000} = \eta_{1000\ \mathrm{bar}}/\eta_{1\ \mathrm{bar}}$. $V^{\neq}$ turns out to be inversely proportional to the thermodynamic quality of the solvent, the pressure dependence becoming a minimum when the solvent power reaches its maximum at athermal conditions [7].

The influence of the shape of the solvent molecules on f_{1000} for a fixed Gibbs energy of mixing can be seen from figure 2 [8]. The comparison is made at the indicated theta temperatures and at the respective critical compositions at 1 bar. Within a given class of solvents (in the present case: substituted *n*-alcanes, alicyclics and carboxylic esters) the f_{1000} values of the pure liquids increase with increasing bulkiness of the molecules; the addition of the polymer raises these values by typically 15%. This means that the *absolute* increase of the pressure dependence of the viscosities of the solutions over that of the solvent reflects the shape of the solvent molecules.

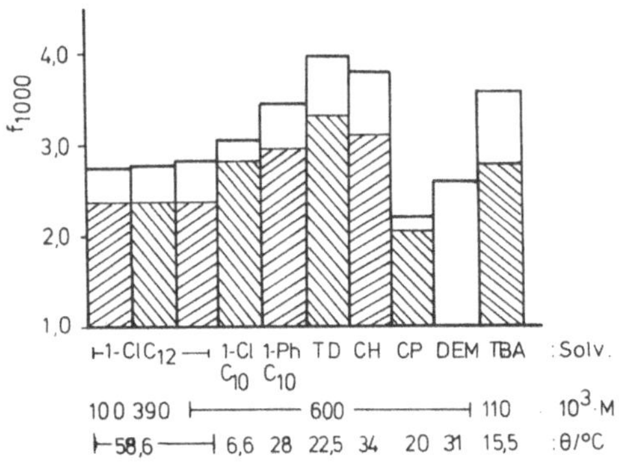

Fig. 2. $f_{1000} = \eta_{1000\ \mathrm{bar}}/\eta_{1\ \mathrm{bar}}$ for the indicated systems at their respective theta-temperatures and critical compositions (1 bar). The hatched parts refer to the pure solvents.
1-ClC_{12}: Chlor-n-dodecane, 1-ClC_{10}: Chlor-n-decane,
1-PhC_{10}: Phenyl-n-decane, TD: trans Decalin,
CH: Cyclohexane, CP: Cyclopentane,
DEM: Diethyl malonate, TBA: tert.-Butyl acetate

3. Discussion

Since even the moderately concentrated solutions normally contain considerably more than 50% solvent, the influence of pressure on the viscosity of the solution primarily reflects the behavior of the pure solvent. Large effects on the latter must be expected for bulky molecules particularly close to the freezing point of the liquid. The present experimental results for the solutions demonstrate that the polymer modifies the characteristics of the solvent in a quite different manner within the two concentration regimes.

In the case of *dilute solutions* the thermodynamic quality of the solvent at 1 bar is of minor importance only, since it merely fixes the hydrodynamic volume of the isolated polymer molecules. It is the pressure dependence of the Flory-Huggins constant χ that governs the effects. The more e.g. the solvent power increases when p is raised ($\partial\chi/\partial p < 0$) the more the coils expand (i.e. $[\eta]$ rises) and the more the behavior of the solutions changes in direction of the polymer, i.e. the more the pressure dependence of the viscosity of the solution exceeds that of the solvent.

In the case of *moderate concentrations*, the number and effectiveness of intersegmental contacts (manifesting themselves in k_H) determines the size of the average flow unit and thus the activation volume of the solution. This means that the χ value at 1 bar already fixes the magnitude of the pressure effects since it determines the statistics of the intersegmental contacts via the Gibbs energy of mixing. The higher χ is at 1 bar, i.e. the worse the solvent, the more these contacts are favored over contacts between segments and solvent molecules and the larger the pressure dependence of η becomes. For moderately concentrated solutions the effect of $\partial\chi/\partial p < 0$ is opposite to that in dilute systems, since the increase of solvent power with pressure loosens the coherence of the solute and shifts the behavior of the solution in direction of the solvent. The above ideas of a thermodynamic "pull-along" effect implies that the enthalpy of mixing should modify the behavior for a given constant χ via the effectiveness of the individual intersegmental contacts. In fact, the present results support this concept [8]. The $V^{\neq}$ values at the theta-temperatures normally increase with the heat of mixing and, furthermore, these heat effects govern the temperature dependence of $V^{\neq}$ as should be the case for the present approach.

References

1. Röntgen, W. C., Wiedemanns Ann. **22**, 518 (1884).
2. Kuss, E., Z. angew. Phys. **8**, 372 (1955).

3. Frisch, H. L., R. Simha, in: F. Eirich (ed.), "Rheology, Theory and Applications", Vol. **1**, Chapt. 12, Academic Press (New York 1956).
4. Schmidt, J. R., B. A. Wolf, Macromolecules, submitted.
5. Wolf, B. A., H. J. Adam, J. Chem. Phys. **75**, 4121 (1981).
6. Wolf, B. A., R. Jend, Macromolecules **12**, 732 (1979).
7. Schmidt, J. R., B. A. Wolf, Makromol. Chem. **180**, 517 (1979).
8. Geerissen, H., J. R. Schmidt, B. A. Wolf, J. Appl. Polym. Sci. **27**, 1277 (1982).

(Received April 15, 1982)

Authors' address:

Prof. Dr. B. A. Wolf, Dipl.-Chem. H. Geerissen
Dr. R. Jend, Dr. J. R. Schmidt
Institut für Physikalische Chemie
Johannes Gutenberg-Universität
Postfach 3980
D-6500 Mainz

Phenomena of high polymer solutions while flowing through capillaries at high rates of shear

K. Kirschke

Bundesanstalt für Materialprüfung, Berlin (West)

Abstract: Generally solutions of high polymers show a shear-rate dependent flow behaviour and so the properties of these fluids have to be measured under conditions of shear corresponding to the practical service. Capillary viscometry is suitable for achieving high rates of shear but relaxation phenomena can effect the results, which is proved experimentally. The flow behaviour of a lubricant blended with a high polymer additive is measured and a graphical representation of the dependence of flow behaviour on temperature and rate of shear is recommended.

Key words: Polymer solution, viscometry, flow behaviour, high rate of shear, lubricant

1. Introduction

Solutions of high polymers generally show a non-Newtonian flow behaviour, if they are flowing at high rates of shear. From this it follows that the viscosity measured in the standardized viscometers at low rates of shear can be quite different from the viscosity which becomes effective in applications which occur mostly at high rates of shear. The multigrade motor oils are an example of this. They contain high polymer additives to improve their viscosity-temperature behaviour. Therefore, the flow behaviour of such fluids has to be measured in viscometers that enable the determination of the dependence of the viscosity on temperature and rate of shear, the latter up to very high values, which are difficult to achieve. A suitable graphic representation of the results, is important for practical purposes.

2. Flow curves over an extensive range of shear rates

The viscosity of solutions is a function of the temperature T, pressure p and shear rate $\dot{\gamma}$:

$$\eta = \eta(T, p, \dot{\gamma})\,.$$

Although the dependence on the pressure can be neglected in many practical cases, the shear rate dependence has to be measured in addition to the dependence of the temperature. It can be proved that for a number of solutions neglecting the shear rate dependence leads to considerable errors which are comparable to those that arise by neglecting the temperature dependence or even substantially worse. For measuring the shear rate dependence we use rotary viscometers (for lower rates of shear) and a capillary viscometer with high driving pressures (for higher rates of shear), both having the possibility of varying the shear rate. At present one problem still remains. Some fluids, for example lubricants, are stressed in several kinds of application with extremely high rates of shear which cannot be simulated in viscometers. Their behaviour in practical applications has to be estimated from their behaviour in the range of shear rates that is accessible for measurement. However, we are interested in extending the measuring range to rates of shear that are as high as possible and investigating the peculiarities of this region. In the following some experimental results are described. Figure 1 shows a result of comparative joint investigations at the German Physical-Technical Federal Institute and the Federal Institute for Materials Testing. The straight lines are the flow curves of two Newtonian calibration oils. The other lines are the flow curves of non-Newtonian mineral oils with

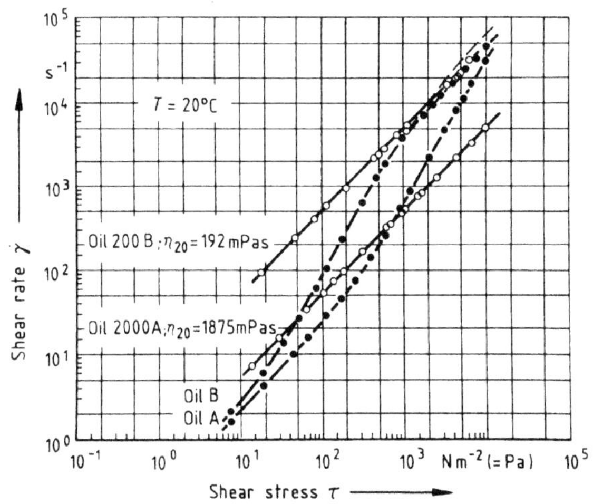

Fig. 1. Flow curves of two Newtonian and two non-Newtonian oils

different high polymer additives. One of these oils (oil B) shows certain deviations from the expected behaviour. This fact has led us to a careful investigation of this phenomenon.

3. Potential phenomena in high shear-rate capillary viscometry

Normally the flow behaviour of solutions of high polymers is represented by flow curves. Figure 2 shows the flow curves of a mineral oil with 4% polyisobutylene 100 at different temperatures. The measured values result in S-curves in log-log plot with some deviations at high rates of shear. As the determination of the flow behaviour in the region of high rates of shear and the possibility of extrapolation to even higher rates of shear is of particular interest for the application of the fluids under consideration we searched for explanations for the occurence of these deviations from the S-shape. For these investigations we used a mineral oil containing polyisobutylene of a very high molecular weight. For this fluid the anomalies of the flow behaviour appear obviously and in a region of the rate of shear that can be covered by measurement. The investigations were carried out with capillaries of the same diameter but different lengths. The result is given in figure 3. Besides the dependence on the rate of shear there is a marked dependence on the length of the capillary at the same

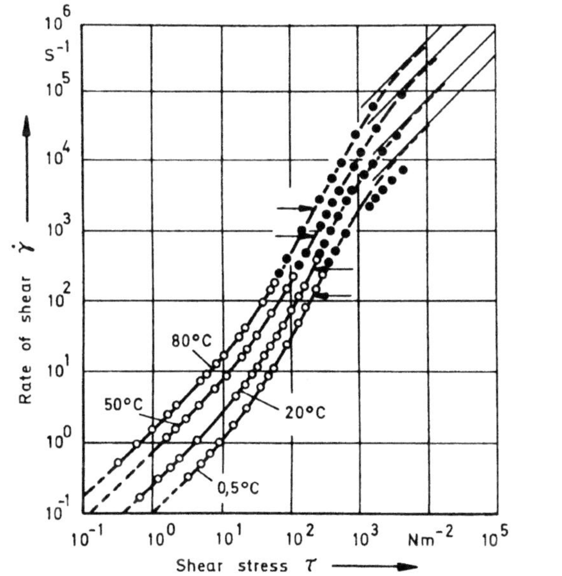

Fig. 2. Flow curves of a mineral oil with 4% polyisobutylene 100 at different temperatures

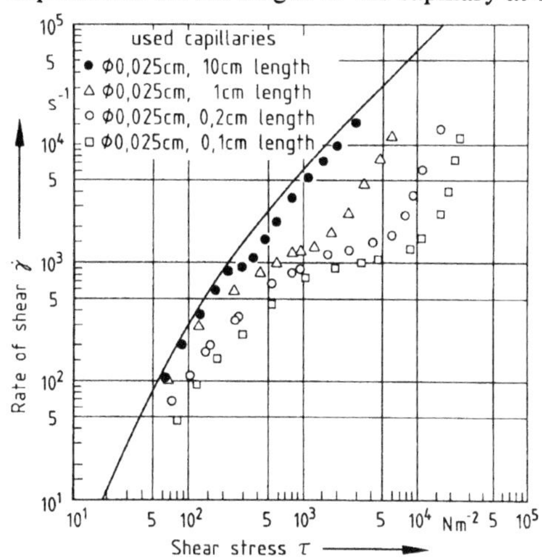

Fig. 3. Investigations with capillaries of the same diameter but different lengths

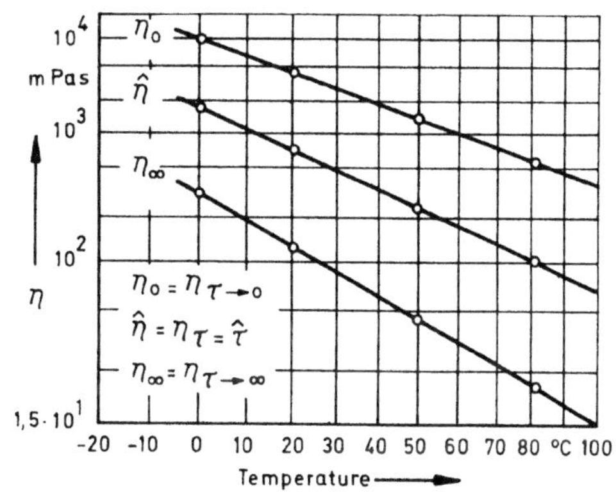

Fig. 4. Dependence of viscosity on temperature based on the data of figure 2. $\hat{\tau}$ is shear stress at the inclination point of the flow curve

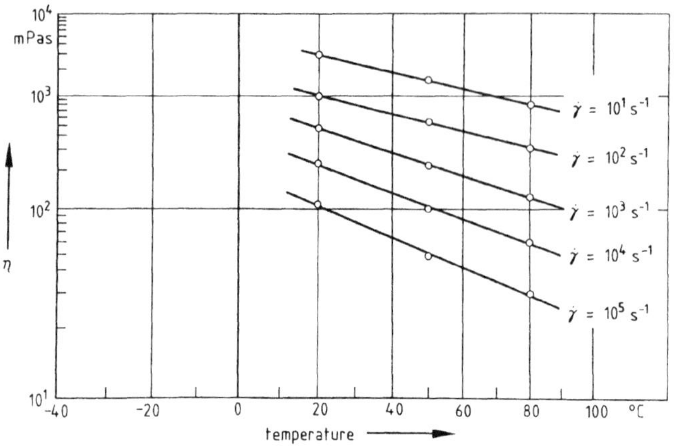

Fig. 5. Mineral oil with 5% styrene-based hydrocarbon

shear rate. This represents a dependence on the duration of working in the capillary. At the same rate of shear the deviation is much larger for shorter working times, i.e. for a shorter duration of stay in the capillary. The reason for this phenomena are probably entrance effects and relaxation effects that occur in these non-Newtonian liquids which are more important the shorter the capillaries are. The correction of the entrance pressure losses and a comparison with the flow behaviour of Newtonian liquids (paraffin oil) under the same conditions showed, that the main part of the deviations is caused by properties of the examined liquid. These effects result in an apparently higher viscosity under shear stresses of very short duration. Extrapolation to sufficiently long capillaries, i.e. sufficiently long times of stay in the capillaries, results in a symmetric extent of the flow curves to the inclination point. We suspect that the degree of asymmetry of some published flow curves can be explained in this way and that they do not reflect the behaviour of the respective fluids during steady-state shear flow.

4. Graphical representation of the dependence on temperature and shear rate of the flow behaviour of solutions, particularly lubricants

In search of a representation of the flow behaviour of solutions $\eta = \eta(T, \dot{\gamma})$ appropriate to practice we found a suitable entry point to be the viscosity-temperature sheet of Ubbelohde with the rate of shear as the parameter. This provides the possibility of interpolation and extrapolation of the results at different rates of shear.

An extrapolation of the flow curves (fig. 2) to the shear stresses $\tau \to 0$ and $\tau \to \infty$ leads to the limiting curves for the non-Newtonian behaviour of the liquid (fig. 4). The curve in between gives the flow behaviour at the inclination point of the flow curves. All these curves coincide in the case of a Newtonian liquid.

Finally, some results of measurements which were carried out in our Institute quite recently shall be given. They refer to a polymer frequently used by one of the major oil companies as an additive for engine lubricants to improve the Viscosity Index of these lubricants. The Viscosity Index characterizes the viscosity-temperature behaviour of lubricants. This oil-soluble additive is a styrene-based hydrocarbon. It is suitable in the formulation of wide-span multigrade motor oils and is superior because of its shear stability. However, apart from the improvement of the viscosity-temperature relation of the blended oil one has to take notice of the non-Newtonian shear-rate-dependent flow behaviour which will be caused. A 5% solution of the additive in mineral oil was prepared and the flow dependence on temperature and shear rate was investigated. The results are shown in figure 5. One can recognize that the variation with the rate of shear within a range of $10\ s^{-1}$ to $10^5\ s^{-1}$ is higher than the variation with the temperature within a range between 20 °C and 80 °C, which is remarkable for the effect of an additive used in practice. The non-observance of the shear rate dependence of the viscosity of such lubricants can therefore lead to mistakes comparable to or higher than the neglect of the viscosity-temperature dependence.

(Received April 29, 1982)

Author's address:

Prof. Dr. K. Kirschke
Bundesanstalt für Materialprüfung
Unter den Eichen 87
D-1000 Berlin 45

Rheologica Acta Rheol. Acta **21**, 511 – 513 (1982)

Some general rheological properties of dilute polymer solutions predicted from intrinsic viscosity measurements

Y. T. Hsü and P. Schümmer

Institut für Verfahrenstechnik der Rheinisch-Westfälischen Technischen Hochschule Aachen

Abstract: Expressions for the rheological properties of dilute polymer solutions at low and moderate deformation rates are established through the computation and analysis of exact Zimm's eigenvalues. It is shown that they can be expressed in terms of measurable parameters from intrinsic viscosity data. Under moderate deformation rates one needs to introduce a "slippage" between the solvent and the "smoothed" polymer to be able to describe shear-thinning behaviour.

Key words: Dilute polymer solution, slipping, Rouse-Zimm model, shear thinning

1. Rouse-Zimm model with slippage

One way to approach this problem is by using the mechanical modeling of polymer molecules based on sound physical considerations. For a dilute polymer solution the most successful model is the well-known Rouse-Zimm model [1, 2]. It should be noted that one of the assumptions of the original Rouse-Zimm model is that the surrounding Newtonian fluid is affinely deformed with an equivalent model continuum within the coils. This would not be the case under moderate shear rates because of the perturbation caused by increased deformation of the whole polymer coil, which has not been considered in the Rouse-Zimm theory.

Following Johnson and Segalman [3], a special modification is introduced, namely, that for the Stokes' force on the i-th bead (a constitutive element of the Rouse-Zimm-coil) the velocity of the solvent in the model will not be identified with the velocity of the "undisturbed" macroscopic continuum while the eigenvalues of the coils remain unchanged.

Our hypothesis is that respective motions should be connected corotationally while the rate of deformation allows for slippage between the two entities:

$$(\nabla \boldsymbol{v}^*) - (\nabla \boldsymbol{v}^*)^\dagger = (\nabla \boldsymbol{v}) - (\nabla \boldsymbol{v})^\dagger, \qquad (1)$$

$$(\nabla \boldsymbol{v}^*) + (\nabla \boldsymbol{v}^*)^\dagger = a_s(D)[(\nabla \boldsymbol{v}) + (\nabla \boldsymbol{v})^\dagger]. \qquad (2)$$

In our approach the slippage function a_s is a function of the rate of deformation tensor itself, i.e. its invariants. The value of a_s tends to one as D tends to zero, giving the original Rouse-Zimm model.

Now we can write down material functions from Rouse-Zimm model (SRZ) with slippage ($s = \sqrt{1 - a_s}$):

Shear viscosity:

$$\eta(\dot{\gamma}) = a_s n k T \sum_{\nu=1}^{N} \frac{t_0 \lambda_\nu^{-1}}{1 + (s t_0 \lambda_\nu^{-1} \dot{\gamma})^2} + \eta_s, \qquad (3)$$

First normal-stress coefficient:

$$\psi_1 = \frac{s_{11} - s_{22}}{\dot{\gamma}^2} = 2 a_s n k T \sum_{\nu=1}^{N} \frac{(t_0 \lambda_\nu^{-1})^2}{1 + (s t_0 \lambda_\nu^{-1} \dot{\gamma})^2}, \qquad (4)$$

Second normal-stress coefficient:

$$\psi_2 = \frac{s_{22} - s_{33}}{\dot{\gamma}^2} = -\frac{1}{2}(1 - a_s)\psi_1, \qquad (5)$$

Dynamic viscosity, real part:

$$\eta'(\omega) = n k T \sum_{\nu=1}^{N} \frac{t_0 \lambda_\nu^{-1}}{1 + (t_0 \lambda_\nu^{-1} \omega)^2} + \eta_s, \qquad (6)$$

Dynamic viscosity, imaginary part:

$$\eta''(\omega) = nkT \sum_{v=1}^{N} \frac{(t_0 \lambda_v^{-1})^2 \omega}{1 + (t_0 \lambda_v^{-1} \omega)^2} . \qquad (7)$$

The model coefficients in these equations consist of the eigenvalues λ_p, the submolecular parameters in natural time $t_0 = m$, and the slippage factor a_s. The number concentration of polymer molecules is n.

It will be shown that all these coefficients can be expressed in terms of measurable parameters.

2. Low-shear rheological properties

The eigenvalue functions

$$s_1 = \sum_{p=1}^{N} \lambda_p^{-1}, \quad s_2 = \sum_{p=1}^{N} \lambda_p^{-2}, \quad s_m = \lambda_m^{-1} \qquad (8)$$

which form basic rheological properties can be related to the molecular weight M and solvent effect through a regression analysis which leads to rather simple expressions instead of the complicated algebraic relations. We shall introduce the results into the material functions. Other eigenvalue functions must be calculated by computer. Because until now no simple analytical approximations for them have been found, our program for the near future is to develop useful simple expressions.

Thus at low shear rates the rheological properties can be rewritten without relaxation times. By simple superposition rules we have been able to take account of the molar mass distribution $\Phi(M)$. A survey of rheological properties at low shear rates is given by the eqs. (9) and (10).

Intrinsic viscosity:

$$[\eta] = K\bar{M}_\eta^a; \; \bar{M}_\eta = (\int_0^\infty M^{a+1} \Phi(M) dM/\bar{M}_n)^{1/a} \quad (9)$$

with

$$a = (11.3\, h^* + 1.10)^{-1/2},$$

$$K = 1.60\, \tilde{N} f b^2 / 6 \eta_s M_s (7.82\, M_s)^a, \quad M_s = M/N,$$

First normal stress coefficient:

$$\psi_1 = K' \int_0^\infty M^\beta \Phi(M) dM; \quad K' = \frac{nE(fb^2)^2}{18\, kTM_s^\beta} \qquad (10)$$

with

$$\beta = 2.06a + 1.90,$$

$$E = 0.105 - 0.0938 \exp(-11.65\, h^*)$$

where $\tilde{N}$ is Avogadro's number and η_s the solvent viscosity.

The submolecular parameters h^*, b, f, M_s cannot be deduced by the model itself. However, by comparison with intrinsic viscosity data [4] we can express them in terms of measurable parameters. By using eq. (9), we have obtained the expression

$$\log \frac{K}{K_\theta} = \left(\frac{1}{2} - a\right) \log(7.82\, M_s) + \left(\frac{h^* \alpha_s^3}{h_\theta^*}\right) \qquad (11)$$

where the intrinsic-viscosity constant K has been related to its value at the θ-temperature. $\alpha_s = b/b_\theta$, K_θ and a are measurable quantities.

Plotting the experimental results of $\log(K/K_\theta)$ as a function of $(a - \frac{1}{2})$, we found that the corresponding data of nearly all flexible polymer solvent systems fall on the same line through zero with a slope equal to 4.0.

If the molecular weight distribution $\Phi(M)$, solvent effect a and chain parameter K_θ are known, all the material functions in low shear rate can be calculated from the following equations (natural time m is given in cgs-units; C is the mass concentration):

$$\tau_p = \frac{m}{\lambda_p} \text{ with } m = 9.634 \cdot 10^{-4} \frac{\eta_s K_\theta}{T}, \qquad (12)$$

$$\tau_{\max} = \frac{m}{\lambda_{\min}} = m E^{1/2} (1.28 \cdot 10^3)^{-\beta/2} M^{\beta/2}, \qquad (13)$$

$$[\eta] = 10^{4(\frac{1}{2} - a)} K_\theta M_\eta^a, \qquad (14)$$

$$\psi_1 = 2.13 \cdot 10^{-14} \frac{C \cdot \eta_s^2 \cdot K_\theta^2 \cdot E}{kT \bar{M}_n (1.28 \cdot 10^3) \beta} \cdot \int_0^\infty M^\beta \Phi(M)\, dM, \qquad (15)$$

$$\eta'(\omega) = \frac{RT}{M} \int_0^\infty \sum_{p=1}^{N} \frac{m \lambda_p}{\lambda_p^2 + m^2 \omega^2} \Phi(M)\, dM + \eta_s, \quad (16)$$

$$\eta''(\omega) = \frac{RT}{M} \int_0^\infty \sum_{p=1}^{N} \frac{m^2 \omega}{\lambda_p^2 + m^2 \omega^2} \Phi(M)\, dM . \qquad (17)$$

3. Rheological properties under moderate shear rates

Now we return to the slippage function a_s. For shear flow we assume that the slippage a_s is a function of the shear rate. A quite good general approximation for *dilute polymer* solutions is found by comparison with published experimental data for variable shear viscosity. We find by regression analysis the following function for $s = (1 - a_s)^{1/2}$:

$$s = \frac{0.9\,\beta^{0.5}}{1 - \beta} \text{ with } \beta = \frac{[\eta]\,\eta_s M \dot{\gamma}}{RT} \tag{18}$$

where β is a dimensionless shear rate. It should be mentioned that this implies that Cox-Merz relation is not valid here. If the intrinsic viscosity is determined, the non-Newtonian properties – following our results – can be estimated for moderate shear rates.

For polydispersed polymer systems one has to calculate the integrals shown in eqs. (16) and (17), which need further effort to obtain manageable expressions.

Acknowledgements

This research was supported by Deutsche Forschungsgemeinschaft (DFG). One author (Y. T. H.) acknowledges support received from the Alexander-von-Humboldt foundation.

References

1. Zimm, B. H., J. Chem. Phys. **24**, 269 (1956).
2. Lodge, A. S., Y. J. Wu, Rheol. Acta **10**, 539 (1971).
3. Johnson, M. W., Jr., D. Segalman, J. Non-Newtonian Fluid Mech. **2**, 255 (1977).
4. Brandrup, F., E. H. Immergut, Polymer Handbook, Interscience (New York 1966).

(Received April 16, 1982)

Authors' address:

Prof. Dr. P. Schümmer, Dr. Y. T. Hsü
Institut für Verfahrenstechnik
RWTH Aachen
Turmstraße 46
D-5100 Aachen

Rheologica Acta Rheol. Acta **21**, 514 – 516 (1982)

Design and operation of the free jet elongational rheometer

P. Schümmer and K. H. Tebel

Institut für Verfahrenstechnik der Rheinisch-Westfälischen Technischen Hochschule Aachen

Abstract: The controlled free jet instability is used for elongational rheometry of polymer solutions. The instantaneous elongational viscosity represents a convenient comparative parameter. Its dependence on the operating parameters and the conditions of operation will be discussed.

Key words: Elongational rheometry, controlled jet breakup, free jet instability, elongational viscosity

1. Introduction

There are several cases of technical application [1, 2] where elongational flow of polymer solutions is of major importance, e.g. polymer flooding in enhanced oil recovery, flow through technical contractions (like granular beds and screen clothes), spinning from solution as well as drag reduction in turbulent flow. Last but not least there is a great interest in finding polymers useful for anti-misting of fuel in aviation.

Besides these cases of technical importance, data on elongational flows are needed for testing continuum models and molecular theories. But useful and handy elongational rheometers are rare and usually of a non-controllable type, where the flow is only approximately a pure uniaxial elongation.

During the investigation of controlled instability and break-up of Newtonian and non-Newtonian fluids we discovered a new method for the realization of a purely uniaxial elongational flow of polymer solutions [3]. Figure 1 shows that the elongational rheometer is based on the behaviour of viscoelastic liquid jets issuing from a harmonically vibrating cylindrical nozzle. As in the Newtonian case there is a regular jet surface with drop regions and filaments inbetween, but due to elastic properties the filaments disappear only asymptotically.

Within an average range of time it is possible to approximate the shape of the filaments by cylinders of local or time dependent radii.

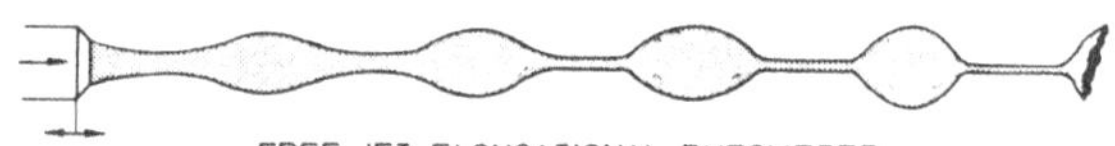

Fig. 1. Controlled free jet instability

This flow situation looks like a purely elongational flow where a cylindrical rod is stretched continuously in the axial direction. Its radius may be considered to vary only with time. The free surface is a coordinate surface in an orthogonal coordinate system [4]. With that the equivalent and generally applicable requirement for a purely elongational flow is fulfilled.

2. Theory

A model of the elongational flow regime containing two ideal geometries (cylinders and spheres) per wavelength λ, which is equal to the ratio of jet velocity and disturbance frequency, leads to the basic equations:

$$S_{zz}(t) - S_{rr}(t) = \frac{\sigma}{R(t)} - p_d\,, \tag{1}$$

$$\dot{\varepsilon}(t) = \frac{2\dot{R}(t)}{R(t)}\,, \tag{2}$$

$$\eta_d(t) = \frac{S_{zz}(t) - S_{rr}(t)}{\dot{\varepsilon}(t)}\,. \tag{3}$$

A simple integral balance applied to the cylindrical part of the filament, eq. (1), in connection with the definition of the strain rate for uniaxial elongation, eq. (2), are the basic equations for estimating the instantaneous elongational viscosity, eq. (3), from the filament profile alone without an additional force measurement.

The pressure p_d is negligible if the drop radius is much larger than that of the filament. In general it is possible to estimate p_d via a mass balance. Hence it follows that p_d can be taken as a constant counter pressure, a result that has been verified experimentally.

The strain rate has to be calculated by computer curve fitting of the filament radius-time relationship followed by differentiation.

3. Experiment

The polymer solutions issue from a short length, cylindrical nozzle of 1 mm in diameter and length into the ambient air while the nozzle vibrates under reproducible conditions in the flow direction, driven by an electromagnetic vibration system. All optical measurements are made using a standard camera system. The negatives have to be analysed with a suitable optical equipment, such as a microscope, a simple projector or a digitizer. A micro computer system is used for curve fitting and other complex computations.

Most of the experimental investigations were carried out with aqueous solutions of polyvinylalcohol (PVA).

4. Results

The experiments have confirmed the suitability of the free jet elongational rheometer. But it is still doubtfull whether the resulting instantaneous elongational viscosity represents a parameter that is independent of the chosen operating parameters, i.e. if there is a general function between the instantaneous values of strain rate and stress. Such a relation exists if the motion is a quasi-steady extension or if there is a unique history for all possible elongations [5]. For this purpose we have compared filament profiles and strain histories for different operating conditions.

In the following the zero-shear viscosity η_0, the surface tension σ and the radius of the first filament R_0 are used for the introduction of dimensionless values:

$$\eta_d^+ = \eta_d/\eta_0, \quad \eta_s^+ = \eta_s/\eta_0, \quad \dot{\varepsilon}^+ = \dot{\varepsilon} \cdot t_s,$$

$$R^+ = R/R_0, \quad p_d^+ = p_d/(\sigma/R_0),$$

where $t_s = 6\eta_0 R_0/\sigma$.

Figure 2 summarizes all elongational viscosity and strain rate data. The operating conditions are indicated as amplitude (in µm), frequency (in Hertz) and velocity (in m/s), e.g. No. 1: 29 µm/400 Hz/ 3 m/s.

In general there is no unique material function and the elongational viscosities do not fall on the same curve. But most of the history curves have shown that with increasing time the experiment becomes a quasi-steady extension. Figure 3 shows an example.

Now some more comments on the computer-curve fitting: Instead of using purely mathematical curve fitting methods, a special material equation of an Oldroyd-type is introduced which leads, in connection with the basic equations, to an ordinary differential equation in the filament radius. By varying the unknown rheological parameter the best fit of the experimental data has to be found. Figure 4 shows a plot of the relaxation time of an upper convected

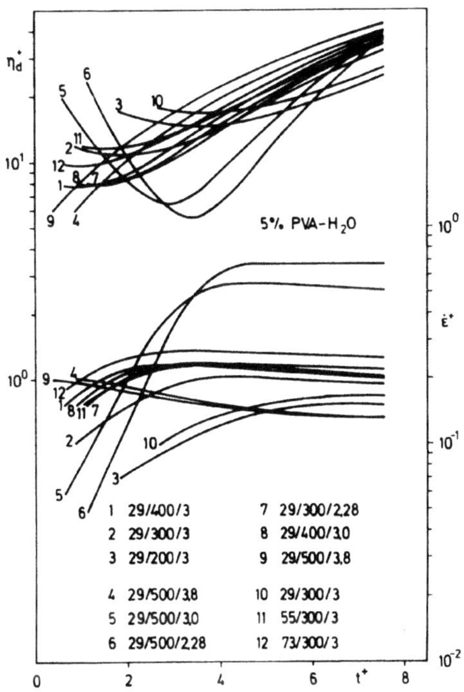

Fig. 2. Elongational data at different operating conditions. The meaning of the subsequent numbers are: disturbance amplitude (in µm), disturbance frequency (in Hz), mean exit velocity (in m/s)

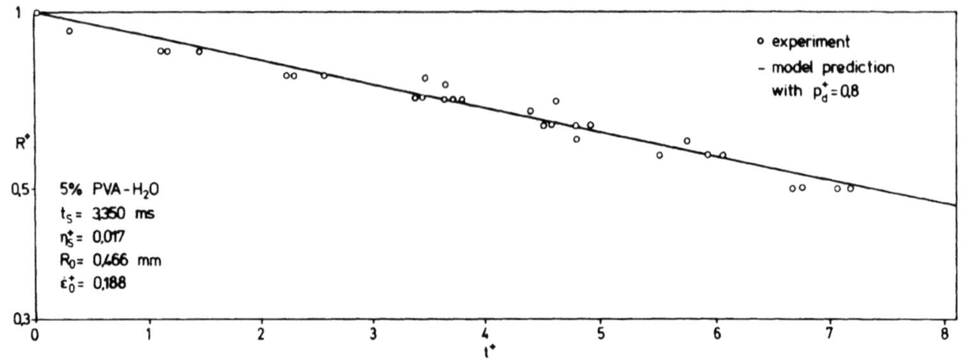

Fig. 3. Quasi-steady extension

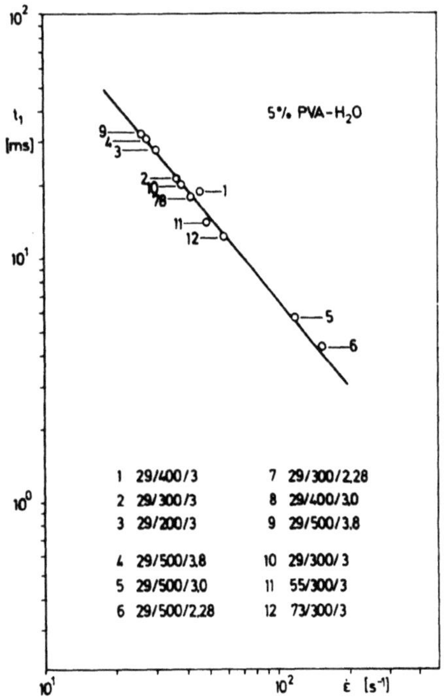

Fig. 4. Relaxation time of an upper convected Jeffreys fluid (for nomenclature see fig. 2)

Jeffreys fluid versus the final strain rate for each operating condition.

5. Conclusions

In conclusion one can say that it is possible to use the flow in the filaments of a harmonically disturbed viscoelastic jet for elongational rheometry of polymer solutions. The frequency of the vibrating nozzle is an additional operating parameter which influences the purely elongational flow. The elongational data follow from the filament profile only, so that no complicated force measurement has to be used. Because of the constant jet velocity a conversion between position and time is easily done.

In general there is no unique deformation history but for suitable operating conditions a transient elongational flow with an approximately constant rate of strain can be realized. In any case convenient parameters can be achieved.

To estimate the elongational viscosity it is not necessary to introduce an explicit material equation. But in the case of only slightly elastic polymer solutions it has been found that a modified upper convected Jeffreys model with a deformation-dependent relaxation time-parameter describes the fluid behaviour in the filament flow quite well. The retardation time is of minor importance.

References

1. Durst, F., R. Haas, W. Interthal, T. Keck, Chem.-Ing. Tech. **54**, 213 (1982).
2. Goren, S. L., M. Gottlieb, J. Fluid Mech. (to be published).
3. Tebel, K. H., Dissertation, RWTH Aachen (to be published).
4. Mac Sporran, W. C., J. Non-Newtonian Fluid Mech. **8**, 119 (1981).
5. Weinberger, C. B., J. D. Goddard, Int. J. Multiphase Flow **1**, 465 (1974).

(Received April 30, 1982)

Authors' address:

Prof. Dr. P. Schümmer, Dipl.-Ing. K. H. Tebel
Institut für Verfahrenstechnik
RWTH Aachen
Turmstraße 46
D-5100 Aachen

Rheologica Acta Rheol. Acta **21**, 517 – 520 (1982)

Measurement of dynamic properties of dilute polymer solutions with a Laser-Doppler velocimeter

P. Schümmer and W. Zang

Institut für Verfahrenstechnik der Rheinisch-Westfälischen Technischen Hochschule Aachen

Abstract: The complex velocity field of an oscillating Couette flow is measured with a Laser-Doppler velocimeter. Different evaluation methods are used for the determination of material functions such as relaxation times and the dynamic complex viscosity.

Key words: Rheometry, highly diluted polymer solution, material function

Even in the linear range of a flow curve highly diluted polymer solutions show elastic behaviour. This may cause important effects even if the system times are only of the same order as the relaxation times of the fluid. Our intention is to determine the material functions of elastic fluids within a viscosity range of 1 – 100 mPa s.

Normally an oscillating fluid shows a position-dependent phase shift and amplitude variation as a function of its properties and the frequency. The local rate of flow can be measured with the aid of a Laser-Doppler velocimeter (LDV) without any mechanical contact.

The test liquid is strained in a cylinder gap-system which is oscillating in the azimuthal direction. In the interior of the gap-system an oscillating flow develops because of inertia. The amplitude variation and the phase shift of the velocity in the middle of the gap as compared to the velocity at the walls and their dependence on the stimulating frequency is found by the previously mentioned technique of LDV and a subsequent Fourier-Analysis.

Figure 1 shows the principle of our test equipment.

The description of the geometrical dimension can be taken from figure 2.

The ϕ-component of the balance of momentum for the laminar shear flow in question is:

$$\rho \frac{\partial u_\phi}{\partial t} = -\frac{1}{r^2} \frac{\partial}{\partial r} (r^2 \sigma_{r\phi}) . \tag{1}$$

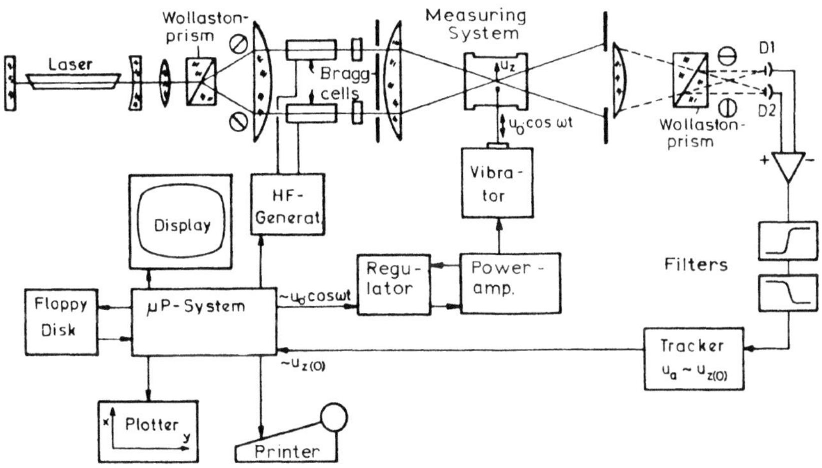

Fig. 1. The test equipment

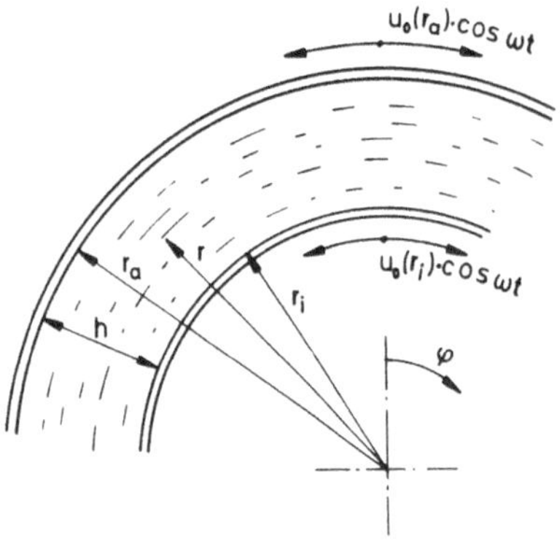

Fig. 2. Geometrical scheme, U_0 = amplitude of the velocity

In the case of small shear rates within steady oscillatory conditions, the theory of linear viscoelasticity can be applied

with

$$\bar{\sigma}_{r\phi}(r, \omega) = \eta^* \frac{\partial \bar{u}_\phi}{\partial r} \tag{2}$$

where

$$\eta^*(\omega) = \eta'(\omega) - i\eta''(\omega) \tag{3}$$

is the complex viscosity.

We thus obtain a homogeneous differential equation of second-order:

$$\frac{d^2\bar{u}_\phi}{dr^2} + \frac{2}{r}\frac{d\bar{u}_\phi}{dr} - i\omega\frac{\rho}{\eta^*}\bar{u}_\phi = 0\,. \tag{4}$$

In the corresponding differential equation for planar parallel plates in cartesian co-ordinates (fig. 3) the term of first-order does not appear.

$$\frac{d^2\bar{u}_z}{dr^2} - i\omega\frac{\rho}{\eta^*}\bar{u}_z = 0\,. \tag{5}$$

This shows that the term of first order can be neglected for a sufficiently large radius.

To simplify matters, the radius of the real gap-system is therefore chosen to be large in comparison to the slit height and the equation is used in the system of cartesian co-ordinates.

As shown in figure 3 the liquid is located between two infinitely extented plates, which both move sinusoidally. From this the following boundary values result:

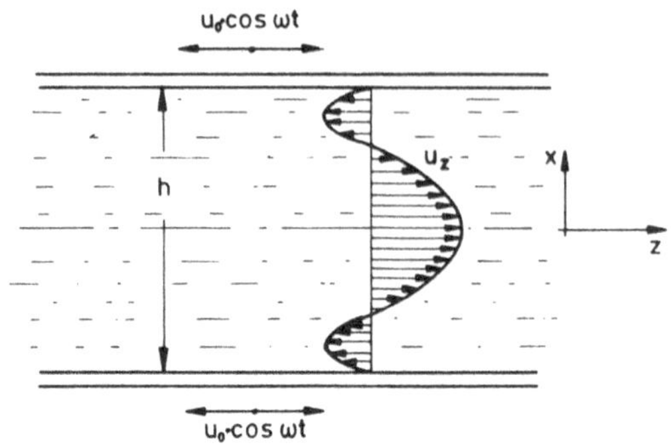

Fig. 3. Geometrical scheme, U_0 = amplitude of the velocity

$$x = 0: \qquad \frac{\partial \bar{u}_z}{\partial x} = 0 \quad \text{(symmetry)}, \tag{6}$$

$$x = \pm\frac{h}{2}: \quad \bar{u}_z = u_0\,. \tag{7}$$

The complex velocity profile resulting from the differential equation is:

$$\bar{u}_z(x) = u_0\frac{\cosh(\alpha x)}{\cosh(\alpha\frac{h}{2})} \quad \text{with} \quad \alpha = \sqrt{i\omega\frac{\rho}{\eta^*}}\,. \tag{8}$$

At the measuring point $x = 0$ the equation simplifies to

$$\bar{u}_z(0) = u_0\frac{1}{\cosh(\alpha\frac{h}{2})}\,. \tag{9}$$

From this the formulas for amplitude variation A_u and phase shift ϕ_u can be calculated:

$$A_u(0) = (\cosh^2(x)\cos^2(y) + \sinh^2(x)\sin^2(y))^{-1/2}, \tag{10}$$

$$\Phi_u(0) = \arctan\frac{\sinh(x)\sin(y)}{\cosh(x)\cos(y)}, \tag{11}$$

with

$$x = \left(\frac{h^2\rho\omega}{8(\eta'^2 + \eta''^2)}\right)^{1/2}\left[(\eta''^2 + \eta'^2)^{1/2} - \eta''\right]^{1/2}, \tag{12}$$

$$y = \left(\frac{h^2\rho\omega}{8(\eta'^2 + \eta''^2)}\right)^{1/2}\left[(\eta''^2 + \eta'^2)^{1/2} + \eta''\right]^{1/2}, \tag{13}$$

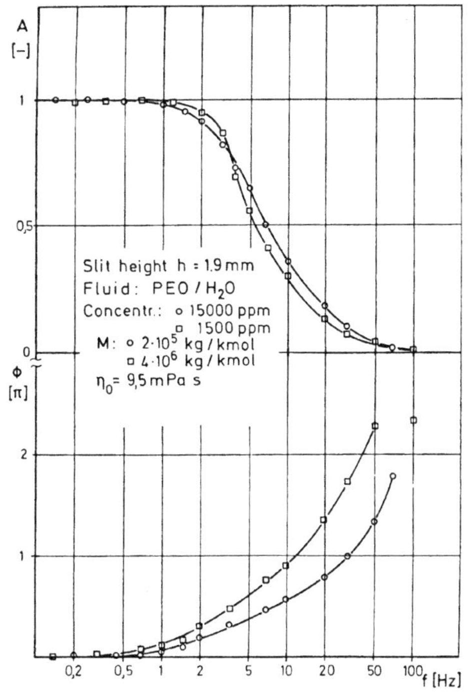

Fig. 4. The measured amplitude variation A and phase shift ϕ as a function of the frequency f

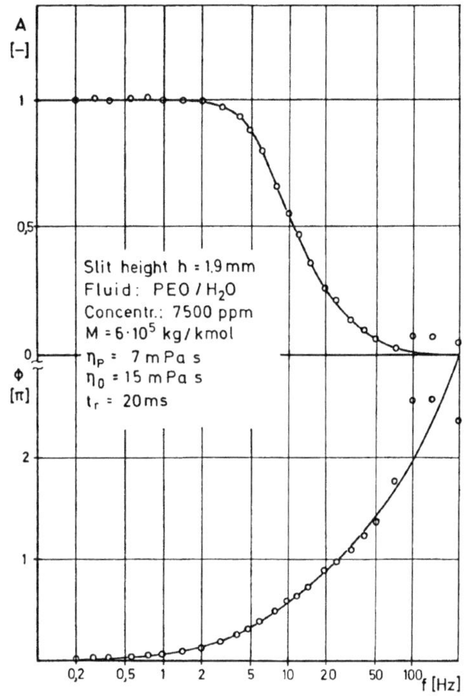

Fig. 6. Comparison of calculated curves with measured values for a slightly elastic fluid

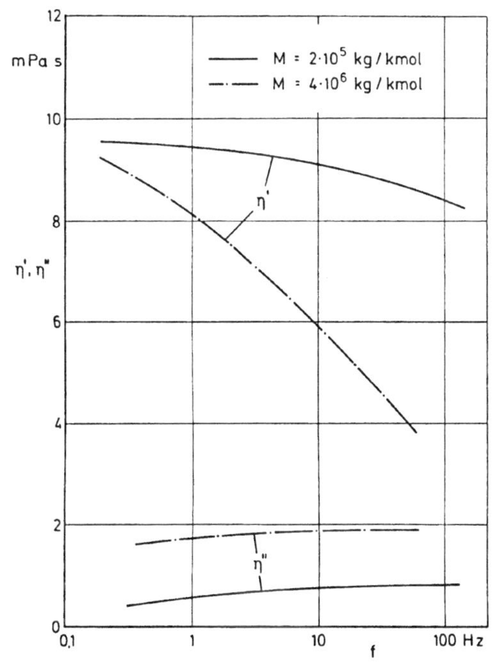

Fig. 5. Real (η') and imaginary part (η'') of the complex viscosity evaluated from the values shown in figure 4

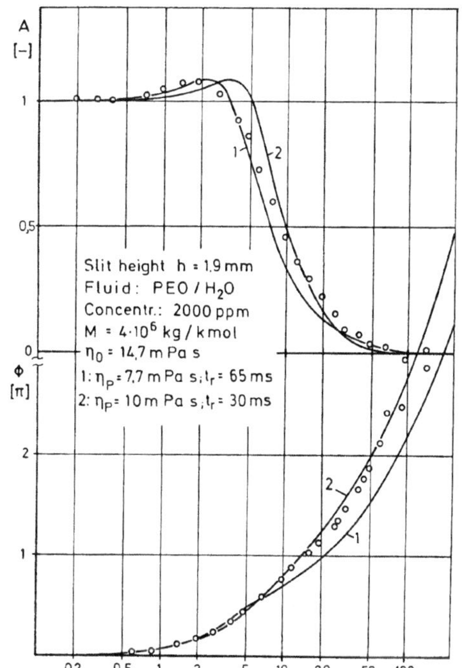

Fig. 7. Comparison of calculated curves with measured values for a moderately elastic fluid

With the aid of these formulae the complex viscosity as a function of the frequency can be found by iteration using a computer program. Figure 4 shows some results for two different solutions of Polyethylenoxide (PEO) in water with different molecular masses but identical zero-shear viscosities.

Figure 5 shows the iteratively calculated curves of the real and imaginary part of the complex viscosity as functions of the frequency. These curves completely describe the viscous and elastic material behaviour under low stress.

For better practical use or in order to derive molecular data from measurements, rheologists use constitutive equations which describe the material response more or less accurately. Such constitutive equations may contain parameters like zero-shear viscosity, polymer viscosity or relaxation times. These parameters can be determined from the complex viscosity. For instance the real and imaginary parts of the complex viscosity for a Jeffreys' constitutive equation are

$$\eta' = \eta_s + \frac{\eta_P}{1 + \omega^2 t_1^2}, \quad \eta'' = \omega t_1 \frac{\eta_P}{1 + \omega^2 t_1^2}, \quad (14)$$

where t_1 is the relaxation time. Combining these formulae for phase shift and amplitude variation the parameters of a constitutive equation can also be found from the phase shift and the amplitude without having to calculate the complex viscosity. In the present case the viscosity η_s is not the viscosity of the solvent of the polymer solution, it is rather a pure fitting parameter.

Figure 6 shows the agreement of the calculated curves with the measured values for the Jeffreys' constitutive equation. Figure 7 shows these curves for a more elastic fluid. It can be seen on these graphs that an increase of the elastic effects coincides with a deterioration of the agreement. This shows that the model hypothesis is no longer valid for highly elastic materials. The instabilities that arise for frequencies larger than a certain critical frequency, which is dependent on the slit height and material parameters, should be mentioned here. Beginning at the critical frequency these instabilities become apparent as strong deviation in the measured values. This means that an analysis in the form presented here is only possible up to this critical frequency.

Nevertheless it is obviously possible to classify the starting point of the instabilities as an extra interesting measurement value. Its analysis is yet to be discussed.

Acknowledgement

This research was supported by the Deutsche Forschungsgemeinschaft (DFG).

(Received May 3, 1982)

Authors' address:

Prof. Dr. P. Schümmer, Dipl.-Ing. W. Zang
Institut für Verfahrenstechnik
RWTH Aachen
Turmstraße 46
D-5100 Aachen

Periodic flows of pseudoplastic fluids in pipes

V. Sobolík, O. Wein, and P. Mitschka

Institute of Chemical Process Fundamentals, Czechoslovak Academy of Sciences, Prague

Abstract: Experimental results on the flow enhancement during flow of clay slurries ($n = 0.15$) through oscillating pipes are compared with theoretical predictions. The agreement is fairly good, especially in the oscillating boundary layer flow regime. Flow enhancement of order 10^8 was found with the slurries used.

Key words: Flow enhancement, oscillating pipe, concentrated suspension, oscillating boundary layer

The rheological behaviour of constitutively nonlinear materials under oscillatory shear stresses or deformations parallely superposed on a steady shear flow differs qualitatively not only from that of Newtonian fluids but also from the prediction of linear viscoelastic theories, as the steady and oscillatory components of the motion influence each other. Even in the case of small oscillatory perturbations the nonlinear operators describing oscillatory response depend on the steady shear rate [1]. In this paper another nonlinearly coupled effect will be treated arising from the case where a superposed oscillatory component influences the relation between the steady characteristics of a flow [2, 3]. The magnitude of this effect, which is called flow enhancement [4] (FE), is controlled by the nonlinearity of the viscosity function [2 – 6] and increases strongly with decreasing flow index, n. With oscillations of sufficiently high frequency it is possible to influence the fluidity, especially of viscoplastic materials, to a sufficient extent for the carrying out of common operations under gravity flow conditions (filtration, degassing, free convection, film flow [5], etc.).

Till now the small-scale studies of both mentioned interactions were mostly confined to nearly-viscometric flows [1 – 4]. It is evident, however, that qualitatively new rheological effects [6], in particular high FE [5] can only be expected in the region of large amplitudes of superposed oscillatory stresses. It seems that memory effects play only a secondary role in the relevant region of high oscillatory stresses and that the process is dominated by purely viscous nonlinear phenomena [2 – 8].

Even if we neglect memory effects, the large amplitudes and frequencies of shear oscillations introduce considerable complications into the mathematical modelling of the problem under consideration, as both the nonlinear viscosity and inertial forces have to be treated in an adequate way. The inertial forces usually dominate in the region of interest and the flow acquires a boundary-layer character [9 – 11]. The value of a single criterion of rheodynamical similarity, Re, is sufficient to distinguish between the boundary-layer or the opposite creeping flow regime. For the flow of power-law fluids, $\tau[\dot{\gamma}] = K\dot{\gamma}^n$ with density ρ in a longitudinally oscillating (amplitude a, angular frequency ω) channel of hydraulic radius R_h the Reynolds number is defined as

$$Re = R_h \rho a \omega^2 / \tau[\omega a / R_h] = R_h^{1+n} a^{1-n} \omega^{2-n} \rho K^{-1}. \tag{1}$$

Under creeping flow conditions [8, 11, 12], $Re \ll 1$, the measure of FE defined as the ratio of volume flow rates under oscillating and steady conditions without oscillation, $I = Q_{osc}/Q_{std}$, can be expressed by the simple relation

$$I = \frac{1}{2\pi} \int_0^{2\pi} [1 + Fr \sin T]^m dT, \quad m = 1/n,$$

$$[\pm x]^m = \pm [x]^m. \tag{2a,b,c}$$

817

At $Re \ll 1$ the criterion of rheodynamical similarity Fr, which together with n and Re fully defines the rheodynamical conditions of superposed flows, represents the ratio of oscillatory and steady stresses. Thus Fr is analogous to the parameter ε introduced in a number of papers [2–4, 6, 7]. In the given case (flow through an oscillating pipe) Fr is defined by the relation

$$Fr = \rho a \omega^2 / (-\partial_z \bar{p} + \rho g_z) \tag{3}$$

where g_z stands for the component of gravity acceleration in the flow direction.

High values of FE can be achieved by increasing a and ω. However, from the definition of Re it is obvious that Re increases with increasing a and ω as well and thus it may reach values $Re \gg 1$, where the flow has a boundary-layer character [8, 9]. The amplitude of the oscillatory stresses is then large only in a thin layer close to the wall of the oscillating pipe and decreases rapidly with increasing distance from it. Assuming the oscillatory stresses in the boundary layer to be greater than the steady stresses, i.e. $Fr\, Re^{-m/(1+m)} \equiv B \gg 1$, we arrive [8, 9, 13] at the following estimate for FE:

$$I = \kappa(m) Fr^{m-1} Re^{-m^2/(m+1)}, \tag{4}$$

$$\kappa(m) = \frac{(m+3)}{(m-1)\sqrt{2}} \left(\frac{m+1}{2}\right)^{2m/(m+1)} \left(\frac{\Gamma(m/2)\, m}{\Gamma(m/2 + 1/2)\sqrt{2}}\right)^{1/(m+1)} \tag{5}$$

Relation (4) represents an asymptote for $Re \to \infty$ and does not yield satisfactory estimates of FE for the common values of $1 < Re < 10$. Some corrections improving the estimates according to (4) down to values of $Re = 1$, $B = 1$ have been given in [13].

As has been shown [14], there is no principal difference between periodic flows generated by a pulsating pressure gradient in a static channel or by oscillations of the channel walls. Both situations can be described by an identical mathematical model provided that the apparent amplitude a is properly expressed in terms of the amplitude of oscillatory pressure drop [8]. Sufficiently high oscillatory stresses can be achieved by technically acceptable means in oscillating channels only. For such a situation Mena and Manero [6] found $I = 10$ for polymer solution with $n = 0.5$ and values of about $I = 100$ were observed [8] for suspensions with $n = 0.33$. In the following we present some results of our experiments with colloidal suspensions having $n = 0.15$ where $I = 10^{12}$ was achieved. Indeed such a high value is not too surprising if we realize the meaning of the definition of I. The value of I may become infinite if the steady stresses are lower than the yield stress.

A very good agreement of experimental I values with the theory according to eqs. (2) and (4) is remarkable for the further mentioned experiments. Reliable data have been obtained using an experimental apparatus (see fig. 1) which enabled exact control and recording of the steady and oscillatory components of the motion investigated. The vibration engine of our own construction [15] made simultaneous measurement of velocity (loudspeaker) and acceleration possible.

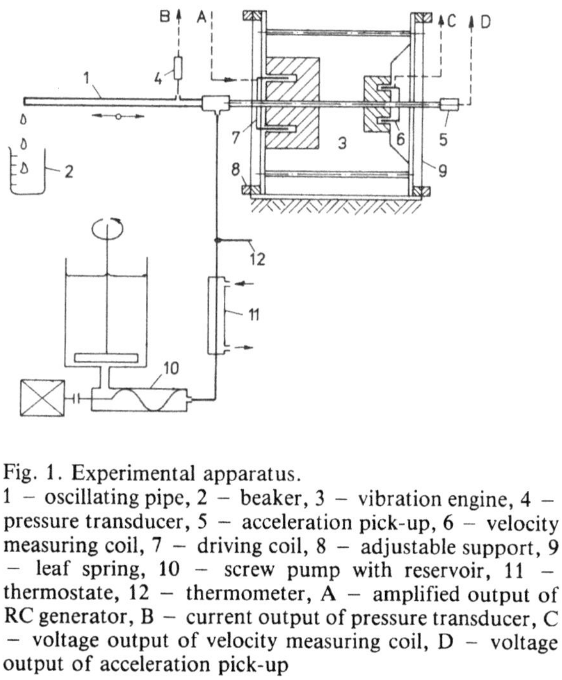

Fig. 1. Experimental apparatus.
1 – oscillating pipe, 2 – beaker, 3 – vibration engine, 4 – pressure transducer, 5 – acceleration pick-up, 6 – velocity measuring coil, 7 – driving coil, 8 – adjustable support, 9 – leaf spring, 10 – screw pump with reservoir, 11 – thermostate, 12 – thermometer, A – amplified output of RC generator, B – current output of pressure transducer, C – voltage output of velocity measuring coil, D – voltage output of acceleration pick-up

A set of kaoline-water suspensions with $n = 0.15$, and consistencies K within the range 22–50 Pa s^n and with densities $\rho \in (1240; 1320)$ kg m^{-3} was used in our experiments. The amplitudes and angular frequencies of the harmonic motion of the capillary tube were adjusted in the range $a \in (0.03; 4)$ mm and $\omega \in (250; 790)$ rad s^{-1}. Three capillaries of radii $R = 2$, 3 and 4 mm and lengths of 300 mm up to 500 mm were used. The flow rates and pressure gradients were varied in the range $Q \in (0.14; 2.4)$ $cm^3\, s^{-1}$ and $\partial_z \bar{p} \in (84; 10^5)$ Pa m^{-1} respectively.

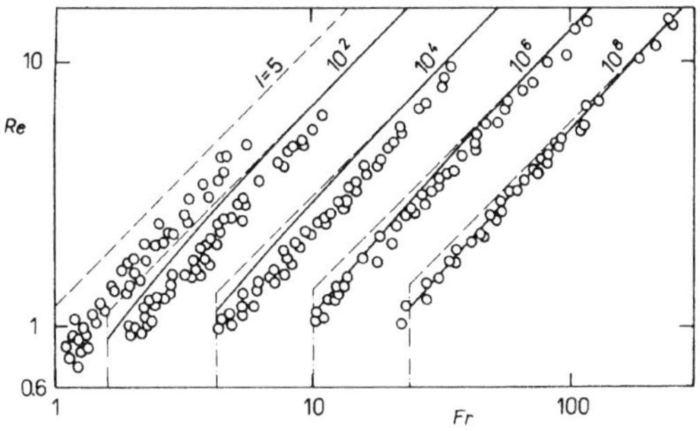

Fig. 2. Experimental data on I in Re-Fr diagram, $n = 0.15$. Dashed lines correspond to the boundary layer theory, eq. (4). Dotted and dashed lines correspond to the creeping flow theory, eq. (2). Full lines represent corrected boundary layer theory [13]

Experimental data concerning FE and the theoretical estimates according to eq. (4) (corrected for finite values of Re and B [14]) are shown in topographic projection $I(Fr, Re)$ = const. in figure 2.

A fairly good agreement of the theory with the experimental results, which is in the proper region $Re \gg 1$, $B \gg 1$ even quantitative, confirms not only the possibility of a realistic mathematical modelling of nonlinear material behaviour at high shear oscillations, but also the necessity of identifying qualitatively different rheodynamical regimes. Anyway the influence of inertial forces, which is likely the cause of discrepancies mentioned in some recent publications [2, 4, 6], has to be considered carefully.

References

1. Powell, R. L., W. H. Schwarz, J. Rheol. **23**, 323 (1979).
2. Barnes, A. H., P. Townsend, K. Walters, Rheol. Acta **10**, 517 (1971).
3. Jones, T. E. R., K. Walters, J. Phys. **A4**, 85 (1971).
4. Phan-Tien, N., Rheol. Acta **19**, 285 (1980).
5. Sobolík, V., O. Wein, P. Mitschka, Rheol. Acta **16**, 394 (1977).
6. Manero, O., K. Walters, Rheol. Acta **19**, 277 (1980).
7. Sundstrom, D. V., A. Kaufman, Ind. Eng. Chem. Proc. Des. Dev. **16**, 320 (1977).
8. Wein, O., V. Sobolík, Coll. Czech. Chem. Commun. **45**, 1010 (1980).
9. Wein, O., Coll. Czech. Chem. Commun. **44**, 2908 (1979).
10. Jones, J. R., T. S. Walters, Rheol. Acta **6**, 240 (1967).
11. Sicardi, S., G. Baldi, A. Gianetto, Ing. Chim. Ital. **11**, 105 (1975).
12. Edwards, M.F., D. A. Nellist, W. L. Wilkinson, Chem. Eng. Sci. **27**, 545 (1972).
13. Wein, O., V. Sobolík, to be published.
14. Wein, O., Coll. Czech. Chem. Commun. **44**, 3231 (1979).
15. Sobolík V., O. Wein, P. Mitschka, Chem. průmysl **27**, 500 (1977).

(Received April 15, 1982)

Authors' address:

Dr. V. Sobolík, Dr. O. Wein, Dr. P. Mitschka
Institute of Chemical Process Fundamentals
Czechoslovak Academy of Sciences
Rozvojová 135
CS-16502 Praha 6 – Suchdol
(Czechoslovakia)

Rheologica Acta Rheol. Acta **21**, 524 – 526 (1982)

Effect of small-amplitude vibrations on viscoelastic flows in a plane slider bearing

S. Zahorski

Institute of Fundamental Technological Research, Polish Academy of Sciences, Warszawa (Poland)

Abstract: The qualitative changes of dynamic lift and friction forces caused by small-amplitude harmonic vibrations superimposed on flows in a plane slider bearing are considered for simple viscous and viscoelastic lubricating fluids. Low- and high-frequency disturbances are analysed in greater detail and the most beneficial situations discussed.

Key words: Vibration, viscoelastic flow, slider bearing

Many previous results on additional vibrations superimposed on various flows of viscoelastic fluids can be found in references [1 – 3]. In the present note we consider the dynamic lift and friction forces caused by small-amplitude harmonic vibrations superimposed on viscous and viscoelastic flows in a plane slider bearing (fig. 1). An effect of low- and high-frequency vibrations (including inertia terms) is discussed in greater detail.

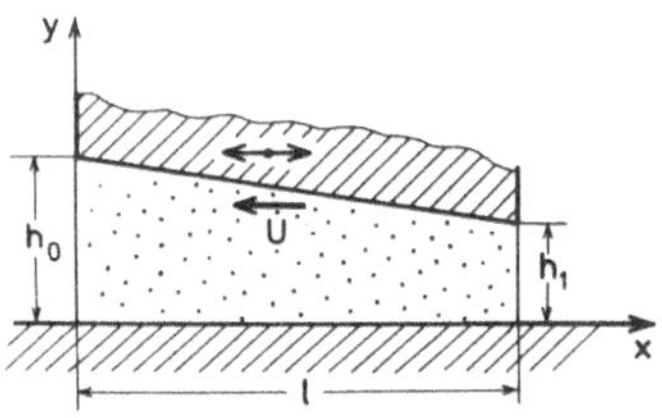

Fig. 1. Geometry of a plane slider bearing

First, we define the following small-amplitude parameter:

$$\varepsilon = \frac{\alpha\omega}{U} \ll 1\,, \tag{1}$$

where U denotes the constant velocity of a fundamental flow, α is the amplitude of vibrations, and ω the angular frequency. In all cases we assume that the "vibrational" Reynolds number

$$Re_v = \frac{\rho\alpha\omega h_0}{\eta} \ll 1\,, \tag{2}$$

where ρ is the density, and η the characteristic constant viscosity of a fluid. For high-frequency vibrations we assume, moreover, that the "frequentative" Reynolds number

$$Re_\omega = \frac{\rho\omega h_0^2}{\eta} \gg 1\,; \tag{3}$$

then, the conditions (2) and (3) can be satisfied simultaneously only for $\alpha/h_0 \ll 1$.

It can be shown that the Navier-Stokes equations take, under the lubrication approximation, the following linearized form:

$$-\frac{\partial p}{\partial x} + \eta\frac{\partial^2 u}{\partial y^2} = \rho\frac{\partial u}{\partial t}\,, \qquad -\frac{\partial p}{\partial y} = 0\,, \tag{4}$$

and the boundary conditions are

$$u(x,0) = 0, \quad u(x,h) = U(1 + \varepsilon\exp i\omega t)\,,$$

$$p(0,y) = p(l,y) = 0\,, \tag{5}$$

where u denotes the horizontal velocity component, p is the hydrostatic pressure, and h refers to current heights of a slit (fig. 1).

For lubricating fluids, we apply either the model of an incompressible second-order fluid, viz.

$$T = -p\mathbf{1} + \eta A_1 + \alpha_1 A_2 + \alpha_2 A_1^2, \quad \mathrm{tr} A_1 = 0\,, \tag{6}$$

where η, α_1, α_2 are the material constants and A_1, A_2 the Rivlin-Ericksen kinematic tensors, or the model of a generalized Newtonian fluid with shear-dependent viscosity, viz.

$$T = -p\mathbf{1} + (\eta - \tfrac{1}{2}\eta_2 \mathrm{tr} A_1^2) A_1\,, \quad \tfrac{1}{2}\mathrm{tr} A_1^2 = \kappa^2\,, \tag{7}$$

where $\eta_2 > 0$ is responsible for a decrease of viscosity with increasing rates of shear κ. One may expect that these models lead to reasonable results, if the numbers:

$$De = -\frac{\alpha_1 U}{\eta h_0}\,, \quad V = \frac{\eta_2 U^2}{\eta h_0^2}\,, \quad De_v = -\frac{\alpha_1 \alpha \omega}{\eta h_0} \tag{8}$$

are sufficiently small.

Introducing the velocity and pressure fields in the form:

$$u = u_N + u' + \ldots\,, \quad p = p_N + p' + \ldots\,, \tag{9}$$

where the subscript N refers to Newtonian solutions, we arrive at

$$u' = 0\,, \quad p' = \frac{\alpha_1}{\eta}\frac{dp_N}{dt} + \left(\frac{3}{2}\alpha_1 + \alpha_2\right)\kappa_N^2\,, \tag{10}$$

since the Tanner theorem [4] holds for second-order fluids. Similarily, for fluids with shear-dependent viscosity, we obtain either

$$p_P' = -\frac{3}{2}\eta_2 \int \left[\left(\frac{\partial u_N}{\partial y}\right)^2 \frac{\partial^2 u_N}{\partial y^2}\bigg|_{y=0} + \left(\frac{\partial u_N}{\partial y}\right)^2 \frac{\partial^2 u_N}{\partial y^2}\bigg|_{y=h}\right] dx + C_P\,, \tag{11}$$

or

$$p_G' = -3\eta_2 \int \left[\frac{1}{2}\left(\frac{\partial u_N}{\partial y}\bigg|_{y=0} + \frac{\partial u_N}{\partial y}\bigg|_{y=h}\right)\right]^2 \cdot \left[\frac{1}{2}\left(\frac{\partial^2 u_N}{\partial y^2}\bigg|_{y=0} + \frac{\partial^2 u_N}{\partial y^2}\bigg|_{y=h}\right)\right] dx + C_G\,, \tag{12}$$

where the approximate intergrations (for $u' \equiv 0$) are performed either at the mean pressure or the mean shear gradient, respectively. In both cases $p_P'(0) = p_P'(1) = p_G'(0) = p_G'(1) = 0$.

In what follows, the perturbation method similar to that presented in reference [3] is applied. We look for solutions in the form:

$$u = u_0 + \varepsilon u_1 \exp i\omega t + \varepsilon^2 (u_2 + \ldots)\,,$$
$$p = p_0 + \varepsilon p_1 \exp i\omega t + \varepsilon^2 (p_2 + \ldots)\,, \tag{13}$$

where u_0, p_0 refer to steady solutions and only real parts of the functions involved are implied. For Newtonian fluids, we arrive at

$$u_0 = u_1 = \frac{U}{h}y + \frac{1}{2\eta}\frac{dp_0}{dx}(y-h)y\,,$$

$$\frac{dp_0}{dx} = \frac{dp_1}{dx} = \frac{6\eta U}{h^2}\left(1 - \frac{H}{h}\right),$$

$$H = \frac{2h_0 h_1}{h_0 + h_1}\,, \tag{14}$$

in the case of low-frequency vibrations ($Re_v = 0$), and at the same expressions for u_0, p_0 but

$$u_1 = Ue^{-k(h-y)} = Ue^{-\nu(h-y)}[\cos\nu(h-y) + i\sin\nu(h-y)]\,,$$

$$p_1 = 0\,, \quad k = (1+i)\nu\,, \quad \nu^2 = \rho\omega/2\eta\,, \tag{15}$$

in the case of high-frequency vibrations ($\nu h \gg 1$ or $Re_\omega \gg 1$).

On the basis of the above equations the dynamic lift and friction forces can be determined. To this end, however, an averaging procedure over one cycle of vibrations should be applied, viz.

$$\langle F \rangle = \frac{\omega}{2\pi}\int_0^{2\pi/\omega} F\left(Reu, Re\frac{\partial u}{\partial y}, \ldots\right) dt\,, \tag{16}$$

where only real parts of velocity derivatives are implied in the integrand. Thus, the averaged dynamic forces up to terms of order ε^2 can be obtained in the form of sums: $\langle F \rangle = F_0 + \langle F_2 \rangle$, since $\langle F_1 \rangle \equiv 0$.

Further numerical calculations have been performed for the ratios:

$$I = \frac{\langle F \rangle - F_0}{F_0} = \frac{\langle F_2 \rangle}{F_0}\,, \tag{17}$$

characterizing relative increases or decreases of the dynamic forces as compared with the corresponding steady values. For example, an enhancement of the high-frequency lift force caused by normal-stress effects is shown in figure 2, for various $\lambda = h_1/h_0$. The low-frequency lift forces behave in a similar way. In figure 3 and 4, relative changes of the dynamic friction forces are plotted for low- and high-frequency vibrations, respectively. It can be concluded that superimposed vibrations are generally beneficial for the lift forces caused by normal stresses and negative for the lift forces caused by shear-dependent (decreasing) viscosities. The dynamic friction forces behave in a quite different way: an enhancement is observed for low-frequency vibrations and a reduction – for high-frequency disturbances. A plot of limit Re_ω, for which the above effects are mutually cancelled, is presented in figure 5. Thus, a range of validity of the methods applied can be estimated.

In a similar way, the case of moving and vibrating base as well as the mixed cases, in which vibrations are superimposed on the parts of bearing remaining stationary in a fundamental flow, can be considered in greater detail. It may be deduced that the most beneficial situations occur when either stationary parts vibrate with low frequencies or moving parts vibrate with sufficiently high frequencies.

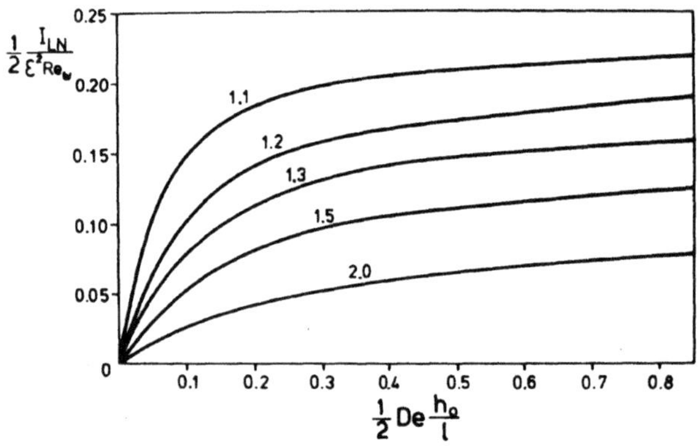

Fig. 2. Relative increase of the dynamic lift force caused by normal stresses for different values of λ. High-frequency vibrations

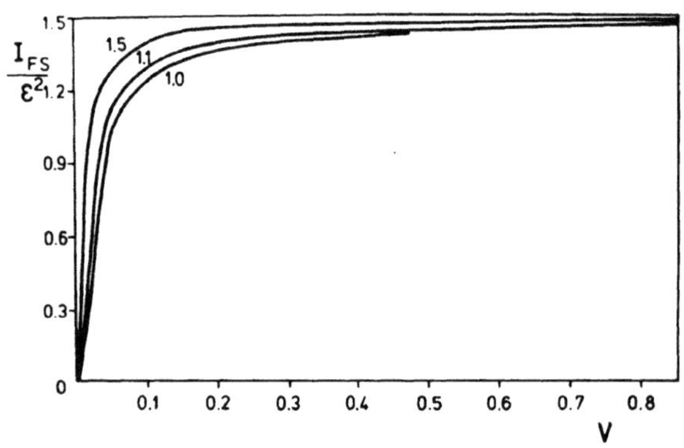

Fig. 3. Relative decrease of the dynamic friction force acting on a slider for different values of λ. High-frequency vibrations

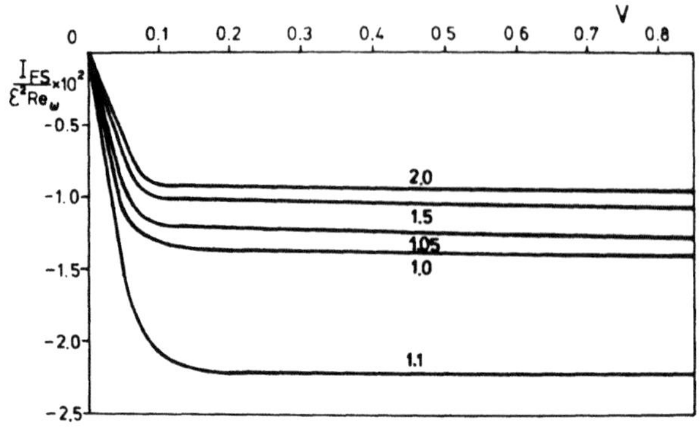

Fig. 4. Relative increase of the dynamic friction force acting on a slider for different values of λ. Low-frequency vibrations

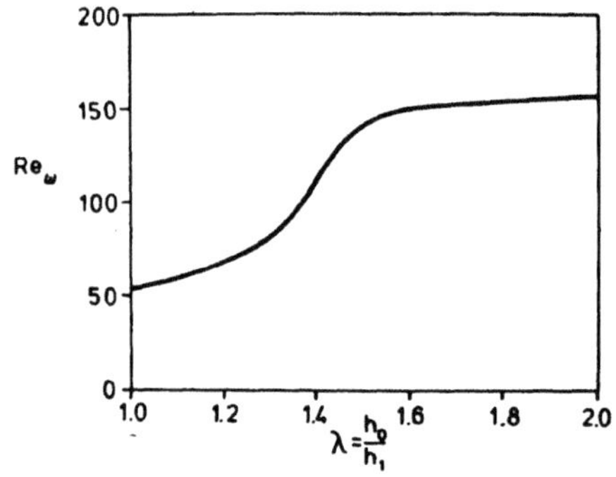

Fig. 5. Limit values of the "frequentative" Reynolds number versus λ

References

1. Rivlin, R. S., J. Non-Newtonian Fluid Mech. **5**, 79 (1979).
2. Manero, O., B. Mena, D. M. Binding, J. Non-Newtonian Fluid Mech. **5**, 427 (1979).
3. Manero, O., K. Walters, Rheol. Acta **19**, 277 (1980).
4. Tanner, R. I., Phys. Fluids **9**, 1246 (1966).

(Received April 15, 1982)

Author's address:

Prof. Dr. S. Zahorski
Institute of Fundamental
Technological Research
Polish Academy of Sciences
ul. Swietokrzyska 21
00-049 Warszawa (Poland)

Rheologica Acta Rheol. Acta **21**, 527–529 (1982)

Enhanced load bearing by polymer solutions lubricating external near-contact of sliding cylinders

D. R. Oliver and S. I. Bakhtiyarov

Department of Chemical Engineering University of Birmingham (U.K.)

Abstract: A small fixed cylinder attached to a load cell almost touches a larger, rotating cylinder (axes parallel). Newtonian liquids in the gap give reasonable loads, but viscoelastic liquids give markedly enhanced load bearing in relation to their apparent viscosities.

Key words: Lubrication, load enhancement, viscoelastic liquid

Notation

F normal load on small cylinder
h minimum film thickness between cylinders
L length of smaller cylinder
R reduced radius, defined by equation $1/R = 1/R_1 + 1/R_2$
R_1 radius of small cylinder
R_2 radius of large cylinder
U surface velocity of large cylinder
α load factor defined by equation $\alpha = Fh/URL\mu$
μ viscosity of liquid

1. Introduction

Doremus and Piau [1] recently demonstrated remarkable load-enhancement effects in cylinder/cylinder sliding experiments using aqueous Polyox solutions. The equipment was first tested using Newtonian liquids, giving loads in approximate agreement with theoretical predictions [2]. When viscoelastic liquids were used, the loads were up to 5000 per cent higher than for a Newtonian liquid of the same viscosity. This greatly exceeds the 800 per cent load enhancement demonstrated by Oliver and Ashton for squeeze film flows, in which aqueous polyacrylamide solutions were used [3]. The work of Doremus and Piau [1] was of such novelty as to merit an independent investigation, which the present paper briefly describes.

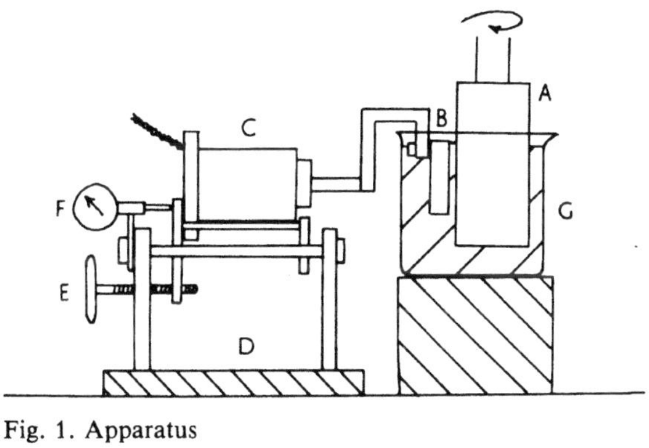

Fig. 1. Apparatus

2. Apparatus (fig. 1)

Rotating cylinder A is the outer cylinder of a "Fann" viscometer, of diameter 38 mm and available speeds of 100, 200, 300 and 600 rpm. The small cylinder B has diameter 10.0 mm (alternatively 3.0 mm) and length 40 mm. This is rigidly attached via a stirrup to the load cell C, which slides on the twin rails of carriage D. The load cell may be moved horizontally by wheel E and the exact position located by dial gauge F with an accuracy of ± 2 μm. Load cell C may also be rotated about a horizontal axis in order to align the cylinders. The load cell is calibrated by placing weights on the stirrup, with the cell in a vertical position.

The relative deflection of the two cylinders under load, caused by the bending of the load cell beam, the stirrup and the mounting of cylinder A is calibrated by first contacting the two cylinders. The load cell is then moved inward in very small steps and the (fairly large) loads detected by the cell are noted. The nominal gap settings used in tests may then be corrected from a knowledge of the individual loads.

3. Liquids used

The Newtonian liquid is a glycerol/water mixture with a viscosity of 16.0 cp (mPa s) at 20 °C. The viscoelastic liquid is an equivolume glycerol/water mixture containing 300 ppm of Magnafloc LT25, an anionic polyacrylamide with a molecular weight of 20 – 25 million (Allied Colloids, Bradford, UK). The viscosity is slightly shear dependent and in the range 15.8 to 18.6 cp (mPa s) at 20 °C.

4. Experimental method

Vertical line contact of cylinders A and B is established with the aid of back illumination and by observing the "kick" of the load cell recorder. The load cell carriage is then moved backwards until the gap is opened to the desired value. Beaker G is then raised and the liquid level brought to the top of the small cylinder with the large cylinder rotating at operating speed. The liquid level is always higher on the entry side of the contraction than on the exit side. Each recorded load, with the large cylinder rotating, is always compared with the "zero" (no rotation) both before and after the measurement. The process is repeated for different gaps and cylinder speeds.

5. Presentation of results

The load factor α is plotted against the surface velocity U of the large cylinder, where $\alpha = Fh/URL\mu$.

All terms are defined under "Notation". The value of α for the Reynolds condition is 2.45 according to Cameron [2], but this involves viscous forces only. Fluid inertia can cause increased values of load (F) at high surface speeds and the magnitude of the correction may be estimated approximately by the use of liquids of low viscosity, such as water. The effect never exceeded 5 per cent of the measured loads in the data presented but this inertia force was always subtracted from "F" before the calculation of α.

6. Results (fig. 2)

The values of α for the Newtonian fluid lie close to 2.45 at low rotational speeds, but are higher at the maximum speed. These data are quite typical for low-viscosity Newtonians (10 – 30 cp) but are not typical

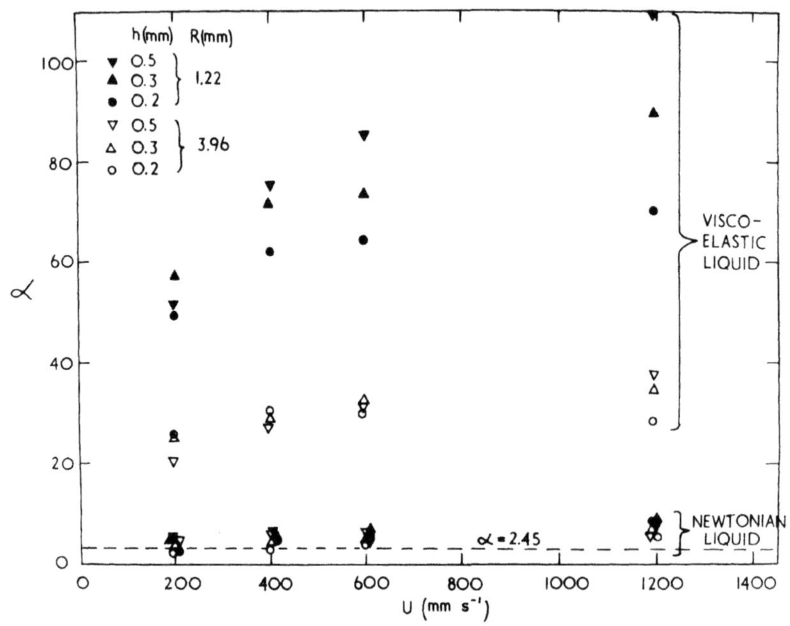

Fig. 2. Relationship between load factor α and surface velocity U of cylinder. Viscoelastic liquid is 50/50 glycerol/water plus 300 ppm Magnafloc LT25

for Newtonian fluids of higher viscosity. For these fluids, α is well below 2.45 at low rotational speeds and approximately equal to 2.45 at high rotational speeds. The drop below 2.45 may be explained on the grounds that the exit side of the contraction runs full at low speeds, and the reduced-pressure part of the Sommerfeld pressure distribution reduces the load. The high values of α for the present glycerol solutions are less easy to explain, because the enhanced load does not increase with ρU^2 as one might expect for effects caused by fluid inertia (including aerodynamic lift). Inertia effects measured using water are too small to account for the observed behaviour. However, there may be some interplay between viscous and inertial effects which cause enhanced loads at high rotational speeds; it may be observed that the smaller fixed cylinder gives more load enhancement than the larger cylinder, possibly because the average flow velocity over the small cylinder is greater.

The viscoelastic liquid gives very large values of α under all experimental conditions, the enhancement effect being of a different order from that discussed above. The smaller fixed cylinder gives the biggest load enhancement (compared with Newtonian fluid behaviour), the relative increase being by a factor of 16 (1500 per cent). This is rather less than the 5000 per cent increase obtained by Doremus and Piau [1], but it must be pointed out that the present cylinder surface velocities reach only 1.2 ms^{-1} compared with 15 ms^{-1} for the earlier workers [1]. The results of Doremus and Piau showed a form of threshold, between 0.3 and 5.0 ms^{-1}, below which no load enhancement was obtained, but above it the load increased very rapidly (or the gap increased at constant load). None of the present results showed this effect, the load being increased by a factor of ten for a surface velocity as low as 0.2 ms^{-1}. This may be caused by the use of a highly elastic liquid (properties not yet determined), since Magnafloc LT25 has been shown to be an outstanding drag reducing polymer [4]. The shear rate in the nip is in the range 380 – 5900 sec^{-1} whilst the maximum extensional strain rate of liquid approaching the nip is 550 sec^{-1}.

It may be concluded that the major finding of Doremus and Piau [1] that viscoelastic liquids, lubricating external sliding contacts between cylinders, give greatly enhanced load bearing in relation to their viscosities has been confirmed. Further work is in progress which deals with the frictional force in addition to the normal force.

References

1. Doremus, P., J.-M. Piau, J. Non-Newtonian Fluid Mech. **9**, 389 (1981).
2. Cameron, A., "Basic Lubrication Theory", 2nd Ed. p. 69. John Wiley and Sons.
3. Oliver, D. R., R. C. Ashton, J. Non-Newtonian Fluid Mech. **7**, 369 (1980).
4. Oliver, D. R., S. I. Bakhtiyarov, J. Non-Newtonian Fluid Mech. (in the press).

(Received April 15, 1982)

Authors' addresses:

Dr. D. R. Oliver
Department of Chemical Engineering
University of Birmingham
Birmingham B15 2TT (U.K.)

Dr. S. I. Bakhtiyarov
Azerbaijan Institute of Oil and Chemistry
Baku, Azerbaijan (U.S.S.R.)

Rheologica Acta Rheol. Acta **21**, 530–533 (1982)

Complex viscosity of viscoelastic damping fluids

R. Andrä and J. H. Spurk

Fachgebiet Technische Strömungslehre, Fachbereich 16, Technische Hochschule Darmstadt

Key words: Vibration, auxiliary mass damper, damping fluid

1. Introduction

The dissipated energy W of the torsional vibration damper depicted in figure 1 using the constitutive model of a second-order fluid is given [1] by the expression

$$\bar{W} = \frac{W}{(\pi/2)\, I_D \omega^2 \phi_G^2} = \frac{2\chi\eta/(I_D\omega)}{(1+\gamma\chi/I_D)^2 + (\chi\eta/(I_D\omega))^2}\,. \quad (1)$$

Here η is the shear viscosity of the fluid being sheared in the clearances between the flywheel (moment of inertia I_D) and the casing, when the latter is vibrating with angular frequency ω and amplitude ϕ_G. χ is a geometry factor and γ is the third coefficient in the constitutive equation of a second-order fluid

$$S + pI = \eta A_{(1)} + \beta A_{(1)}^2 + \gamma A_{(2)}\,.$$

It is evident from eq. (1) that a frequency-independent optimal damping, defined by the tuning $1 = -\gamma\chi/I_D$, can be achieved by properly selecting χ and I_D, since for a given second-order fluid, γ is a negative constant. Eq. (1) is also valid for a damper using general viscoelastic fluid at least for sufficiently small amplitudes; however the material properties η and γ which may be combined to a complex viscosity $\eta^* = \eta + i\omega\gamma$ are now functions of frequency and the tuning is than no longer frequency-independent. Since the frequency independence is of course a desirable feature in damper applications, measurements of frequency dependence of the viscosity η^* of viscoelastic fluids commonly used in damper application have been made, in order to assess the effect of frequency dependence on damper performance [2].

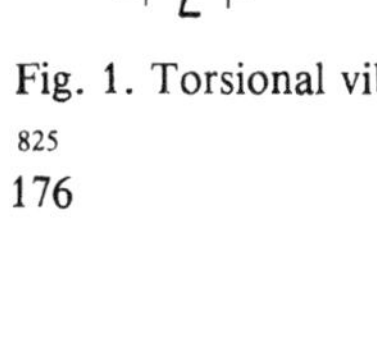

Fig. 1. Torsional vibration damper

Measurements of the fluid properties at 22 °C for five polydimethylsiloxanes ranging in molecular weight from 10000 to 100000 have shown that the frequency dependence in the optimal tuning does not significantly affect damper performance but the maximum damping $\bar{W}$ obtainable with the combination of γ/η particular to polydimethylsiloxanes, is restricted to a maximum value of about 2, i.e. the maximum damping is about twice as large as the maximum damping in an untuned damper using Newtonian damper fluid [3].

The present paper extends the work reported in [2] in three ways: (i) we show the behaviour of material properties for four equilibrated polydimethylsiloxanes at elevated temperatures; (ii) we give results for mixtures of polydimethylsiloxanes having different molecular weights, and (iii) we present the performance of dampers using a relatively concentrated solution of high-molecular-weight polyisobutylene in low-molecular-weight polyisobutylene as the damping fluid and also of dampers using a dilute solution of high molecular-weight-polyisobutylene in mineral oil.

2. Results

Figures 2a–c show the ratio of shear viscosities $\eta(\omega)/\eta_0$ for four siloxanes as function of the reduced frequency $\omega\eta_0/\bar{\Gamma} = \omega\tau$. Here η_0 is the viscosity at zero-frequency, $\bar{\Gamma}$ is the effective shear modulus introduced by Michele [4], and τ the relaxation time that may be formed with these quantities. Using the shear stress S_{12} and the first normal-stress difference N_1, the effective modulus is defined by the equation $\bar{\Gamma} = 2S_{12}^2/N_1$. In steady shear experiments on siloxanes [4], the effective shear modulus was found to be nearly independent of molecular weight with a value of $\bar{\Gamma} = 1.28 \cdot 10^4\ \mathrm{N/m^2}$. The corresponding quantity in oscillatory shear is $-\eta^2(\omega)/\gamma(\omega)$ and is shown made dimensionless with Micheles value in Figure 2a–c.

This figure shows, that $\eta(\omega)/\eta_0$ is weakly frequency-dependent and that the "effective modulus" $-\eta^2(\omega)/\gamma(\omega)$ reaches Micheles steady-state value at low frequencies, while at high frequencies it is about twice as large as Micheles value, but here again almost frequency independent. Comparison of figures 2a–c among each other show that the quantity $-\eta^2(\omega)/(\gamma(\omega)\bar{\Gamma})$ and the ratio $\eta(\omega)/\eta_0$ are independent of temperature and almost independent of molecular weight. Dependence on temperature and molecular weight enters only through the zero-viscosity η_0.

Figure 3 shows as an example the dissipated energy $\bar{W}$ of equilibrated siloxanes in comparison with the theoretical prediction for a Newtonian fluid ($\gamma = 0$) and a second-order fluid. The condition for optimal tuning $-\gamma_m\chi/I_D = 1$ is based on average values of η_m and $\gamma_m = -\eta_m^2/\bar{\Gamma}_m$. The experimental points include data up to reduced frequencies $\omega\tau = 6$. The data follow the optimal curve for second-order fluids quite closely up to reduced frequencies of about $\omega\tau = 2$. Above this value the frequency dependence of γ does not longer allow the condition for optimal tuning to be satisfied. The maximum energy $\bar{W}$ is seen to be about 2 and this corresponds to a maximum of $|\omega\gamma/\eta|$ of about 1; because we find from eq. (1)

$$\bar{W}|_{\gamma\chi/I_D=-1} = -2\omega\gamma/\eta = 2\omega\eta/\bar{\Gamma}. \tag{2}$$

Higher values of dissipated energy $\bar{W}$ require damping fluids with larger values of $|\omega\gamma/\eta|$ or, equivalently, with smaller effective shear modulus. Mixtures of siloxanes are known to give lower values of effective shear modulus in steady shear experiments [4], and as figure 4a and b show, this behaviour is also found in our oscillatory-shear experiments.

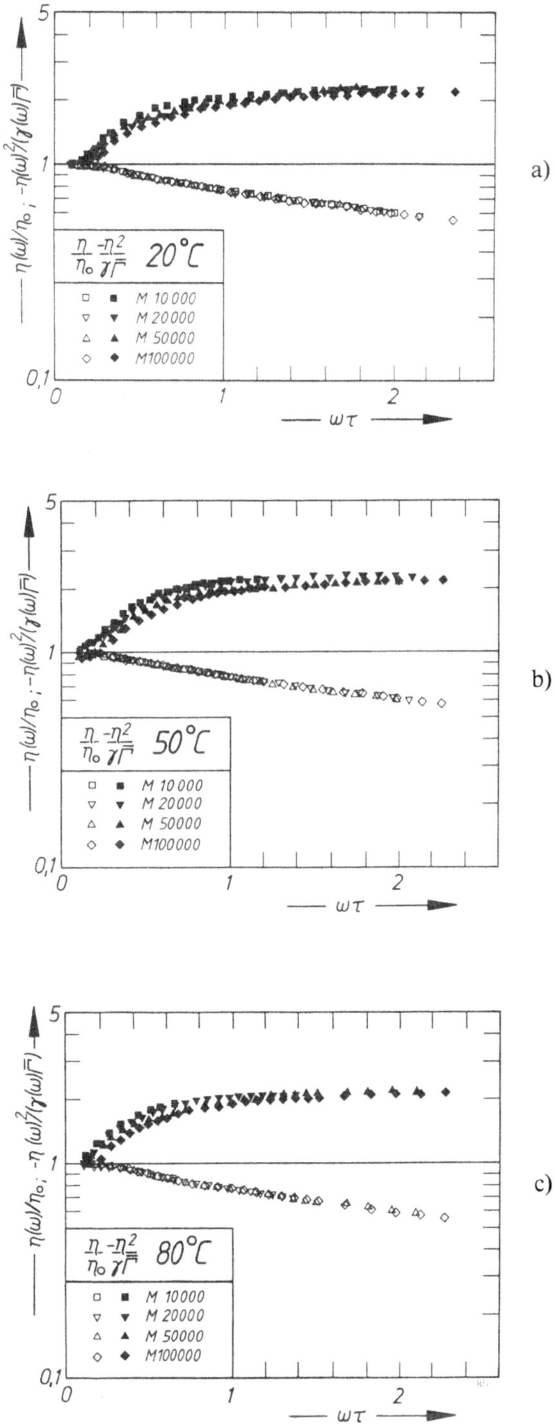

Fig. 2. Experimental values of shear viscosity $\eta(\omega)/\eta_0$ and effective shear modulus $-\eta^2(\omega)/(\gamma(\omega)\bar{\Gamma})$ for equilibrated polydimethylsiloxanes at different temperatures: (a) 20°C, (b) 50°C, (c) 80°C

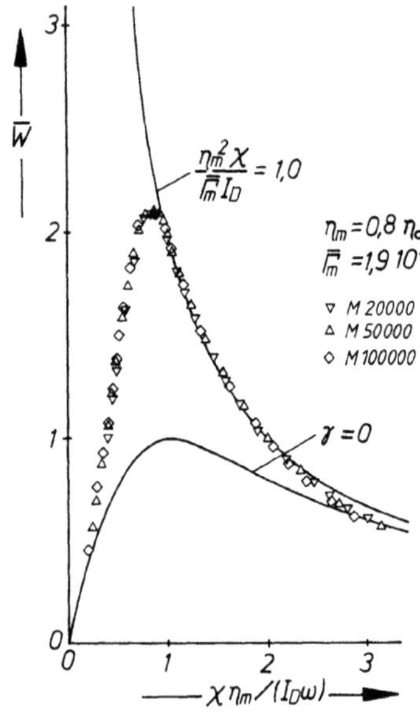

Fig. 3. Dissipated energy for polydimethylsiloxanes

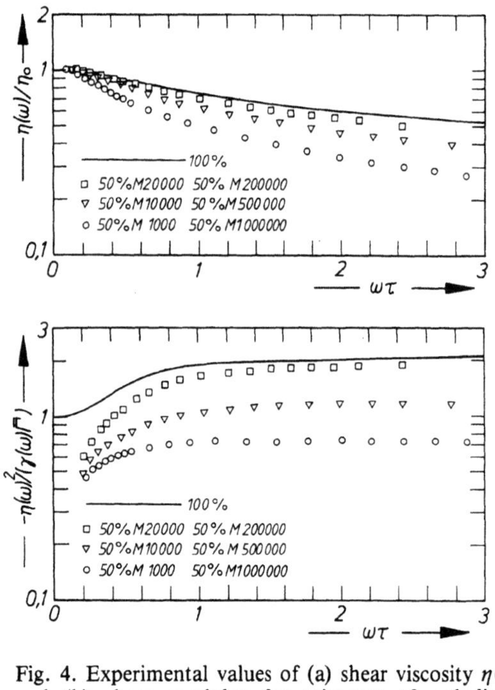

Fig. 4. Experimental values of (a) shear viscosity $\eta(\omega)/\eta_0$ and (b) shear modulus for mixtures of polydimethylsiloxanes

However, as may be seen from figure 5, only a modest increase in dissipated energy results, since the decrease in modulus is almost offset by the decrease in shear viscosity.

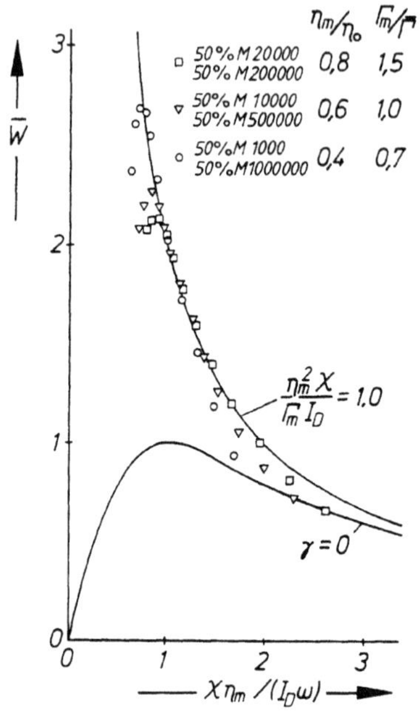

Fig. 5. Dissipated energy for mixtures of polydimethylsiloxanes

Eq. (2) may also be written as

$$\bar{W}|_{\gamma\chi/I_D = -1} = -2\omega\gamma/\eta = 2G'/G'' \quad (3)$$

where G' and G'' are the storage and the loss modulus respectively. Eq. (3) and the fact that G' as a function of ω is equal to $N_1/2$ as a function of $\dot{\gamma}$ for small $\dot{\gamma}$ and ω and G'', under the same circumstances, equal to S_{12} suggest that one examines liquids, which in steady-state shear flow have $N_1/2 > S_{12}$. Liquids exhibiting this property are, e.g., solutions of polyisobutylene in mineral oil [5] and solutions of high-molecular-weight polyisobutylene in low-molecular-weight polyisobutylene.

Figure 6 shows the dissipated energy that may be achieved with a dilute solution of polyisobutylene in mineral oil. In the same figure theoretical curves for a "frequency-tuned" damper using a Newtonian damping fluid [6] are also shown.

By means of a torsional spring with stiffness k, these dampers are tuned to a particular frequency by the condition $k/I_D = \omega^2$. These dampers are effective only over a narrow band width as compared to untuned viscous dampers ($\gamma = 0$). It is easy to show, that the optimal damping curve $\gamma\chi/I_D = -1$ (γ = const) is the envelope of all tuned dampers. The comparison shows that dampers using solutions of iso-

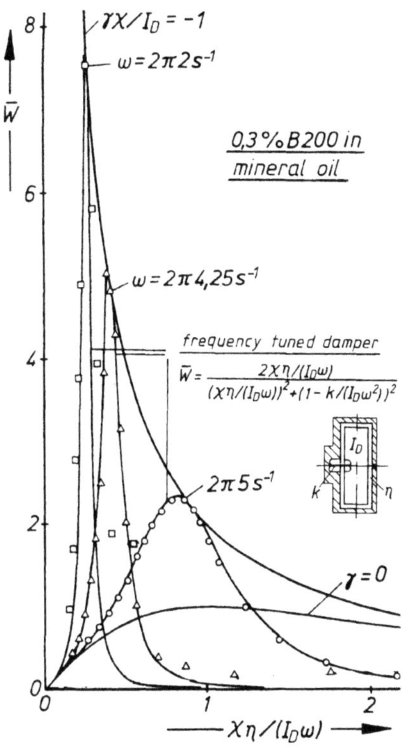

Fig. 6. Dissipated energy for solution of polyisobutylene ("Oppanol B 200") in mineral oil

butylene in mineral oil behave even quantitatively just as frequency tuned dampers and give approximately the same band widths at the same dissipated energies. Figure 7 shows, however, that solutions of moderately concentrated high-molecular-weight polyisobutylene in low-molecular-weight polyisobutylene lead to equally high dissipated energies at substantially increased band width and are therefore preferable in damper applications.

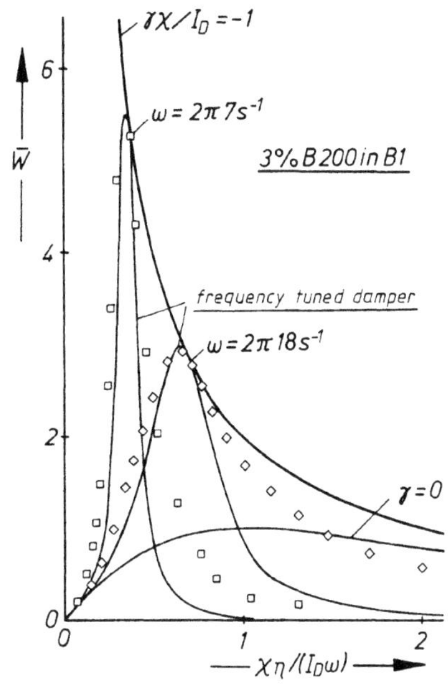

Fig. 7. Dissipated energy for solution of high-molecular-weight polyisobutylene ("Oppanol B 200") in low-molecular-weight polyisobutylene ("Oppanol B 1")

References

1. Spurk, J. H., Ingenieur-Archiv **48**, 121 – 127 (1979).
2. Andrä, R., J. H. Spurk, J. sound and vibration **82**, 465 – 472 (1982).
3. Den Hartog, J. P., Mechanical vibrations 4. ed., McGraw-Hill Book Company (New York-Toronto-London 1956).
4. Michele, J., Rheol. Acta **15**, 15 – 22 (1976).
5. Michele, J., Rheol. Acta **17**, 42 – 58 (1978).
6. Harris, C. M., C. E. Crede (eds.), Shock and vibration Handbook. 2. ed., McGraw-Hill Book Company (New York 1976).

(Received July 1, 1982)

Authors' address:

Dipl.-Wirtsch.-Ing. R. Andrä, Prof. Dr.-Ing. J. H. Spurk
Fachgebiet Technische Strömungslehre
Fachbereich 16
Technische Hochschule Darmstadt
Petersenstr. 30
D-6100 Darmstadt

Rheologica Acta Rheol. Acta **21**, 534 – 536 (1982)

A finite element code for initial value problems with a Maxwell model

O. Hassager and C. Bisgaard

Instituttet for Kemiteknik, Danmarks Tekniske Højskole, Lyngby (Denmark)

Key words: Sphere, wall effect, finite element method, Lagrangian description

1. Introduction

Substantial efforts have recently been made in the development of numerical methods for the simulation of nonhomogeneous flow of viscoelastic fluids described by models of the Oldroyd-Maxwell type [1]. Most of the models used may be formulated either in differential form or in integral form. The differential and integral forms of the models are equivalent, but the numerical methods used in the two situations are quite different. Most simulation programs developed to this day use the differential form (see e. g. [2, 3]), but simulations with integral models have also been performed [4 – 6].

The present simulation program is based on the integral form of the contravariant convected Maxwell model [1]. A novel feature of the program is the use of a Lagrangian description of the flow. This means that the simulation of a flow situation is formulated as an initial value problem, where the initial position of all particles must be specified, and the simulation program then follows the motion of the particles forward in time.

2. Numerical method

The liquid that one wishes to consider is divided into a number of finite elements. We use linear quadrilateral elements [7] with element coordinates (ξ, η). Each element node then designates a material particle, and the flow of the liquid is specified by following the space coordinates of each node as a function of time. The motion of an arbitrary particle is then given by the specification that the particle remains in the same element with unchanged element coordinates (ξ, η) at all times. If the space coordinates of a particle at a past time t' are denoted x_i' $(i = 1, 2, 3)$ and the coordinates at the present time are denoted x_i $(i = 1, 2, 3)$ the contravariant convected Maxwell model gives the following explicit expression for the stress tensor $\tau_{ij}(x, t)$ at the present time:

$$\tau_{ij}(x, t) = -\frac{\eta}{\lambda^2} \int_{t'=-\infty}^{t} e^{-(t-t')/\lambda} E_{im} E_{jm} dt' \tag{1}$$

where

$$E_{ij}(x, x', t') = \frac{\partial x_i}{\partial x_j'} \tag{2}$$

and the two parameters in the model are the zero-shear-rate viscosity η and the time constant λ.

At each time during the forward integration the equations to be solved for the current space coordinates of each nodal point are formulated with the use of a variational principle developed recently [8], and solved with Newton iteration.

3. Results and discussion

The element program has been used to simulate the creeping motion around a solid sphere of radius R_s moving axially in a cylinder of radius R_c filled with liquid. The sphere is initially at rest and suddenly at time $t = 0$ it is acted upon by a force F in the axial direction. The subsequent motion of the sphere including the rapid approach to steady state motion is followed.

An example of a finite element mesh is shown in figure 1. The mesh is shown at two successive instants

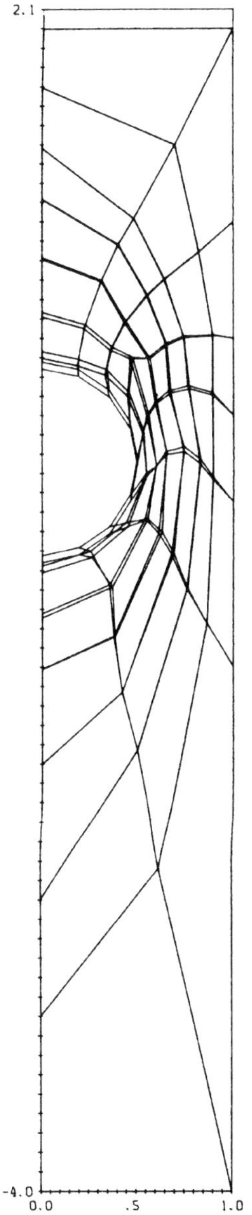

Fig. 1. Finite element mesh for the axial motion of a sphere in a cylinder with $R_s/R_c = 0.5$. The mesh is drawn at two successive time steps to give an indication of the magnitude of the deformations obtained

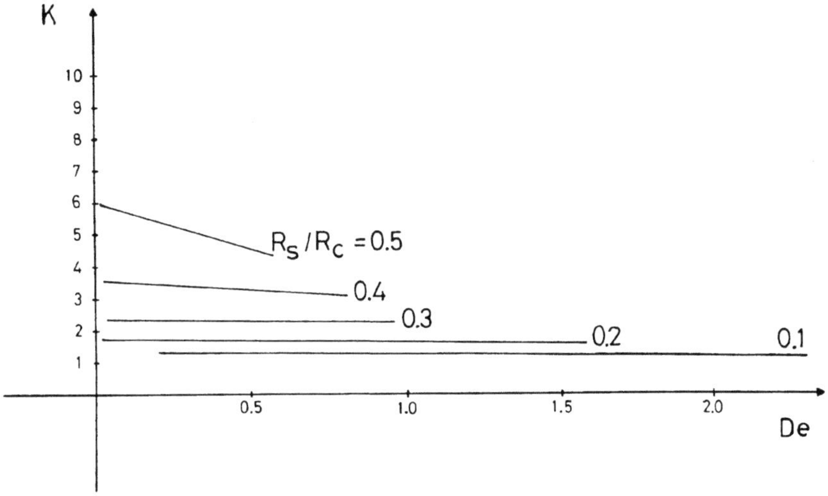

Fig. 2. Simulation results for $K = F/6\pi\eta R_s v_0$ as function of Deborah number $De = \lambda v_0/R_s$ for $R_s/R_c = 0.1, 0.2, 0.3, 0.4$ and 0.5

thus giving an indication of the magnitude of deformations between successive time steps.

The steady state velocity is denoted v_0 and results are displayed in the form of values of $K = F/6\pi\eta R_s v_0$ as function of the ratio R_s/R_c and the Deborah number $De = \lambda v_0/R_s$ as shown on figure 2.

The predictions for K at very low Deborah numbers agree within 1% with the values for the motion of a sphere in a cylinder filled with a Newtonian fluid [9].

It is evident from figure 2 that the presence of the cylinder wall has a dramatic influence on the K-factor. For $R_s/R_c = 0.1$ the value of K at a Deborah number of 2.3 has dropped by only about 9% from the Newtonian value, whereas for $R_s/R_c = 0.5$ the value of K at a Deborah number of 0.53 has dropped by 27% from the Newtonian value.

The maximum Deborah number for which it has been possible to reach a steady state motion (2.3) is well above that which has been reached with steady-state simulations of the Maxwell model on differential form [10, 11]. We feel that the method proposed here shows promise as an improvement over available numerical simulation techniques for a number of flow situations.

Acknowledgements

One of the authors (C.B.) wishes to acknowledge support from the Danish Council for Scientific and Industrial Research.

References

1. Bird, R. B., R. C. Armstrong, O. Hassager, Dynamics of Polymeric Liquids, Vol. I, Fluid Mechanics, Wiley (New York 1976).
2. Crochet, M. J., R. Keunings, J. Non-Newtonian Fluid Mech. 7, 201–218 (1980).

3. Coleman, C. J., J. Non-Newtonian Fluid Mech. **8**, 261 – 270 (1981).
4. Viriyayuthakorn, M., B. Caswell, J. Non-Newtonian Fluid Mech. **6**, 245 – 268 (1980).
5. Bernstein, B., D. S. Malkus, Proc. IUTAM Symp. Var. Meth. Mech., Northwestern U., Evanston, Il. (1978).
6. Court, H., A. R. Davies, K. Walters, J. Non-Newtonian Fluid Mech. **8**, 95 – 117 (1981).
7. Zienkiewicz, O. C., The Finite Element Method in Engineering Science, McGraw-Hill (London 1911).
8. Hassager, O., J. Non-Newtonian Fluid Mech. **9**, 321 – 328 (1981).
9. Happel, J., H. Brenner, Low Reynolds Number Hydrodynamics, p. 320, Noordhoff (Leyden 1973).
10. Crochet, M. J., The flow of a Maxwell fluid around a sphere, in: Gallagher (ed.), Finite Elements in Fluids IV, Wiley (Chichester 1981) to be published.
11. Tiefenbruch, G., L. G. Leal, J. Non-Newtonian Fluid Mech. **10**, 115 – 155 (1982).

(Received April 14, 1982)

Authors' address:

Dr. O. Hassager, C. Bisgaard
Instituttet for Kemiteknik
Danmarks Tekniske Højskole
Bygning 229
DK-2800 Lyngby

An experimental investigation of velocity fields around spheres and bubbles moving in non-Newtonian liquids

C. Bisgaard and O. Hassager

Instituttet for Kemiteknik, Danmarks Tekniske Højskole, Lyngby (Denmark)

Key words: Sphere, bubble, non-Newtonian liquid, negative wake, Laser-Doppler anemometry

1. Introduction

Spheres and bubbles moving in non-Newtonian liquids are known to show several interesting phenomena. For example Hassager [1] found that the liquid velocity behind the bubbles rising in a non-Newtonian liquid is in the downward direction away from the rising bubble. This phenomenon was called the "Negative wake". Sigli and Coutanceau [2] reported the existence of a negative wake behind spheres moving in a non-Newtonian liquid, by using a visualization technique with tracer particles.

In the present study we have measured the axial component of the velocity on the axis of gas bubbles and solid spheres moving in a non-Newtonian liquid. The most interesting feature of the observed flow fields is the occurrence of oscillations in the wake behind the bubbles and spheres, implying the existence of instabilities in the flow field. The physical constants of the test liquid are determined by comparing experimental data for spheres moving in the test liquid to the results of finite element calculations from Hassager and Bisgaard [3].

2. Experimental equipment

The experimental apparatus consists of a vertical 4.5 inch diameter glass cylinder filled to a depth of 2 m with test fluid. Gas bubbles are injected by a syringe near the bottom of the tube, go through a centering device and proceed axially up through the test section. Solid spheres are released near the top on the axis of the cylinder and proceed down through the test section. In the test section two photo arrays are used to measure the velocity of the object. A Laser-Doppler anemometer (LDA) is used to measure the axial component of the liquid velocity at a fixed point on the cylinder axis as a function of time. Having crossed the "measuring volume" one of the laser beams impinges on a photo cell. This photo cell is used to register the exact time at which the object enters into and exists from the measuring volume and keep the LDA frequency tracker in "hold" during the passage of the object. The photo cells and the LDA output are sampled and stored by a microprocessor. For measuring the force acting on the spheres, when it is moving in the test fluid, is used a Westfalsk balance, which measure the difference between the gravity and the bouyancy forces with four significant digits.

771

3. Experimental results

The test fluid used was a 1% polyacrylamide solution in glycerol. For determining the zero-shear-rate viscosity η_0 and a characteristic time λ for the fluid, three different spheres were used. The ratio between sphere radius and cylinder radius were: $R_k/R_c = 0.5$, 0.4, 0.3, where R_k is the sphere radius and R_c is the cylinder radius. It was possible to separate the spheres and add small weights to increase the density of the spheres. The finite element calculations give the wall correction factor K_f's dependence of the Deborah number $De = v_k\lambda/R_k$, where v_k is the sphere velocity. The K_f is defined by:

$$K_f = \frac{F}{6\pi R_k \eta_0 v_k}$$

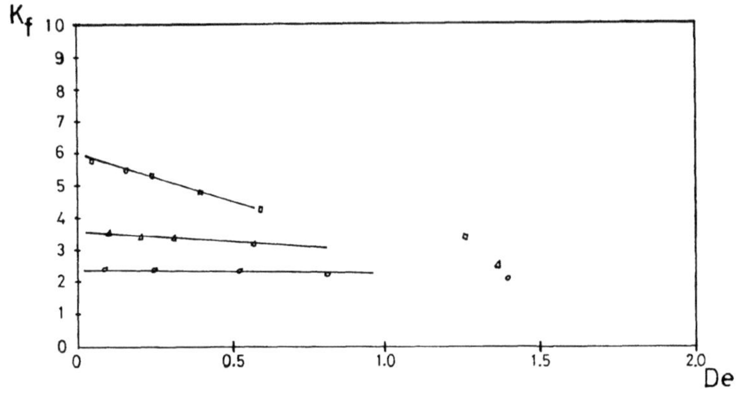

Fig. 1. Calculations of K_f as a function of *De* are indicated by straight lines for three different values of R_k/R_c. The experimental data are indicated by □ for $R_k/R_c = 0.5$, △ for $R_k/R_c = 0.4$ and ○ for $R_k/R_c = 0.3$. $\eta_0 = 21.0$ Pa s, $\lambda = 12.0$ s

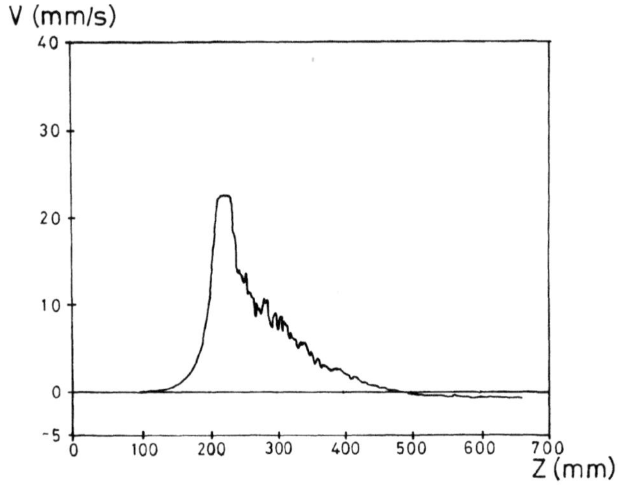

Fig. 2. Axial component of liquid velocity on the axis of a steel sphere. $V_k = 22.3$ mm/s, $t = 20\,^\circ$C, $R_k = 6$ mm, $R_k/R_c = 0.12$, $De = 44.6$, $7.9 \cdot 10^{-3} \leqslant Re \leqslant 1.1 \cdot 10^{-1}$. The abscissa, z is the distance along the axis of the sphere

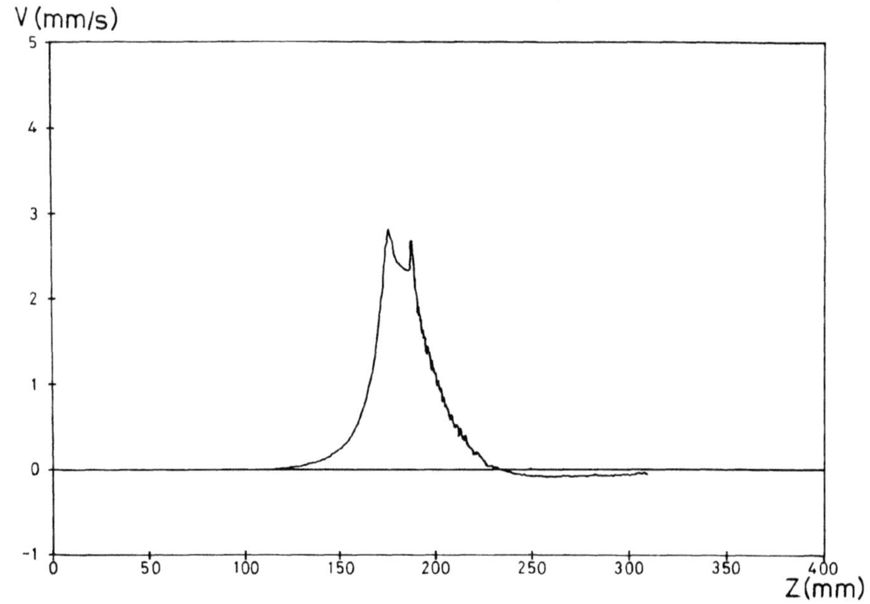

Fig. 3. Axial component of liquid velocity on the axis of a PTFE sphere. $V_k = 2.9$ mm/s, $t = 19.9\,^\circ$C, $R_k = 6$ mm, $R_k/R_c = 0.12$, $De = 5.8$. $1.0 \cdot 10^{-3} \leqslant Re \leqslant 1.4 \cdot 10^{-2}$. The abscissa, z is the distance along the axis of the sphere

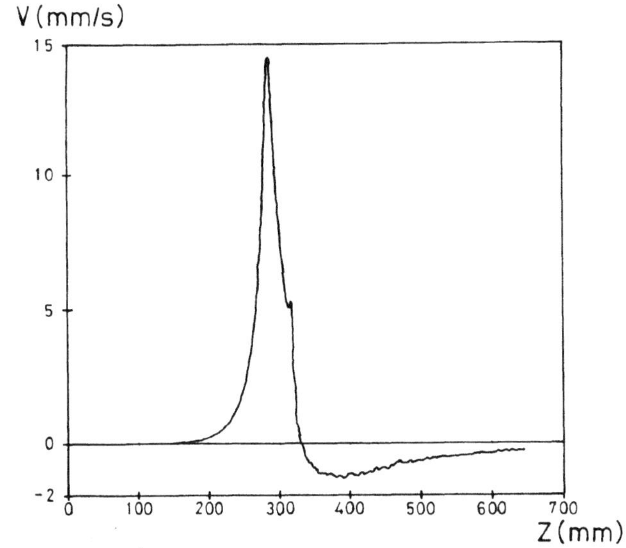

Fig. 4. Axial component of liquid velocity on the axis of an air bubble. V_b = 14.4 mm/s, t = 19.9 °C, R_b = 7.7 mm, R_b/R_c = 0.154, De = 22.4, $6.54 \cdot 10^{-3} \leqslant Re \leqslant 9.22 \cdot 10^{-2}$. The abscissa, z is the distance along the axis of the bubble

where F is the force acting on the sphere. A number of experiments were performed, with the three different values of R_k/R_c. The experimental values of $F/6\pi R_k v_k = K_f \eta_0$ and $v_k/r_k = De\,\lambda$ were fitted to the calculations of $K_f = f(De, R_k/R_c)$ by the least squares method, to calculate the values of η_0 and λ, which gave the best fit of the experimental data to the finite element calculations. The results are shown in figure 1. The values of η_0 = 21.0 Pa s and λ = 12.0 s were used to calculate the Reynolds number and the Deborah number for the Laser-Doppler measurements.

The Laser-Doppler anemometer measured the velocity as a function of time. The constant velocity of the moving objects allows us to transform the experimentally obtained velocity vs. time records into velocity vs. space records that show the axial component of liquid velocity on the axis of the objects. Figure 2 shows the velocity on the axis of a steel sphere falling in the test fluid. The figure clearly shows the negative wake behind the sphere, but close behind the sphere, there seems to be some rapid oscillations in the velocity, indicating instabilities in the flow field behind the sphere. The oscillations can not be explained by turbulence, because the Reynolds number is very low. The minimum value of the Reynolds number is calculated by using η_0, and the maximum value of the Reynolds number is calculated by using the viscosity of glycerol. The Reynolds number is defined by: $Re = v_k R_k \rho / \eta$, where ρ is the density for the test fluid.

Figure 3 shows the velocity on the axis of a PTFE sphere falling in the test fluid. For this sphere the Deborah number is 5.8, but there are still some weak oscillations present in the wake behind the sphere.

Consider now in figure 4 the velocity on the axis of an air bubble rising in the test fluid. The figure clearly shows the negative wake behind the bubble, but there also seem to be oscillations in the negative wake for the bubble. Another striking feature is that the negative wake is closer to the bubble than for the spheres. The Reynolds number and the Deborah number are calculated by using the radius R_b. R_b is calculated from the bubbles volume, assuming the shape of the bubble to be spherical.

4. Discussion

The present experiments provide information that improves our knowledge about the motion of isolated objects in non-Newtonian liquids. The most important result is probably the existence of instabilities in the wake behind the objects at rather low Deborah number.

Acknowledgements

The authors gratefully acknowledge the financial assistance provided by the Danish Council for Scientific and Industrial Research. The DISA Company has generously placed a 55L35 frequency tracker at our disposal. Finally we wish to thank Mr. B. Johnsen for his invaluable aid with electronics and programming.

References

1. Hassager, O., Nature **279**, No. 5712, pp. 402–403 (1979).
2. Sigli, O., M. Coutanceau, J. Non-Newtonian Fluid Mech. 2, 1–21 (1977).
3. Hassager, O., C. Bisgaard, Rheol. Acta **21**, 534–536 (1982).

(Received April 14, 1982)

Authors' address:

C. Bisgaard, Dr. O. Hassager
Instituttet for Kemiteknik
Danmarks Tekniske Højskole
Bygning 229
DK-2800 Lyngby

Rheologica Acta Rheol. Acta **21**, 540 – 542 (1982)

Natural convection of a viscoelastic fluid

K.-J. Röpke and P. Schümmer

Institut für Verfahrenstechnik der Rheinisch-Westfälischen Technischen Hochschule Aachen

Abstract: The influence of elasticity and shear thinning viscosity on the temperature distribution and heat transfer in natural thermal convection is discussed. The numerical investigations are based on a four-parameter Oldroyd constitutive equation, which represents the typical fluid response of dilute solutions and melts. It was found that especially the second normal-stress difference affects the heat transfer mechanism.

Key words: Numerical investigation, transient natural convection, viscoelasticity, heat transfer

1. Introduction

The hydrodynamics of thermal natural convection is based on the temperature dependent density variation and the resulting buoyancy forces in the gravitational field. It is observed in nature as well as in technical applications and its principal mechanism has been investigated by several scientists. But nearly all papers published have dealt with the steady state of Newtonian fluids [1]. For technical problems however, for example transient heating/cooling processes in chemical engineering as well as sophisticated heat storage tanks or artificial oil reservoirs, it becomes more and more important to predict the heat exchange of fluids with complex response or to optimize "conventional" processes by special additives. From the rheological point of view this complex behaviour can for example be shear-thinning viscosity and elasticity.

2. Constitutive equation

The constitutive equation on which our investigation should be based has to describe such viscoelastic fluid response. Moreover it should be material objective and convenient from the numerical point of view. Therefore a reduced four-parameter Oldroyd fluid (ROF) was chosen:

$$s^{+} + t_0 \overset{\nabla}{s}{}^{+} + \kappa t_0 (s^{+} \cdot d + d \cdot s^{+}) = 2\eta_0 \left[d + t_0 \frac{\eta_s}{\eta_0} [\overset{\nabla}{d} + 2\kappa d^2] \right]. \qquad (1)$$

This equation couples the extra stress tensor s^{+} with the rate-of-deformation tensor $\boldsymbol{d}$ and qualitatively describes well dilute and very dilute polymer solutions as well as polymer melts. The value t_0 is a characteristic time of the fluid, a measure for the elastic response; the parameter κ describes the shear-rate-dependency of elasticity and viscosity as well as the level of the second normal-stress difference. η_0 is the fluid viscosity at zero-shear-rate, while η_s describes the influence of the solvent viscosity. The superscripts "∇" and "." denote the upper convected and material derivatives, respectively. Furthermore this model shows a second Newtonian regime and overshooting for $\eta_s \neq 0$.

3. Physical model

In order to investigate the principal influence of such a complex fluid on a transient cooling/heating process in natural thermal convection a simple physical model was chosen: a rectangular enclosure with adiabatic top and bottom. The whole system is held at temperature T_0 for all times t less than 0. Then at $t = 0$ the temperatures of the left and right wall jump to T_H and T_C respectively: $T_H > T_0 > T_C$. These conditions initiate a clock-wise motion which is free of physical instabilities.

4. Theory

The governing differential equations describing this transient problem generally are those of mass, energy and motion. Here the usual assumptions about the heat conduction and inner energy can be introduced. Furthermore when sufficiently small derivations from the state of equilibrium are considered, one can neglect the dissipation term in the energy balance and apply the Boussinesq-approximation. The latter means that the density variation is only considered in the buoyancy term of the balance of momentum. All other properties are considered to be constant. Introducing the quasi-incompressibility because of the Boussinesq-approximation, the general stress tensor s can be split into an isotropic pressure $-pI$ and an extra stress tensor s^+.

Taking the curl of the equation of motion eliminates the unknown isotropic pressure. The resulting differences of the local velocity gradients can be abbreviated by the vorticity function ω. Thus the two components of the equation of motion are reduced to one equation of vorticity. Together with the stream function ψ, ω alters the outer appearance of the equation of mass.

For simplicity non-dimensional substitutes are introduced, in the following marked by capital letters.

Thus one gets the non-dimensional equations of flow (2), energy (3) and vorticity (4) [1]:

$$\frac{\partial^2 \Psi}{\partial X^2} + \frac{\partial^2 \Psi}{\partial Y^2} = -\Omega, \tag{2}$$

$$\frac{\partial \theta}{\partial \tau} + U\frac{\partial \theta}{\partial X} + V\frac{\partial \theta}{\partial Y} = Pr^{-1}\left[\frac{\partial^2 \theta}{\partial X^2} + \frac{\partial^2 \theta}{\partial Y^2}\right], \tag{3}$$

$$\frac{\partial \Omega}{\partial \tau} + U\frac{\partial \Omega}{\partial X} + V\frac{\partial \Omega}{\partial Y} = \frac{\partial}{\partial X\,\partial Y}(S^+_{YY} - S^+_{XX}) + \frac{\partial^2 S^+_{YX}}{\partial X^2} - \frac{\partial^2 S^+_{YX}}{\partial Y^2} - Gr\frac{\partial \theta}{\partial Y}, \tag{4}$$

with X, Y as coordinates; U, V as velocities; θ as excess temperature; τ as time; Prandtl-number Pr and Grashof-number Gr as measures for heat transfer properties and bouyancy forces respectively; all other quantities have already been discussed.

It must be emphasized that for the transformation ν_0, the kinematic viscosity at zero-shear rate and reference temperature T_0 is used.

This set of six coupled differential equations, i.e. (2), (3), (4) and XX-, YY-, XY-components of the ROF (1) were solved numerically.

5. Results and conclusions

The Nusselt-number $\overline{Nu}$ can be taken as a measure of heat transfer:

$$\overline{Nu} = \frac{b}{h}\int_0^{h/b} \left.\frac{\partial \theta}{\partial Y}\right|_{\text{wall}} dX \quad (h \text{ length, } b \text{ width of enclosure}).$$

Figure 1 shows the slope of $\overline{Nu}$ versus the dimensional time t for viscoelastic fluids which differ only

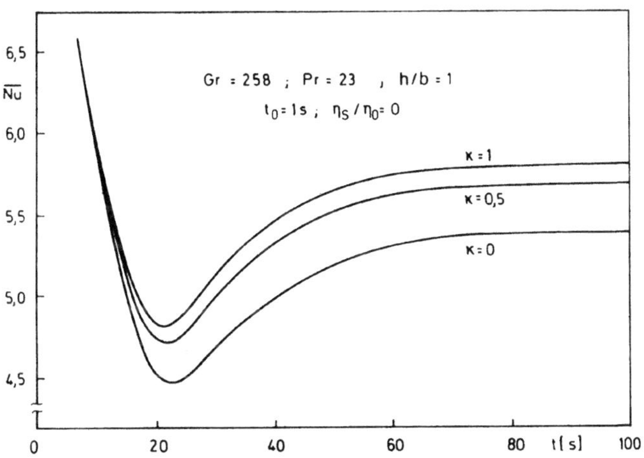

Fig. 1. Global Nusselt-number $\overline{Nu}$ as a function of the dimensional time t

in the value of the parameter κ; $\kappa = 0$ for stress-independent properties, disappearing second normal-stress difference, and $\kappa < 0 \leqq 1$ for stress-dependent properties, non-disappearing second normal-stress difference.

So far the investigations have shown

- that increasing elasticity also increases the heat transfer. Hereby one should consider that the level of the second normal-stress difference is the relevant feature,
- and that the stress-dependence of the fluid properties can be neglected, i.e. for example, that the Nusselt-power-law for Newtonian fluids can also be applied to predict heat transfer rates of purely viscous fluids.

The question however as to whether these power-laws are also valid for a wide range of realistic viscoelastic fluids will be answered in future.

Acknowledgement

The authors gratefully acknowledge the Deutsche Forschungsgemeinschaft for the financial support of this research.

Reference

1. Newell, M. E., F. W. Schmidt, J. Heat Transfer, 159–168 (1970).

(Received April 14, 1982)

Authors' addresses:

Dipl.-Ing. K.-J. Röpke
Rütgerswerke AG
Varziner Straße 49
D-4100 Duisburg 12

Prof. Dr. P. Schümmer
Institut für Verfahrenstechnik
RWTH Aachen
Turmstraße 46
D-5100 Aachen

Rheologica Acta Rheol. Acta **21**, 543 – 546 (1982)

Numerical simulation of elongational flow behaviour in the cross cell experiments

M. G. N. Perera, A. Lyazid, and O. Scrivener

Institut de Mécanique des Fluides – ERA CNRS, Strasbourg

Abstract: In this work, numerical simulation of an elongational flow of polymer solutions is tried. The flow field is computed in a cell of simple geometry for newtonian, non-newtonian inelastic and non-newtonian elastic fluids. For the latter, an Oldroyd four-constant-model was adopted. The results are qualitatively in good agreement with the experimental results.

Key words: Elongational flow, polymer solution, finite difference method

1. Introduction

The purpose of our investigation is to study the dynamical behaviour of macromolecules submitted to an elongational flow. Previous studies have shown [1] that macromolecules can extend in given regions of the flow. Extension was visualized by a flow birefringence technique, for several polymer-solvent pairs and in cells of different shapes [2]. In our experiments we tried to relate the appearance of extension to the flow field and to compute the flow so as to try to explain the abnormal velocity profiles observed in the extensional part of the flow. For numerical simulation the Oldroyd four-constant-model has been used together with a finite difference technique.

2. Experiments

The elongational flow is performed in the gap of a cell, termed "cross cell", designed jointly with the University of Metz [1]. Four right-angled corner pieces are placed in a 30 mm of depth chamber leaving 2 mm width flow passages arranged as a cross (fig. 1). The fluid comes into the cell through two opposite slits and flows out through the two others. The center of the cell is a stagnation point and the fluid is strongly accelerated in the outlet channels. The flow is laminar and the appearance of instabilities was only observed at high flow rates. The velocities were measured in the gap of the cell by Laser-Doppler anemometry.

For newtonian fluids like water or water-glycerol mixtures, both in the inlet and the outlet channels, velocity profiles were found to be not very different from Poiseuille flow-profiles.

Moreover, in the inlet channels, for PEO WSR 301 solutions (c = 0.1 or 0.4%), the profiles were not found to be significantly different from the newtonian (fig. 2), while in the outlet channels the profiles have show an unexpected shape. Before rearranging the profiles to parabolic form they present an

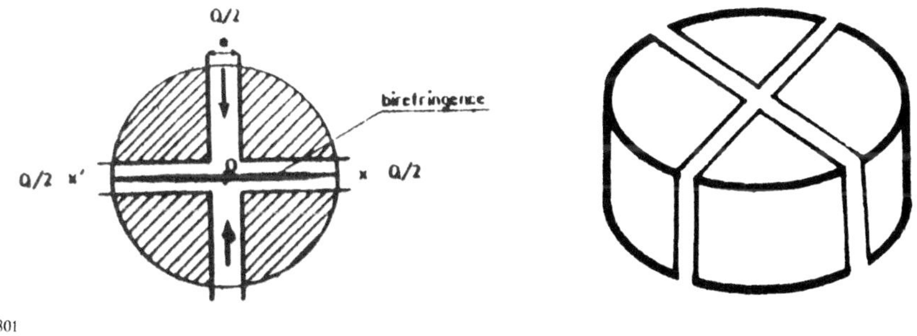

Fig. 1. Experimental device: the "cross cell"

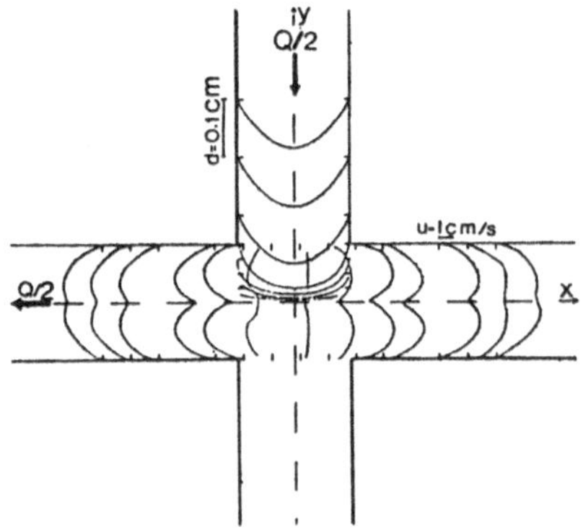

Fig. 2. Velocity measurements in the inlet and outlet channels with PEO WSR 301 solutions, $C = 0.4\%$, $Q = 2.63\ \mathrm{cm^3/s}$

important depression on the axis. This depression is related both to the polymer concentration and to the flow rate.

The calculations of the longitudinal velocity gradients in the central gap and in the outlet channels shows that the condition of extension $T_1 \dot{\gamma} > 1$ [3] is not reached even though in the same region extension can be visualized by flow birefringence.

3. Mathematical formulation

To compute the flow we used the equation of continuity together with the general form of the equation of motion where the total stress is developed into the contribution of the pressure and of the viscous stresses P'_{ik}. In this equation the expression of the stress tensor depends on the choice of the liquid model. For this work we used inelastic liquid and elastic liquid models.

For non-newtonian inelastic liquids, as for newtonian liquids the stress-strain relation is expressed by:

$$P'_{ik} = 2\eta e^{(1)}_{ik} \tag{1}$$

where the viscosity η is a constant for newtonian fluids and a function of the second invariant I_2 of the deformation tensor for non-newtonian fluids. The shear thinning viscosity is given either by a power-law model or by:

$$\eta(I_2) = \eta_0 \frac{(1 + \mu_0 \lambda_2 I_2)}{(1 + \mu_0 \lambda_1 I_2)}. \tag{2}$$

For elastic liquids, the Maxwell-Oldroyd models, compared to simple fluid models, exhibit elastic properties in larger amounts if necessary and are valid under all conditions of motions and stresses:

$$P'_{ik} + \lambda_1 \frac{\mathfrak{D} P'_{ik}}{\mathfrak{D} t} + \mu_0 P'^{j}_{j} e^{(1)}_{ik} = 2\eta_0 e^{(1)}_{ik} + \lambda_2 \frac{\mathfrak{D}}{\mathfrak{D} t} e^{(1)}_{ik} \tag{3}$$

where $\mathfrak{D}/\mathfrak{D}t$ is the upper convected Oldroyd time derivative and λ_1, λ_2, μ_0 are time parameters and η_0 the limiting zero-shear viscosity. Other liquid models, like convected Maxwell model with viscosity dependent on second invariant or Phan-Thien-Tanner model using a series of relaxation times are expected to model even better real liquids, but were not tried here.

4. Numerical simulation

Equations of the form (3) are implicit in the stresses and as such have to be solved separately as a set of simultaneous partial differential equations before their values are fed into the equation of motion. The simplicity of the cross cell geometry enables all the advantages of finite difference techniques to be used. The details of the solution method of the equations are described in recent literature [4–7]. The essence of the method is to discretise all equations in diagonal dominant form to ensure local convergence and help in the global convergence to as large values of the elastic parameters as possible.

Classical boundary conditions are used for the velocities and the stresses: no-slip condition at the boundaries, open boundaries far enough so that on them fully developed conditions can be used, zero-shear stress on the line of symmetry and Von-Neumann conditions for normal stresses.

To solve equations of continuity and of motion the stream function Φ and the vorticity ω formulation are used. In the equation of motion the following substitution has been found useful:

$$P'_{ik} = 2\eta_0 e^{(1)}_{ik} + \bar{P}_{ik}. \tag{4}$$

The resulting equation of motion becomes:

$$\rho \left(\frac{\partial \Phi}{\partial x} \frac{\partial \omega}{\partial y} - \frac{\partial \Phi}{\partial y} \frac{\partial \omega}{\partial x} \right) + \eta_0 \nabla^2 \omega = \frac{\partial^2 \bar{P}_{xx}}{\partial x \partial y} + \frac{\partial^2 \bar{P}_{yx}}{\partial y^2} - \frac{\partial^2 \bar{P}_{xy}}{\partial x^2} - \frac{\partial^2 \bar{P}_{yy}}{\partial x \partial y}. \tag{5}$$

The vorticity equation and the stress component are discretised using central differences in such a way that all discretized equations have diagonally dominant iterative matrices associated with them. One global iteration consists of three iteration loops, that is for the stream function, stress and vorticity. Successive overrelaxation methods are used to solve each set of equations.

Numerical solutions are obtained in two forms. Firstly a sort of non-dimensionalising is achieved by using values, like half width of the channel, flow rate, viscosity equal to one. When, for example, the viscosity varies throughout the flow field, non-dimensionalising is in the strict sense not possible. So as a second we used, as much as possible, values existing in the experimental set-up. This was found to be a better way to simulate experimental situations.

The newtonian solution can either be obtained by the viscoelastic program using $\mu_0 = \lambda_1 = \lambda_2 = 0$ or by using the explicit values of the stresses issued from Newton's law.

The non-newtonian inelastic equivalent of the Oldroyd four-parameter-model can be shown to have the form (1) with the viscosity expressed as (2). This solution which involves the explicit equation (2) can also be solved using a simple version of the program.

The non-newtonian elastic solution is obtained with the complete version of the program. For the non-dimensional set-up of results mentioned above solutions are available for values of λ_1 given by 0.06 sec, 0.12 sec, 0.24 sec, 0.36 sec, 0.48 sec and 0.60 sec. For the solution with experimental conditions, solutions are obtained up to a value of $\lambda_1 = 0.06$ sec. The Deborah number corresponding to this value is $De \simeq 1$.

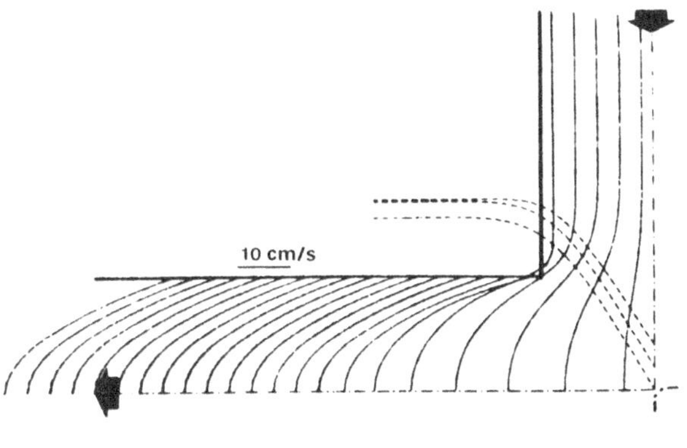

Fig. 3. Simulation of the flow (quarter of the cell) with a newtonian solution: $\eta_0 = 0.5$ poise, $Q = 4.5$ cm^3/s, $e = 0.2$ cm; ---- magnitude of the longitudinal velocities at different distances from the axis

In all the cases $\mu_0\lambda_1$ and λ_1/λ_2 are kept constant:

$$\mu_0\lambda_1 = 0.003 \text{ sec}^2; \quad \lambda_1/\lambda_2 = 1/6 .$$

5. Analysis of results

Using numerical methods it has been possible of simulate qualitatively the experimentally observed flow patterns (fig. 2). Figure 3 shows the numerically simulated newtonian solution and figure 4 the inelastic non-newtonian solution. Comparison of the fully developed flow region shows the large amount of shear-thinning. Figure 5 shows the elastic solution. Here the depressions can be clearly seen. Comparison of the figures 4 and 5 shows that only elasticity has been responsible for these velocity depression on the center-line. It is also to be noted that the elastic solution produces these depressions only for values of the relaxation parameter λ_1 which are rather high,

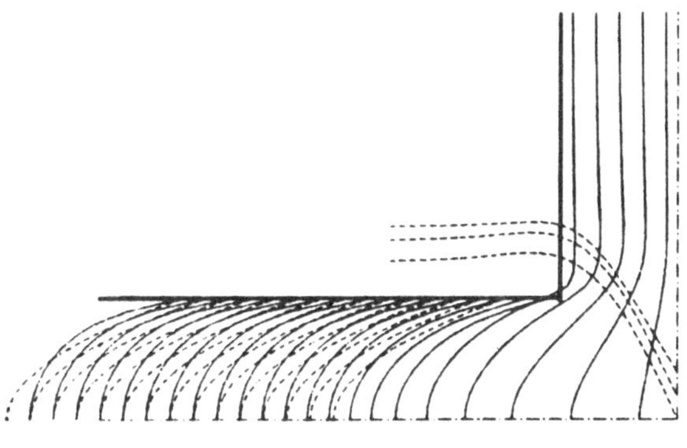

Fig. 4. Non-newtonian inelastic solution: $\eta_0 = 0.5$ poise, $Q = 4.5$ cm^3/s, $e = 0.2$ cm; ---- comparison with the newtonian solution

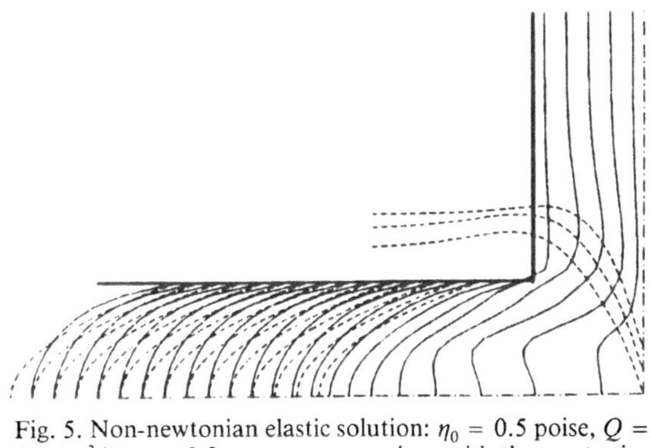

Fig. 5. Non-newtonian elastic solution: $\eta_0 = 0.5$ poise, $Q = 4.5$ cm^3/s, $e = 0.2$ cm; ---- comparison with the newtonian solution

with which values simulation becomes difficult due to numerical instabilities.

6. Conclusions

Numerical results so far obtained are essentially in qualitative agreement with experimental results. To obtain better simulations it is realised that a more complete rheological model has to be used.

References

1. Cressely, R., R. Hocquart, A. Lyazid, O. Scrivener, Proc. IUPAC Symp. "Makromainz" (Mainz 1979).
2. Cressely, R., R. Hocquart, O. Scrivener, Optica Acta **20**, 1173 (1979).
3. Mackley, M. R., J. Non-Newtonian Fluid Mech. **4**, 111 (1978).
4. Perera, M. G. N., K. Walters, J. Non-Newtonian Fluid Mech. **2**, 49 (1977).
5. Perera, M. G. N., K. Strauss, J. Non Newtonian Fluid Mech. **5**, 269 (1979).
6. Davies, A. R., K. Walters, M. F. Webster, J. Non-Newtonian Fluid Mech. **4**, 325 (1977).
7. Perera, M. G. N., PhD Thesis, Univ. Coll. of Wales, Aberystwyth (1976).

(Received April 15, 1982)

Authors' addresses:

Dr. M. G. N. Perera
EVT Energie- und Verfahrenstechnik GmbH
Postfach 395
D-7000 Stuttgart 1

Dr. O. Scrivener, Dr. A. Lyazid
Institut de Mécanique des Fluides
2, rue Boussingault
F-67083 Strasbourg Cedex

On some contraction flows for Boger fluids

K. Walters and D. M. Rawlinson

Department of Applied Mathematics, University College of Wales, Aberystwyth (U.K.)

Key words: Contraction flow, Boger fluid

1. Introduction

The contents of this lecture are a part of a general research programme which seeks to investigate flows for which elastico-viscous effects are a dominating influence. The basic experimental work is supported by attempts to simulate numerically the observed flow characteristics using finite-difference techniques (see, for example, Cochrane, Walters and Webster [1], Walters and Webster [2]).

Briefly, existing work has shown that

(i) slight asymmetries in the flow geometry which have a small effect on flow characteristics for Newtonian liquids can have an exaggerated influence in the case of elastico-viscous liquids. For example, the *precise* shape of re-entrant corners is an important factor in determining the size of the recirculating vortices found in many contraction flows involving highly elastic liquids. In some cases, slight changes in corner geometry can affect the flow in regions at a significant distance from the corner.

(ii) Fluid memory can sometimes result in flow characteristics which are *qualitatively* different to those found in the corresponding Newtonian situations (Walters and Webster [2]).

(iii) Elastico-viscous Boger fluids [3] show an expected vortex enhancement in *circular* contraction flows, but corner vortices are virtually absent in the corresponding *planar* contraction case (Walters and Webster [2]). Since, in a steady simple shear flow, Boger fluids have a reasonably linear shear stress τ and quadratic first normal-stress difference ν_1 over a significant shear rate range q, one is tempted to appeal to the Tanner-Giesekus (T-G) theorem (see, for example, Tanner [4]) to support such unusual behaviour in the plane contraction case. However, fluid inertia is important in the experiments referred to and conditions near the important corner regions are certainly outside the "second-order" range required for the application of the T-G theorem. So, the complete difference between the vortex behaviour for Boger fluids in circular and planar contraction flows remains a provocative experimental observation which as yet is not fully understood.

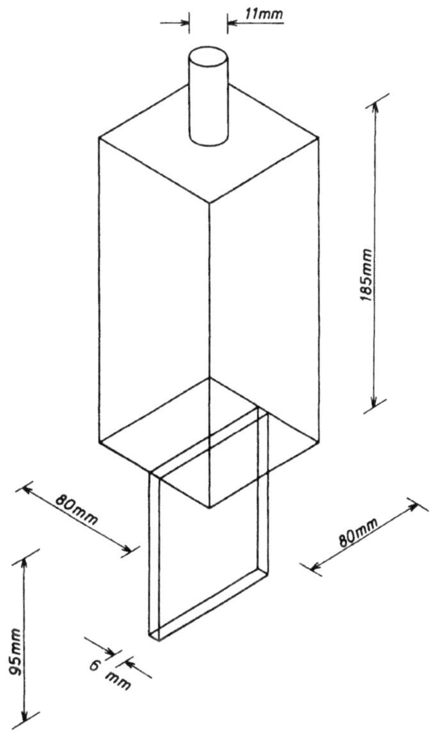

Fig. 1. The planar contraction (slit) geometry

The contraction flow experiments already published are for contraction ratios of 4:1. In the present lecture, attention is drawn to experiments on a planar 13.3:1 contraction of the sort shown in figure 1, the third dimension being clearly such as to give the wide channel a square cross-section. Also, in place of the usual axisymmetric circular contraction, we consider the square-square contraction shown in figure 2.

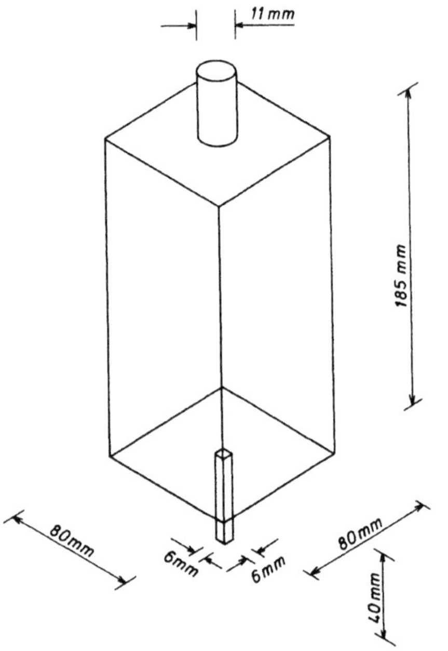

Fig. 2. The square-square contraction geometry

2. Flow visualization

The Newtonian liquids used for comparison purposes in the present experiments were mixtures of maltose syrup (C.P.C., U.K.) and water. The non-Newtonian Boger test liquids were solutions of polyacrylamide (N100, Cyanamid, Great Britain) in a maltose syrup/water base. Measurements of τ and ν_1 at different shear rates q were obtained on a Weissenberg R16 Rheogoniometer and these allowed the flows to be characterized by means of a Reynolds number R and elasticity number W through (cf. Walters and Webster [2])

$$\tau = \eta_0 q\,, \quad \nu_1 = 2\eta_0 \lambda q^2\,,$$

$$R = \frac{\rho \bar{U} L}{\eta_0}\,, \quad W = \frac{\lambda \bar{U}}{L}\,, \tag{1}$$

η_0 and λ being material constants, ρ the density, $\bar{U}$ the mean velocity and L a characteristic length. In the

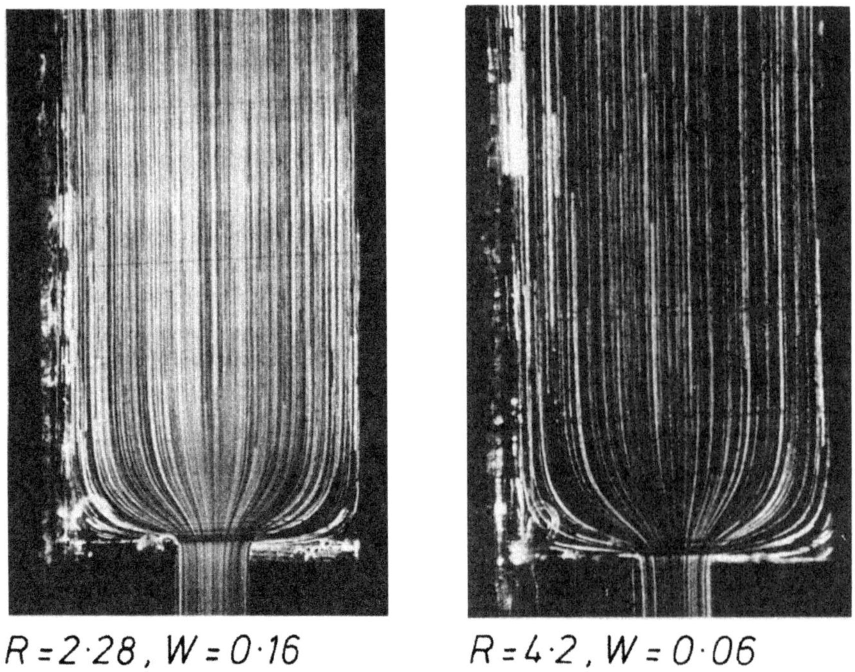

Fig. 3. Flow visualization pictures for B12 (left) and B14 (right) in a 4:1 planar contraction geometry

Table 1. Estimated density, viscosity and relaxation times for liquids B12 – B17

Liquid	ρ(gcm^{-3})	η_0(Pa s)	λ(s)
B12	1.3	0.6	0.1
B13	1.3	1.15	0.16
B14	1.3	0.15	0.07
B15	1.3	0.7	0.08
B16	1.3	1.3	0.36
B17	1.3	0.75	0.35

present experiments, we used conditions in the wide channel to determine R and W with L being the width of the side (i.e. 80 mm). All the experiments were carried out at room temperature (20 °C).

The relevant values of the physical variables for the Boger fluids used in the experiments are given in table 1.

The flow visualization apparatus has been described in detail in earlier papers (Cochrane, Walters and Webster [1], Walters and Webster [2]). The test fluids are seeded and a parallel beam of laser light makes flow visualization possible. The use of a conventional camera, used with different exposure times, permits a photographic record of the flow field. Figure 3 contains an example of the flow field for liquids B12 and B14 flowing through a 4:1 planar contraction (cf. Walters and Webster [2]).

In the experiments reported in this lecture for the 13.3:1 contractions, the beam of laser light entered one of the faces normally with the obvious symmetrical positioning. (It was not possible for technical reasons to visualize the flow with the laser beam shining along a diagonal).

Figure 4 contains the flow structure for the slit geometry for various flow rates. *No* significant vortex

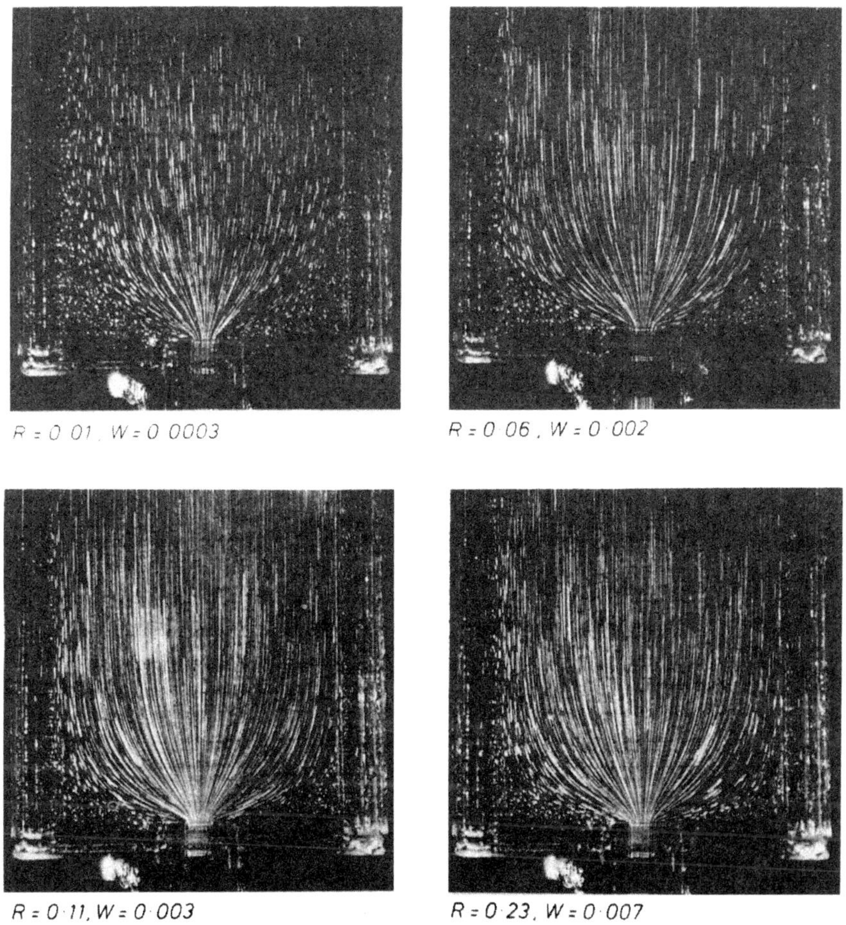

Fig. 4. Flow visualization pictures for liquid B13 in the planar contraction geometry of figure 1

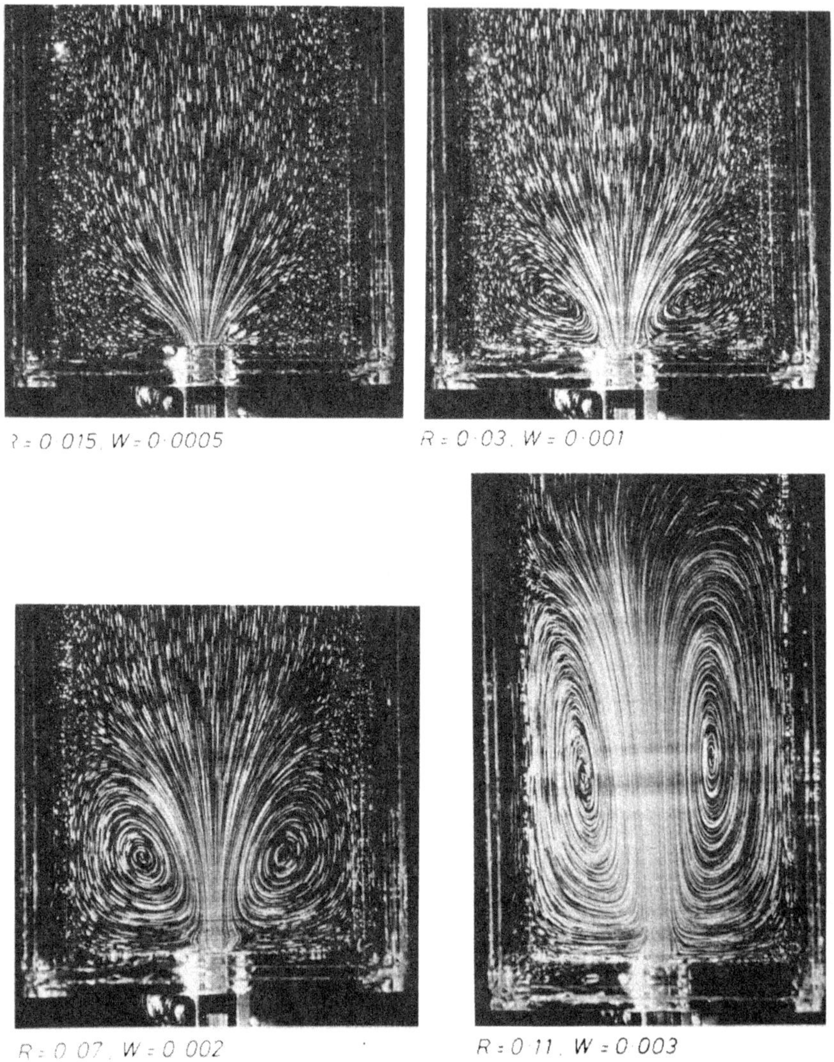

Fig. 5. Flow visualization pictures for liquid B13 in the square-square geometry of figure 2

activity is evident (in agreement with our earlier experiments with a 4:1 contraction), and no amount of experimentation with different Boger fluids and different flow conditions gave rise to any change in the basic flow structure.

This situation contrasts dramatically with the corresponding square-square contraction case (see fig. 5). Here, significant recirculating vortices are already in evidence at the lowest flow rate and, as the flow rate is increased, vortex enhancement dominates the whole of the available upper channel. Especially worthy of note in the rather abrupt expansion in the central jet as the flow is just about to enter the contraction.

The complete difference between the flow structure in figures 4 and 5 led Mr. F. N. Cogswell to suggest to us that there may be some merit in designing a square-square geometry with an intermediate slit (see fig. 6), the extra resistance from the addition of an intermediate slit being (possibly) compensated by a change in the flow structure. There is therefore the *possibility* of an increased flow rate through the square-slit-square geometry, other things being equal, although the introduction of an extra slit would certainly be expected to lead to a decreased flow rate *in normal circumstance.*

Accordingly, we constructed the "inserted-slit" geometries with dimensions as given in figure 6.

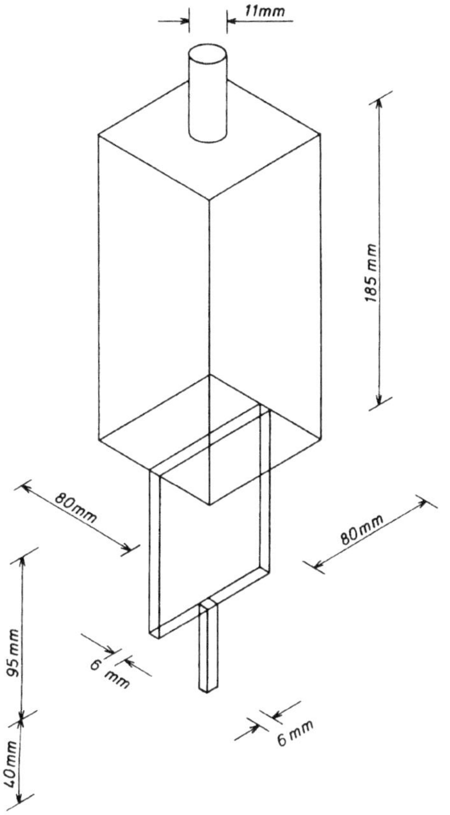

Fig. 6. The inserted-slit geometry. Dimensions are for long slit. Length of slit for short slit is 25 mm

Fig. 7. Flow visualization picture for top section of short inserted-slit geometries. Liquid B13

Flow visualization experiments with the inserted-slit geometries were notable for the strong tendency towards asymmetrical flow. (Admittedly, such asymmetries were also evident in the square-square geometry (see fig. 5) *but at much higher flow rates*). Figure 7 for liquid B13 in the top square section of the *short* slit geometry is a typical example of the strong vortex flow which was consistently observed for this geometry in both the square and slit sections. The figure also shows the strong asymmetry referred to above. Clearly, the inserted slit in this case is too short to generate the sought-for vortex-free flow characteristic of the basic slit geometry of figure 1. In contrast, the long slit geometry did permit a greater degree of vortex-free flow (see fig. 8). In this case, the low flow rate regime has *no* strong vortex structure, and the top section is free of such behaviour until an asymmetric vortex activity arises at the highest flow rate. In contrast, the vortex free flow in the lower slit section quickly degenerates into an unstable asymmetrical vortex flow.

3. Flow rate measurements

We have recently begun a detailed study of the pressure-flow rate behaviour of Boger fluids in the various geometries discussed in section 2. As a means of comparison, we intend to study the flow rates for the various geometries for the same value of the pressure at the entrance to the geometries. Our preliminary experiments have not indicated any significant trends but the work is continuing.

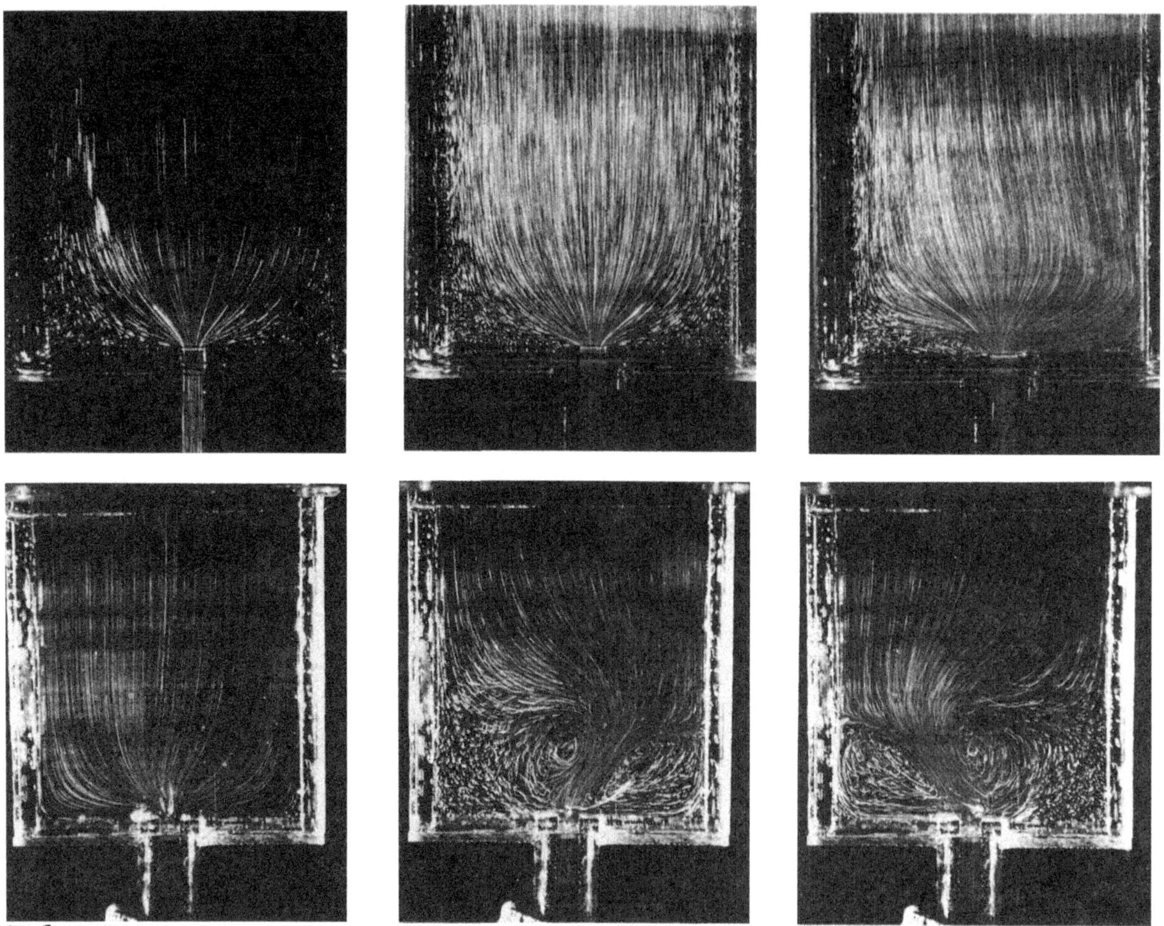

(a) $R = 0.17$, $W = 0.001$ (b) $R = 0.45$, $W = 0.003$ (c) $R = 0.69$, $W = 0.004$

Fig. 8. Flow visualization pictures for the long inserted-slit geometry. Top pictures for top section. Bottom pictures for slit section. Liquid B15

Acknowledgements

We have benefited from helpful discussions with Professor R. S. Rivlin and Mr. F. N. Cogswell. We are also indebted to Mr. Robin Evans for help with the photography.

References

1. Cochrane, T., K. Walters, M. F. Webster, Phil. Trans. Roy. Soc. London **301**, 163 (1981).
2. Walters, K., M. F. Webster, to appear in Phil. Trans. Roy. Soc. London (1982).
3. Boger, D. V., J. Non-Newtonian Fluid Mech. **3**, 87 (1977/78).
4. Tanner, R. I., Phys. Fluids **9**, 1246 (1966).

(Received July 6, 1982)

Authors' addresses:

Prof. K. Walters, D. M. Rawlinson
Department of Applied Mathematics
The University College of Wales
Physical Sciences Building
Penglais, Aberystwyth, SY23 3BZ (U.K.)

Rheologica Acta Rheol. Acta **21**, 553 – 555 (1982)

Structural changes of polyisobutylene solutions induced by orifice flow

G. Dembek

Lehrstuhl für Strömungsmechanik, Abteilung Chemietechnik der Universität Dortmund

Key words: Orifice flow, flow-induced structural change, viscometry, polyisobutylene solution

1. Introduction

Orifice flow of polymer melts and solutions is of great interest as it occurs in a wide variety of applications e.g. extrusion and viscometry. It is assumed in most cases that the polymer molecules are partially orientated in the core of the flow region in front of the orifice. This leads to storage of elastic energy in the fluid observable as die swell at the exit. Recent experiments on polyisobutylene (PIB) solutions in orifice flow indicate that there is not only a partial orientation but structural change is also possible in the fluid.

2. Experimental set-up

The examined solutions consist of high molecular weight PIB Oppanol B 200[1]) dissolved in low molecular weight PIB B 1[1]). The solution is pumped through a glass tube with an inner diameter of 32 mm by a gear pump. The orifice is mounted in the glass tube and can be changed. The flow rate and the pressure difference over the glass tube are measured. The diameter of the orifices ranges from two to six and the length from one to twenty millimeters.

The flow is visualized by illuminating a plane upstream of the orifice to record the movement of small air bubbles sucked in on the low pressure side of the gear pump. To ensure that the movement of the air bubbles is not very different from that of the fluid, a photochemical coloration method developed by Hummel et al. [1] and described elsewhere [2] is used.

3. Experimental results

The PIB solutions are characterized by recording their zero-shear viscosity as a function of concentrations as shown in figure 1. At concentrations up to 0.1% there is only a small dependence of the zero-shear viscosity on concentration, whereas at higher concentrations the influence is very strong. Straight lines drawn through both regions on the double logarithmic plot cross at a concentration of 1.5 %. Small changes of the concentration in this transition region have important consequences for orifice flow.

At concentrations higher than 1.9% the fluid behaves upstream of the orifice as sketched in figure 2 from a photograph and well known for many polymer

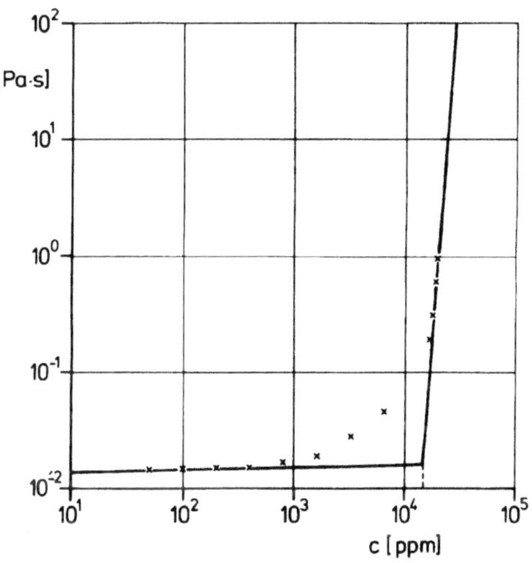

Fig. 1. Concentration dependence of the zero-shear viscosity for solutions of PIB Oppanol B 200 in Oppanol B 1

[1]) Oppanol B 1, Oppanol B 200; trademark of BASF AG, Ludwigshafen.

770

Fig. 2. Flow pattern of a 2% PIB solution streaming from a tube of d = 32 mm into an orifice of d = 6 mm (sketched from a photograph)

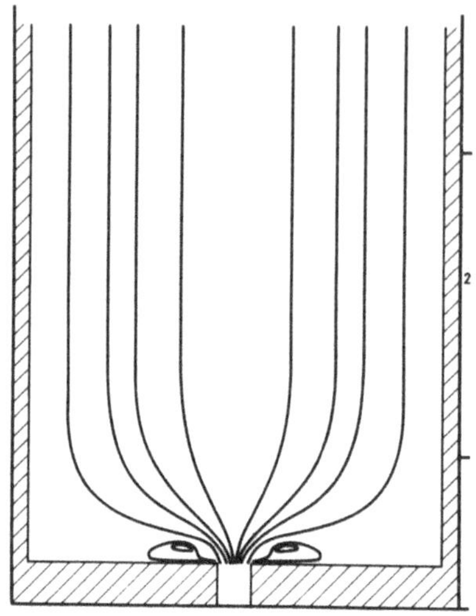

Fig. 3. Comma-like vortices formed by a 1.7% PIB solution flowing into an orifice with d = 3 mm (sketched from a photograph)

melts and solutions. A converging zone is formed, the size of which increases with increasing flow rate, while the convergence angle decreases until the flow becomes asymmetric and unstable, first swinging then collapsing. Remarkable is an indication of bulging at the border of the orifice.

At a concentration of 1.7% the entrance flow pattern is quite different. At low flow rates there are almost no vortices at the corners of the contraction and the pressure loss at the orifice is nearly proportional to the discharge. With increasing flow rate comma-like vortices occur at the border of the orifice as shown in figure 3. They are accompanied by a strongly increasing pressure drop. For further increases in flow rate these vortices grow forming a hose through which the fluid flows into the orifice

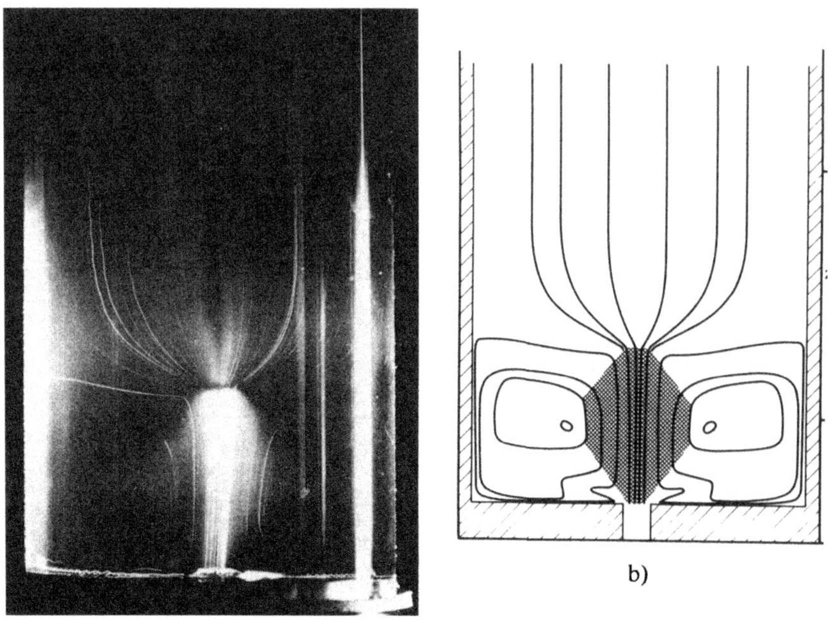

a) b)

Fig. 4. Hose formed by a 1.7% PIB solution flowing into an orifice with d = 3 mm. The silky appearance of the hose and the neighbouring region indicates a structural change of the fluid. a) Photograph; b) Sketch from the photograph

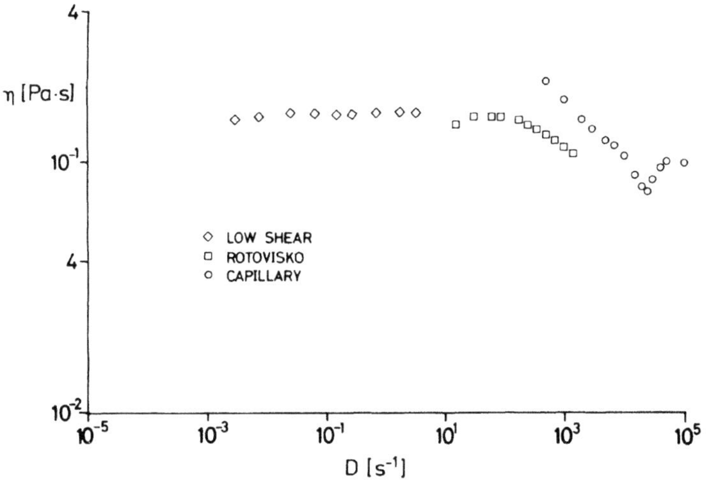

Fig. 5. Shear-rate dependence of the shear viscosity of a 1.7% PIB solution as measured by rotational and capillary viscometers

(fig. 4). The hose is similar to that observed by Giesekus [3] for aqueous polyacrylamide solutions. It can reach a length several times the orifice diameter. The outer part of the hose is bulged significantly just in front of the orifice so that at the border of the orifice the flow is the same as for the comma-like vortices at lower flow rates.

The hose and neighbouring regions flowing parallel to it show a silky appearance when illuminated perpendicular to the viewing direction. This difference in the optical properties compared to the rest of the fluid provides evidence of a change of the fluid in this region.

The viscosity plot measured for this fluid (fig. 5) is quite different from those of more highly concentrated PIB-solutions, for which measurements made using rotational and capillary viscometers overlap. For the 1.7% solution the values obtained from measurements in a capillary viscometer[2]) with a diameter of 0.156 mm are nearly twice as high as those from rotational viscometers and even higher than the zero-shear viscosity. This gives rise to the supposition that the fluid is changed by the extension in the orifice flow present also at the inlet of the capillary.

The valley in the viscosity plot of the capillary-data is identical with the shear-rate region in which the hose occurs. Here the Bagley correction has to be taken into account as it is up to 25% of the total pressure loss at a capillary of l/d = 1000. The following constant viscosity is measured at shear-rates where the orifice flow becomes unstable. However the unstable flow of the 1.7% solution is quite different from that of more highly concentrated solutions. The main disturbances are in the direction of flow whereas in more highly concentrated solutions it is a motion swinging perpendicular to it.

[2]) These measurements were kindly carried out at the BASF AG, Ludwigshafen, by Dr. Münstedt.

References

1. Popovich, A. T., R. L. Hummel, Chem. Engng. Sci. **22**, 21–25 (1967).
2. Dembek, G., Flow visualization of newtonian and non-newtonian media through contractions by photochromic dyes. Prepr. Int. Symp. Flow Visualization 508–512, Bochum (1980).
3. Giesekus, H., Rheol. Acta **8**, 411–421 (1969).

(Received May 10, 1982)

Author's address:

G. Dembek
Lehrstuhl für Strömungsmechanik
Abteilung Chemietechnik
Universität Dortmund
Postfach 500500
D-4600 Dortmund 50

Rheological interpretation of pressure anomalies of aqueous dilute polymer solutions (ADPS) in orifice flow

D. H. Fruman and M. Barigah

Laboratoire de Biorhéologie et d'Hydrodynamique Physico-Chimique, Université Paris VII

Abstract: The response of a considerable number of solutions of several polymers (PEO, HPAM, PAM) with concentrations of less than 100 ppm in orifice flow has been investigated. It is shown that the excess pressure (difference between the ADPS and the solvent total pressure drop) behaves linearly as a function of a superficial strain rate (ratio between a velocity and a length scale). In rheological terms this behaviour is interpreted as the result of a constant elongational viscosity whose values are two to three orders of magnitude larger than the shear viscosity. A formal approach to this phenomenological interpretation is suggested.

Key words: Elongational viscosity, dilute polymer solution

1. Introduction

It is well known that concentrated, semi-dilute and dilute polymer solutions display anomalous behaviour in extensional flow situations, such as those occurring in the upstream converging region of a sink flow. Considerable attention has therefore been devoted to investigate these anomalies in abrupt entry geometries and to interpret pressure drop measurements in terms of the rheological properties of the polymer solutions. For dilute solutions, Giles [1] originally demonstrated that pressure drops significantly exceeding those of the solvent (water) did exist for orifice and slit flows provided their geometric dimensions were small enough. These results were later confirmed and extended by Bilgen [2] to a variety of concentration of PEO compounds. Also, he succeeded in correlating the excess pressure (difference between the ADPS and the solvent total pressure drop) with the relaxation time of the ADPS computed using molecular theories. Piau [3] performed experiments with orifices of different diameters and suggested that the onset of the excess pressure was related to a critical superficial strain rate, ratio between a velocity U and a length scale D of the flow. For ADPS discharged into air from a small orifice, Balakrishnan and Gordon [4] computed elongational viscosities from pressure drop entrance measurements. James and Saringer [5] measured in a small angle converging nozzle the wall pressure upstream of the exit. They showed that a radial normal stress component equal to a third of the axial normal stress is responsible of the observed decrease of the wall pressure.

In all of the above mentioned works but [2] the total pressure drop ΔP was used in the analyses and correlations instead of the excess pressure ΔP_e, difference between the ΔP of an ADPS and the ΔP of a solvent of equal shear viscosity. A correlation of data based on the ΔP_e was presented by Fruman and Tulin [6] for two extensional flow situations: Pitot tube and orifice. They found that, in the case of PEO solutions, the ΔP_e increases linearly with U/D provided a critical (onset) value was exceeded. This method of analysis was then used by Ouibrahim [7] and Ouibrahim and Fruman [8] to investigate the anomalous pressure of PEO and HPAM solutions respectively. In this paper our previous work is extended to HPAM solutions with concentrations of up to 100 ppm in solvents of different pH and ionic strengths. It is shown that, provided the values of U/D are small and moderate, the ΔP_e of *all* ADPS tested is a linear function of U/D. The slope of the straight lines is a measure of the elongational viscosity

η_E of the polymer solutions. Moreover, for ADPS behaving as Newtonian in shear, excess pressures only occur beyond a critical value of U/D, $(U/D)_c$. If the viscous behaviour is non-Newtonian (shear thinning), $(U/D)_c$ coincides with the origin. Both, the non-Newtonian behaviour and the absence of a non-zero onset value appear to be directly related to the conformation of the polymer molecules.

2. Experimental

The experimental set-up was described in [9] in detail. Simply, it consists of a plastic cylindrical cell containing a 0.112 mm diameter thin-walled orifice drilled in a 0.4 mm thick stainless steel sheet. Pressure drop between the upstream and downstream chambers were measured with the aid of pressure differential gauges, while the flow rate was determined by weighting procedures. The calibration of the orifice with a Newtonian fluid (water-glycerol-mixtures) allows the computation of the Newtonian ΔP to be subtracted from the ΔP measured during polymer solution flow at equal Reynolds number and U/D.

Several polymer molecules were used throughout the test program under a variety of conditions affecting the concentration, the pH and the salt (NaCl) content. In this short communication we will only consider the results obtained with a particular type of hydrolyzed polyacrylamide, Pusher 700, usually employed in improved oil recovery operations. The polymer solutions were prepared with extreme care in order to reduce or eliminate mechanical and/or biological degradation. Demineralized water filtered through a 5 – 15 μ sintered glass filter, brought to pH 7 by either adding HCl or NH_4OH and stored in closed containers was employed. All solutions were obtained by manually agitating the required quantity of polymer and water until visual observation indicated a complete dispersion of the polymer. This dispersion was then kept during 24 hours in a refrigerator (4 °C) and, once removed, the pH was adjusted to 8 before leaving the solution for another 24 hours at room temperature. Prior to the tests the pH was again controlled and brought to 8 or to another value if desired. For the solutions prepared with water containing salt the procedure was the same besides the fact that the salted water was filtered, brought to pH 7 and used as a solvent. Viscosity data for shear rates comprised between 10^{-2} and $10^3\ s^{-1}$ were obtained from Couette and capillary tube viscosimeters. Density measurements were made using a standard 25 cm^3 pycnometer and a 10^{-4} g precision balance.

3. Interpretation of orifice flow data

Several attempts to interpret orifice flow data in terms of the elongational viscosity of the polymer solutions have been made [10, 4]. Assuming a conical flow of angle Φ upstream of the orifice, considering that the total pressure drop is the sum of inertial, viscous and elastic effects and that the viscous contribution is the same for a Newtonian fluid of equal shear viscosity than for the ADPS, it can be shown [9] that,

$$\Delta P_e = -\rho U^2\left(\frac{D}{D_j}\right)^2\left[1-\left(\frac{D_{jN}}{D_{jNN}}\right)^2\right] + \frac{2}{3}\int_{\dot\varepsilon_0}^{0}(\tau_{rr}^{el}-\tau_{\theta\theta}^{el})\frac{d\dot\varepsilon}{\dot\varepsilon} \tag{1}$$

where D_j is the diameter of the jet issuing into the downstream chamber, $\dot\varepsilon$ is the strain rate, subscripts N and NN refer to Newtonian fluid and ADPS respectively and subscript o refers to the orifice plane. τ_{ij}^{el} are the components of the stress tensor associated with the elasticity of the ADPS. Since little is known about the changes in the diameter of the jets as a result of elasticity it can be further assumed that $D_{jN} \approx D_{jNN}$ and, based on previous and present data,

$$\tau_{rr}^{el}-\tau_{\theta\theta}^{el} = \sigma_e(\dot\varepsilon_o)(\dot\varepsilon/\dot\varepsilon_o)^n \tag{2}$$

where $\sigma_e(\dot\varepsilon_o)$ is the first normal-stress difference at $\dot\varepsilon = \dot\varepsilon_o$. Replacing (2) into (1) and integrating we obtain,

$$\Delta P_e(\dot\varepsilon_o) = -\sigma_e(\dot\varepsilon_o)\frac{2}{3n} \tag{3}$$

and, with $n = 1$,

$$\eta_E = -\frac{\sigma_e(\dot\varepsilon)}{\dot\varepsilon} = \frac{3}{2}\,\frac{\Delta P_e(\dot\varepsilon_o)}{\dot\varepsilon_o}\,. \tag{4}$$

The strain rate at the orifice can be evaluated considering a sink flow and

$$\dot\varepsilon_o = 2\frac{U}{D}\sin\Phi(1+\cos\Phi)\,. \tag{5}$$

For ADPS and Φ small enough (~15°), we finally obtain

$$\eta_E \approx \frac{3}{2}\,\frac{\Delta P_e(U/D)}{U/D-(U/D)_c} \tag{6}$$

which relates the elongational viscosity to the slope of the excess pressure versus U/D data.

4. Results

Figure 1 shows typical results for HPAM solutions containing 2% of NaCl and concentrations ranging from 10 to 100 ppm. As indicated the slopes increase and the $(U/D)_c$ decrease with concentration increasing. These solutions display a Newtonian behaviour in shear and have an elastic behaviour very similar to PEO solutions [6]. For unsalted solutions, whose

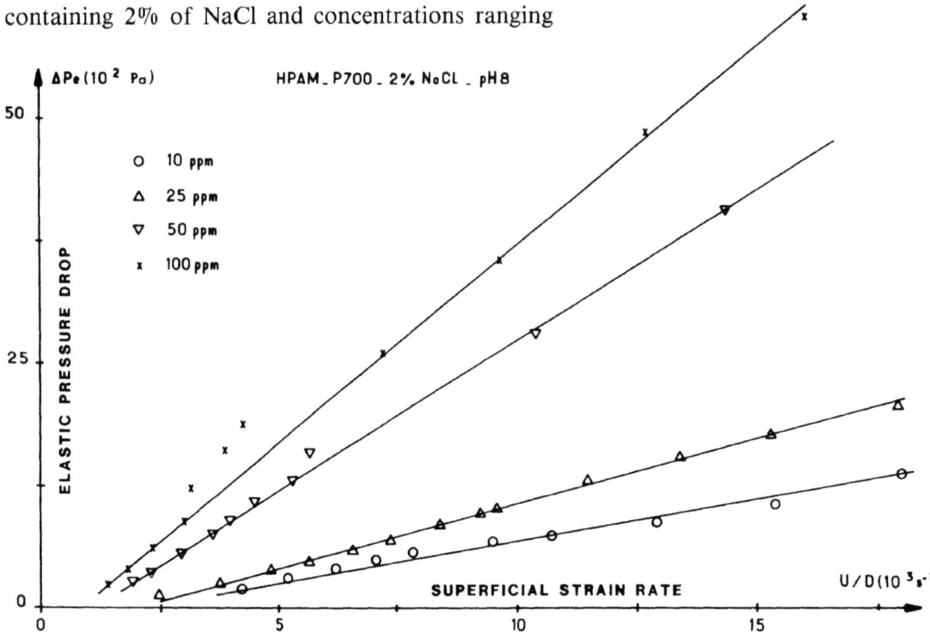

Fig. 1. Elastic pressure drop versus superficial strain rate for HPAM solutions containing 2% of NaCl, pH = 8

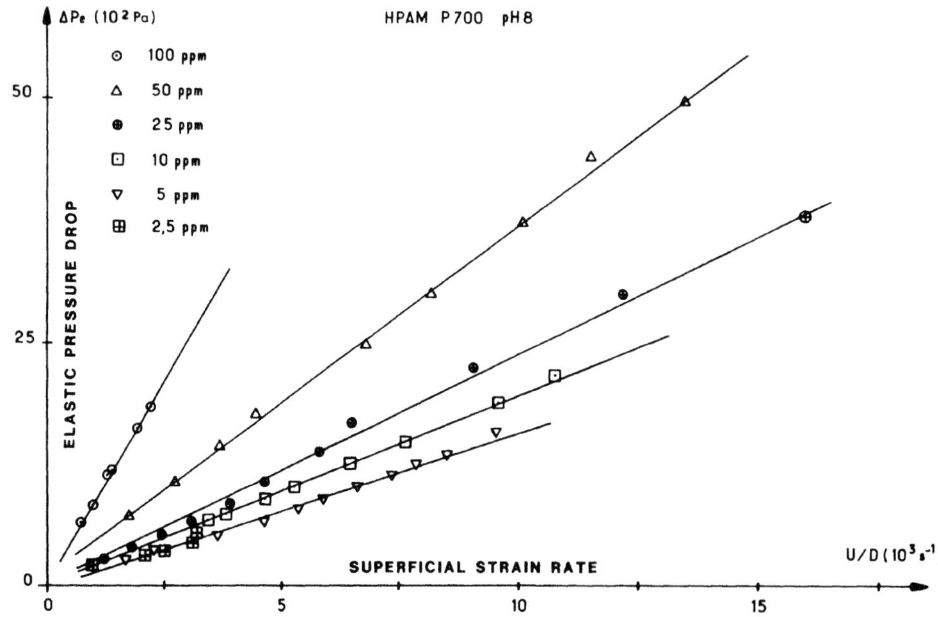

Fig. 2. Elastic pressure drop versus superficial strain rate for HPAM solutions without NaCl, pH = 8. Solutions prepared from a 1000 ppm master solution

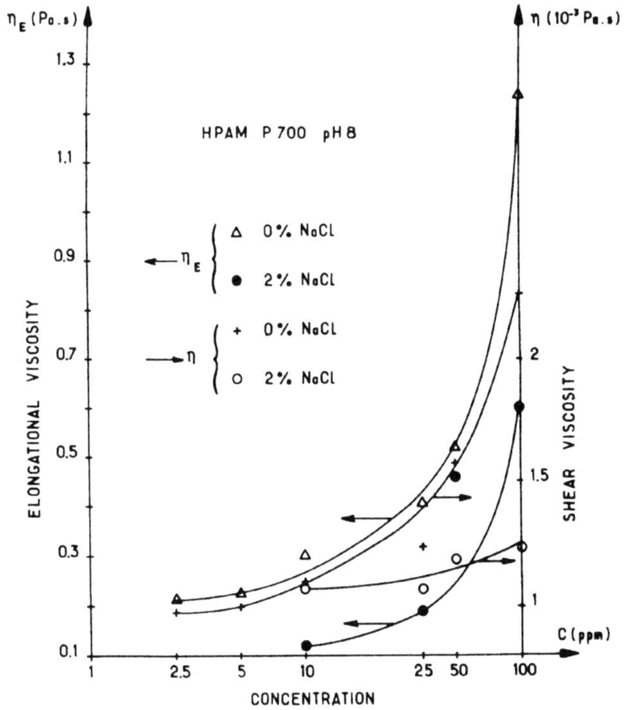

Fig. 3. Elongational and shear viscosity versus concentration for salted and unsalted HPAM solutions

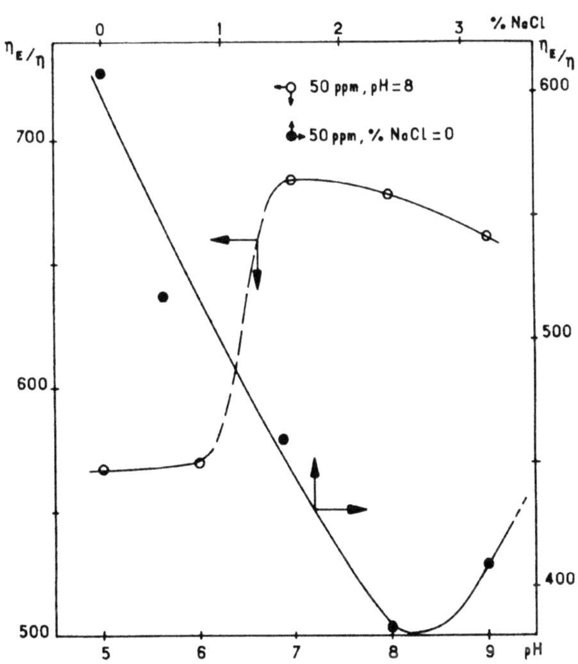

Fig. 4. Ratio between elongational and shear viscosity versus pH and salt content for HPAM solutions at constant concentration, 50 ppm

behaviour is shear thinning, ΔP_e increases also linearly but $(U/D)_c = 0$ as shown in figure 2, even for concentrations as low as 2.5 ppm. In Figure 3 we have summarized the results concerning these solutions by plotting η_E computed using eq. (6) and the shear viscosity, η, as a function of the concentration. As shown, the former can be several orders of magnitude larger than the latter. It is interesting to note that η_E is more sensitive than η to changes in polymer concentration. For example, in the case of salted solutions a tenfold increase in concentration leads to an increase of 5.07 of η_E and only 1.15 of η. For unsalted solutions and a fortyfold augmentation of the concentration the factors are respectively 5.7 and 2.3. Also, for very dilute solutions, such as those containing only few ppm, η_E tends to a non-zero limiting value which can be interpreted as a property of the polymer molecule in the absence of strong molecular interactions. Figure 4 shows the values of η_E/η for different pH and salt contents and a constant concentration, 50 ppm. These results indicate that the consequence of an increase in salinity is to reduce the elastic effects of the HPAM solutions and are in agreement with those of Durst et al. [11] for flow through porous media. Indeed, these authors observe a pronounced shift of the onset towards larger strain rates (Reynolds numbers) as already shown in our case in figures 1 and 2, and a decrease of the pressure drop (resistance coefficient) at equal strain rate. Moreover, comparison of figures 9(a) and 15(c) of [11] shows a drastic reduction in elastic properties for pH dropping from 8.8 to 3.3 in agreement with our own observations, figure 4. The pH and the ionic strength of the solvent affect the conformation of the dissolved molecules and hence their elastic contribution to the flow in terms of changes of the onset strain rate and the pressure drop magnitude. Drag reduction [12] and porous media data [11] for HPAM solutions lead to the same interpretations.

The existence of a constant elongational viscosity, constant slope of the ΔP_e versus U/D curves, is however limited to U/D such that the ratio between inertial and elastic forces is not too large. Analysis of a large number of results indicates that for ADPS there exists an upper bound of U/D given by

$$(U/D)^{+} < \eta_E/\rho D^2 . \tag{7}$$

For values of U/D larger than the above the behaviour of ADPS is no longer linear and the elongational viscosity computed using eq. (6) is no longer valid. Relation (7) indicates that transition into an inertially dominated flow will occur sooner for very dilute solutions, for large orifice diameters and for solutions with limited intermolecular interaction.

5. Conclusions

Pressure anomalies observed during the flow of ADPS through a small orifice can be interpreted in terms of an elongational viscosity and an onset strain rate below which no elastic effects are displayed. For a single polymer compound (HPAM) the above parameters vary with the concentration and the conformation of the molecules as determined by the pH and the ionic strength of the solvent. Our results are in general agreement with the conclusions drawn from drag reduction and porous media data. Due to its simplicity, orifice flow experiments associated with the proposed method of analysis provide a practical way of characterizing the viscoelastic properties of dilute polymer solutions.

References

1. Giles, W. B., Nature **224**, 584 – 585 (1969).
2. Bilgen, E., J. Mécanique **12**, 375 – 391 (1973).
3. Piau, J. M., Mécanique des Solutions de Polymères. Application à la Réduction de Frottement, Colloques Internationaux du C.N.R.S., N° 233, 225 – 240, Brest 20 – 23 Mai 1974.
4. Balakrishnan, C., J. R. Gordon, Proc. VII. Intern. Congr. Rheology, pp. 265 – 267. Gothenburg, Sweden (1976).
5. James, D. F., J. H. Saringer, J. Fluid Mech. **97**, 655 – 671 (1980).
6. Fruman, D. H., M. P. Tulin, Phys. Fluids **21**, 1896 – 1899 (1978).
7. Ouibrahim, A., Phys. Fluids **22**, 784 – 785 (1979).
8. Ouibrahim, A., D. H. Fruman, J. Non-Newtonian Fluid Mech. **7**, 315 – 331 (1980).
9. Barigah, M. D., Caractérisation Rhéologique des Solutions de Polymères en Ecoulement Elongationnel, Thèse de 3° Cycle, Université Paris VI, Décembre 1981.
10. Metzner, A. B., A. P. Metzner, Rheol. Acta **9**, 174 – 181 (1970).
11. Durst, F., R. Haas, B. U. Kaczmar, J. Appl. Polym. Sci. **26**, 3125 – 3149 (1981).
12. Frommer, M. A., D. H. Fruman, J. E. Santo, The Influence of the Chemical Nature of Polymers on their Drag Reduction Characteristics, Hydronautics, Incorporated, Technical Report N° 7213 – 2, January 1974.

(Received April 16, 1982)

Authors' address:

Dr. D. H. Fruman, M. Barigah
L.B.H.P.
Université Paris VII
2, Place Jussieu
F-75251 Paris Cedex 05

Rheologica Acta Rheol. Acta 21, 561 – 563 (1982)

Shear induced phase transitions in highly dilute aqueous detergent solutions

H. Rehage and H. Hoffmann

Institut für Physikalische Chemie der Universität Bayreuth

Abstract: Aqueous solutions of some cationic detergents show rheopectic behaviour at very low concentrations. Rheological measurements and the electrical anisotropy of the streaming solutions indicate that the rheopectic behaviour of the solutions is due to a shear induced phase transition.

Key words: Rheopexy, electrical anisotropy, stress relaxation, phase transition, detergent solution

1. Introduction

Highly dilute solutions of ionic and non-ionic detergents behave usually as Newtonian liquids. Only a few systems are known which have a more complicated rheological behaviour. Solutions of these systems are elastic. This striking phenomenon is relatively rare and it is not completely understood in spite of the effort that has recently been undertaken to explain it [1]. For a few systems the phenomenon has been observed in very dilute solutions with volume fractions of the detergent of the order of 10^{-3} [2].

The aggregates that are present in these solutions have been characterized by static and dynamic light scattering, electric birefringence, NMR, rheological, kinetic and neutron scattering experiments [3]. The results of these investigations indicate that non-spherical rodlike micelles are present in these solutions. The axis of the rods are randomly oriented when the solutions are at rest. The lengths of the rods increase linearly with the detergent concentration [4].

The present investigations have been performed in order to obtain more informations about the visco-elastic behaviour of the detergent solutions. The measurements were carried out on the system Cetylpyridiniumsalicylate (CPySal) which had been well characterized in previous investigations.

2. Preparation of the solutions

The CPySal solutions were prepared as previously described by ion exchange procedure from CPyCl solutions or by dissolving CPySal which had been synthesized before. Both methods gave identical results. The solutions were left standing for two days in order to reach equilibrium [1].

3. Experimental

Three types of viscometers were employed for measuring the rheological properties of the detergent solutions. The measurements of viscosity changes with time and the dynamic experiments have been carried out with a rotary viscometer (Contraves Low Shear 30 sinus).

The viscosities at low detergent solutions were determined with a Zimm-Crothers viscometer at a constant shear stress of $1.27 \cdot 10^{-4}$ Pa.

Measurements of superposed sinusoidal oscillations onto steady shear have been performed with a rotary viscometer (Haake RV 100/CV 100). In order to measure the anisotropy effect of the electrical conductivity we have constructed a Couette-type apparatus which permits measurements of the electrical conductivity in three orthogonal directions [5].

4. Results and discussion

It was previously reported that highly dilute solutions of the amphiphilic system CPySal shows a rheopectic rheological behaviour [1]. This effect is demonstrated in figure 1.

It shows the transient viscosity calculated from the measured shear stress as a function of the shear time.

790

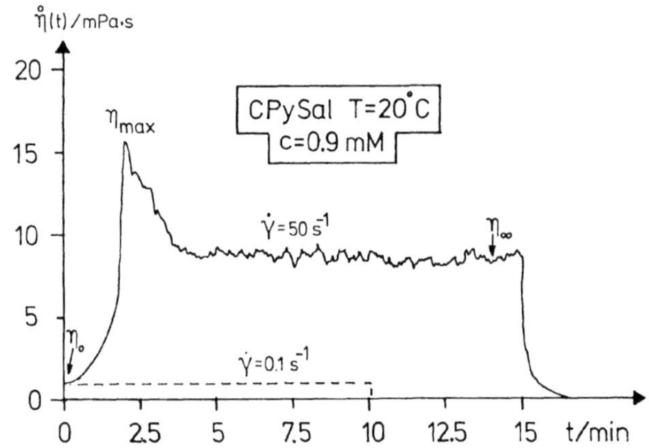

Fig. 1. Dependence of the transient viscosity on time of shearing at $\dot{\gamma}$ = 0.1 s^{-1} and $\dot{\gamma}$ = 50 s^{-1}

The rate of deformation is constant and it has been applied at t = 0. The parameter of each curve is the shear rate $\dot{\gamma}$. At low shear rates the transient viscosity is constant. The flow behaviour of the solution is that of a Newtonian liquid. At some higher shear rates however, there is a pronounced increase in viscosity which reaches a maximum value at about 2 min and then decreases again. After about 4 min the transient viscosity reaches a steady-state value. The latter part of the curve shows the relaxation process after the flow has been stopped. The rheopectic behaviour of the detergent solution can be explained by postulating the formation of a supermolecular structure during flow.

Additional informations about this structure can be obtained by measuring the anisotropy effect of the electrical conductivity. It is based on the phenomenon that the electrical conductivity of a solution of charged anisometric particles depends on the direction of measurement if the particles are aligned in the streaming solution. The deviations from the average conductivity in three orthogonal directions of measurements are related to the dimensions of the anisometric polyelectrolyte [6]. The results of the conductivity measurements indicate that the supermolecular structure must be rod-shaped and that it is aligned at a shear rate of about 50 s^{-1}.

The rheopectic behaviour of the solutions can be measured up to a detergent concentration of 7.7 mM. Above this concentration the viscous resistance increases rapidly and the solutions show a thixotropic behaviour. In our previous studies of the amphiphilic system several methods have been used to determine the lengths of the cylindrical micelles [4]. These investigations have shown that at exactly 7.7 mM the rodlike micelles start to overlap. At this concentration the lengths of the rods are equal to the spacing between the aggregates. In this situation the rodlike micelles can make contact and they form a new structure, a three-dimensional network. The rheological properties above concentrations of 7.7 mM are typical of a network solution and they are not the result of hydrodynamically interacting separate cylinders [5].

Observations with the unaided eye suggest that the shear induced structure has many properties of the network solution which is formed at higher concentrations. In trying to elucidate the mechanism behind the rheopectic behaviour it is necessary to compare the shear-induced supermolecular structure with the network structure. Five different types of experiments have been performed in order to get informations on the shear induced structure:

(i) Stress relaxation experiments.
(ii) Relaxation experiments of the electrical anisotropy effect.
(iii) Investigations of the thermal behaviour of the two phases.
(iv) Parallel superposition of sinusoidal oscillations onto steady shear in order to determine the elastic moduli of the shear-induced structure.
(v) Investigations of the ageing process of the network solutions.

The results indicate that the shear-induced structure and the network structure have similar properties [5].

5. Conclusions

The above observations suggest that the shearing of the detergent solutions can result in reversible formation of a three-dimensional network structure.

Acknowledgement

Financial support of this work by a grant of the "Deutsche Forschungsgemeinschaft (DFG)" and the "Fonds der Chemischen Industrie" is gratefully acknowledged.

References

1. Hoffmann, H., G. Platz, H. Rehage, W. Schorr, W. Ulbricht, Ber. Bunsenges. Phys. Chem. **85**, 255 (1981).

2. Gravsholt, S., J. Colloid Interface Sci. **57**, 576 (1976); J. Ulmius, H. Wennerström, L. B. A. Johansson, G. Lindblom, S. Gravsholt, J. Phys. Chem. **83**, 2232 (1979).
3. Kalus, J., H. Hoffmann, K. Reizlein, W. Ulbricht, Ber. Bunsenges. Phys. Chem. **86**, 37 – 42 (1982).
4. Hoffmann, H., G. Platz, H. Rehage, W. Schorr, Ber. Bunsenges. Phys. Chem. **85**, 877 – 882 (1981).
5. Hoffmann, H., H. Rehage, in preparation.
6. Götz, K. G., J. Colloid Sci. **20**, 289 – 314 (1965).

(Received April 16, 1982)

Authors' address:

Dipl.-Chem. H. Rehage, Prof. Dr. H. Hoffmann
Institut für Physikalische Chemie
Universität Bayreuth
Postfach 3008
D-8580 Bayreuth

Use of polymers in oil recovery processes

J. F. Stanislav

Department of Chemical an Petroleum Engineering, The University of Calgary, Calgary, Alberta (Canada)

Key words: Polymer, mobility control, enhanced oil recovery, waterflooding, thermodynamics

Water-soluble polymers are used extensively in various stages of gas and oil production operations, typical examples being enhanced oil recovery, water production control and well drilling. A variety of polymeric materials, both naturally occurring and synthetic ones, are currently used: guar and cellulose derivatives, xanthan gum, polysaccharides, polyacrylamides and others. In this paper only the application of polymeric materials to enhanced recovery processes is discussed.

Waterflooding is a typical secondary recovery scheme. It involves a continuous injection of large quantities of water into the oil bearing formation. Areas of high water saturation which are initially formed around each injection well are moving through the reservoir towards the pressure sinks, that is producing wells. As a result of this water bank motion through the reservoir, additional oil is displaced to the wellbores where it can be recovered. Oil production begins to increase over the level of the recovery by primary depletion (by natural forces) until the moment when water breaks into the producers; at that time water production rises rapidly and ultimate recovery of secondary oil is reached.

The amount of oil which can be efficiently displaced by the waterflooding process depends on the degree to which the properties of the reservoir rock are uniform. In general, oil reservoirs are never homogeneous.

In regard to the oil recovery processes, the distribution of the rock permeabilities is the most important formation characteristics. The injected water is following the path of the least resistance, that is, it preferentially flows through the high permeability zones; the oil in the low permeability stratas is then bypassed and remains unrecovered. The matter is further aggrevated by the effective permeability increase to water in high water saturation layers. This phenomenon of nonuniform water invasion of the rock formation is known as a "fingering effect" and accounts for early water breakthrough into the producers.

The reduction of the water fingering and thus enhancement of oil recovery can be achieved by improving the water-oil mobility ratio which is defined as:

$$M = \frac{k_w/\mu_w}{k_0/\mu_0}. \tag{1}$$

The mobility ratio, M, less or equal to one ensures a piston-like displacement whereas at values greater than one the more mobile water will finger through the oil zone leaving some oil unrecovered.

In theory, the mobility ratio can be controlled by either reducing the oil viscosity, or, increasing the water viscosity or, through permeability changes. Virtually all efforts have been aimed at increasing the viscosity of the displacing phase. Water-soluble polymers proved to be so far more efficient that any other materials considered: large size of the macromolecules suggests a significant energy dissipation under the flow conditions and thus a substantial viscosity increase even at low polymers concentration.

Currently, the high-molecular-weight polymers are being used as mobility control agents in both miscible and immiscible forms of oil displacements; both polyelectrolytes (such as partially hydrolyzed polyacrylamides) and nonionic materials (such as polyethyleneoxides), and biopolymers should be mentioned [1].

Initially, the polymer additives were considered to act only as viscofiers; therefore, the data on shear-rate-dependent viscosity were used to determine which material could perform as an efficient mobility control agent. In this respect the partially hydrolyzed polyacrylamide is the best candidate: in a dissolved state it expands to a larger spherical form, 2000–5000 Å, (due to the electrical repulsion of the like charges along the polymer chain), so that under the flow conditions the solution viscosity is increased. However, the experimental investigation clearly indicated that the pressure loss of the polymeric fluid flow through porous medium cannot be predicted on the basis of the laboratory measurement of the shear viscosity only. Obviously, other factors are involved which determine the polymer behaviour under the reservoir conditions. Needless to say a great deal of further research endeavour is needed to fully understand all aspects of the process.

The pressure loss calculation is usually based on the assumption of laminar flow and an inelastic behaviour. However, this assumption may not always be justified. At high rates inertial effects become noticeable and under certain conditions even elastic effects may occur. Both cases would result in pressure drop increase at the fixed flow rate. It also should be noted that the viscosity value used in calculation is obtained under viscometric conditions. However, the fluid flow in porous medium is likely to be of a non-viscometric nature.

Further investigation suggested that the polymeric materials may reduce the mobility of the displacing phase not only through viscosity increase but also by lowering the effective permeability. The polymer-rock interaction is featured by polymer retention. Although the actual mechanism is not yet well understood the polymer adsorption, entrapment and plugging are likely involved.

Interestingly enough, none of the papers dealing with the subject looks into the process thermodynamics. Yet such an approach promises to improve the understanding of a role of polymeric materials in oil recovery processes. To illustrate this point the matter of polymer retention is approached from the thermodynamical viewpoint [2, 3].

Consider two regions in a porous system: one where polymer solution flows (R1) and another one where the fluid is stagnant (R2). Under the equilibrium conditions there is a zero free energy difference between the two regions and the following equation may be written:

$$RT\ln\left(\frac{C_2}{C_1}\right) = -\left\{\frac{1}{2V_2}\int_{V_2}\rho(v_2-\bar{v}_2)^2 dV_2 + \frac{1}{V_2}\int_{V_2}(p_{11}-p_{22})dV_2\right\} + \left\{\frac{1}{2V_1}\int_{V_1}\rho(v_1-\bar{v}_1)^2 dV_1 + \frac{1}{V_1}\int_{V_1}(p_{11}-p_{22})dV_1\right\}. \quad (2)$$

In the region R2 the velocity is zero and the first two integrals thus disappear from the above equation. The remaining two terms are positive and so must be the left-hand side of the equation. Polymer concentration is therefore, higher in dead end pores compared to pores where the solution is flowing. The phenomenon of retention can therefore, be interpreted strictly in thermodynamical terms only.

In conclusion, I would like to emphasize that a better understanding of the polymer behaviour under the reservoir conditions not only helps to achieve more efficient use of polymers in oil recovery processes but also could lead to development of new materials featured by properties specifically suitable to the needs of enhanced oil recovery.

References

1. Stanislav, J. F., C. S. Kabir, J. Canadian Petroleum Technology **17**, 1–7 (1977).
2. Hlavacek, B., J. F. Stanislav, Proc. VII. Intern. Congr. Rheology, Gothenburg (Sweden) 1976, pp. 296.
3. Unpublished discussions with B. Hlavacek.

(Received May 21, 1982)

Author's address:

Dr. J. F. Stanislav
Department of Chemical and Petroleum Engineering
The University of Calgary
2500 University Drive N.W.
Calgary, Alberta T2N 1N4 (Canada)

Rheologica Acta Rheol. Acta **21**, 566–571 (1982)

Viscoelastic flow of dilute polymer solutions in regularly packed beds

R. Haas and F. Durst

Institut für Hydromechanik der Universität Karlsruhe und Lehrstuhl für Strömungsmechanik, Technische Fakultät der Universität Erlangen-Nürnberg

Key words: Porous media flow, viscoelasticity, polymer solution, dumbbell model, elongational flow

1. Introduction

It is well known that dilute polymer solutions may exhibit strong viscoelastic flow effects when exposed to elongational flow fields with elongation rates that exceed certain "Onset"-conditions. Such flow fields occur in the converging-diverging flow passages of a porous matrix but also in regularly packed beds of spheres.

2. Theoretical approach

Concerning the extensional flow behaviour of dilute high-molecular weight polymer solutions, the authors refer to the work of Bird et al. [1] and Marrucci [2] who show that major properties of linear, flexible polymers or macromolecules are best described by simple dumbbells, which are suspended in a Newtonian solvent with viscosity η_s. In a fluid at rest, the single dumbbells are characterized by a finite end-to-end distance R determined by the motion of the free ends of the molecule. These are separated by forces caused by Brownian motion and by an internal elastic force F. The simplest dumbbell model available in literature takes the spring connector between the spheres of the dumbbell to be Hookean, so that the intramolecular force F is linearly related to the molecular extension R. H is the spring constant describing the simple spring given in figure 1.

It has been found that the linear force-displacement law is not capable of describing a wide range of experimental data, especially those obtained for elongational flow of dilute polymer solutions. To overcome these shortcomings a finite-extensible, nonlinear-elastic dumbbell model was proposed by Warner [3]. His ideas are summarized and shown on the right-hand side of figure 1. Warner's dumbbell model with the nonlinear force-displacement law reacts linearly for small extensions but will get stiffer and stiffer as the spring is extended beyond small values of R. Furthermore, the spring cannot be extended beyond a maximum value R_0, which means a completely stretched polymer chain in its zig-zag conformation.

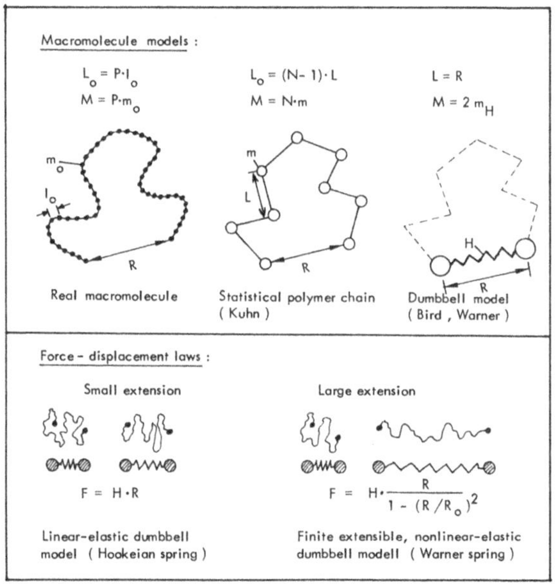

Fig. 1. Molecular picture of a macromolecule

If such finite extensible, nonlinear-elastic dumbbells are suspended in a Newtonian solvent with viscosity η_s, one can speak of a model liquid for the polymer solution. One obtains a model of a viscoelastic liquid. This "dumbbell solution" is characterized by the elastic nature of the dumbbells (H) and by the viscous properties of the solvent (η_s). These two properties can be combined by defining a relaxation time τ for the solution:

$$\tau = \frac{[\eta]\,\eta_s M}{A k T}\,. \qquad (1)$$

In this equation $[\eta]$ is the intrinsic viscosity, M the molecular weight, A the Avogadro constant, k the Boltzmann constant and T the absolute temperature. Eq. (1) holds fairly well for finite-extensible nonlinear-elastic dumbbell solutions, if the stretching parameter $b = R_0^2/R^2$ can be assumed to be very large.

If the dumbbell suspension is in a flowing state, the "Onset" of viscoelastic effects is usually characterized by the well-known Deborah number, which is defined as the ratio of the above defined relaxation time of the solution to the characteristic flow time:

$$De = \frac{\text{Relaxation time of the liquid}}{\text{Characteristic time of the flow}} = \frac{\tau}{(1/\dot{\varepsilon})} = \tau \cdot \dot{\varepsilon}\,. \qquad (2)$$

For elongational flow, the characteristic time of the flow may be expressed by the inverse of the elongation rate $\dot{\varepsilon}$, so that the Deborah number is obtained as the product of the relaxation time and the elongation rate.

As it has been shown by Bird et al. [1], the extensional viscosity of a dumbbell solution is increased above the Newtonian value $3\eta_s$, if the aforementioned model solution is subjected to pure elongational flow. In [1] a reduced extensional viscosity is plotted versus the Deborah number and one can see, that for large stretching parameters b, the extensional viscosity increases at a Deborah number of 0.5, reaching an asymptotic value at higher Deborah numbers. The maximum value of λ^* is characteristic of the nonlinear force law of the dumbbell model.

To test these results, the elongational flow nature of a porous media was used by Durst and Haas [4], to experimentally investigate the characteristic features of dilute polymer solutions and to compare these with the corresponding predicted values.

As it was shown in [4], a large number of fluid-mechanic and physico-chemical parameters are involved to describe the "Onset" of viscoelastic polymer effects. This is clearly seen if a modified porous-media Deborah number

$$De_e = \frac{k_1 \eta_s^2 (1-n)(1+c[\eta])[\eta] M}{\rho_s d^2 A k T} Re \qquad (3)$$

with $\eta_p = \eta_s(1 + [\eta]c)$ and $\rho_s \approx \rho_p$ is used to characterize the state of the solution instead of the generally used Reynolds number

$$Re = \frac{\bar{v} d \rho}{\eta(1-n)}\,. \qquad (4)$$

In eq. (3), the elongation rate $\dot{\varepsilon}$ of the porous media flow is replaced by the expressions:

$$\left(\frac{\bar{v}}{d}\right) = \frac{(1-n)\eta}{\rho d^2} Re\,; \quad \dot{\varepsilon}_e = k_1 \cdot \left(\frac{\bar{v}}{d}\right), \qquad (5)$$

where $\bar{v}$ is the superficial flow velocity, d the sphere diameter of the porous matrix, n the porosity of the porous bed and k_1 a parameter, which describes the average strength of the elongation rate within a porous media channel. It is one of the aims of the present paper to compute the k_1-values for regularly packed beds and to deduce from these studies suitable values for randomly packed beds.

3. Extensional flow geometry in regularly packed beds

To carry out the above mentioned study, the authors use two simple sphere arrangements, i. e. the

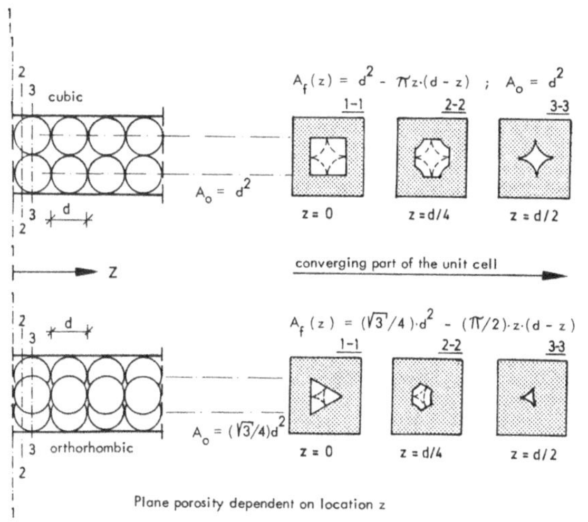

Fig. 2. Sphere arrangement and flow geometry of regularly packed beds

simple cubic sphere arrangement and the orthorhombic sphere structure. For simplification of the experiments only one single channel of each regular matrix is used, as shown in figure 2.

As this figure indicates, the two flow channels have periodically varying cross-sectional areas A_f, i.e. converging-diverging cross-sections. A_0 is the cross-sectional area corresponding to the superficial flow velocity $\bar{v}$. Investigating only the converging part of the unit cells, the plane porosity $n_f(z) = A_f(z)/A_0$ can be estimated from simple geometrical relationships:

Cubic sphere arrangement

$$n_f(z^*) = 1 + \pi(z^{*2} - z^*), \quad 0 \leqslant z^* \leqslant 0.5, \tag{6}$$

Orthorhombic sphere arrangement

$$n_f(z^*) = 1 + (2\pi/\sqrt{3})(z^{*2} - z^*), \quad 0 \leqslant z^* \leqslant 0.5, \tag{7}$$

where $z^* = z/d$ is the dimensionless location in the mean direction of the flow. The function $n_f = f(z^*)$ is shown in figure 3 for the two sphere arrangements. The dimensionless elongation rate $\dot{\varepsilon}^*$ in the converging part of the unit cells can be given in the general form:

$$\dot{\varepsilon}^*(z^*) = \frac{\partial v^*}{\partial z^*} = (-1) \cdot n_f(z^*)^{-2} \cdot \frac{\partial n_f(z^*)}{\partial z^*}, \tag{8}$$

where $v^*(z^*) = v(z^*)/\bar{v} = 1/n_f(z^*)$ is the dimensionless flow velocity. With eqs. (6) and (7), one gets for $0 \leqslant z^* \leqslant 0.5$:

Cubic sphere arrangement

$$\dot{\varepsilon}^*(z^*)_c = (-1) \cdot [1 + \pi(z^{*2} - z^*)]^{-2} \cdot \pi(2z^* - 1), \tag{9}$$

Orthorhombic sphere arrangement

$$\dot{\varepsilon}^*(z^*)_{or} = (-1) \cdot [1 + (2\pi/\sqrt{3})(z^{*2} - z^*)]^{-2} \cdot (2\pi/\sqrt{3})(2z^* - 1). \tag{10}$$

These functions are also shown in figure 3. One can clearly see that the dimensionless elongation rate $\dot{\varepsilon}^*$ is

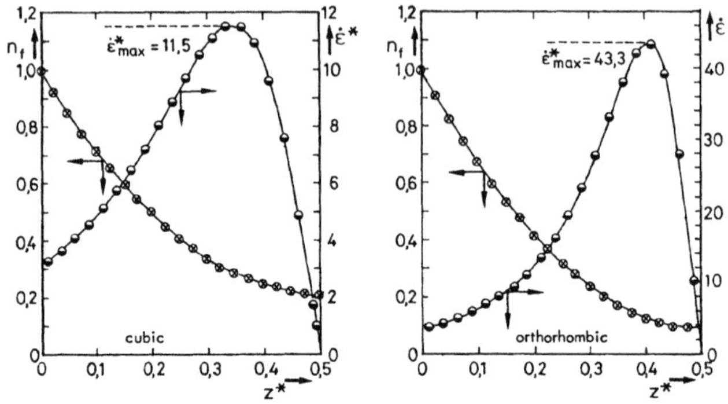

Fig. 3. Plane porosity n_f and dimensionless elongation rate $\dot{\varepsilon}^*$ as function of the dimensionless location z^* in the converging part of the unit cells

not constant along the mean direction of the flow and thus, the elongation rate has a defined distribution in the converging part of the unit cells. Eq. (5) takes the following general form:

$$\dot{\varepsilon}_e = \dot{\varepsilon}^*_e(z^*) \cdot (\bar{v}/d) = k_1 \cdot (\bar{v}/d). \tag{11}$$

In this sense, the k_1-value is the effective value $\dot{\varepsilon}^*_e$ of the elongation-rate distribution $\dot{\varepsilon}(z^*)$. k_1 can be determined from experiments as outlined below.

4. Experimental results

Beside the usually employed test sections for randomly packed beds, cf. [4], two simple flow geometries of regularly arranged spheres were used in the experiments. These are shown in figure 4 and consisted of 15 unit cells. In some experiments, a smaller number of segments was employed to study the influence of segment number w on the viscoelastic flow behaviour. In the experiments, the pressure drop was measured for both the solvent and the viscoelastic polymer solution. The varying parameter was the superficial flow velocity $\bar{v} = Q/A_0$, where Q is the measured flow rate. To plot the data, the authors employed dimensionless parameters, i.e. a resistance coefficient and the Reynolds number. The detailed definitions of these numbers may be obtained from [4].

In figure 5 the influence of segment number w of unit cells on the viscoelastic flow behaviour of a 50 ppm polyacrylamide solution ($M = 18.21 \cdot 10^6$ g/mol) is shown for the cubic sphere arrangement employing a sphere diameter $d = 5$ mm. The specific resistance increase $\Lambda_{sp} = (\Lambda_p - \Lambda_s)/\Lambda_s$ due to the

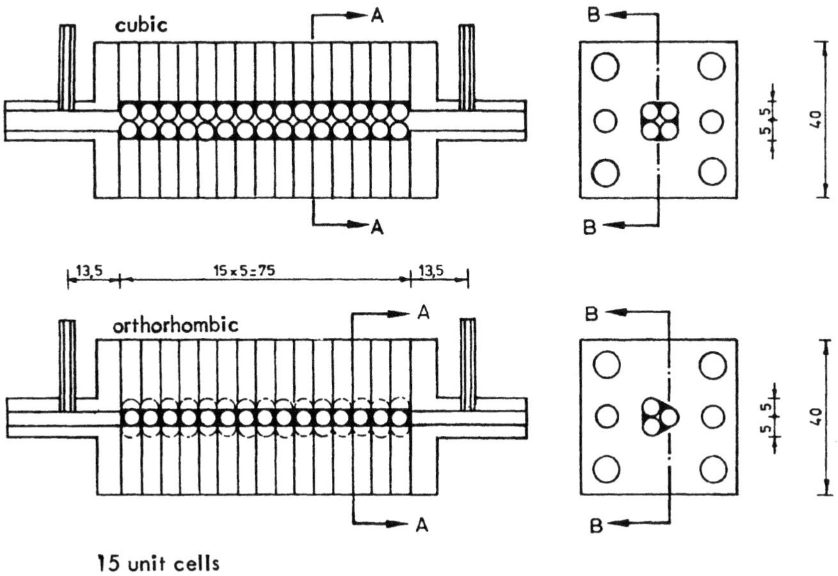

Fig. 4. Test sections for regularly packed beds

polymer action is plotted versus the Reynolds number. The indices p and s correspond to the polymer and pure solvent, respectively.

From this plot, it is obvious that the viscoelastic flow behaviour starts at a constant Reynolds number of the flow, i.e. at a certain superficial velocity within the cubic channel, independent of the number of unit cells. We can assume that the macromolecules are stretched at an "Onset" flow-rate, but the extent to which they are stretched, is dependent on the number of contractions they have passed. A maximum increase of the resistance coefficient is obtained after passing a number of about 4 segments followed by an subsequent decrease after the passage of 8 and 15 segments. This variation of the effect with increasing segment number can be explained partly by variations in the flow field caused by the stretched molecules and by degradation of the macromolecules after they have reached a stretched state after 4 contractions. The initially observed increase in the resistance coefficient with segment number indicates that the molecules are stretched step by step in subsequent contractions. One single contraction cannot fulfil the condition request-

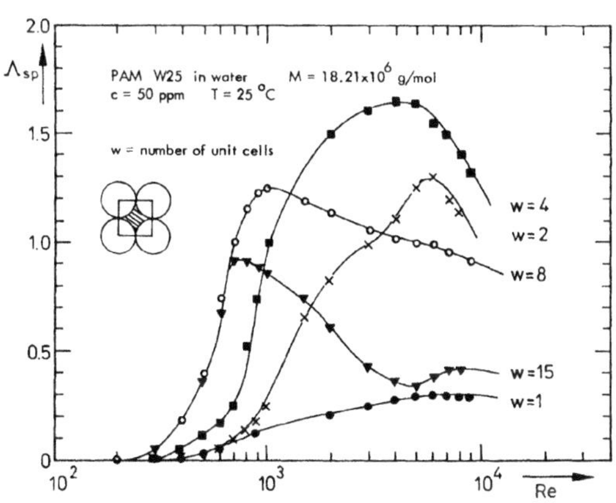

Fig. 5. "Onset"-behaviour and specific resistance increase of a polymer solution in a cubic channel dependent on the segment number w of unit cells

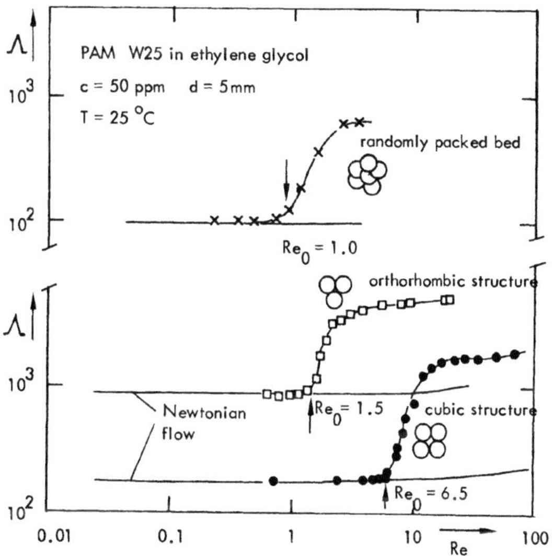

Fig. 6. "Onset" Reynolds-numbers for a 50 ppm PAM-solution in various bed structures (d = 5 mm)

Table 1. "Onset"-conditions in various bed structures for PAM W25 in ethylene glycol ($[\eta]$ = 1132 cm³/g, M = 18.21 · 10⁶ g/mol, T = 25 °C)

	Relaxation time (s)	Re_0 –	$(\bar{v}/d)_0$ (1/s)	$\dot{\varepsilon}_{e,0}$ (1/s)	$k_1 = \dot{\varepsilon}^*_{e,0}$ –	Δt_{ex} (s)	x_{ex} –
cubic sphere structure	0.138 0.058	6.5	2.154	3.623 8.627	1.682 4.000	0.110 0.082	0.398 0.708
orthorhombic sphere structure	0.138 0.058	1.5	0.574	3.623 8.627	6.310 15.030	0.190 0.082	0.688 0.708
randomly packed bed	0.138 0.058	1.0	0.398	3.623 8.627	9.100 21.680	– –	– –

ing the macromolecule to be a limit time under strain to be fully stretched.

To determine the effective k_1-value for the different structured beds, the authors employed the experimental results obtained for polyacrylamide (M = 18.21 · 10⁶ g/mol) in ethylene glycol. These data are shown in figure 6. For the cubic sphere arrangement, an "Onset" Reynolds-number of about 6.5 is used. For the orthorhombic sphere arrangement, the corresponding value resulted in 1.5 and for the randomly packed bed, a value of 1.0 was found, if a sphere diameter of d = 5 mm and a segment number of w = 15 are used. For the randomly packed bed the pressure drop was measured with a length of $\Delta L = 15\ d$. In table 1 the characteristic "Onset"-data are summarized. In the first column the relaxation time of the polymer solution is given. The simple dumbbell relaxation time resulted in a value of τ = 0.138 s, computed from the formula given in eq. (1). The largest Zimm-relaxation time, which is 0.42 times the simple dumbbell relaxation time, has a value of τ_z = 0.058 s. The above mentioned theory yields an "Onset" of viscoelastic effects at a Deborah number of 0.5. Utilizing this information an "Onset" elongation-rate $\dot{\varepsilon}_{e,0}$ can be determined. The computed values for $\dot{\varepsilon}_{e,0}$ are given for the different sphere structures in column 4 of the table. From these values the effective dimensionless elongation rates $k_1 = \dot{\varepsilon}^*_{e,0}$ are obtained and are given in column 5 of table 1.

5. Stretching-conditions for flexible macromolecules in elongational flow

It has been pointed out by Marrucci [2] that the request for the Deborah number to exceed a value of 0.5 in order to obtain viscoelastic effects is only a necessary but not a sufficient condition. He pointed out, that the elongation rate corresponding to Deborah number of 0.5 has to be applied for a sufficiently long time in order to ensure that the actual stretching and alignment of the macromolecules can take place irrespective of the superimposed random Brownian motion of the parts of the molecules. For a constant elongation rate, the condition of Marrucci reads:

$$\dot{\varepsilon}_0 \cdot \Delta t \geqslant x\,, \tag{12}$$

where Δt is the time needed for stretching and alignment of the macromolecule and x exceeds a number of the order of 4 or larger.

The considerations of Marrucci apply to a constant elongation rate $\dot{\varepsilon}$ along the mean direction of the flow. As indicated in figure 3, this is not the case for the converging part of the unit cells, where a defined elongation-rate distribution is obtained. If the real elongation rate exceeds the "Onset" elongation-rate in a test section, one would expect the time Δt, which a molecule has to spend in the elongational flow field, would be reduced. Computing the time Δt_{ex} which a molecule spends between fields in the unit cell geometries, where a certain elongational rate is exceeded, one obtains:

$$\Delta t_{ex} = \frac{\Delta z}{v(z)} = \frac{1}{\bar{v}/d} \int_{z_1^*}^{z_2^*} n_f(z^*)\,\partial z^*\,. \tag{13}$$

In figure 7, the lower and upper locations z_1^* and z_2^* show that between these two locations the "Onset" elongation-rate is exceeded. Table 1 shows also the times Δt_{ex} computed in this way and the value x_{ex} indicates that the times are smaller than those requested by the Marrucci condition (x = 4). The reason for this is the higher elongation rate that is experienced by the molecules between z_1^* and z_2^* at "Onset"-condition.

Finally, we may take the k_1-values either for the simple dumbbell relaxation time or for the Zimm-relaxation time to compute the corresponding

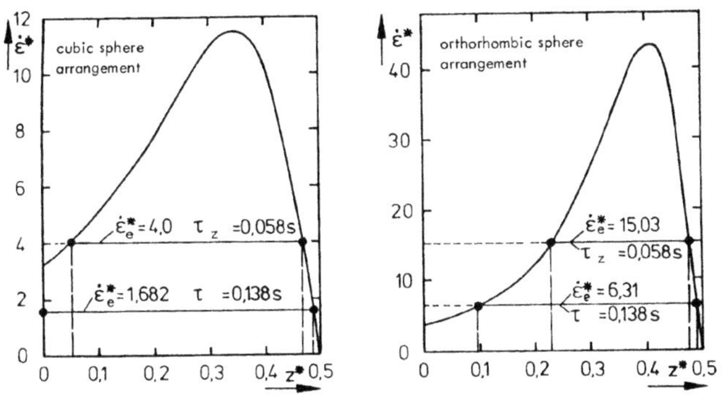

Fig. 7. Critical dimensionless elongational rate $\dot{\varepsilon}^*$ dependent on the choice of the relaxation time (τ or τ_z)

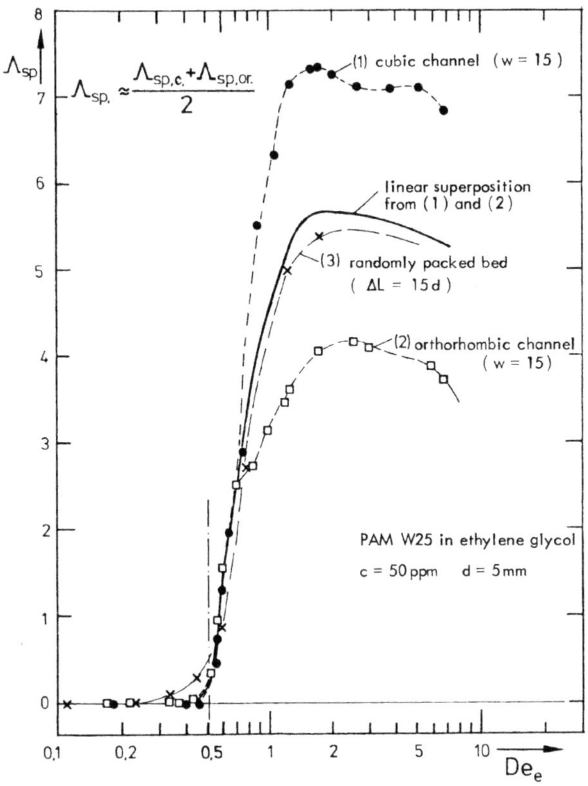

Fig. 8. Linear superposition of the resistance increase

Deborah number for the data given in figure 6. If we here again plot the specific resistance increase Λ_{sp} as a function of the Deborah number, the graphs in figure 8 are obtained. These graphs are in remarkable agreement to the theoretical results of Bird et al. [1], at least qualitatively. Under the assumption that a randomly packed bed is composed of regularly structured unit cells, which was shown by Franzen [5, 6] for the flow of Newtonian fluids, we may take a superposition of the viscoelastic polymer effects in regularly packed beds to get the corresponding effects for the random package. This is also shown in figure 8.

References

1. Bird, R. B., O. Hassager, R. C. Armstrong, C. F. Curtiss, Dynamics of Polymeric Liquids, Vol. 2, J. Wiley & Sons (New York 1977).
2. Marrucci, G., Pol. Eng. Sci. **15**, 229 (1975).
3. Warner, H. J., Jr., Ind. Eng. Chem. Fundam. **11**, 379 (1972).
4. Durst, F., R. Haas, Rheol. Acta **20**, 179 (1980).
5. Franzen, P., Rheol. Acta **16**, 548 (1977).
6. Franzen, P., Rheol. Acta **18**, 392, 518 (1979).

(Received May 17, 1982)

Authors' addresses:

Dr. R. Haas
Institut für Hydromechanik
Universität Karlsruhe
Kaiserstraße 12
D-7500 Karlsruhe

Prof. Dr. F. Durst
Lehrstuhl für Strömungsmechanik
Universität Erlangen-Nürnberg
Egerlandstraße 13
D-8520 Erlangen

Rheologica Acta Rheol. Acta 21, 572 – 577 (1982)

Laminar and turbulent flows of dilute polymer solutions: a physical model

F. Durst, R. Haas, and W. Interthal

Lehrstuhl für Strömungsmechanik, Technische Fakultät der Universität Erlangen-Nürnberg,
Institut für Hydromechanik der Universität Karlsruhe und Abt. Angewandte Physik der Hoechst AG,
Frankfurt/Main-Höchst

Key words: Drag reduction, porous media flow, elongational viscosity, dilute polymer solution, macromolecule

1. Introduction

It is well known that diluted high-molecular polymer solutions give rise to drastic flow effects; for example, in laminar flow through porous media, a pressure drop increase of up to thirty times the value of the flow of the pure solvent is observed, whereas in turbulent wall bounded flow, a pressure drop decrease or drag reduction of up to 80% or five times lower than the Newtonian value is measured. Although both these dramatic influences on the flow are caused by the same polymers, dissolved in the same solvent and employed in the same concentration range, the effects are usually interpreted by different models in the literature. The reason for this is the apparent contrariety of the observed polymer effects in laminar and turbulent flows.

In the following we advance a unified theory that is generally valid for both effects. Hence, the present presentation is an extension of the authors' work [1, 2]. In the description of the polymer molecule, we start from a simple dumbbell model for the macromolecule and explain the interaction of suspended dumbbell molecules with elongational flow fields as a common cause of these intrinsically opposite effects.

Elongational flow fields occur in porous media flow in constricted passages between the spheres which form the porous matrix and in turbulent flows, for example, near the wall in wall boundary layer flows between counter-rotating eddy pairs. This is schematically shown in figure 1 which also indicates the postulated deformation of the polymer molecules suspended in the flowing fluid. The indicated stretching of the molecules will occur if the strength of the elongational flow field and its spatial extension fulfil certain "Onset"-requirements. Under these conditions, a marked increase in the flow resistance or in the corresponding elongational viscosity is caused by the flow-induced molecule stretching.

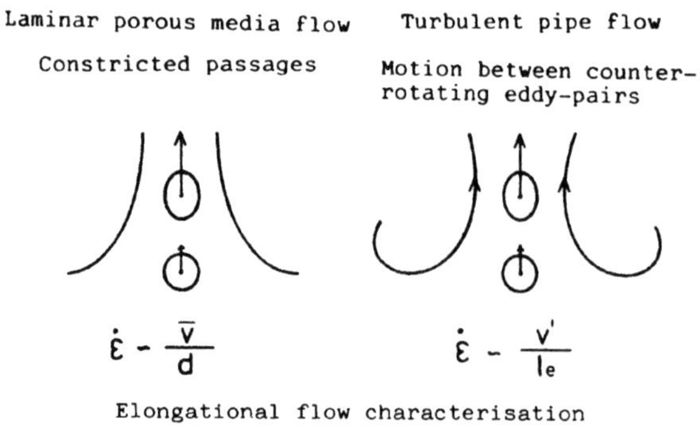

Fig. 1. Sketch of flow fields and elongation of macromolecules between spheres of porous media and counter-rotating eddy pair

Due to this stretching, one observes:

(i) an increase in apparent pressure loss in porous media flow or a reduction in flow rate when compared, at the same pressure loss per unit length, with the flow rate of the solvent only,
(ii) a decrease in turbulent cross velocity fluctuations and, hence, a reduction of turbulent energy extraction from the mean flow.

This clearly indicates that the same molecule action can explain both the observed increase in pressure loss in porous media flow and the well-known drag reduction caused by polymers added to wall boundary layer flow. Polymer actions in other turbulent flows can also be explained if these contain elongational flow fields that are sufficiently strong to elongate the suspended macromolecules. In this way, the energy transfer from the mean motion of the fluid to the dissipative turbulent flow field is reduced, yielding a reduction in flow losses.

2. Molecular Model

For a quantitative description of the observed molecule action, a macromolecule model must first be constructed. Following the derivations in [1 – 4] a finite extensible, nonlinear-elastic dumbbell model is employed with properties in accordance to those described by Warner [5] and Bird et al. [6]. This dumbbell model consists of two spheres representing the total polymer mass of a macromolecule. These spheres are connected by means of a massless spring with a nonlinear force-displacement law. The law proposed by Warner [5] is given in figure 2 together with the molecule simplifications adopted in the present study.

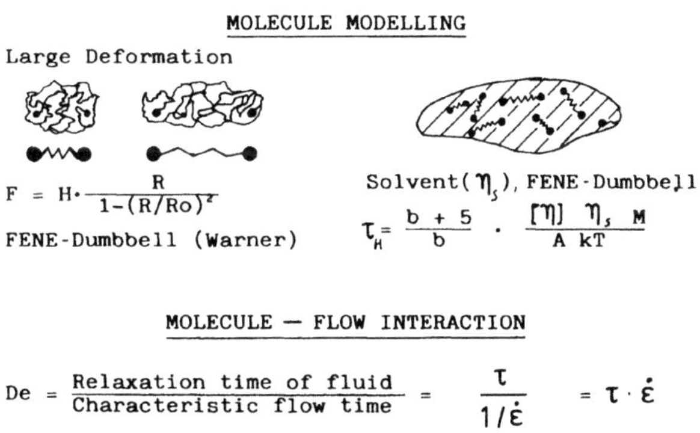

Fig. 2. Major equations to describe molecule properties by the FENE-dumbbell model

These are the FENE-dumbbells that are thought to be suspended in the solvent. The suspended dumbbells interact with the surrounding fluid and are found to respond to fluid deformations with a typical time constant. It is shown by Bird [6] that this time constant can be given in terms of a stretching coefficient b, the intrinsic viscosity $[\eta]$, the solvent viscosity η_s, the molecular weight M, the Avogadro number A, the Boltzmann constant k and the absolute temperature T. The definition of this time constant is given in figure 2 together with the force-displacement law introduced by Warner [5]. Figure 2 also indicates the definition of a Deborah number describing the ratio of the time constant of the dumbbell model fluid and a characteristic flow time. In accordance with Bird et al. [6], the latter is introduced as the inverse of the elongation strain rate of the flow field. Using this characteristic flow time, Bird et al. have derived asymptotic expressions for the elongational viscosity of FENE-dumbbell solutions that are exposed to pure elongational flow fields. The details of the extensive theoretical derivations are not repeated here, only the final expressions are given in figure 3 together with

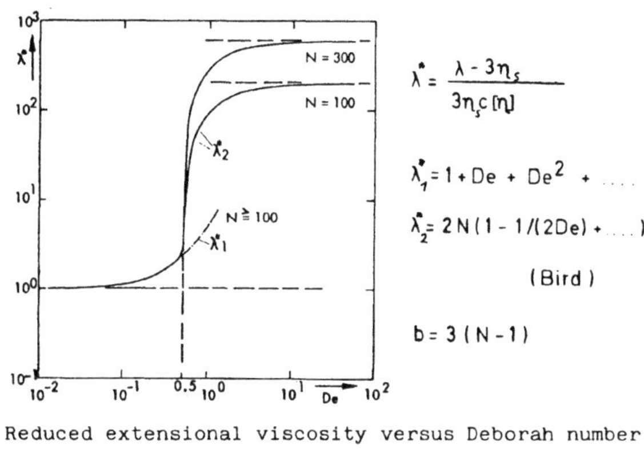

Fig. 3. Predicted variation of elongation viscosity for a FENE-dumbbell solution subjected to pure elongational flow

their graphical representations, e.g. $\lambda^* = f(De)$. λ_1^* applies to small De-numbers and λ_2^* is the asymptotic solution for large De-numbers. N is the number of statistical segments of the polymer chain which is related to the stretching parameter b as given in figure 3. The diagram in this figure indicates that at low De-number, the polymer solution shows only a small increase of elongational viscosity. However, at a characteristic De-number of 0.5, a dramatic elongational viscosity increase occurs. At higher De-numbers, a saturation is reached which depends on N (or b).

3. Porous media flow

The similarity between the above mentioned theoretically predicted increase in elongational viscosity and data obtained experimentally in porous media is immediately apparent for the data given in figure 4. The data of a 50 ppm solution of a polyacrylamide with a molecular weight of $18.3 \cdot 10^6$ g/mol dissolved in ethylene glycol have been plotted. Different bead diameters for the porous matrix have been employed and in figure 4 the experimental data are given in terms of modified Ergun-coordinates. As theoretically predicted, at low Reynolds numbers the flow is not influenced by the macromolecules. However, in each case, from a certain "Onset" Reynolds-number onwards, the flow resistance increases suddenly and dramatically.

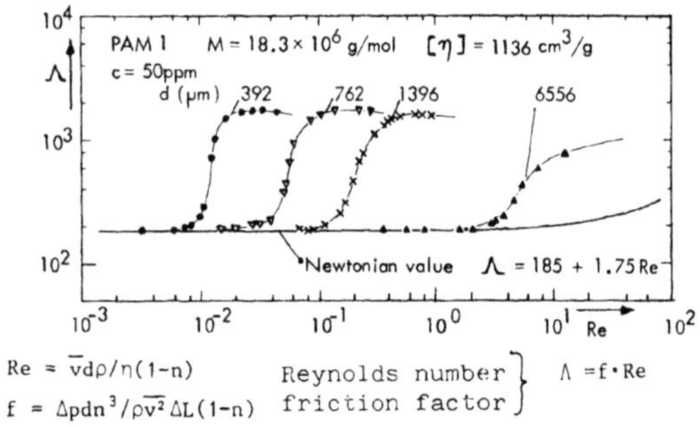

Fig. 4. Experimental results of resistance coefficient versus Reynolds-number for different sphere diameters of the porous matrix

For comparison of the experimental results of figure 4 with the theoretically derived extensional viscosity, the experimental data are replotted in a more general form, namely in terms of experimentally determinable values of an extensional viscosity η_e^* and a Deborah-number De_e. The experimental data obtained for the matrices with different sphere diameters fall together, e.g. see figure 5, and are well represented by the theoretical line of figure 3.

Only three parameters N or b, τ_H and k_1 are necessary, of which two, k_1 and τ_H are independently determinable. k_1 takes a value of 8.5 which has been calculated by two of us from the flow paths through regularly packed beds. With these equations, we can predict the influence of all these parameters on the elongational viscosity increase. More details can be obtained from Refs. [1] to [4] which provide information on other parameter variations and which also show that these are well described by the theoretical results of Bird et al.

4. Turbulent wall flow

As mentioned earlier, local elongation strain rates also occur in turbulent flows; for example, particularly intensely between counter-rotating eddies in the wall-bounded buffer layer. In analogy to porous media flow, we suggest that a mean local elongation strain rate v'/l_e be introduced to treat turbulent flows analogously to porous media flow; l_e is the diameter of the most energetic eddy, as introduced by Townsend [7] and v' the mean value of the cross-flow velocity fluctuations. Introducing l_e is equivalent to the assumption that, if at all, the most energetic eddies create the strongest elongation strain rates that are large enough to exceed the "Onset"-condition and to stretch the polymer molecules.

In Newtonian turbulent flow, kinetic energy is usually withdrawn from these eddies within the so-called sweep-burst mechanism resulting in small dissipative eddies. This energy transfer and loss is

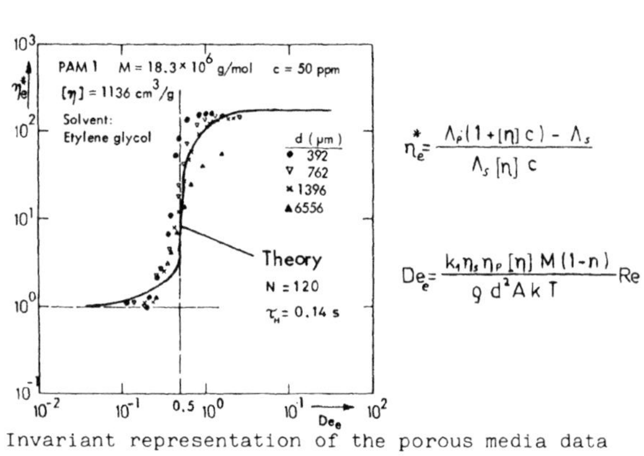

Fig. 5. Comparison of elongational viscosity and predicted variation with Deborah number

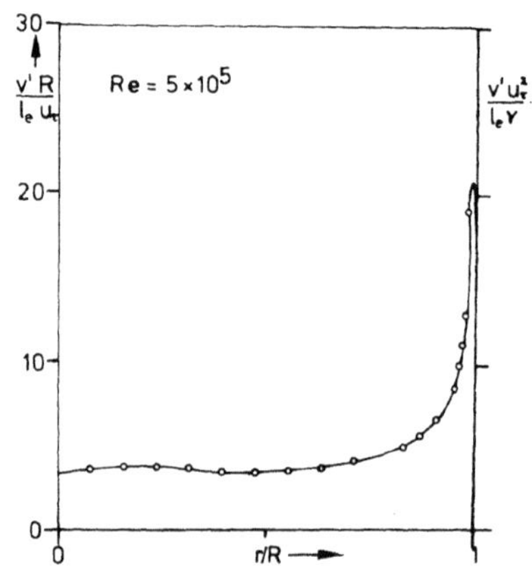

Fig. 6. Computed distribution of mean elongation strain rate using experimental data of Laufer [8]

reduced by the polymer action because the formation of regions with strong elongational viscosities yields a decrease of the very small dissipative eddies and, hence, a reduction of the turbulent energy losses. For a quantitative description information on v' and l_e is needed.

In figure 6, the curve of the time-averaged local strain rate that is produced by the most energetic eddy is plotted against the pipe radius. Measured data of Laufer [8] were employed to calculate this distribution. For further details, the reader is refered to [2]. In this paper, only the shape of the curve is discussed, which remains almost constant from the center of the pipe until near to the wall, where a sharp increase is measured. Closer to the wall, the elongation strain rate decreases again due to the strong viscous effects in the sublayer.

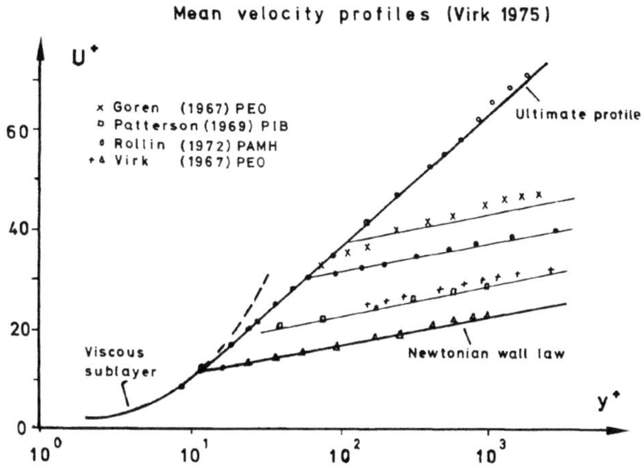

Fig. 8. Shift of log-region of normalized velocity indicates finite region of molecule action

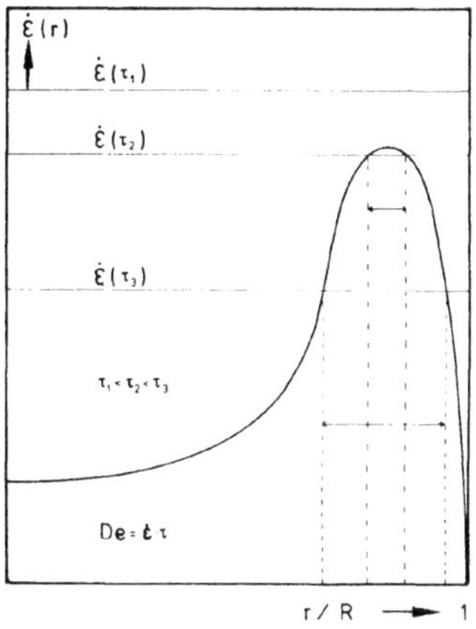

Fig. 7. Schematics of distribution of mean elongation strain rate and "Onset" elongation strain rate for various molecules

We can now use such data and predict the region across a pipe in which the molecules act. In figure 7, these elongation strain rate data are replotted schematically. If we now introduce a macromolecule with a low time constant τ_1 which is due to a low molecular weight, the product of $\dot{\varepsilon} \cdot \tau_1$ takes a value which exceeds at no point in the pipe cross-section the "Onset"-condition $De \geqslant 0.5$. Hence this polymer is ineffective. By increasing τ, the "Onset"-condition $De \geqslant 0.5$ can be fulfilled in a small region away from the wall. The higher τ, the lower the corresponding $\dot{\varepsilon}$-line lies and the broader the region of the buffer layer which is influenced by the polymer action until most of the pipe cross-section is covered. This predicted behaviour is in excellent agreement with drag reduction data available in literature. As a matter of fact, the present model is the only one that gives a correct description of the polymer action in a finite region indicated by the shift of the log-region of normalized velocity profiles given in figure 8.

As stated above the polymer will effect the flow if a critical elongation strain rate is exceeded. As a consequence of the assumed similarity between elongational flow through porous media and the cross motion between the vortices in turbulence, experimentally obtained data should scale with the proposed elongation strain rate. In figure 9 the normalized drag reduction function RV' is plotted against the De-number. The data for different pipe diameters

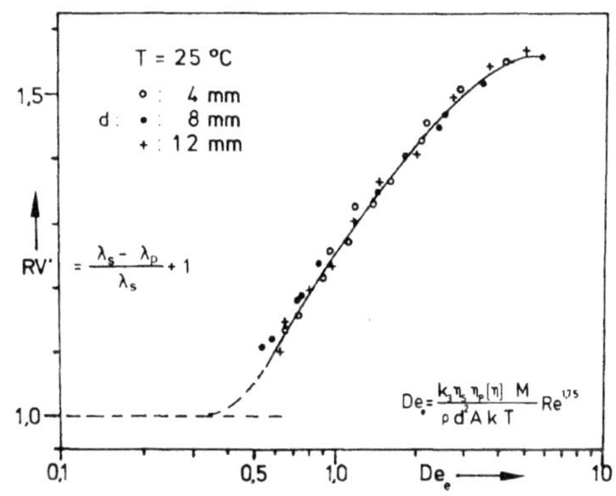

Fig. 9. Normalized drag reduction data for different pipe diameters plotted versus Deborah number

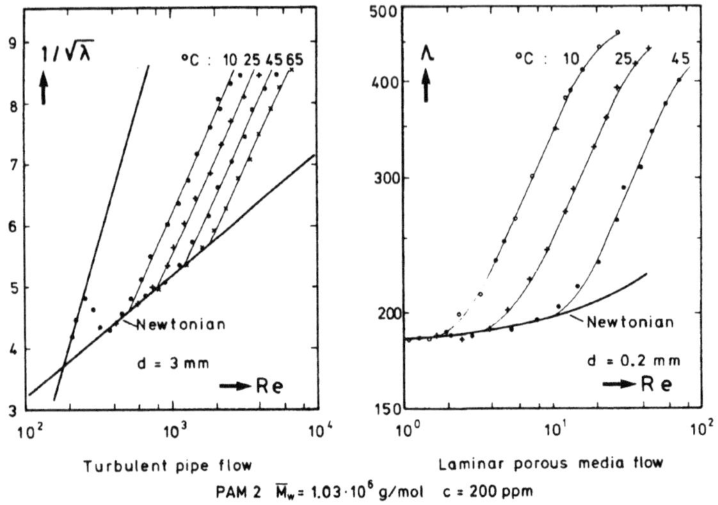

Fig. 10. Flow behaviour of turbulent pipe flow and laminar porous media flow for a polyacrylamide solution at different temperatures

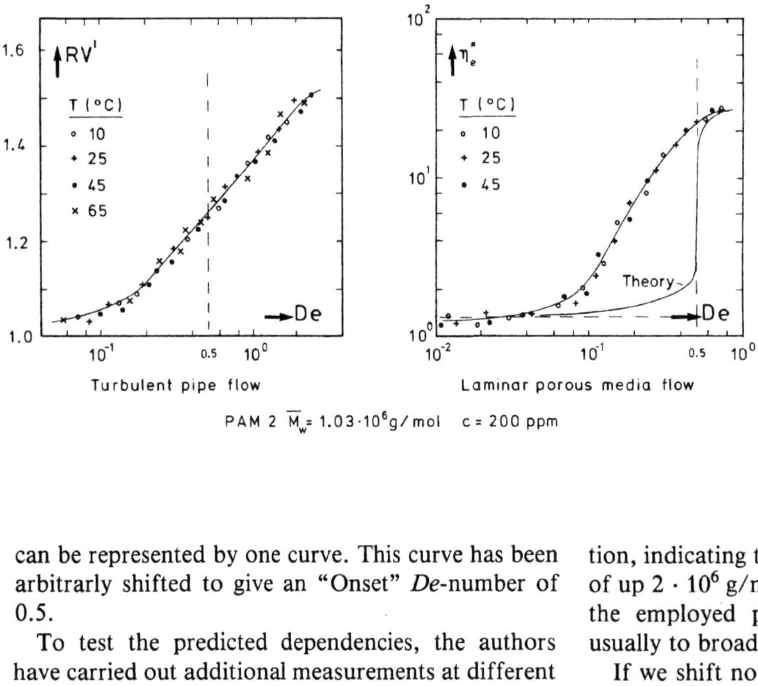

Fig. 11. Invariant representation of the data from figure 10

can be represented by one curve. This curve has been arbitrarily shifted to give an "Onset" *De*-number of 0.5.

To test the predicted dependencies, the authors have carried out additional measurements at different temperatures as well in laminar porous media flow as in turbulent pipe flow. The same polyacrylamide with an average $M_w = 10^6$ g/mol, determined by light scattering, has been used. In figure 11, the actual data of figure 10 are shown in the invariant representations. In the laminar porous-media case, the "Onset" starts earlier at $De \approx 0.1$ and the slope is more flat. This is assumed to be due to a broader M_w-distribution, indicating that parts of the molecules have a M_w of up $2 \cdot 10^6$ g/mol. This assumption is supported by the employed polymerisation process which leads usually to broader M_w-distributions.

If we shift now the turbulent invariant data in the same manner at "Onset" *De*-number 0.1 due to a molecular weight of $2 \cdot 10^6$ g/mol, we can recalculate the constant k_2 which is then 8.5. That is the same value as in the porous media case. This is another strong hint for the validity of these relationships and the chosen parameter. In the meantime, these relationships were proved to hold with different types of polymers and under various solvent conditions.

References

1. Durst, F., R. Haas, Rheol. Acta **20**, 179 (1981).
2. Durst, F., R. Haas, W. Interthal, T. Keck, Chem.-Ing.-Tech. **54**, 213 (1982).
3. Interthal, W., R. Haas, Proceedings of Euromech 143, Flow and Transport in Porous Media, A. A. Balkema (Rotterdam 1981).
4. Warner, H. J., Jr., Ind. Eng. Chem. Fundam. **11**, 379 (1972).
5. Bird, R. B., O. Hassager, R. C. Armstrong, C. F. Curtiss, Dynamics of Polymeric Liquids 2, J. Wiley & Sons (New York 1977).
6. Townsend, A. A., The Structure of Turbulent Shear Flow, Cambridge University Press (New York 1956).
7. Laufer, J., NACA Technical Note, No. 2954 (1953).

(Received May 17, 1982)

Authors' addresses:

Prof. Dr. F. Durst
Lehrstuhl für Strömungsmechanik
Universität Erlangen-Nürnberg
Egerlandstraße 13
D-8520 Erlangen

Dr. R. Haas
Institut für Hydromechanik
Universität Karlsruhe
Kaiserstraße 12
D-7500 Karlsruhe

Dr. W. Interthal
Abt. Angewandte Physik
Hoechst AG
Postfach 800320
D-6230 Frankfurt 80

Rheologica Acta Rheol. Acta **21**, 578 – 581 (1982)

Experimentelle Untersuchung turbulenter Rohrströmungen hochreiner wäßriger Polyacrylamid-Lösungen

B. Gampert und P. Wagner

Fachbereich 12 – Strömungslehre, Universität Essen – Gesamthochschule

Zusammenfassung: Es werden Molekulargewichtsfraktionen vergleichsweise niedriger Dispersität von im Labor synthetisiertem, hochreinem Polyacrylamid untersucht. Die Widerstandsverminderung wurde in einem einfachen Rohrsystem ohne Umlauf gemessen. Aus der Annahme, daß der Toms-Effekt auf die im Vergleich zum reinen Lösungsmittel stark erhöhte Dehnviskosität zurückzuführen ist, welche die Zahl und die Intensität der Bursts im turbulenten Strömungsfeld verringert, wurden eine Deborah-Zahl *De* und ein Ausdruck für die Widerstandsverminderung *WV** hergeleitet. Es werden Meßergebnisse angegeben, die den Zusammenhang zwischen *De* und *WV** darstellen.

Abstract: The polymeric material investigated consisted of laboratory synthesized polyacrylamide molecular weight fractions of relatively low dispersity. Drag reduction measurements were performed in a single pass pipe flow system. A Deborah number *De* and an expression for the drag reduction *WV** were derived from the hypothesis that the Toms effect results from an increase in the elongational viscosity which reduces the number and the intensity of bursting periods. Figures presenting *WV** as a function of *De* are shown.

Key words: Polymer solution, highly clean polyacrylamide, turbulent pipe flow, elongational viscosity, Deborah number

1. Einleitung und Aufgabenstellung

Der Toms-Effekt [1, 2], d. h. die Widerstandsverminderung in turbulenter Strömung infolge Beigabe polymerer Additive wird wesentlich beeinflußt durch die molekularen Parameter der beigesetzten Polymere. Bei der Untersuchung des Zusammenhangs zwischen der Molmasse der Polymersubstanz, z. B. von Polyacrylamid (PAA), und der Widerstandsverminderung in einer von wäßriger Lösung durchströmten Rohrmeßstrecke ist es erforderlich, mit hinreichend engen Molmassenverteilungen zu arbeiten, da nur dann die mittlere Molmasse ein aussagekräftiger Parameter ist [3].

Da eine präparative, z. B. gelchromatographische Fraktionierung technischer wasserlöslicher Polymere, insbesondere für den interessierenden hohen Molmassenbereich zur Zeit noch mit großen Unsicherheiten und Schwierigkeiten behaftet ist, wurde PAA in Zusammenarbeit mit dem Institut für Chemische Technologie der Technischen Universität Braunschweig (Prof. Dr. Klein, Priv. Doz. Dr. Kulicke) im Labor polymerisiert [4]. Durch eine entsprechende Führung des Prozesses gelang es, bereits bei der Herstellung hinreichend enge Molmassenfraktionen zu erzeugen.

Die Experimente haben das Ziel, zunächst den Einfluß der Molmasse von Polymeren auf globale Größen, wie den Strömungswiderstand und die mechanische Degradation, zu studieren.

Bei letzterem Vorgang, der mit einem Wiederanstieg des Druckabfalls verbunden ist, handelt es sich um das Aufbrechen der Moleküle im Strömungsfeld. Die mechanische Degradation wurde in der Literatur bisher im wesentlichen durch die Untersuchung ihrer Wirkung, d. h. den veränderten Strömungswiderstand, studiert. Die im Labor hergestellten hochreinen Substanzen bieten den Vorteil, daß im Gegensatz zu technischen Produkten eine sichere viskosimetrische

Bestimmung der mittleren Molmasse möglich ist, so daß deren Änderung infolge Degradation direkt verfolgt werden konnte.

Die Grenzviskosität wurde im Zimm-Crothers Viskosimeter bestimmt. Anschließend konnte das Viskositätsmittel der Molmasse über die $[\eta]$-M-Beziehung von Scholtan berechnet werden.

Die Rohrmeßstrecke bestand aus einem Rohr von 6,1 mm Durchmesser. Auf einen geschlossenen Umlauf wurde verzichtet, um unkontrollierte Degradationswirkungen zu vermeiden. Der Durchfluß wurde durch eine Zeitmessung und Auswägen der zugehörigen Fluidmenge festgestellt. Die Druckmeßdosen zur Aufnahme des Druckabfalls befanden sich in einer Entfernung von 1,8 m und 2,3 m vom Rohreingang. Mit den geometrischen Daten des Rohres und der im Ubbelohde-Viskosimeter bestimmten Viskosität der Lösung konnte anschließend der Widerstandsbeiwert $\lambda_L = [2\Delta p/(\rho u_m^2)] \cdot d/L$ als Funktion der Reynolds-Zahl $Re = u_m \cdot d/\nu_L$ bestimmt werden (Index L: Lösung).

2. Diskussion der Ergebnisse

Durch die Beigabe von PAA erhöht sich die Scherviskosität einer wäßrigen Lösung nur unwesentlich, während die Dehnviskosität sehr stark ansteigt. Entsprechend beobachtet man in laminaren Dehnfeldern, daß der Strömungswiderstand deutlich erhöht wird, wie dies z. B. bei Sickerströmungen der Fall ist [5]. Dieser Effekt tritt allerdings nur ein, wenn ein hinreichend großer Wert für das Verhältnis zwischen „charakteristischer Molekülzeit" und „charakteristischer Strömungszeit" erreicht wird.

In turbulenter Strömung sind polymere Additive insbesondere dann wirksam, wenn sie in Wandnähe vorliegen. Dort wird über den sogenannten Burst-Zyklus die kinetische Energie der Turbulenz erzeugt. In der Abfolge von Strömungsvorgängen, die zum Burst-Zyklus gehören, treten lokale Dehnfelder auf. Die Beigabe polymerer Additive erniedrigt die Burst-Rate [6].

Die eben angesprochenen Tatsachen wollen wir zu der Hypothese zusammenfassen, daß die Wirkung der Polymere darin besteht, über die erhöhte Dehnviskosität den Burst-Zyklus und so die mit dem turbulenten Impulsaustausch verbundenen Verluste zu dämpfen, sofern ein charakteristisches Zeitverhältnis erreicht wird.

Wir beziehen die in üblicher Weise definierte Widerstandsverminderung WV

$$WV = \left.\frac{\lambda_w - \lambda_L}{\lambda_w}\right|_{\dot{V} = \text{const}} \tag{1}$$

(wobei λ_w und λ_L den Widerstandsbeiwert des reinen Wassers bzw. der Lösung und $\dot{V}$ den Durchfluß bedeuten) auf die Polymerkonzentration c und die Grenzviskosität $[\eta]$ und erhalten eine reduzierte Widerstandsverminderung

$$WV^* = \frac{\lambda_w - \lambda_L}{\lambda_w} \cdot \frac{1}{c \cdot [\eta]} . \tag{2}$$

Wir definieren jetzt eine Deborah-Zahl De als das Verhältnis von charakteristischer Polymerzeit t_p und charakteristischer Strömungszeit t_s. Die charakteristische Polymerzeit werde gleich der in [7] angegebenen Relaxationszeit

$$t_p = \frac{[\eta] \cdot \eta_s \cdot M}{A k T} \tag{3}$$

gesetzt (η_s = Scherviskosität, M = Molmasse, A = Avogadrosche Konstante, k = Boltzmann-Konstante, T = Temperatur). In Strömungen newtonscher Flüssigkeiten gilt nach [6] und [8] für die Burst-Zeit

$$t_B' \sim 1/u_\tau^2 . \tag{4}$$

Nach Definition einer dimensionsrichtigen Burst-Zeit

$$t_B = K_1' \nu_L/u_\tau^2 = K_1^*/u_\tau^2 \tag{5}$$

($u_\tau = \sqrt{\tau_w/\rho}$ = Schubspannungsgeschwindigkeit, ν = kinematische Viskosität) wird diese der Strömungszeit t_s zugeordnet. Damit erhalten wir für die Deborah-Zahl unter Berücksichtigung einer der Einsteinschen Viskositätsgleichung analogen Beziehung für die kinematische Viskosität

$$De = K_1 \frac{M \cdot \rho_L \cdot u_\tau^2 \cdot [\eta]}{A k T \cdot (1 + c \cdot [\eta])} , \tag{6}$$

worin ρ_L die Dichte der Lösung bedeutet.

Abbildung 1 zeigt die reduzierte Widerstandsverminderung als Funktion der Deborah-Zahl für wäßrige Lösungen aus PAA-Proben mit den mittleren Molmassen (Viskositätsmittel) $\bar{M}_\eta$ = 3.24 · 10^6, 4.9 · 10^6 und 5.03 · 10^6 g/mol. Im Ausdruck für die Deborah-Zahl wurde $K_1 = 1$ gesetzt. Die Konzentration ist in allen drei Fällen c = 50 ppm. Man erkennt, daß die Meßwerte um weniger als 10% von der eingezeichne-

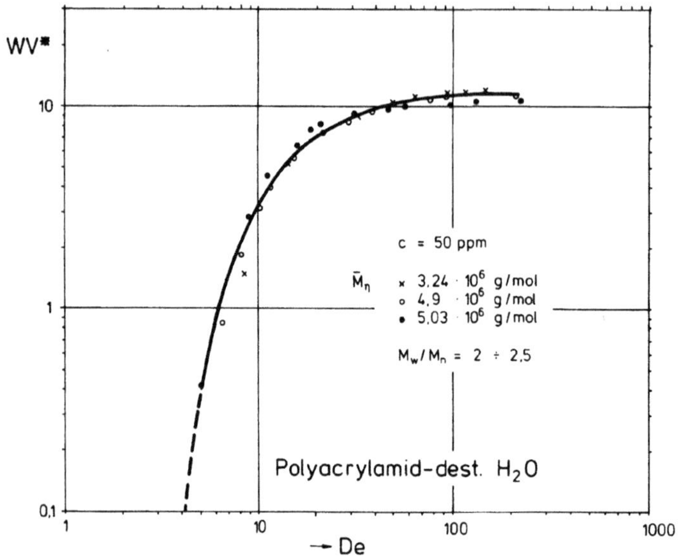

Abbildung 1. Die reduzierte Widerstandsverminderung WV^* als Funktion der Deborah-Zahl für wäßrige Polyacrylamidlösungen aus PAA-Proben unterschiedlicher mittlerer Molmasse

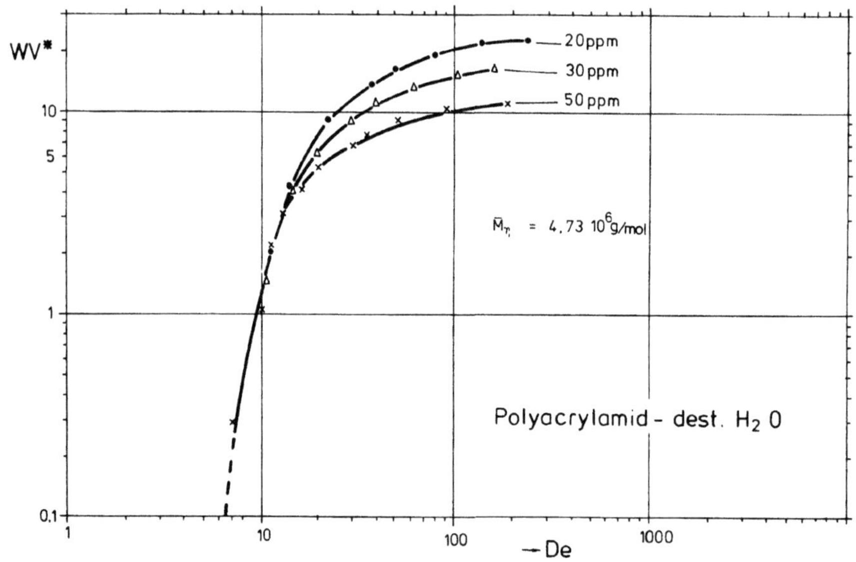

Abbildung 2. Die reduzierte Widerstandsverminderung WV^* als Funktion der Deborah-Zahl für wäßrige Polyacrylamidlösungen unterschiedlicher Konzentration

ten Kurve abweichen und die Auftragung $WV^* = WV^*(De)$ von der Molmasse weitgehend unabhängig ist.

Qualitativ entspricht der Verlauf der Kurve dem von Kurven $\eta_D^* = \eta_D^*(De)$, was man als Hinweis auf die Wirksamkeit der Dehnviskosität auffassen könnte.

Für unterschiedlich konzentrierte PAA-Lösungen einer bestimmten PAA-Probe mit der mittleren Molmasse $\bar{M}_\eta = 4.73 \cdot 10^6$ g/mol erhält man $WV^* = WV^*(De)$-Verläufe entsprechend Abbildung 2. Die reduzierte Widerstandsverminderung hängt in systematischer Weise von der PAA-Konzentration ab.

Die Onset-Deborah-Zahl ist für alle Konzentrationen (c = 20 ppm, 30 ppm, 50 ppm) die gleiche und gegenüber dem von Abbildung 1 zu einem höheren Wert verschoben. Messungen weisen darauf hin, daß die PAA-Probe mit $\bar{M}_\eta = 4.73 \cdot 10^6$ g/mol eine etwas engere Molmassenverteilung besitzt. Die in die Gl. (3) eingegangenen theoretischen Ergebnissen nach [7] wurden für monodisperse Polymere hergeleitet. Die Verschiebung des Onset-Punktes sollte daher mit der unterschiedlichen Molmassenverteilung der in den Abbildungen 1 und 2 untersuchten Proben zusammenhängen. Die Untersuchung besonders eng verteilter Polystyrol-Standards scheint diese Vermutung zu bestätigen.

Zur Untersuchung des Zusammenhanges zwischen der Molmasse der Polymere und der Degradation wurden wäßrige Lösungen aus PAA-Proben unterschiedlicher Molmasse ($\bar{M}_\eta = 3.24, 4.6$ und $5.03 \cdot 10^6$ g/mol) bei n Rohrdurchläufen dem gleichen Druckgradienten $\Delta p/L = 0.45$ bar/m ausgesetzt und die veränderte Molmasse $\bar{M}_\eta$ nach Aufkonzentrieren der Lösung viskosimetrisch aufgenommen. Es zeigte sich, daß eine hohe Molmasse die mechanische Degradation begünstigt und die stärksten Änderungen der Molmasse während der ersten Rohrdurchläufe auftreten.

Anschließend wurde für die in gleicher Weise degradierten Lösungen die Änderung der Widerstandsverminderung bei unterschiedlicher *Re*-Zahl bestimmt. Bei höherer *Re*-Zahl sind auch kürzere, noch nicht degradierte Moleküle widerstandsvermindernd wirksam. Die Auswirkung einer gegebenen Degradation ist daher für kleinere *Re*-Zahlen größer.

Danksagung

Die Verfasser danken Herrn Prof. Dr. J. Klein und Herrn Priv.-Doz. Dr. W.-M. Kulicke vom Institut für Chemische Technologie der TU Braunschweig für die großzügige Hilfe bei der Herstellung der PAA-Proben sowie Herrn cand. ing. W. Brauck und Herrn cand. ing. E. Freesemann für die einsatzfreudige Unterstützung bei der Durchführung der Messungen.

Literatur

1. Toms, B. A., Proc. 1st. Int. Congress on Rheology, Scheveningen (Holland) 1948, Vol. II, pp. 135 – 141.
2. Virk, P. S., E. W. Merrill, H. S. Mickley, K. A. Smith, E. L. Molle-Christensen, J. Fluid Mech. **30**, 305 – 328 (1967).
3. Gampert, B., P. Wagner, ZAMM **62**, 999 – 1000 (1982).
4. Gampert, B., P. Wagner, Turbulent flow with polymer additives, erscheint in Arch. Mech. (1982).
5. Durst, F., R. Haas, Rheol. Acta **20**, 179 – 192 (1981).
6. Kim, H. T., S. J. Kline, W. C. Reynolds, J. Fluid Mech. **50**, 133 – 160 (1971).
7. Bird, R. B., P. J. Dotson, N. L. Johnson, J. Non-Newtonian Fluid Mech. **7**, 273 – 235 (1980).
8. Achia, B. U., D. W. Thompson, J. Fluid Mech. **81**, 439 – 464 (1971).

(Received April 15, 1982)

Anschrift der Verfasser:

Prof. Dr.-Ing. B. Gampert, Dipl.-Ing. P. Wagner
Fachbereich 12 – Strömungslehre
Universität Essen – Gesamthochschule
Postfach 103764
D-4300 Essen 1

Influence of drag reducing additives on the structure of turbulence in a mixing layer

M. Hibberd, M. Kwade, and R. Scharf

Lehrstuhl für Strömungsmechanik, Abteilung Chemietechnik der Universität Dortmund

Abstract: The structure of a free turbulent mixing layer in Newtonian and drag reducing fluids is discussed using results from visualization and LDA-techniques. Particular emphasis is placed on results for the turbulent shear stress and the time correlation function.

Key words: Drag reduction, free mixing layer, coherent structure, flow visualization, Laser-Doppler anemometry

1. Introduction

Many investigators have considered the drag reduction effect, which appears in a number of flow fields, to be a phenomenon that is closely related to a modification of the turbulent structure in the flow. It is well known, for example, that in pipe flows the addition of small amounts of long chain polymer material to a low viscosity Newtonian fluid strongly influences the buffer layer near the wall to produce drag reduction [1]. However, changes in the flow structure in the core region may also cause a significant friction loss. Indeed, injection of a polymer thread into the turbulent core region has been found to produce marked drag reduction [2, 3] although almost no polymer reaches the wall region, at least not sufficient to produce the measured drag reduction. In order to be able to study the influence of drag reducing additives on the structure in a flow undisturbed by wall influences, a free mixing layer between two plane parallel fluid streams (one of the most fully explored types of free turbulent flow) was chosen for the experiments described here.

2. Experimental set-up

The two streams flow together with constant velocity profiles (1 and 2 m/s) and a suppressed turbulence level after having passed the rear edge of a separating plate. The mixing layer develops down a 40 × 20 mm rectangular glass walled test channel about one metre in length. Several precautions were taken to ensure steady flow conditions for the continuously circulating fluid such as the use of flow regulators and the suppression of vibrations. This is particularly important because of the time dependent changes in the viscosity of the polymer solutions (mainly caused by degradation effects) and because, as is well known, the developing mixing layer structure is very sensitive to small disturbances in the entrance region. Further details of the flow system are given by Kwade [4].

Two techniques, flow visualization and Laser-Doppler anemometry (LDA), were used to study the changes in the flow structure in the channel when 50 ppm of Separan AP 30 (a partly hydrolized polyacrylamide of high molecular weight) was added to water. Pressure measurements along the 32 mm diameter return pipe showed that the polymer addition produced drag reduction of approximately 60% at the experimental flow rates.

3. Flow visualization experiments

For the flow visualization experiments, the slower stream was heated slightly before it entered the channel to provide sufficient variation of the refractive index in the mixing layer for the Toepler

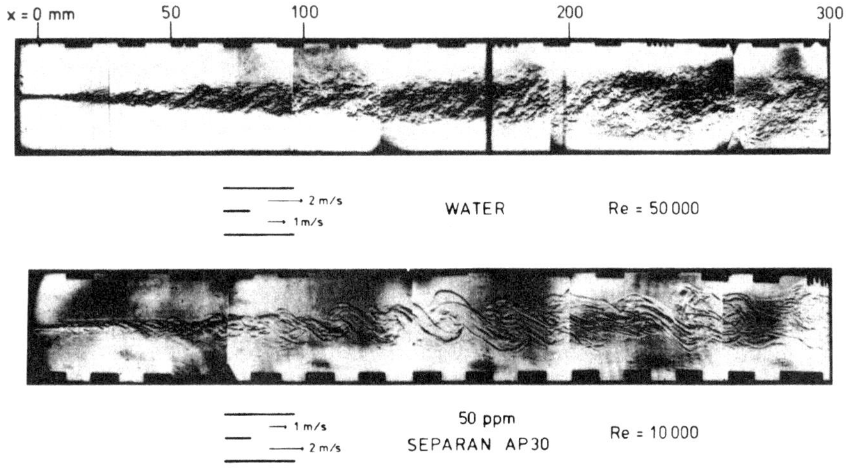

Fig. 1. Influence of polymer additive (50 ppm Separan AP30) on the structure in a free mixing layer

schlieren method to be used. The small temperature difference did not influence the mixing layer. The diameter of the lenses in the optical system did not allow the whole channel to be photographed at one time and so the following pictures were made by overlapping a sequence of shots taken at different times.

Despite shot to shot variations in the positions of structure within the flow, the composite pictures provide some valuable information. The appearance of large scale vortices and a reduction in the amount of small scale turbulence in the Separan solution can be clearly seen in figure 1. These pictures demonstrate that the addition of polymer has a large effect on the structure of the mixing layer. The diminished micro-turbulence is similar to that observed in gases by Brown and Roshko [5] for decreasing Reynolds numbers. The small regular oscillations up to x = 20 mm are caused by interactions of the boundary layers on each side of the splitter plate, and they give way to the much larger vortices, the size of which increases with increasing x. This suggests that a vortex pairing process similar to that clearly observed by Winant and Browand [6] is occurring and, indeed, examples of vortex pairing have occasionally been observed in the present experiments. That it is only occasionally observed confirms that the pairing process occurs relatively quickly, as found by Brown and Roshko.

4. Statistical analysis by Laser-Doppler anemometry

A much more detailed quantitative study of the properties of these structures was provided by statistical evaluation of measurements made using the LDA method. This technique is necessary because of the irregular behaviour of thermal probes, e.g. hot wires, in turbulent flows of polymer solutions. Additionally, for spatial correlation measurements marked errors occur with two hot wire probes when one is placed in the wake of the other. Thus two modified forward scatter arrangements were used to prevent mechanical collision of the two devices, see [4]. A further modification allowed the determination of the Reynolds-stress, proportional to $\overline{u'v'}$. This special set-up was described in [7].

At first sight the coherent structures shown in figure 1 appear to be quite regular. However their characteristic features such as diameter and lifetime are not in fact uniform but are distributed statistically. To obtain quantitative information the correlation method, first described by G. I. Taylor in 1935, was used. The visualization results of the coherent structures led to a coarse model for the vortex-like structure: Favourable positions in the flow field for detecting maximum oscillation of the fluctuation velocities are for the longitudinal component the boundary and for the transversal fluctuations the centre of the mixing layer. It is clear that these fluctuations will also appear in the space-time correlation function, whose normalized form, e.g. for the longitudinal components, is given by:

$$R_{xx}(\tau) = \frac{\overline{u_1'(t)\,u_2'(t-\tau)}}{(\overline{u_1'^2}\,\overline{u_2'^2})^{1/2}},$$

where $u_1' = u'(r,t)$ and $u_2' = u'(r + \Delta r, t)$ are two longitudinal fluctuation components of the flow field,

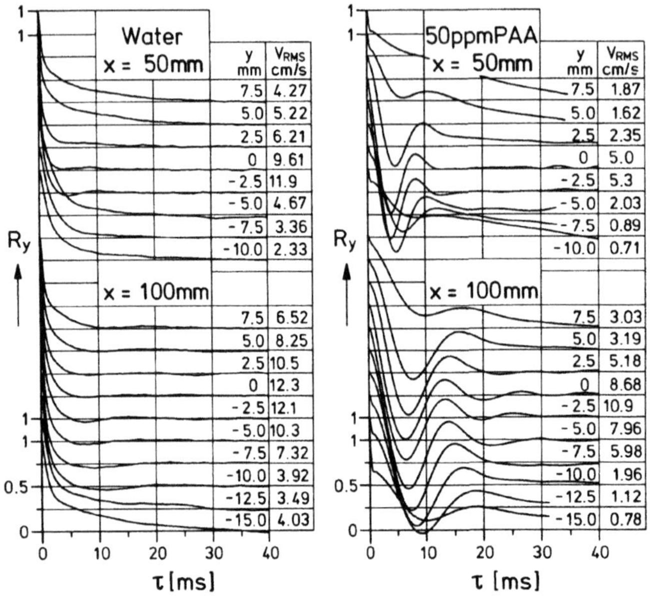

Fig. 2. Transversal time correlations across the developing mixing layer

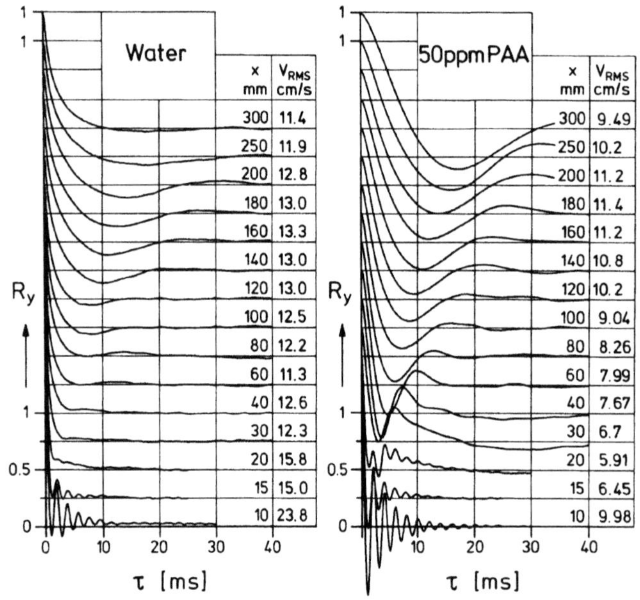

Fig. 3. Transversal time correlations along the middle axis of the mixing layer

measured at two positions $\boldsymbol{r} = \{x, y, z\}$ and $\boldsymbol{r} + \Delta \boldsymbol{r} = \{x + \Delta x, y + \Delta y, z + \Delta z\}$.

The most significant longitudinal vibrations were indeed found at the boundary plane of the mixing zone; not only for the space-time correlations but also for $\Delta \boldsymbol{r} = 0$, i.e. for the time correlations $R_x(\tau)$, see [4]. Similarly a strong oscillation of the transverse time correlation $R_y(\tau)$ was observed at the middle axis of the developing mixing layer, see figure 2. The slower decay of $R_y(\tau)$ in Separan and the more distinct oscillations whose frequency is greatest at the central region ($y \approx 0$) for both x-positions, should be noticed.

Figure 3 shows the development of the transverse time correlation down the channel. The strong oscillations of constant frequency for water and polymer

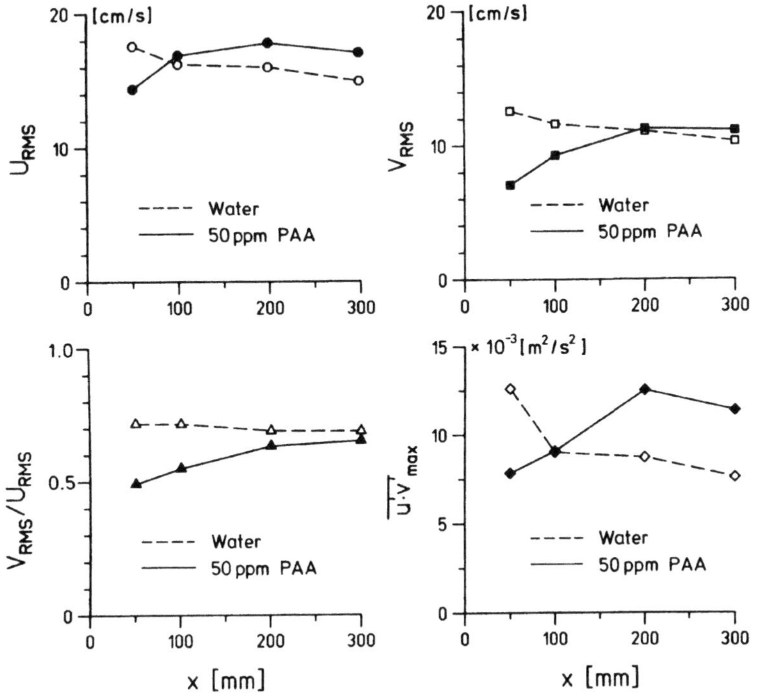

Fig. 4. x-dependence of typical turbulence quantities in the mixing layer at $\overline{u' v'}_{\text{MAX}}$-positions

solution behind the splitter plate are caused by the small vortices seen in figure 1. At $x \approx 20$ mm these disappear and are replaced by oscillations, the period of which increase continuously due to the enlargement of the mixing layer.

The normalized correlations provide information about length and time scales characterizing the vortex elements, but not directly about the intensity of each element, i.e. its contribution to the total energy content. For this information we turn to the fluctuation intensities and the Reynolds-stresses, the latter characterizing virtual stresses arising from the fact that there is a mean transport of momentum due to velocity fluctuations between adjacent fluid layers.

The upper part of figure 4 shows the x-dependence of the longitudinal (U_{RMS}) and the transverse (V_{RMS}) fluctuation intensities and the lower part shows the ratios $V_{\text{RMS}}/U_{\text{RMS}}$ (left side) and the maximum Reynolds-stresses, which are proportional to $\overline{u' v'}_{\text{MAX}}$ (right side). To obtain a direct comparison, the RMS-quantities were measured at y-positions at which the Reynolds stresses show their peak values.

The $V_{\text{RMS}}/U_{\text{RMS}}$-ratios, an indication of the relative strength of the transverse fluctuations, become constant and nearly equal for water and the Separan solution when the mixing layer is fully developed, i.e. $x \geqslant 200$ mm. However, at the origin of the mixing zone the remarkably lower polymer values provide evidence for a smaller width of the mixing layer, possibly caused by an orientation of the long chain molecules at the inlet of the test section.

The x-dependence of the maximum turbulent shear stress $\overline{u' v'}$ shows a behaviour similar to the RMS-values for both liquids: The polymer peaks are first lower, and grow till further downstream they become higher than those for water; whereas the water peaks remain nearly constant in the developed mixing layer, i.e. $x \geqslant 100$ mm. The correlation quotient

$$R_{xy}(\Delta r = 0, \quad \tau = 0) = \frac{\overline{u' v'}_{\text{MAX}}}{U_{\text{RMS}} \cdot V_{\text{RMS}}},$$

calculated for each x-position from the data of figure 4, is nearly constant in the developed mixing layer ($x \geqslant 100$ mm) for water and for Separan solution; the values are about 0.49 for water and about 0.60 for the polymer solution.

5. Conclusions

The addition of some 50 ppm of PAA-material to water enhances coherent structures in a fully turbulent mixing layer, as shown by photographic and statistically evaluated results. The diminution of small-scale turbulence, a feature closely related to drag reduction, does not necessarily lead to a general attenuation of turbulent intensities and shear stresses.

Acknowledgement

One of the authors (R.S.) gratefully acknowledges the financial support of the Deutsche Forschungsgemeinschaft.

References

1. Virk, P. S., AIChE J. **21**, 625 (1975).
2. Vleggaar, J., Diss. Universiteit Amsterdam (1973).
3. Bewersdorff, H. W., K. Strauß, Rheol. Acta **18**, 104 (1979).
4. Kwade, M., Rheol. Acta **21**, 120 (1982).
5. Brown, G., A. Roshko, J. Fluid Mech. **64**, 775 (1974).
6. Winant, C. D., F. K. Browand, J. Fluid Mech. **63**, 237 (1974).
7. Giesekus, H., M. Kwade, R. Scharf, Paper presented at the VIIIth Int. Congress on Rheology, Naples, Sept. 1–5, 1980.

(Received May 28, 1982)

Authors' address:

M. Hibberd, Ph. D., Dr.-Ing. M. Kwade,
Dipl.-Phys. R. Scharf
Lehrstuhl für Strömungsmechanik, Abt. Chemietechnik
Universität Dortmund
Postfach 500500
D-4600 Dortmund 50

Rheologica Acta Rheol. Acta **21**, 587 – 589 (1982)

Effect of a centrally injected polymer thread on drag in pipe flow

H. W. Bewersdorff

Lehrstuhl für Strömungsmechanik, Abteilung Chemietechnik der Universität Dortmund

Key words: Drag reduction, tube flow, polyacrylamide

Drag reduction is obtained if the friction loss in turbulent flow is reduced by adding a substance to the flow at constant flow rate. The magnitude of this effect may be expressed in terms of the friction factors of the solvent f_s and the drag reducing fluid f_p as

$$DR = 1 - \frac{f_p}{f_s}. \tag{1}$$

Although an overwhelmingly large number of articles and reports have appeared in which the effect was explained, no interpretation was complete enough to explain all features characteristic of drag reduction.

Drag reduction is accompanied by a change in the intensity of the turbulent fluctuations and in the velocity profile. This change can be clearly seen in the universal velocity profile which shows the relation between the fluid velocity $\bar{u}$ and the distance from the wall y. The dimensionless coordinates are:

$$u^+ = \bar{u}/u_\tau, \quad y^+ = y(u_\tau/\nu) \tag{2}$$

where ν is the kinematic viscosity and

$$u_\tau = \sqrt{\tau_w/\rho} \tag{3}$$

designates the wall friction velocity, τ_w and ρ being the wall shear stress and the density of the fluid. In this universal velocity profile (usually a semilogarithmic plot) a Newtonian fluid shows two regions, called the viscous sublayer and the turbulent core.

For polymer solutions it has been found that there is a third intermediate region called the elastic sublayer. In this case the velocity profile of the core shows a parallel displacement. As the level of drag reduction increases the parallel shift also increases. It is clear that this is accompanied by a thickening of the elastic sublayer. For maximum drag reduction Virk [1] showed that the elastic sublayer extends to the center line, i.e. the turbulent core vanishes.

In the elastic sublayer the turbulence production and dissipation reaches a maximum. The addition of polymers expands the elastic sublayer and moves the location of the greatest production and dissipation closer to the pipe center as drag reduction is increased [2 – 4]. These are the fundamental results for a homogeneous distribution of the polymer over the cross section of the tube, i.e., if there is no concentration gradient.

Now the question arises, what will happen if a polymer solution with a higher concentration is injected into the center of the pipe. To answer this an experimental set-up was designed of which a sketch is shown in [5].

For determining the drag reduction from pressure drop measurements the test pipe of acrylic glass was provided with 14 pressure taps spaced 1 m apart. A sufficient entrance pipe length was provided to ensure fully turbulent flow in the measuring section. The water was pumped by a Mohno pump. A dosing pump was used for the desired flow rate of the polymer solution. The polymer solution was injected with a needle into the center of the tube. The diameter of the needle was chosen so that the bulk velocities of the water and the polymer solutions were nearly identical. Colouring the polymer solution provided the possibility of looking for turbulent diffusion of the polymer solution. At the end of the pipe, which was 14 meters from the injection point, the velocity profile was measured with a two-colour Laser-Doppler Anemometer (LDA) working in the back-scattering mode. Because of the curvature of the tube and the resulting reflections of the laser-beams a flat window was built in the tube. Velocity profile measurements showed that this did not influence the flow at the opposite part of the tube.

If drag reduction is plotted against the distance from the injection point, it is seen to be a function of the following parameters:

$$DR = f(Re, c_R, c_p, d^+, P) \quad (4)$$

where Re is the Reynolds number, c_R the polymer concentration averaged over the cross section of the tube, c_p the polymer concentration of the injected polymer solution, d^+ the dimensionless tube diameter formed as in eq. (2), and P designates the polymer type.

If a certain value of c_p is exceeded the injected polymer solution forms a long thread (fig. 1) which appears to remain intact over the entire length of the tube [6]. Thus it can be concluded that at least a significant part of the injected polymer is not dispersed while passing down the tube. This is in contrast to a glycerol solution with the same zero-shear viscosity, which was dispersed completely after a few diameters at relatively low Reynolds numbers. Although there is no important mixing of the injected polymer solution, a remarkable drag reduction is found [6]. Drag reduction as a function of the distance from the injection point shows an increase at the beginning and a nearly constant value at the end of the measuring section (fig. 2). Comparing these ultimate values of the injection experiments with those of experiments with a homogeneous polymer solution at the same concentration averaged over the cross section it is found that there are differences in the measured friction factor.

In a plot of the friction factor versus Reynolds number (fig. 3) all values measured at one concentration of the injected polymer solution lie on a straight line. The dashed line belongs to the measured values of a homogeneous polymer solution. Especially at high Reynolds numbers, the differences between the injection experiments and the experiments with a homogeneous polymer solution are remarkable. The onset-point for drag reduction depends on the concentration of the injected polymer solution. Most effective for drag reduction is a concentration range between 0.15 and 0.3%.

Velocity profile measurements for water at various Reynolds numbers show very similar lines. These are in good agreement with the theoretical profiles and thus demonstrate the accuracy of the LDA-measurements.

Fig. 4 shows velocity profiles from the injection experiments. The full line represents the region where measurements were taken. The dashed line is the extrapolation to the viscous sublayer. In contrast to

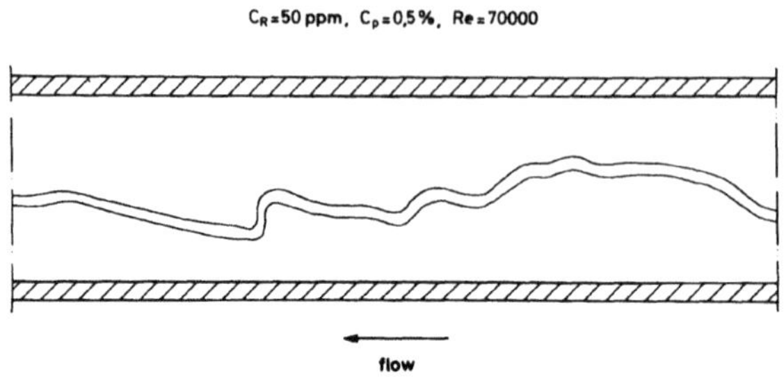

Fig. 1. Polymer thread (drawing from a movie taken with a high-speed camera)

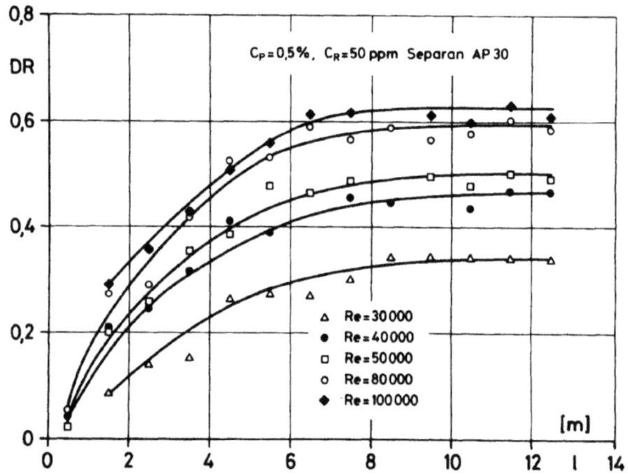

Fig. 2. Drag reduction versus the distance from the injection point

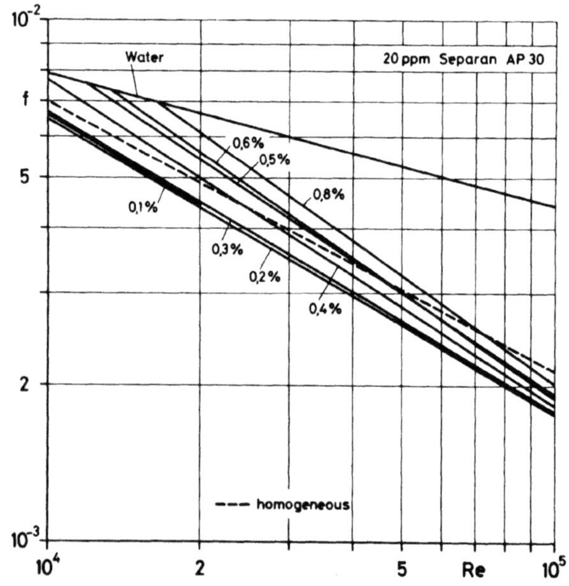

Fig. 3. Friction factor for homogeneous and heterogeneous drag reduction

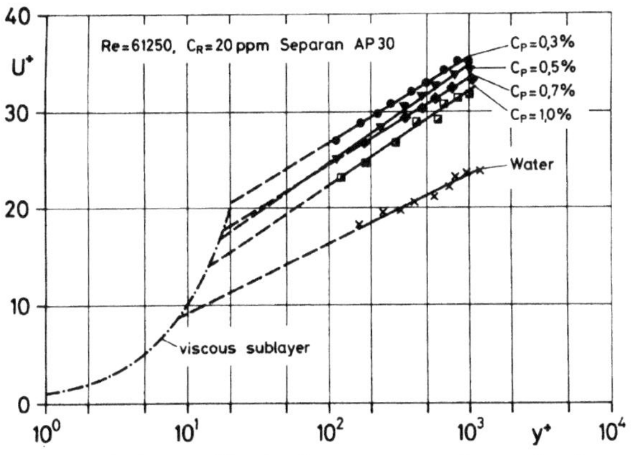

Fig. 4. Velocity profiles for heterogeneous drag reduction

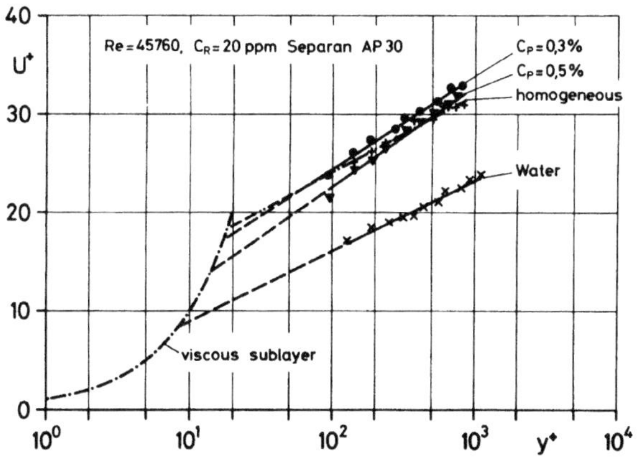

Fig. 5. Velocity profiles for homogeneous and heterogeneous drag reduction

the velocity profile measurements of homogeneous polymer solutions there is no longer a parallel shift of the velocity profile of the turbulent core, but the slope of the velocity profile is changed. It is interesting to note that all velocity profiles of the injection experiments are nearly parallel if the same polymer concentration averaged over the cross-section is chosen but different concentrations of the injected polymer solution are used.

Differences in the slope of the velocity profile of the turbulent core between homogeneous and heterogeneous drag reduction are shown in figure 5. With the same concentration averaged over the cross-section and the same Reynolds number, homogeneous drag reduction shows a velocity profile, which is parallel to the Newtonian profile, whereas heterogeneous drag reduction shows a different slope.

As the Reynolds number increases, drag reduction also increases. For the injection experiments the velocity profiles show a parallel shift with increasing

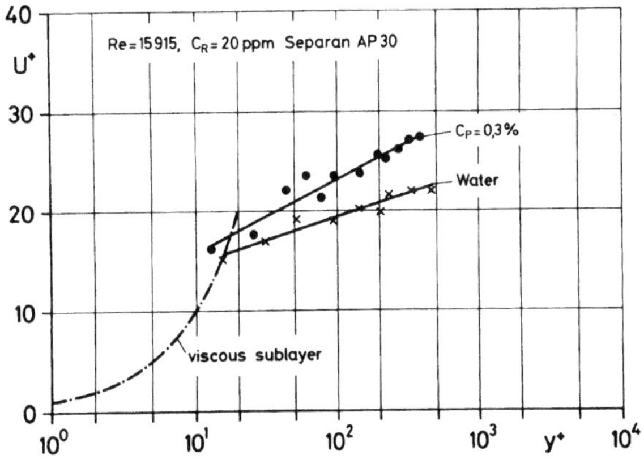

Fig. 6. Velocity profile for heterogeneous drag reduction at a low Reynolds number

Reynolds number, but they are not parallel to the Newtonian line.

All velocity profiles presented above had to be extrapolated to the viscous sublayer and thus could not answer the question whether there is a two- or a three-layer velocity profile for heterogeneous drag reduction. Therefore one velocity profile measurement (fig. 6) was taken at a lower Reynolds number. Because of the limits of accuracy of the tracker of the LDA-equipment, errors are here larger than at higher Reynolds numbers. This velocity profile leads to a two-layer model with a thickened viscous sublayer for heterogeneous drag reduction, but, of course, further measurements are needed to confirm or to reject this result.

Acknowledgement

The author thanks the Deutsche Forschungsgemeinschaft (DFG) for providing the LDA-equipment.

References

1. Virk, P. S., AIChE J. **21**, 625 (1975).
2. Mizushina, T., H. Usui, Phys. Fluids **20**, S 100 (1977).
3. Berman, N. S., Ann. Rev. Fluid Mech. **10**, 47 (1978).
4. Lumley, J. L., Polym. Sci. Macromol. Rev. **7**, 263 (1973).
5. Bewersdorff, H. W., K. Strauß, Rheol. Acta **18**, 104 (1979).
6. Vleggaar, J., Diss. Universiteit van Amsterdam (1973).

(Received May 6, 1982)

Author's address:

Dipl.-Ing. H.-W. Bewersdorff
Lehrstuhl für Strömungsmechanik
Abteilung Chemietechnik
Universität Dortmund
Postfach 500500
D-4600 Dortmund 50

Rheologica Acta Rheol. Acta **21**, 590 – 592 (1982)

The rheological behaviour of $Ca(OH)_2$ suspensions in water

A. J. G. van Diemen and H. N. Stein

Laboratory of Colloid Chemistry, Eindhoven University of Technology, Eindhoven (The Netherlands)

Abstract: The rheological characteristics of $Ca(OH)_2$ suspensions are investigated in order to check the "elastic floc" model for energy dissipation during flow. It is found that the energy, required to overcome the viscous drag experienced by particles moving within flocs, can account for the total energy dissipation.

Key words: Energy dissipation, $Ca(OH)_2$ suspension, elastic floc model, suspension in water

The rheological behaviour of $Ca(OH)_2$ suspensions in water was investigated in order to elucidate one aspect of the "elastic floc" model for energy dissipation during stationary flow in coagulated suspensions. This model was developed by Hunter et al. [1 – 3] and van de Ven and Hunter [4]. According to it, energy dissipation during flow occurs only to an insignificant extent through formation and subsequent breaking of bonds between the suspended particles; the energy dissipation is caused in the main by

(i) energy required to stretch bonds between primary particles,
(ii) energy required to overcome the viscous drag experienced by particles moving within flocs,
(iii) energy required for the internal movement of liquid in the flocs when they change shape and/or volume.

It is, however, not clear whether the third energy dissipation term is really an independent one: the work required to move an object through a liquid as calculated, e.g., for isolated spheres by the Stokes equation, includes the work for displacing part of the surrounding liquid. Thus, the third energy dissipation term may already be comprized in the second.

With regard to the testing of this model, aqueous $Ca(OH)_2$ suspensions offer special opportunities since $Ca(OH)_2$ is soluble enough to employ its dissolution rate after diluting the suspension with water, as a measure for the degree of aggregation of the solid [5]; thus one of the parameters which influence the rheological behaviour of a suspension is accessible to independent check.

On the other hand, the solubility is small enough to let the liquid phase retain the characteristics of a dilute electrolyte solution with a viscosity not largely different from that of water.

Since it is known [5] that the properties of $Ca(OH)_2$ suspensions are dependent on the time during which the systems are subjected to shear, we investigated the rheology by registering the torque experienced by the inner cylinder of a rotation viscometer (Epprecht 15T-FC, Contraves A.G., Zürich), as a function of time at constant shear rate; values obtained at one particular shearing time were then combined into a torque-shear rate diagram. Primarily the diagrams obtained by extrapolation to shearing time zero (where the structure not yet broken down by shear is approached) and for long shearing times (where the final structure is approached) are of interest.

The torque-shear rate diagrams were used to calculate a linear relation between shear rate and shear stress; thus, a τ_0 and an η were calculated as for a Bingham liquid, both for shearing time → 0 and for a shearing time of 50 min. The use of a Bingham equation implies in the context of the present work only that the torque-$\dot{\gamma}$ relationship can, within the $\dot{\gamma}$ region investigated and within the accuracy of the measurements, be represented by a linear relation; it does not

798

imply that the liquid really behaves as a Bingham liquid down to very low values of the shear rate.

Predominantly suspensions were investigated at a solid volume fraction (c_v) of 0.25, in which sedimentation did occur only very slowly.

τ_0 decreases whereas η increases with increasing shearing time; the decrease of τ_0 is pronounced (a factor 0.3 – 0.5) whereas the η increase is less (a factor 1.2). Shearing did not, however, effect any change in the dispersion characteristics: it did not influence the dissolution rate of $Ca(OH)_2$ after dilution of the suspension, and the stack-wise aggregates of plate-like $Ca(OH)_2$ crystals observed by SEM in the original $Ca(OH)_2$ samples were found to a similar extent in samples isolated from sheared suspensions. Neither did shearing affect the sediment volume of more dilute suspensions (c_v = 0.082), or the ζ potential of the $Ca(OH)_2$ as determined by electro-osmosis or electrophoresis.

It is concluded from these results that the stack-like aggregates are persistent to shear; thus, one of the energy dissipation mechanisms recognized by van de Ven and Hunter, viz. that connected with the stretching of bonds between particles, can be neglected.

Since the unsheared suspension does not significantly settle, it consists of one single floc. A shear imposed on such a system will not occur uniformly over the volume but will be realized in shear planes dividing this floc into domains. An equation was derived for the energy dissipation during stationary flow caused by the viscous drag experienced by particles moving within these domains:

$$E = 12\pi\eta_0 f \frac{\delta_0^2 A}{\Delta^3 \overline{\cos u}} \cdot \frac{\dot{\gamma}^2}{1-(q-1)l^2} \tag{1}$$

where E is the energy dissipation per unit of volume and time; η_0 the viscosity of the suspension medium; f the quotient of friction experienced by an aggregate moving in a domain and the friction experienced by an isolated sphere of comparable radius; δ_0 the average distance over which an aggregate bordering a shear plane is displaced on encountering another aggregate; A the average distance between shear planes; Δ^2 the average area which can be assigned to an aggregate bordering a shear plane, in this plane; $\Delta\,\overline{\cos u}$ the average distance travelled by an aggregate bordering a shear plane between two successive encounters; q the average number of directly connected neighbours of an aggregate; and l the fraction of displacement communicated (on the average) by an aggregate to its direct neighbours.

In stationary flow, $E = \tau\dot{\gamma}$. Thus, if $\tau = \tau_0 + \eta\dot{\gamma}$,

$$\eta = 12\pi\eta_0 f \frac{\delta_0^2 A_1}{\Delta^3 \overline{\cos u}} \cdot \frac{1}{1-(q-1)l^2} \tag{2}$$

where A_1 is the average distance between shear planes for $\dot{\gamma} \to \infty$.

An estimate of f was made on the basis of results obtained by Brenner [6], Roeberson [7], Honig et al. [8] and Frankel and Acrivos [9]. On moving two spheres (of radius c) such that the distance between their centers increases from r_0 to $r_0 + \delta$, the friction factor was found to be:

$$f' = 1 + \frac{1}{6}\frac{c}{\delta}\ln\frac{r_0+\delta-2c}{r_0-2c} + \frac{1}{3}\frac{c}{\delta}\ln\frac{r_0+\delta-\frac{4}{3}c}{r_0-\frac{4}{3}c}. \tag{3}$$

f' is determined primarily by c/δ; it significantly surpasses one only for the case of spheres which are very close to each other either at the start or at the end of the movement.

For aggregates moving in domains during flow this means, that f is determined predominantly by the ratio of r_0 and a, the equivalent aggregate radius.

With A_1/Δ = 1 and δ_0/Δ = 0.2 at c_v = 0.25, the experimental values of η can be accounted for with reasonable values for r_0/a (about 2.02), if the motion of aggregates within the domains takes part in the total energy dissipation to a considerable degree, i. e. if the motion of aggregates from their positions within the domains is not restricted in the main to aggregates bordering a shear plane.

It is concluded that there is no need for the introduction of an additional energy dissipation term by flow movement inside the domains.

A more extended report on this work will be published soon [10].

References

1. Firth, B. A., R. J. Hunter, J. Coll. Int. Sci. **57**, 248, 266 (1976).
2. Firth, B. A., J. Coll. Int. Sci. **57**, 257 (1976).
3. Hunter, R. J., J. Frayne, J. Coll. Int. Sci. **71**, 30 (1979); **78**, 107 (1980).
4. Ven, T. G. M. van de, R. J. Hunter, Rheol. Acta **16**, 534 (1977).

5. Ohnemüller, W., Schriftenreihe des Bundesverbandes der Deutschen Kalkindustrie **10**, 101 (1970); Tonind. Ztg. u. Keram. Rundschau **89**, 197 (1965).
6. Brenner, H., Chem. Eng. Sci. **16**, 242 (1961).
7. Roebersen, G. J., Ph. D. Thesis (Utrecht 1974).
8. Honig, E. P., G. J. Roebersen, P. H. Wiersema, J. Coll. Int. Sci. **49**, 28 (1974).
9. Frankel, N. A., A. Acrivos, Chem. Eng. Sci. **22**, 847 (1967).
10. van Diemen, A. J. G., H. N. Stein, Rheol. Acta (in the press).

(Received May 17, 1982)

Authors' address:

Dr. A. J. G. van Diemen, Prof. Dr. H. N. Stein
Laboratorium voor Colloidchemie
Technische Hogeschool Eindhoven
Postbus 513
NL-5600 MB Eindhoven

Behaviour of double emulsions in shear flows

J. J. Ulbrecht, P. Stroeve, and P. Prabodh

Department of Chemical Engineering, State University of New York at Buffalo, and University of California at Davis

Abstract: The stability behaviour of heterogeneous drops indicates that, for liquid-liquid dispersions, we may have to modify the stability criterion to incorporate either the rheological characteristics of the dispersion or its cohesion or both.

Key words: Shear deformation, stability, heterogeneous drop, liquid membrane

1. Introduction

When a water-in-oil (W/O) emulsion is resuspended in an aqueous phase a new emulsion is formed in which the drops are heterogeneous. The two aqueous phases, the continuous one and the one contained in the sub-drops within the oily drops, are separated by a liquid (oil) membrane. Inversely, an O/W/O double emulsion can be also formed by redispersing an O/W emulsion in a hydrocarbon solvent. The foundations of the technology of liquid membranes were laid by Li [1, 2] and the industrial applications were recently reviewed by Marr and Kopp [3]. Since reagents can be introduced into the subdrops which will provide a sink for species which diffuse from the continuous phase across the liquid membrane and because of the high interfacial area per unit volume, the short diffusion distances and the ability to modify the membrane permeability characteristics, double emulsions have been shown to be efficient media for many separation processes. However, the industrial applications of double emulsions to separation processes is still limited due to the instability of the drops to breakup in flow. In stirred tanks, motionless mixers, pipelines, pumps and other process equipment, the double emulsion drops are exposed to varying degrees of viscous and inertial forces. The breakup of these drops due to these forces is presently not fully understood. However, it is known that during breakage some of the subdrops are expelled into the continuous phase with the release of the species to be separated, which gives rise to a reduced selectivity (Hochhauser and Cussler [4]). The understanding of the mechanisms of breakage is necessary to design process equipment or to set operating conditions to minimize the breakage (Haugh [5]).

In some other applications of double emulsions, the breakage of drops and release of subdrops filled with reagents can be utilized to obtain changes in physical properties and or initiation of chemical reactions. For example, blow-outs in oil well pipes can be repaired by forcing a reactive double emulsion through the fissure, with subsequent polymerization of reagents released from the subdrops to form a polymer plug.

Although the stability of double emulsions could be, to some extent, controlled by the proper choice of surfactants, it is the stability in shear which will ultimately determine the feasibility of the industrial applications of these systems.

2. Stability and deformation of homogeneous drops

The first concepts and the terminology were developed by Taylor [6, 7] who investigated, both experimentally and theoretically, drops in both plane, hyperbolic, and in simple shear flow. By applying Lamb's hydrodynamic theory [8] for the slow motion of viscous fluids, i.e. ignoring inertial effects, Taylor derived a relationship for the pressure difference across the drop interface

$$\Delta P = -4G\mu_c f(p) + \frac{2\sigma}{a} \qquad (1)$$

where G is the velocity gradient in the continuous phase, μ_c is the continuous phase viscosity, p is the viscosity ratio between drop and continuous phase, σ is the interfacial tension, a is the droplet radius and the function $f(p)$ is

$$f(p) = \left(\frac{19p + 16}{16p + 16}\right). \tag{2}$$

The first term in eq. (1) is due to the normal stresses, and the second term is the contribution of ΔP due to the pressure difference across a spherical interface as given by the Young-Laplace equation. Deformation of the drop is given by Taylor as

$$d = \frac{L - B}{L + B} = \frac{\mu_c G a}{\sigma}\left(\frac{19p + 16}{16p + 16}\right) \tag{3}$$

where L and B are the major and minor axes of the deformed drop as shown in figure 1. Taylor proposed that when the pressure difference due to the normal stresses exceeds the pressure difference due to the surface tension, the drop will break up so that (ignoring deviation from spherical form)

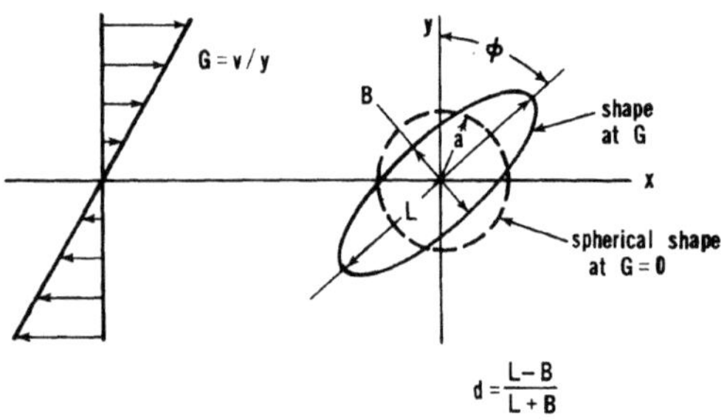

Fig. 1. Drop deformation in simple shear

$$4G\mu_c\left(\frac{19p + 16}{16p + 16}\right) \geq \frac{2\sigma}{a}. \tag{4}$$

The critical deformation d_B at which point the drop would burst is then

$$d_B = \tfrac{1}{2} \tag{5}$$

and the critical ratio of the applied viscous force to the surface tension force multiplied by $f(p)$ is

$$\frac{\mu_c G_B a}{\sigma} f(p) = \frac{1}{2}. \tag{6}$$

The ratio of the applied viscous force to the surface tension force will be called here the deformation parameter

$$\varepsilon = \frac{\mu_c G a}{\sigma} \tag{7}$$

and d the deformation. (Hinze [9] has given the name "generalized Weber number" to ε, but traditionally the Weber number is used for the ratio of the inertial force to the surface tension force.) Eq. (6) now becomes

$$\varepsilon_B f(p) = \tfrac{1}{2} \tag{8}$$

where ε_B is the critical deformation parameter for breakup. The function $f(p)$ is not very sensitive to the viscosity ratio as the value of $f(p)$ varies from 1 to 1.187 when p goes from zero to infinity.

Since the work of Taylor, the subject area of drop stability and deformation has been extensively researched but a full account of this work is beyond the scope of this presentation. An interested reader may wish to consult the recently published work of Grace [10], of Torza et al. [11] and that of Karam and Bellinger [12].

Experimental results obtained by Torza et al. [11] for the critical deformation parameter as a function of the viscosity ratio are shown by curve C in figure 2. For viscosity ratios greater than 3, drops could not be broken by shear flow indicating asymptotic behavior.

Karam and Bellinger [12] studied drop breakup in a Couette device, cf. figure 3C. They utilized a wide

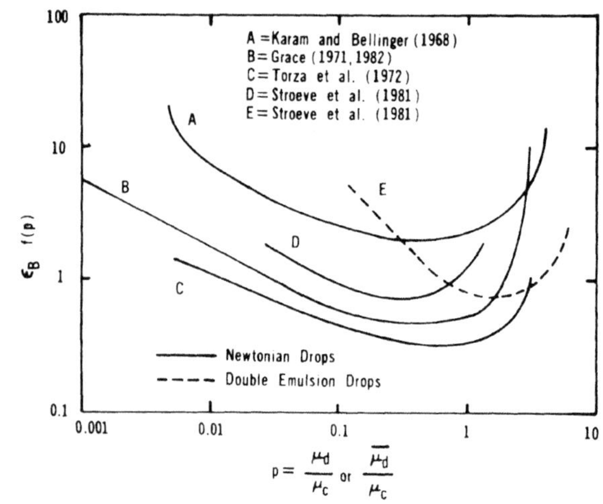

Fig. 2. Breakup of homogeneous drops, as observed by different authors, and of double emulsion drops

range of viscosity ratios and their data are given by curve A in figure 2. Karam and Bellinger concluded that the viscosity range must be within $0.005 \leq p \leq 4$ for a drop to break up. The presence of a second asymptote, below which drops do not break, was not observed by Grace [10] who conducted the most extensive study of the drop deformation and breakup to date.

Stroeve et al. [13] measured drop breakup in a cone-and-plate device where the cone and plate are driven at equal but opposite speed, cf. figure 3D. Breakup of small drops (20 – 80 μm) were measured and the data are shown as Curve D in figure 2. The data are in close agreement with that of Torza et al. and Grace, who studied drops of size 200 μm to 1 mm. Considering the variety of equipment used and the numerous liquid-liquid systems employed, there is good quantitative agreement between curves B, C and D. The higher data by Karam and Bellinger are possible due to their use of the equilibrium interfacial tension rather than the dynamic interfacial tension obtained at zero time (Grace [10]).

3. Stability and deformation of heterogeneous drops

The first report on the stability and deformation of heterogeneous drops in a shear field was presented by Stroeve et al. [13] who used a counter-rotating cone-and-plate device.

The cone and plate had a diameter of 5 cm and gap angles were 2 and 3 degrees (fig. 3D). The device was mounted on an inverted microscope as shown on figure 3D, and individual drops could be observed as these were being sheared. A similar device has been used by Schmid-Schönbein and colleagues (Fisher et al. [14]). The breakup of drops was followed both by photography or by automated size analysis with the Quantimet System 23 image analyser.

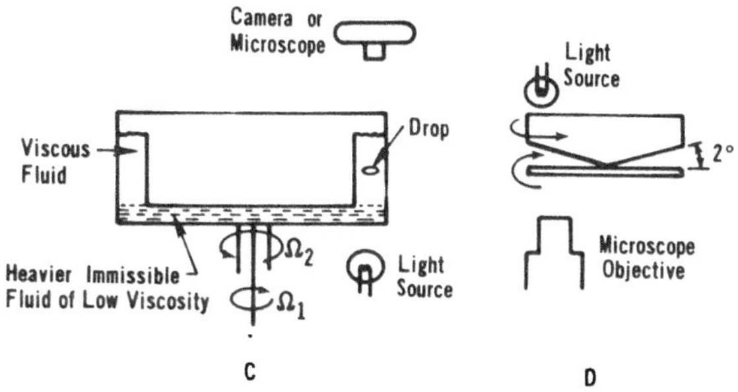

Fig. 3. Sketch of experimental devices for the observation of drop breakup: C – Couette device as used by Karam and Bellinger [12], D – Cone-and-plate device as used by Stroeve et al. [13]

The use of an image analyser speeds up particle size analysis considerably. Breakup of both single drops of known initial diameter and dilute suspensions of drops (<0.5 vol-%) was measured as a function of shear rate and system properties. For the case of dilute suspensions, the maximum drop size that could exist at a particular shear rate was found by performing size analysis on at least 500 drops. Both techniques gave identical results regardless of the analysis (photography or image analysis) method used.

The breakup data of double emulsion drops, as shown by curve E in figure 2, were obtained for oil drops fully packed with aqueous subdrops. When fully packed, the volume fraction of subdrops was 62 percent. The subdrop phase was made up from different proportions of water and glycerin with a very small amount of methylene blue dye added to color the particles. The subdrops had diameters from approximately 1 μm to 7 μm with a number average of about 2.5 μm and a Sauter mean diameter of about 4 μm. The subdrops are smaller than the size of the oil drop containing them. The oil phase was a mixture of an oil soluble surfactant (sorbitan monooleate) and mineral oil. The surfactant is practically insoluble in the aqueous phase. Double emulsion drops were suspended in a variety of corn syrups so that the total system was of the W/O/W type. The drop size range was identical to that employed for the homogeneous Newtonian drops (20 – 80 μm). The mode of the drop breakup was similar to that found for homogeneous drops. Figure 4 shows a photograph of the breakup of a heterogeneous drop. Satellite drops (from one to three) usually formed between the remaining globules of the original drop. No dependency of ε_B on the drop radius was observed, i.e. for a given viscosity ratio large of small drops had the same ε_B.

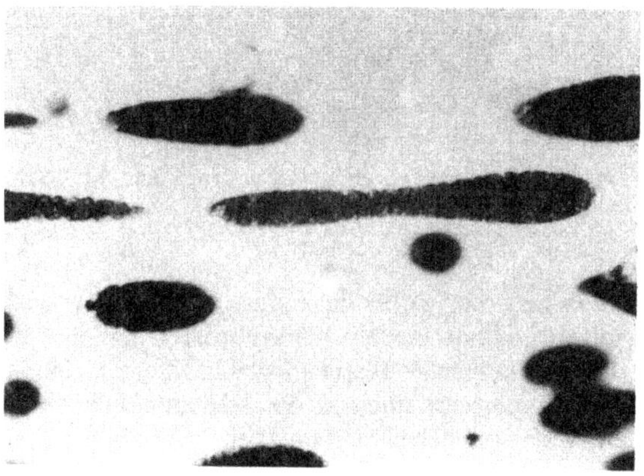

Fig. 4. Micro-photograph of the breakup of a heterogeneous drop

The W/O drops exhibited shear thinning behavior. Viscosity measurements of the W/O emulsion were made in a Haake viscometer as a function of shear rate. The viscosity was independent of the proportion of glycerin to water, indicating that the subdrops act like solid spheres in the shear rate range investigated.

The initial work reported by Stroeve et al. [13] has now been extended to include the breakup of double

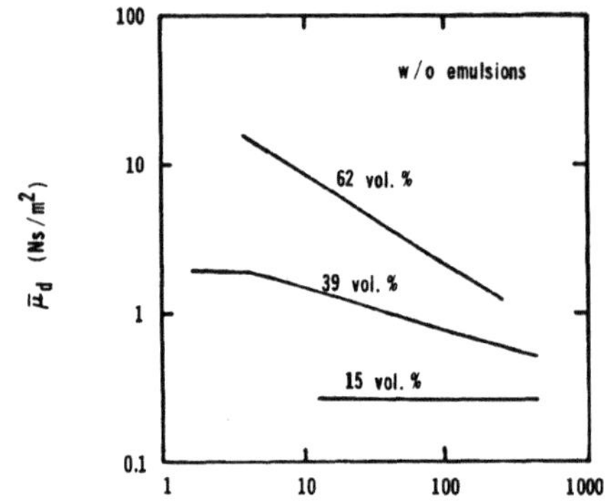

Fig. 5. Flow curves of W/O emulsions with different content of the discontinuous phase

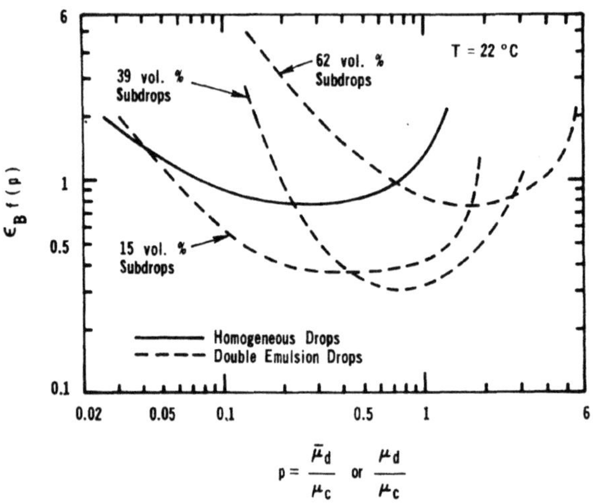

Fig. 6. Breakup of homogeneous drops and of double emulsion drops with different subdrops volume content

emulsion drops with subdrop volume fractions of 39 and 16 volume percent. The viscosity data for the three W/O emulsions are shown in figure 5. The 39 volume percent emulsion is still non-Newtonian, although less dependent on shear rate than the 62 volume percent emulsion, whereas the 15 volume percent emulsion is Newtonian over the shear rate investigated. Breakup data for the three systems are shown in figure 6 and are compared to homogeneous drops. As the volume fraction of subdrops decreases, the breakup curves clearly shift back to the left, or to lower viscosity ratios. The 15 volume percent emulsion drops approach the breakup behavior of the homogeneous (Newtonian) drops. At zero volume percent subdrops, the data do fall on the solid line. The decrease in the minimum of the deformation parameter can be interpreted by advancing one or more of the three following hypotheses.

The existence of the subdrops will have a bearing on the cohesion of the drop phase. At the low concentrations of 15% and 39% the cohesion of the drops themselves is not likely to be affected but, as soon as the drop is drawn out into a long thread, the thickness of it obviously cannot become less than is the average size of the subdrops. Thus, for a given viscosity ratio the critical value of ε_B is lower. The highest concentration of 62% represents the maximum volume concentration of randomly packed spheres. Under this condition, the cohesion of the drop is greatly increased due to the cohesive forces among the subdrops themselves. The consequence of this is a shift to higher values of ε_B.

Although the argument based on the variable cohesion may be a perfectly valid one, it may not tell the whole story. The curves in figure 6 were plotted using an apparent viscosity of the W/O emulsion of the shear rate of the macro-flow between the cone and the plate. It is immediately obvious that this is not the shear rate inside the drop. Let us assume that the boundary of the drop, the characteristic size of which is δ, moves around the drop centre with a lateral velocity v_0 so that the velocity gradient inside the drop is of the order of $(2v_0/\delta)$. If the drop is sufficiently small and prolate and we may assume that the velocity field of the continuous phase is essentially undisturbed, then the velocity gradient surrounding the drop is of the order of $(\Delta v/\delta)$ where Δv is the velocity difference in the continuous phase on both sides of the drop. Thus, the ratio of the two velocity gradients will be of the order of $(2v_0/\Delta v)$. We have not succeeded yet in measuring v_0 accurately but it is clear from our observations that v_0 is at least ten times higher than Δv or more. Since the emulsion inside the drop is shear-thinning the apparent viscosity of it is lower at higher shear rates and this will result into a shift of the curves in figure 6 towards lower values of p.

Finally, we must not overlook that the emulsion inside the drops may be viscoelastic. This has been not only predicted by Schowalter et al. [15] (their analysis was subsequently refined by Cox [16] and by Frankel and Acrivos [17]) but experimentally observed by Han and Funatsa [18]. The rotational flow inside an ellipsoid was analysed by Astarita and Marrucci [19] who recognized that this flow is opposite in nature to pure stretch flows. If a Maxwell liquid (which has a shear independent shear viscosity) is subjected to the rotational flow inside an ellipsoid it dissipates less energy than it would under the conditions of a viscometric flow. Phenomenologically it appears that such a liquid has a reduced viscosity. This indeed would contribute to a further shift of the broken curves in figure 6 towards lower values of p.

Acknowledgement

The authors wish to acknowledge the helpful discussion which one of us (J. J. U.) had with Professor Astarita.

References

1. Li, N. N., AIChE J. **17**, 459 (1971).
2. Li, N. N., J. Membr. Sci. **3**, 265 (1978).
3. Marr, R., A. Kopp, Int. Chem. Eng. **22**, 44 (1982).
4. Hochhauser, A. M., E. L. Cussler, AIChE Symp. Ser. **71**, 136 (1976).
5. Haugh, T. M., Science **193**, 134 (1976).
6. Taylor, G. I., Proc. Roy. Soc. A **146**, 501 (1934).
7. Taylor, G. I., Proc. 11th Int. Congr. Appl. Mech., Munich (1964).
8. Lamb, H., Hydrodynamics, 6th ed., Dover Reprint, (New York 1945).
9. Hinze, J. O., AIChE J. **1**, 289 (1955).
10. Grace, H. P., Chem. Engng. Commun. **14**, 225 (1982).
11. Torza, S., R. G. Cox, S. G. Mason, J. Coll. Interf. Sci. **38**, 395 (1972).
12. Karam, H. J., J. C. Bellinger, Ind. Engng. Chem. Fund. **7**, 576 (1968).
13. Stroeve, P., P. V. Prabodh, T. Elias, J. J. Ulbrecht, "Stability of Double Emulsion Droplets in Shear Flow", 53rd Annual Meeting of the Society of Rheology, Louisville, Kentucky (1981).
14. Fisher, T. M., M. Stohr, H. Schmid-Schönbein, AIChE Symp. Ser. **182** (74), 38 (1978).
15. Schowalter, W. R., C. E. Chaffey, H. Brenner, J. Coll. Interface Sci. **26**, 152 (1968).
16. Cox, R. G., J. Fluid Mech. **37**, 601 (1969).
17. Frankel, N. A., A. Acrivos, J. Fluid Mech. **44**, 65 (1970).
18. Han, C. D., K. Funatsa, J. Rheol. **22**, 113 (1978).
19. Astarita, G., G. Marrucci, First Intern. Congr. on Theor. Appl. Mechan. Udine 1971, pp. 21 – 43.
20. Hinch, E. J., A. Acrivos, J. Fluid Mech. **98**, 305 (1980).

(Received June 24, 1982)

Authors' addresses:

Prof. Dr. J. J. Ulbrecht
Department of Chemical Engineering
State University of New York
Clifford C. Furnas Hall
Buffalo, NY 14260 (U.S.A)

Dr. P. Stroeve, Dr. P. Prabodh
Department of Chemical Engineering
State University of California
Davis, CA 95616 (U.S.A)

Rheologica Acta Rheol. Acta 21, 598 – 604 (1982)

Behavior of viscoelastic glass bead suspensions under superposed flows

P. Masi, L. Nicodemo, L. Nicolais, and G. Taglialatela

Polymer Engineering Laboratory, Institute of Chemical Engineering Principles,
Engineering School of the University of Naples

Abstract: Parallel superposed flows may be useful in order to explain the rheological behavior of viscoelastic suspensions. It has been shown that relaxation times of molecules are affected by a combined flow, and the rheological behavior of glass bead suspensions in a viscoelastic medium is explaned in terms of such variations.

Key words: Suspension, combined flow, dynamic viscosity

1. Introduction

The rheological behavior of fluids undergoing parallel superposed steady and oscillatory shear flows, has become of increasing interest in the scientific field. The popularity of this situation is mainly due to the fact that such an experiment, which is easily carried out, is at the same time a crucial test of any rheological equation and offer the possibility to

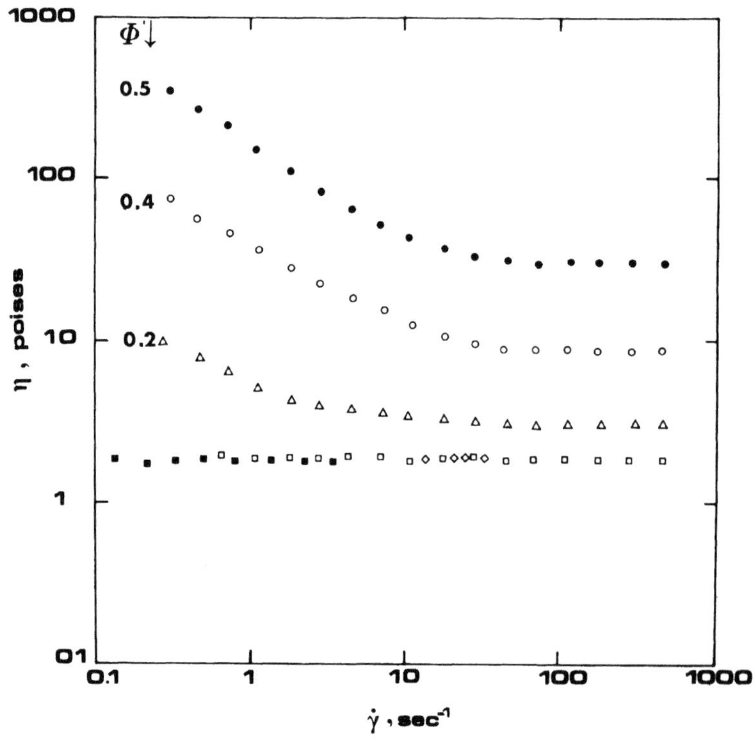

Fig. 1. Shear viscosity η vs. shear rate $\dot{\gamma}$ of glass bead suspensions in vaseline oil at various filler content

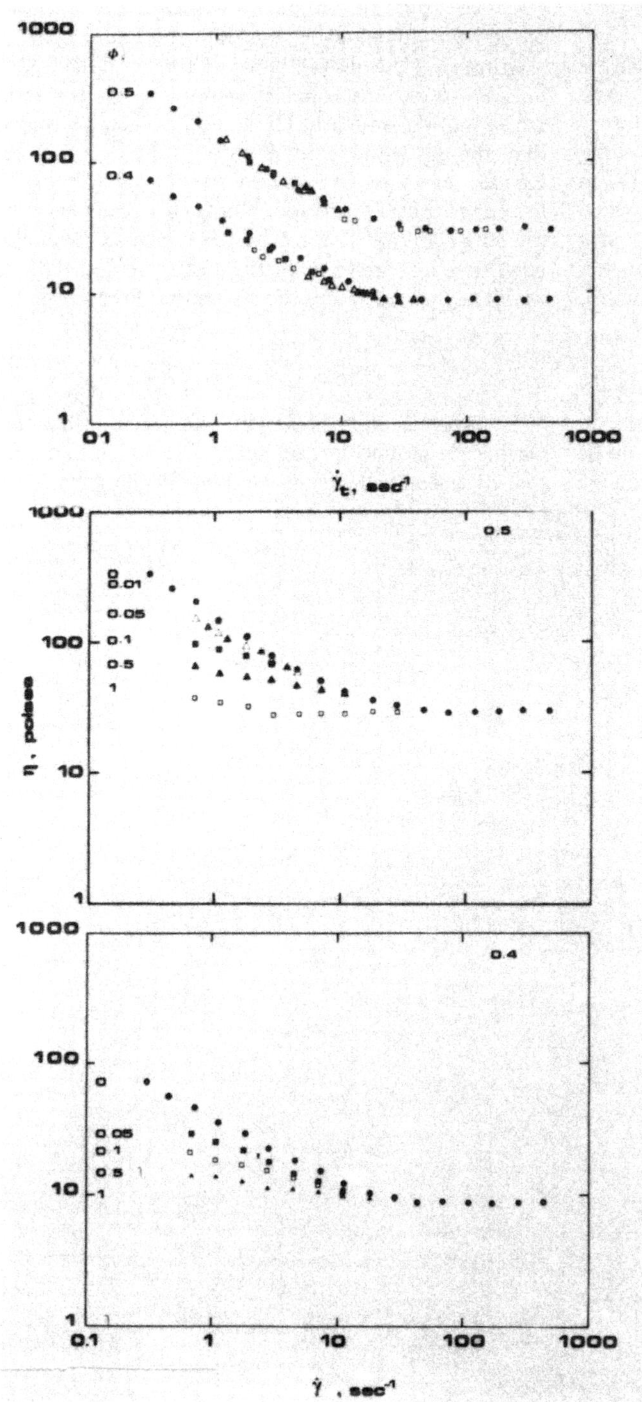

Fig. 2. Mean viscosity vs. shear rate of glass bead suspensions (ϕ = 0.4 and ϕ = 0.5) in vaseline oil at various oscillation frequencies

get information about structural parameters of nonlinear fluids [1].

The peculiarity of parallel superposition is that the trajectories of the particles due to the viscometric motion are parallel to those due to the oscillatory shear flow. The theoretical analysis of the resulting flow has been extensively studied by Pipkin and Owen [2]. According to their theory, the total stress τ_t which arises from the combined motion is composed of a stationary contribution due to the steady shear τ, and a fluctuating part due to the oscillatory motion τ_0^0. The total stress can be thus written in the form:

$$\tau_t = \eta\dot{\gamma} + \frac{\eta'\dot{\gamma}_0^0}{\sin\delta}\sin(\omega t + \delta) \tag{1}$$

where η is the fluid viscosity which may or may not be dependent on the shear rate $\dot{\gamma}$, η' is the dynamic viscosity, $\dot{\gamma}_0^0$ the amplitude of the shear wave and δ the phase lag between stress and deformation.

From eq. (1) follows that η' can be calculated with the same procedure which is used when only the oscillatory motion is present [3]. In addition eq. (1) shows that the unsteady contribution to the total stress vanishes at values of ωt equal to $n\pi - \delta$, n being an integer number. Correspondingly, η is calculated by dividing the total stress τ_t by $\dot{\gamma}$. Of course under this condition τ_t is the value of the mean stress τ_m defined as $\frac{1}{2}(\tau_{t\max} + \tau_{t\min})$ and the viscosity is then given by:

$$\eta = \tau_m/\dot{\gamma}\,. \tag{2}$$

This paper is concerned with the effect which the oscillatory motion has on shear viscosity η and with that of a steady shear on the dynamic viscosity η' of

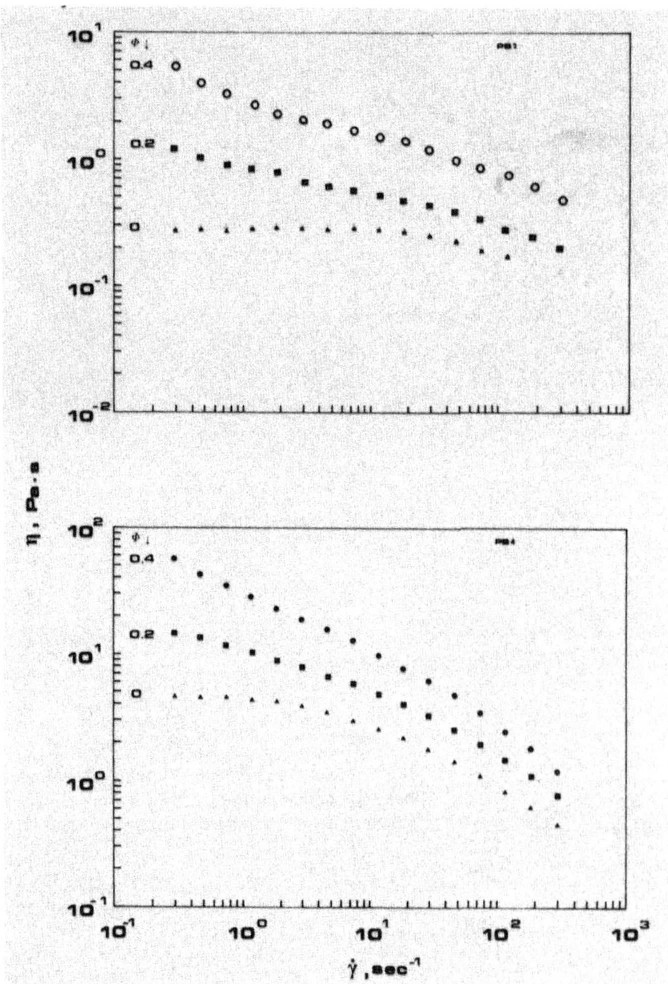

Fig. 3. Shear viscosity η vs. shear rate $\dot{\gamma}$ of glass bead suspensions in solutions of polyisobutylene in dekaline

suspensions of rigid microspheres having diameter <30 μm in Newtonian and non-Newtonian liquids.

2. Results and discussion

The suspending media and their suspensions were tested in parallel superposed flows at 25 °C in a Weissenberg Rheogoniometer R 18, using a cone-and-plate geometry with 7.5 cm diameter and 1° angle cone.

Figure 1 shows the viscosity η in steady shear vs. the shear rate $\dot{\gamma}$ for vaseline oil suspensions at a volumetric concentration of beads equal to 0.2, 0.4 and 0.5 and the viscosity curve of pure vaseline. While the oil exhibits a Newtonian behavior, the suspensions show a thinning effect approaching a constant value at high shear rate.

In a previous paper [4], this effect is attributed to the presence of bead aggregates which trap some liquid, thus making the effective particles concentration larger than the nominal one.

Figure 2 reports viscosity curves of the suspensions with volumetric bead concentration of 0.4 and 0.5 respectively, obtained by superimposing to the steady shear an oscillatory motion of 10^{-3} rad of amplitude and various frequencies. At a given shear rate, the viscosity η decreases with increasing the oscillation frequency, and the lower $\dot{\gamma}$ of the steady shear the more the viscosity decreases. Such effect disappear at high $\dot{\gamma}$ where all curves approach the same asymptotic value.

If a sinusoidal perturbation of amplitude $\dot{\gamma}_0^0$ and frequency f is superposed to a steady shear, the shear rate will change on time between $\dot{\gamma} - \dot{\gamma}_0^0$ and $\dot{\gamma} + \dot{\gamma}_0^0$. If the aggregation and desaggregation of the particles

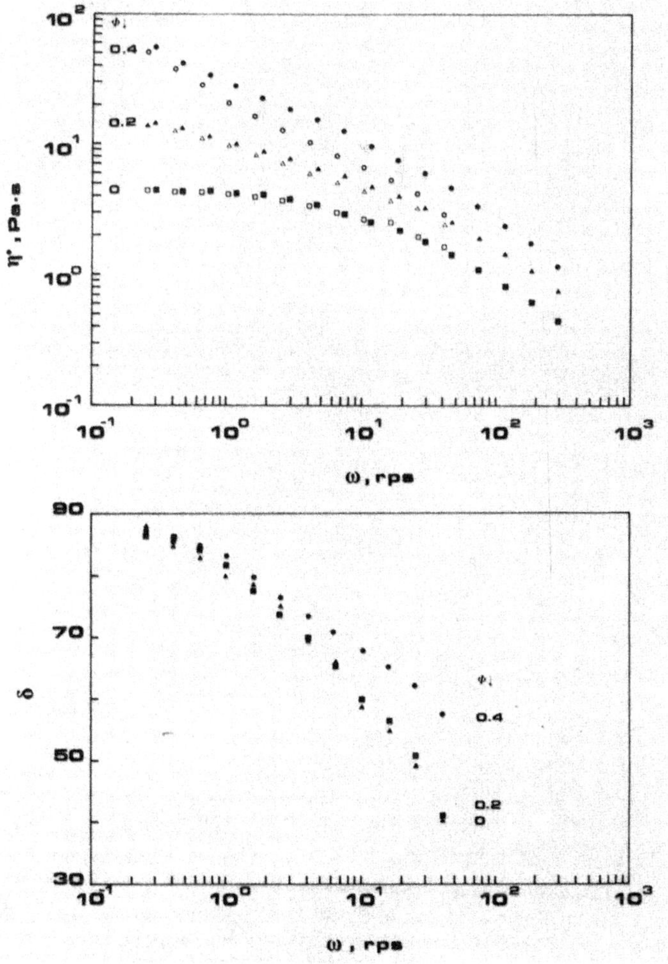

Fig. 4. Complex viscosity η^* and phase lag δ of glass bead suspensions (ϕ = 0.4 and 0.2) in solutions of 4% (w/w) of polyisobutylene in dekaline at various angular frequencies. Full symbols refer to shear viscosity η vs. shear rate $\dot{\gamma}$ plotted on the same ω scale

were not time dependent and reversible, the viscosity would describe on time the part of viscosity curve corresponding to $\dot{\gamma}(t)$, and the viscosity measured according to eq. (2) would be independent on frequency. However this is not our case. We suggest that aggregation and desaggregation phenomena have different rate and that the aggregation rate is much smaller than the desaggregation rate. If it is so, the viscosity we measure is the one which correspond to the highest shear rate $\dot{\gamma} + \dot{\gamma}_0^0$. The detection of a single curve obtained replotting the viscosity data as a function of the total shear rate $\dot{\gamma}_t = \dot{\gamma} + \dot{\gamma}_0^0$ confirm such hypothesis.

Figure 3 shows the viscosity η in steady shear vs. shear rate $\dot{\gamma}$ for suspensions of glass beads at $\phi = 0.2$ and $\phi = 0.4$ in solutions of polyisobutylene in dekaline at 2 and 4% in weight of polymer. The viscosity of the suspensions increases with decreasing shear rate depending on the polymer and filler content; at fixed filler content the viscosity increases with decreasing shear rate and such effect is stronger in the shear rate value corresponding to the constant viscosity zone of the suspending medium and there is a change in the curvature. This behavior was interpreted [4] supposing that the viscosity decreases in the Newtonian region because of the particles desaggregation and in the non-Newtonian region because of an increase of the relaxation time of the macromolecules.

Figure 4 reports the complex viscosity η^* and the phase lag δ, for PB4 suspensions at various filler content, as a function of ω; in the same figure also the viscosity curves η vs. $\dot{\gamma}$ are reported. In the low $\dot{\gamma}$

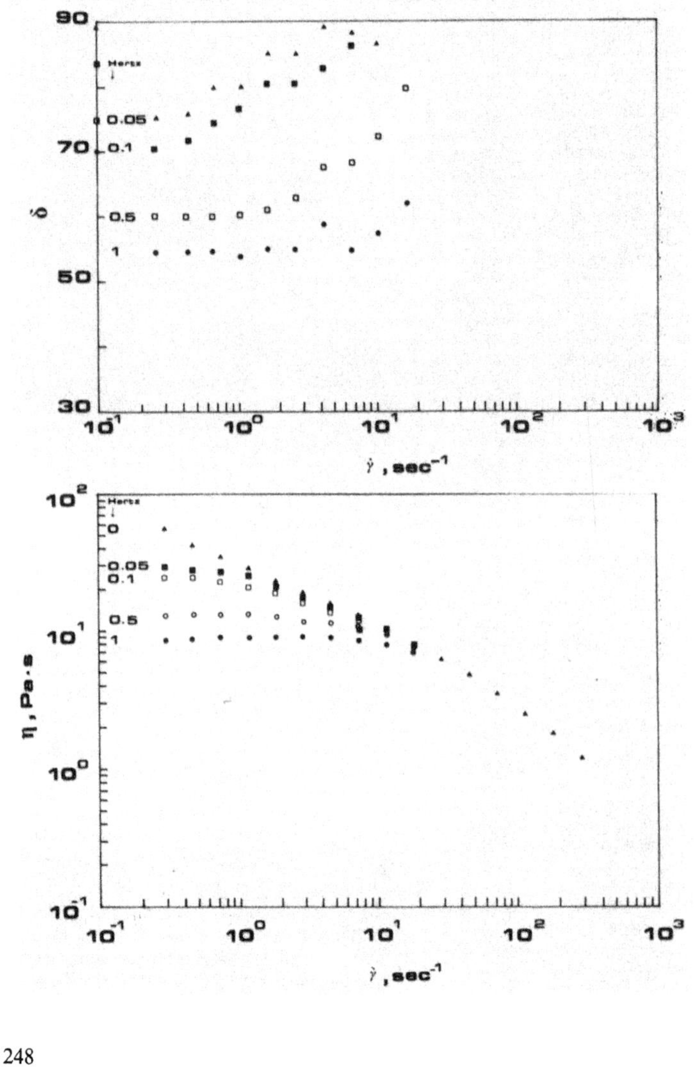

Fig. 5. Mean viscosity and phase lag vs. shear rate of glass bead suspensions ($\phi = 0.4$) in solutions of 4% (w/w) of polyisobutylene in dekaline at various frequencies

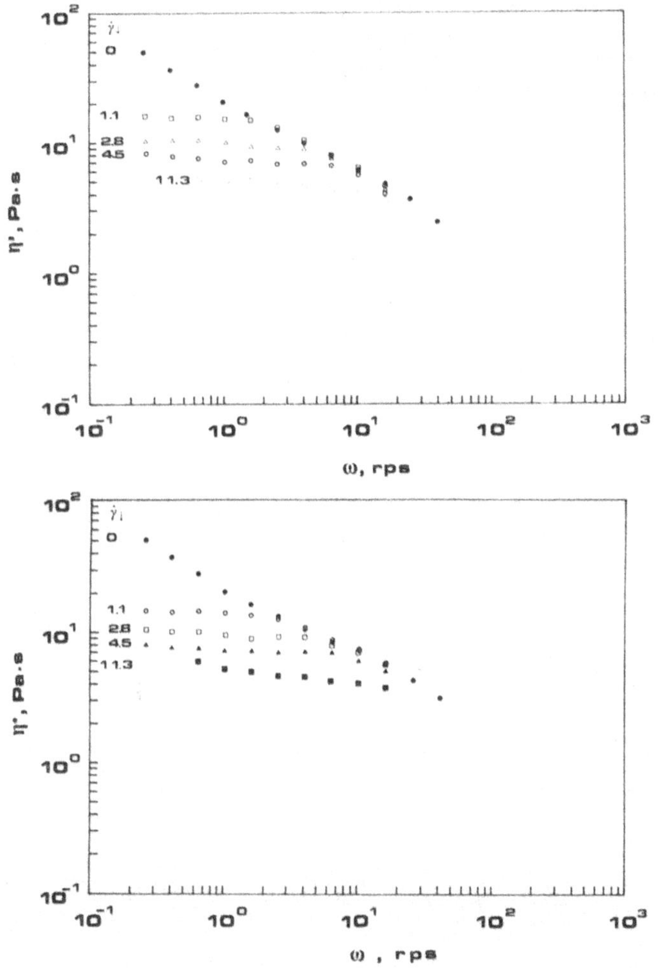

Fig. 6. Complex viscosity η^* and dynamic viscosity η' vs. angular frequencies ω of glass bead suspensions ($\phi = 0.4$) in solutions of 4% (w/w) of polyisobutylene in dekaline at various shear rates

region the fluids response is mainly viscous ($\delta \simeq 90°$), vice versa with increasing $\dot{\gamma}$ the viscoelastic behavior of both suspensions and suspending medium becomes more relevant, this in agreement with the hypothesis of two different mechanism responsables of the viscosity decrement.

Figure 5 shows the effect of the oscillatory perturbation on shear viscosity and figure 6 the effect of steady shear on the dynamic viscosity. When oscillatory motion and stationary shear are simultaneously acting, both dynamic viscosity and steady shear viscosity curves, exhibit a constant viscosity region corresponding to low ω and $\dot{\gamma}$ values respectively. Figure 5 shows that with increasing $\dot{\gamma}$ the viscosity decreases and approaches the curve of pure steady shear. The departure from newtonian behavior is realized first with decreasing frequency. This behavior suggests that when $\dot{\gamma}_0^0$ is larger than the shear rate of the steady shear flow (low $\dot{\gamma}$ region) the molecules have a constant relaxation time, vice versa if the shear rate of the steady shear is predominant, the relaxation times will change, as the data of phase lag reported in figure 5 indicate.

Acknowledgements

This work was supported by C.N.R., Grant n. 8100674 Progetto finalizatto Chimica fine e secondaria, sottoprogetto materiali polimerici.

References

1. Osaki, K., M. Tamura, M. Kurate, T. Kotaka, J. Phys. Chem. **69**, 4183 (1965).
2. Pipkin, A. C., D. R. Owen, Phys. Fluids **10**, 836 (1967).
3. Powell, R. L., W. H. Schwarz, J. Polym. Sci., Polymer Phys. Ed. **17**, 969 (1979).
4. Nicodemo, L., L. Nicolais, R. F. Landel, Chem. Eng. Sci. **29**, 729 (1974).

(Received April 14, 1982)

Authors' address:

Dr. P. Masi, Prof. Dr. L. Nicodemo, Prof. Dr. L. Nicolais, Mr. G. Taglialatela
Istituto di Principi di Ingegneria Chimica
Facoltà di Ingegneria dell'Università di Napoli
Piazzale Tecchio
I-80125 Napoli

Rheologica Acta Rheol. Acta **21**, 605 – 606 (1982)

Biorheological matching: muco-ciliary interaction and epithelial clearance

A. Silberberg

Polymer Research Department, Weizmann Institute of Science, Rehovot (Israel)

Key words: Muco-ciliary clearance, muco-ciliary interaction, mucus viscoelasticity, biorheological matching

1. Introduction

Respiratory epithelia such as cover the surface of the nose, the sinus, the Eustachian tube, the upper airways of the lung, etc. are designed to protect the underlying tissue and to keep the surface free from air-borne pollutants and from tissue generated particulate waste.

This is done as follows: Some of the cells, which constitute the epithelium, are provided with cilia, i.e. fine, closely spaced, mobile protusions, which execute a flail-like beat, in some or other form of synchrony, across the surface (table 1). The cilia move in a low viscosity, aqueous, periciliary fluid, presumably of Newtonian character. Their tips, however, interact directly with a layer of mucus, which is a gel-like, viscoelastic, aqueous system based on a special mucus glycoprotein as gel substance. The ciliary tips propel this mucus layer in a specified direction over the exposed surface. Mucus and entrapped pollutants and waste are then eliminated together. Mucus is secreted in situ from cells (most often organized in the form of glands) and is supplied locally as the need arises, i.e. to the extent that there are particle loads which must be cleared.

Table 1. Some Characteristics of Muco-Ciliary Epithelia

Cilia	Frequency of beat	: 12 – 20 Hz
	Effective Stroke	: $\frac{1}{4} - \frac{1}{3}$ of cycle
	Duration of effective stroke	: $\sim 10^{-2}$ sec
	Number	: $6 - 11 \times 10^8/cm^2$
	Length	: $6 - 7 \times 10^{-4}$ cm
	Diameter	: $0.16 - 0.2 \times 10^{-4}$ cm
	Tip-to-tip distance	: $0.3 - 0.4 \times 10^{-4}$ cm
	Wave length	: $10 - 12 \times 10^{-4}$ cm
	Tip velocity	: $\sim 3 \times 10^{-2}$ cm/sec
	Force/cilium	: $\sim 3 \times 10^{-7}$ dynes
Periciliary fluid	Thickness	: $5 - 6 \times 10^{-4}$ cm
	Viscosity	: $\sim 10^{-2}$ poise (?)

The details of this mechanism emerged when it was shown [1] that the presence of mucus is essential for transport, i.e. for clearance, to be effected. It was shown to be possible to excise an epithelium, deplete it of its locally stored supply of mucus while studying the transport rate of applied particle loads. When the epithelium was totally depleted of mucus, particles ceased to be moved even though ciliary beat was seen to continue. Only when mucus was resupplied, along with the particle, was transport re-instated. The particle velocity measured after mucus was restored matched that found originally on the excised tissue exactly.

Mucus harvested from other epithelia can also function to restore transport on a depleted epithelium, but the velocity will not necessarily match the original velocity. It could be shown in effect [2] that any biochemically not-incompatible aqueous gel system is able to restore transport. The highest velocities were found when systems in the gelation region were tested. For example an aqueous Guar Gum gel, in its gelation region, gave velocities even higher than those found with indigenous mucus. However, both below the gelation point, in the sol region, i.e. at lower concentrations, and above the gelation concentration, i.e. with stiffer gels, the transport velocity is less, showing that a biorheological requirement exists which has to be matched optimally for good clearance. It has been proposed that this optimum arises because of conflicting requirements: The ciliary tip must both be able to penetrate the mucus layer and then, in a sufficiently dissipationless manner, transfer

765

its force laterally [3]. The rheological properties of mucus in other words must be attuned to achieve biorheological matching.

2. Biorheological matching

The rheological properties of freshly secreted mucus have been characterized in an experiment which tests the system *in situ* on the epithelium using a fine glass rod as probe [4]. The arm holding the probe is mechanized and its exact location and the forces which act upon it can be measured due to the incorporation of force and displacement transducers into the design. The device was shown to give excellent results when tested with high viscosity Newtonian fluids. When mucus layers, as they are freshly secreted onto the epithelium were tested, it was found that the stress generated is a non-linear function of the rate of shear $\dot{\gamma}$. At low rates of shear (less than 1 sec^{-1}) the stress is constant and equal to about 10^3 dynes/cm^2. Above 1 sec^{-1}, however, the stress increases linearly with rate of shear along a straight line which (extrapolated to $\dot{\gamma} = 0$) passes through the zero of the coordinate system and corresponds to an effective Newtonian viscosity of about 10^3 poise. We may expect that weaker mucus gels would give a plateau stress less than 10^3 dynes/cm^2, and would possess an effective Newtonian viscosity less than 10^3 poise. Exactly the opposite would happen with stiffer mucus gels.

A cilium which begins to penetrate the mucus layer moves at high velocity normal to the mucus/periciliary fluid interface but at essentially zero relative velocity parallel to this surface. Hence as penetration occurs, the cilium takes advantage of the large plateau modulus at low rates of shear to generate, or rather maintain, the motion of the mucus sheet. The whole event – penetration of the cilium, lateral shift of the tip, extraction of the tip-takes only some 10^{-2} sec (table 1). Mucus will thus react elastically, but if, on a depleted epithelium, the mucus substitute is too weak a gel, it will relax in this time span. The cilium will tend to stay "stuck" and most of the energy will be dissipated instead of being transferred to the mucus as kinetic energy of translation. If the mucus substitute is too stiff a gel, on the other hand, only a minimal depression will be created at the periciliary fluid/mucus interface, there will be only inadequate penetration, the ciliary tip will slip and the force transferred to the mucus will be minimal. In either of these extreme cases the ciliary beat only serves to generate a shear flow pattern in the periciliary fluid layer. This has some component tending to move the mucus but is, by itself, inadequate to the task [5].

Inadequate clearance will also arise if in disease mucus rheology is altered. Mucuciliary epithelia thus represent a very clear example of the importance of proper biorheological matching between the anatomical and structural features of certain organs and their physiological function.

References

1. Sadé, J., N. Eliezer, A. Silberberg, A. C. Nevo, Am. Rev. Resp. Disease **102**, 48 – 52 (1970).
2. King, M., A. Gilboa, F. A. Meyer, A. Silberberg, Am. Rev. Resp. Disease **110**, 740 – 745 (1974).
3. Meyer, F. A., A. Silberberg, Biorheology **17**, 163 – 168 (1980).
4. Gilboa, A., A. Silberberg, Biorheology **13**, 59 – 65 (1976).
5. Yates, G. T., T. Y. Wu, R. E. Johnson, A. T. W. Cheung, C. L. Frand, Biorheology **17**, 151 – 162 (1980).

(Received April 16, 1982)

Author's address:
Prof. A. Silberberg
Polymer Research Department
Weizmann Institute of Science
Rehovot 76100 (Israel)

Rheologica Acta Rheol. Acta **21**, 607 – 610 (1982)

The rheological properties of the arterial wall

K. Pascale, H. W. Weizsäcker, A. Gauster, and G. Dohr

Physiologisches Institut (T. Kenner) und Institut für Histologie und Embryologie (W. Burkl) der Universität Graz

Key words: The rheological properties of rat arteries were studied in vitro. Using intact arterial segments undergoing large deformations in longitudinal and circumferential direction, both the stress-strain relationship and the corresponding histological changes with deformation are presented. Correlations between the mechanical properties and the histological structure are discussed.

Key words: Rat carotid, stress-strain relation, arterial wall structure

1. Introduction

The mechanical properties of blood vessels are an important determinant of hemodynamics. Large arteries, as chief site of vascular impedance, determine the pulsatile load on the heart. They fulfill a "windkessel"-function in smoothing out the pulsatile flow stemming from the heart in a continuous flow through the capillaries. Because of its mechanical function and because of its proneness to disease, the arterial wall is object of intensive research.

From a rheological point of view the vessel wall is highly deformable, inhomogeneous, multiphase, anisotropic, non-linear, and, to a certain extent, viscoelastic [3].

Arterial tissue is a highly organized composite material. It contains mainly elastin, collagen and smooth muscle, that are arranged in concentric layers with fibers interconnecting them [3]. The majority of these structures certainly have a mechanical function. With their position, orientation and interaction they account for the passive and active performance of the vessel under physiological loading. Thus knowledge of the relation between the overall-properties of the artery and the architecture of its tissue is important for a deeper understanding of vascular function both under normal and pathologic conditions. It is furthermore useful for a proper mathematical description of the mechanical behavior in form of constitutive equations.

In order to find out the characteristic properties of arteries under physiological loading and to correlate them to their structure, both experiments in vivo and in vitro were carried out.

2. Methods

The investigations were performed on 9 segments of rat carotid arteries under passive smooth muscle conditions. The experimental protocol consisted of three parts.

First, during dissection, length, outer diameter and pressure of the intact segment were recorded in situ, using a pressure transducer (P23Db, Statham) and a movie camera (Beaulieu R16).

After excision, the stress-free dimensions of the segment were measured and its mechanical properties tested under physiological loading conditions: The vessel was cannulated at one end, inserted in a tensile testing machine (Vibrodyn, Paar), stretched to various levels of longitudinal strain and inflated, at each level, from pressure 0 to $26.66 \cdot 10^4$ Pa. During deformation both the longitudinal force F and the intraluminal pressure p were measured by transducers while changes in length l and diameter d were filmed with the above mentioned camera (fig. 1a). A detailed description of the experimental set-up and the method used is given in [5] and [7].

After completion of the mechanical test, the specimen was inserted, between two glass holders, in the apparatus for fixation shown in figure 1b.

For control purposes, first a well defined state of deformation was restored in physiological solution

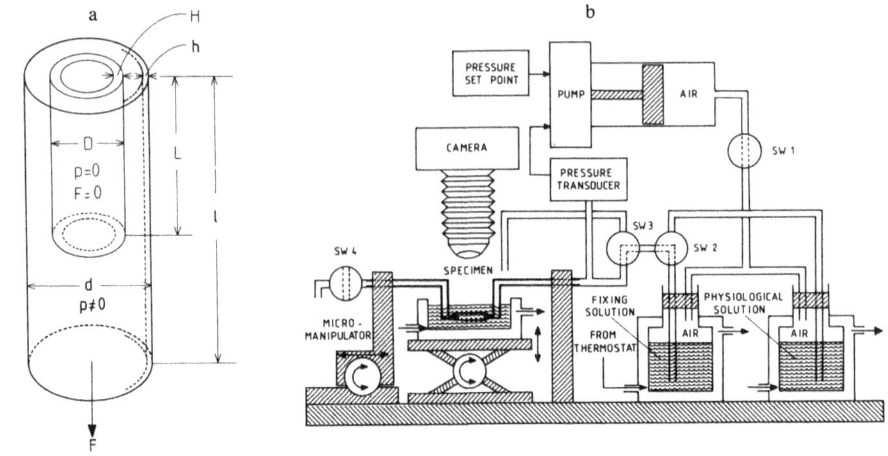

Fig. 1. Sketch of an artery segment (a) and of the apparatus used for fixation (b)

(Ca^{++}-free + EGTA) using a micromanipulator and a pump (lower left and upper right in figure 1b, respectively). Then, by turning the hydraulic switches SW2, SW3 and SW4 to the position shown in the figure, the physiological solution both inside and outside the vessel was replaced by fixative. During fixation the intraluminal pressure was held constant by means of the feed-back loop formed by the pressure transducer and the Harvard pump (Harvard Apparatus: Infusion/Withdrawal Pump Model 997 and Servo Control Amplifier Model 990). After fixation for 1 h in glutaraldehyde the specimen was rinsed, dehydrated and embedded in EPON 812. Then, using an ultramicrotome, the carotid segment was cut in 1 μ thin longitudinal, transverse and tangential slices.

These sections were stained using the method of Humphrey and Pittman [1] and Unna [4].

Corresponding to the 9 vessels investigated, 9 states of deformation were analysed under a microscope. These states of deformation corresponded to 3 different longitudinal strains and 3 different intraluminal pressures, chosen to be smaller, equal or larger then the corresponding in vivo values.

3. Results

Figure 2 shows the typical deformation behavior of carotid segments, when they are stretched longitudinally without torsion and inflated by internal pressure.

In this figure the longitudinal force F (fig. 2a) and the outer diameter d (fig. 2b) are plotted vs. the longitudinal strain ε. This is defined as $\varepsilon = (l/L) - 1$ (see fig. 1a). In both diagrams a and b each single curve corresponds to a different constant internal pressure p.

The most striking feature of figure 2a is that all the force vs. strain curves pass through a common point

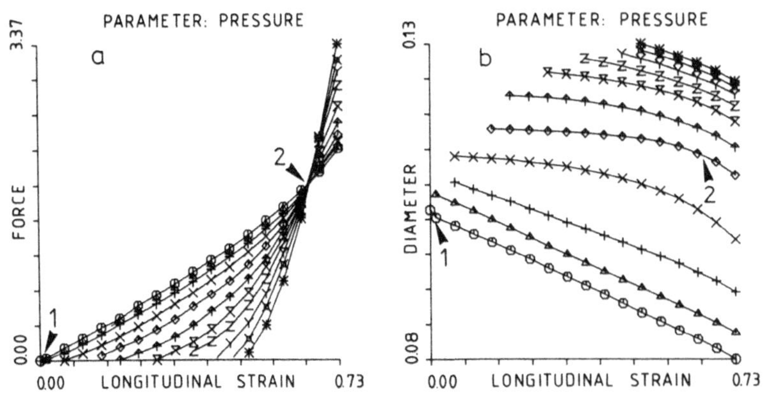

Fig. 2. Deformation of carotid segments (see also fig. 1a): Force (in 10^{-2} N), diameter (in 10^{-2} m), pressure (in 10^4 Pa). ⌽ corresponds to $p = 0$, △ to $p = 26.66 \cdot 10^3$ Pa and $*$ to $26.66 \cdot 10^4$ Pa

[5]. This highly characteristic crossing pattern seems to be common to all large blood vessels [6, 2].

Diagram 2a states that at the strain where the curves intersect each other (and only at this very strain), the longitudinal force does not depend on pressure [5, 2]. Comparison of this strain with the in vivo strain prior to dissection shows that, for each single vessel, the pairs of values are nearly congruent [7, 2]. Diagram 2b suggests that at the in vivo prestretch and near to the average physiological pressure, carotids display a maximum distensibility in circumferential direction. This findings raise the interesting question which structural principle, common to all arteries, generates this characteristic behavior.

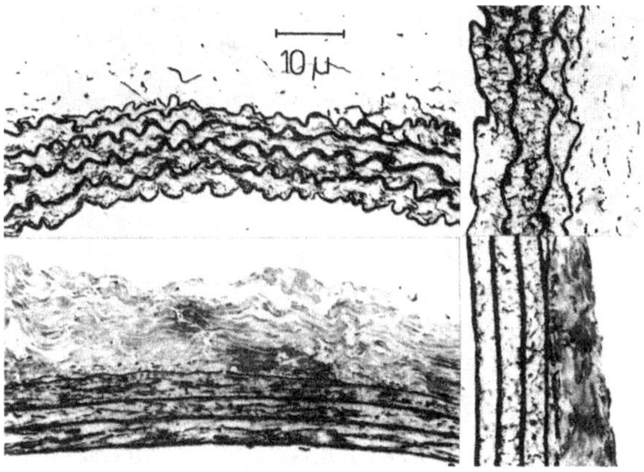

Fig. 3. Cross sections (left) and longitudinal sections (right) of rat carotid artery wall fixed under (a) stress-free and (b) loading conditions

Figure 3 shows photomicrographs of representative sections of carotid walls, fixed under stress-free conditions (a) and under physiological loading conditions (b). The straightening of the elastic lamellae (dark bands) due to physiological deformation is evident. Examination of several sections like in figure 3 indicate that the carotid tunica media appears as a series of 4–5 approximately concentric, thick, fenestrated, fibrillar elastic laminae, connected by an intricate network of elastin fibrils with interspersed collagen fibers and smooth muscle cells. The tunica intima consists of endothelial cells whereas the adventitia is mainly built by bundles of collagen fibers [3]. In the stress free state of deformation the elastic laminae are folded and the interlamellar fibers seem to be randomly oriented. On distension, however, a regular structure becomes more evident: All elastic laminae throughout the wall straighten simultaneously, the unfolding transition being rather abrupt and occurring around the average physiological state of deformation ($\varepsilon = 0.7$, $p = 10.67 \cdot 10^3$ Pa: Figure 2: 1 and 2). The fibers in the interlamellar space become more oriented, running obliquely between adiacent thick lamellae, to form a characteristic herringbone pattern. Adventitial collagen while unfolding somewhat, does not seem to completely straighten even at pressure of $26.66 \cdot 10^4$ Pa.

4. Discussion

The results presented in this paper show that there is a characteristic deformation pattern common to all arteries (and not displayed, e.g., by rubber tube) which is highly correlated with the conditions of loading that occur in vivo. That is, at in vivo strain the longitudinal force is independent of intraluminal pressure. Thus the perivascular tissue, while holding the vessel prestretched, is not likely to undergo large shear fluctuations during transition of the pulse wave. Stated more generally, the artery appears to be stably inserted into its surrounding tissue.

While at this stage of histological investigation any statement on the structural mechanism underlaying the above exposed behavior would be highly speculative, the results gained confirm the practicability of this approach to histomechanics. The findings are in good agreement with literature [8], the fixation procedure proved to be highly reproducible, and several structural observation (such as waviness of the laminae and pitch of the fibers) are enough marked to be put on a quantitative basis. Worthwhile of further investigation seems the evident herringbone pattern displayed by the interlamellar structures. This suggests a helicoidal arrangement of the fibers and possibly a mechanism of successive recruitment of elastin and collagen. A simple structural model consisting of such helices and displaying the basic features of figure 2 has already been proposed [5]. This model suggests, that taking into account the characteristic structures of a biotissue can be useful in reducing the constitutive parameters necessary to describe its mechanical behavior.

References

1. Humphrey, C. D., F. E. Pittman, Stain Technol. **49**, 9–14 (1974).
2. Loon, van P., W. Klip, E. L. Bradley, Biorheology **14**, 181–201 (1977).

3. Patel, D. J., R. N. Vaishnav, Basic Hemodynamics and its Role in Disease Processes. University Park Press (Baltimore 1980).
4. Unna, P. G., Mon. prakt. Dermat. **12**, 394–396 (1891).
5. Weizsäcker, H. W., K. Pascale, Basic Res. Cardiol. **72**, 619–627 (1977).
6. Weizsäcker, H. W., K. Pascale, Pflügers Archiv **391**, R10 (1981).
7. Weizsäcker, H. W., K. Pascale, Cardiovascular System Dynamics. Models and Measurements. (Edited by T. Kenner et al.) pp. 347–362. Plenum Press (New York and London 1982).
8. Wolinsky, H., S. Glagov, Circulation Research **14**, 400–413 (1963).

(Received April 24, 1982)

Authors' addresses:

Dr. K. Pascale, Dr. H. W. Weizsäcker
Physiologisches Institut

A. Gauster, Dr. G. Dohr
Institut für Histologie und Embryologie

Universität Graz
A-8010 Graz

Rheologica Acta Rheol. Acta **21**, 611 – 613 (1982)

Viscoelastic parameters of blood in patients with myocardial infarction compared to normal donors

I. Anadere, H. Chmiel, and P. Heimburg

Institut für Grenzflächen- und Bioverfahrenstechnik der Fraunhofer Gesellschaft
und Zentrum für Innere Medizin, Robert-Bosch-Krankenhaus, Stuttgart

Abstract: The viscoelastic measurements of whole blood in regard to blood and plasma is a sensitive method for the acquisition of pathological changes in flexibility and aggregation tendency of erythrocytes. This could be demonstrated at the example of patients with myocardial infarction.

Key words: Blood viscoelasticity, hematocrit, myocardial infarction

1. Introduction

Clinical investigation of blood was restricted until recently either to the investigation of corpuscles with filtration methods or to blood measurements under steady flow conditions. The filtration method alone cannot discriminate between the effects of altered flexibility or altered aggregation tendency of the erythrocytes. This disadvantage in fact is circumvented when steady-state viscosity measurements are performed with Couette devices at very low and at high shear rates [1]. Figure 1 shows the comparison of these measurements between healthy controls and patients with cardiac infarction measured at a shear rate of 0.1 s^{-1}. The difference between the two groups is strongly marked. However, due to the well know surface effects on the boundary layer between blood – air or blood plasma – air, the reproducibility of these measurements is unsatisfactory. Moreover, in the native human circulation, blood is never put under such steady flow conditions. For this reason, we are using a rheometrical method which considers the pulsating flow of blood [2]. In this case the viscoelastic properties of blood will be of more importance and they must be taken into consideration with the technique of the apparatus.

The observation of blood from patients with cardiovascular disorders with a light microscope shows that these have a marked increase of erythrocyte aggregation tendency. If such an aggregate is put in

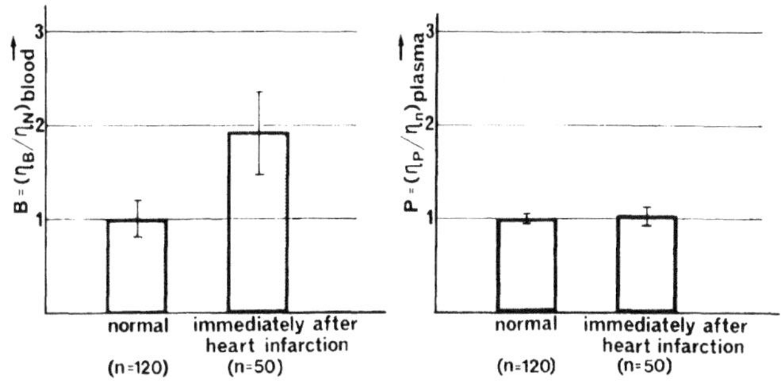

Fig. 1. Comparison of blood and plasma viscosities, shortly after heart infarction, with corresponding normal values measured with a Couette device at a shear rate of 0.1 s^{-1}

oscillating motion under very low shear rates, it will be deformed purely elastically. At higher shear rates, it will be partly elastically deformed and partly sheared. These latter two effects are analysed separately with our method. For this purpose the linear viscoelastic theory was applied for blood at low shear rates and frequencies, which assumes a complex coefficient of viscosity

$$\eta^* = \eta' - i\eta''$$

that consists of the viscous η' and elastic η'' components [2]. At very high oscillating flow no aggregates exist, then one can measure the flexibility of the erythrocytes. In addition to the determination of plasma viscosity and hematocrit, i.e. the volume concentration of blood corpuscles, all parameters that influence blood rheology are included in this program.

2. Material and methods

The following parameters were measured: a) hematocrit, b) the viscoelasticity of whole blood, c) plasma viscosity, d) whole blood and plasma density.

Blood for establishing normal values was obtained from healthy donors from the blood bank. The normal donor group was comprised of both sexes. The pathological group consisted of 11 in-patients (9 men and 2 women) with fresh myocardial infarct from Robert Bosch Hospital, Stuttgart. In the preparation of whole blood, 5 ml of blood were anticoagulated with edetic acid (EDTA-K). 2.5 ml were retained for the measurement of viscoelasticity, density and hematocrit of whole blood. Plasma was separated from the rest after 15 min centrifugation at 2000 xg. Hematocrit values were obtained using a micro-hematocrit centrifuge with double determinations after 15 min centrifugation. Determination of whole blood and plasma density was performed with a precision densitometer (Paar DMA 55) with a resolution of 10^{-5} g/cm^3. For the determination of the viscous and elastic components of complex viscosity an oscillating capillary viscometer was used [2]. The frequency of the oscillation applied was 2 Hz which is near to the in vivo pulsation of vessels. The results of several viscoelasticity measurements showed that the reproducibility of the instrument was better than 3% for consecutive measurement of blood samples.

In this study, a single shear rate of $\dot{\gamma} = 10\ s^{-1}$, which is near to that in the arterioles, was used for the comparative evaluation of the data. All measurements were performed at 23 °C and the results were analysed statistically using Student's *t*-test.

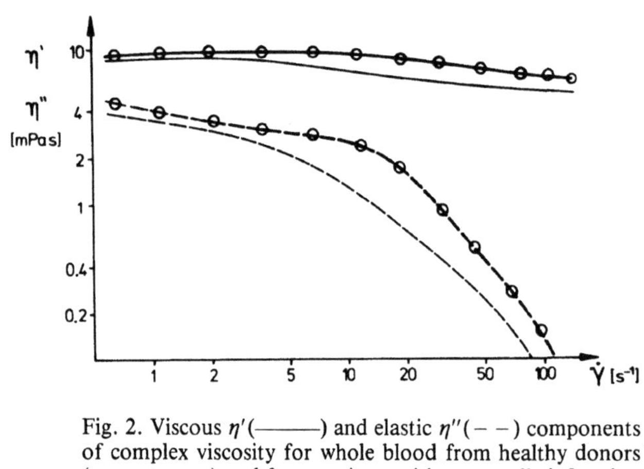

Fig. 2. Viscous η'(——) and elastic η''(– –) components of complex viscosity for whole blood from healthy donors (——, – –) and from patients with myocardic infarction (—⊖—, – ⊖ –) (HK_z = 44%, F = 2 Hz, T = 23 °C)

3. Results

Figure 2 shows typical plots of the viscous and elastic components of viscoelasticity as a function of shear rate for a normal donor and for an infarct patient, both with a hematocrit of 44%. The flow curves for both viscous and elastic components of complex viscosity lie evidently higher than those for the normal donor. One of the major factors affecting blood viscosity is the hematocrit. The tendency of read blood cells (RBC) to form aggregates can lead to a falsification of the hematocrit experimentally determined using the micro-hematocrit centrifuge. The closest packing of the cells is not achieved because of plasma being caught between the aggregates. A more precise method of hematocrit determination is the

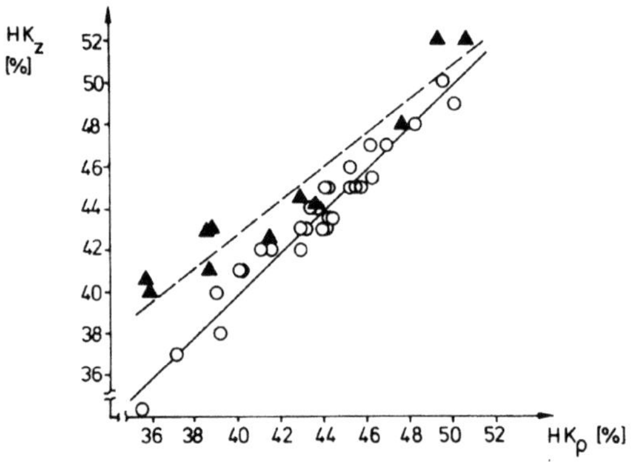

Fig. 3. Hematocrit experimentally determined (HK_z) vs. calculated by means of density measurements (HK_ρ) for healthy donors (0) and for patients with myocardiac infarction (△)

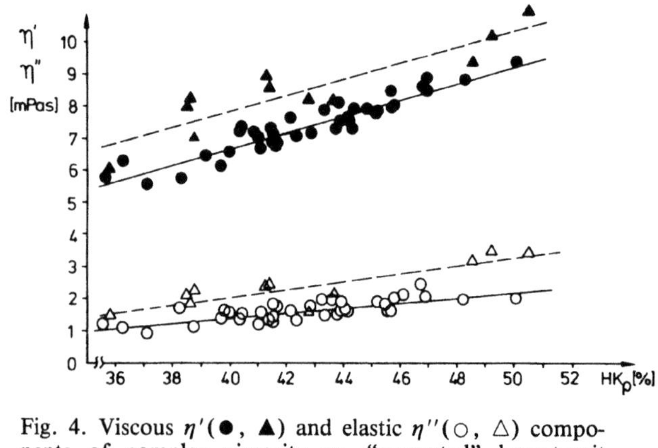

Fig. 4. Viscous η'(●, ▲) and elastic η''(○, △) components of complex viscosity vs. "corrected" hematocrit (HK_ρ) for healthy donors (●, ○) and for patients with myocardiac infarction (▲, △) (F = 2 Hz, T = 23 °C, $\dot{\gamma}$ = 10 s^{-1})

determination of whole blood and plasma density in order to calculate an "aggregation independent" hematocrit which we call HK_p [3] figure 3. Shows the relationship between this aggregation independent hematocrit and hematocrit determined with the microcentrifuge. For the healthy group the relationship is linear, but the values for the group of patients with cardiac infarction diverge strongly from the normal curve. The relationship of both viscous and elastic components of viscoelasticity as a function of "corrected" hematocrit (HK_p) for both healthy donors and patients with cardiac infarction, is demonstrated an figure 4. Both viscoelastic parameters lie for the patients significantly higher than the normal donor group.

To compare two groups of donors with different mean hematocrit values is not relevant. A possibility to circumvent this problem is the calculation of the viscoelasticity data from a standard hematocrit of 45% which is near to the mean hematocrit value of male healthy donors. For this purpose the correlation between hematocrit and viscoelastic parameters must be known for al large random population. As shown figure 5, within a certain range of hematocrit (about 35 – 52%) this correlation can be assumed as linear. Hematocrit values from healthy donors and patients lay within this range in this study.

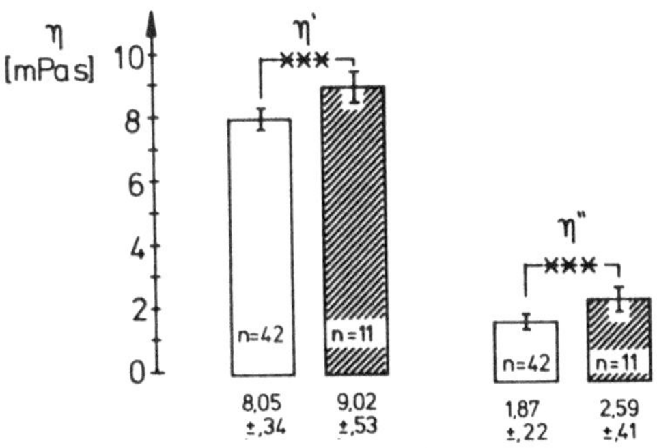

Fig. 5. Mean viscoelasticity data for healthy donors (□) and patients with cardiac infarction (▨), calculated from a corrected standard hematocrit of 45% (HK_{p45}) (F = 2 Hz, T = 23 °C, $\dot{\gamma}$ = 10 s^{-1})

Figure 5 shows the transformed viscoelasticity data of healthy donors and patients with cardiac infarction calculated from a corrected standard hematocrit of 45%. The mean viscous component of complex viscosity is 12%, and the elastic component 39% higher in patients with cardiac infarction compared to normal donors. These results confirm that the patients with myocardial infarction have an increased tendency of erythrocyte aggregation. The correlation with the laboratory findings is very high. The patients with the highest viscoelastic parameters showed also the highest values of creatine phosphokinase, the most specific enzyme available for the determination of myocardial tissue necrosis. Thus the determination of the viscoelastic parameters can be a further diagnosis aid and expecially can allow possible indications for the application of drugs and their effects during the therapy.

References

1. Chmiel, H., E. Walitza, On the Rheology of Blood and Synovial Fluids, Research Studies Press (Chichester 1980).
2. Thurston, G. B., Biophys. J. **12**, 1205 (1975).
3. Anadere, I., H. Chmiel, H. Hess, G. B. Thurston, Biorheol. **16**, 171 (1979).
4. Chmiel, H., I. Anadere, P. Heimburg, La Ricerca Clin. Lab. **11** (Supl. 1) 189 (1981).

(Received June 28, 1982)

Authors' addresses:

Dr. I. Anadere, Prof. Dr.-Ing. H. Chmiel
Fraunhofer-Institut IGB
Institutszentrum Stuttgart
Nobelstraße 12
D-7000 Stuttgart 80

Prof. Dr. P. Heimburg
Zentrum für Innere Medizin
Robert-Bosch-Krankenhaus
Auerbachstraße 110
D-7000 Stuttgart 50

Preliminary experiments for the preparation of hyaluronic acid from synovial fluid without rheological changes

S. Altmann and H. Zeidler

Abteilung für Krankheiten der Bewegungsorgane und des Stoffwechsels (Leiter: Prof. Dr. med. F. Hartmann), Zentrum Innere Medizin und Dermatologie der Medizinischen Hochschule Hannover

Abstract: Hyaluronic acid (HA) isolated from synovial fluid by CsCl density-gradient ultracentrifugation show a decrease in the viscoelasticity. Metrizamide, a substance used recently for non-ionic density-gradient ultracentrifugation, do not change the rheology of synovial fluid and HA. Thus it may be possible to isolate HA without altering rheological properties.

Key words: Synovial fluid, viscoelasticity, hyaluronic acid, isopycnic ultracentrifugation, metrizamide

1. Introduction

Hyaluronic acid (HA) is generally thought to be the polymer responsible for the rheological properties of the normal and pathological synovial fluid [7, 10, 11, 16, 21–23]. Chemically it belongs to the glycosaminoglycans and is a linear unbranched polymer of D-glucuronic acid and D-glucosamine which are linked to one another by glycoside bonds. In addition the synovial fluid contains a large number of dissolved organic and inorganic substances as well as cellular and solid elements [1]. Synovial fluid is believed to be a dialysate of the blood serum, with an electrolyte content and other lower molecular weight substances more or less corresponding to that of the serum. Besides this there are other special products formed by the synovial membrane, namely hyaluronic acid and chondroitin sulfate.

The question arises as to whether the HA alone is responsible for the viscoelastic properties of the normal and pathological synovial fluid or whether protein also participates through complex formation. An answer to this question is only possible by isolating pure HA from the synovial fluid and comparing the rheological values measured with those of the natural synovial fluid.

Previous investigations by other authors on HA, which was extracted by ultrafiltration, suggest a complex formation with proteins [4, 5, 11, 12, 17] whereas Balazs and co-workers [2] using HA isolated by electrodeposition could find no decrease in viscosity when compared with the natural synovial fluid.

2. Experiments and results

Because caesium-chloride ultracentrifugation was reported to be the best method for isolating HA at the end of the sixties and in the early seventies [18–20], we decided to use this method for the separation of synovial fluid as described earlier [1]. The molecular weight and intrinsic viscosity values, the flow curves and normal forces of the isolated HA were markedly lower at comparable concentrations than the original synovial fluid (fig. 1). These findings let us first conclude that other macromolecules besides the HA in synovial fluid are responsible for the viscoelasticity and network structure [1].

Provided that the HA isolation does not involve an alteration of the tertiary and quaternary conformations and that the reduction in viscoelasticity is caused only by the prevention of the complex formation with proteins, then a complete reproduction of the rheological behavior should be expected in recombination experiments.

However, corresponding investigations with normal and pathological synovial fluid samples ($n = 7$)

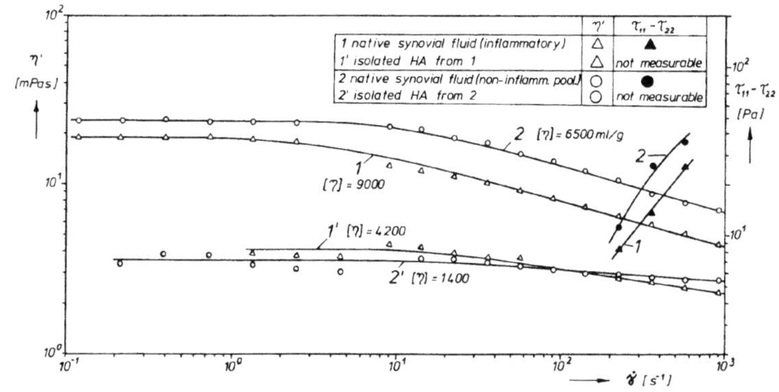

Fig. 1. Flow curves and normal forces of native human pathological synovial fluid and isolated hyaluronic acid ($\vartheta = 37\,°C$)

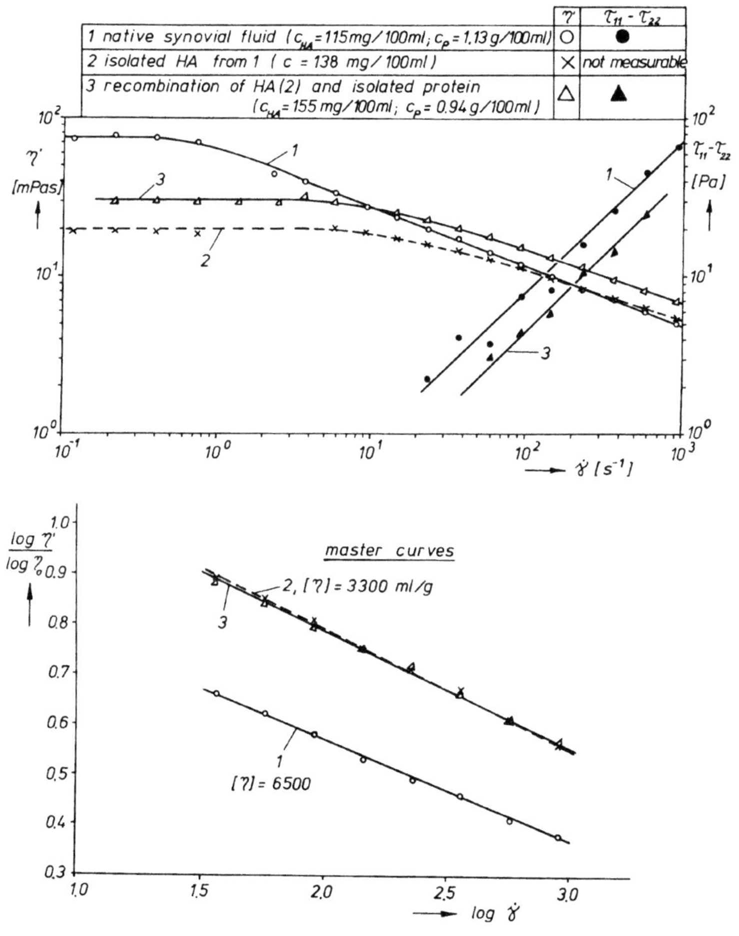

Fig. 2. Flow curves, normal forces and master-curves of native ox synovial fluid, isolated hyaluronic acid and recombination ($\vartheta = 37\,°C$)

did not bring about a restoration of the flow curves and normal forces of the original sample, which is shown in figure 2 for healthy ox synovial fluid, when the separated protein was added to isolated HA. The influence of the HA-concentration differences on the position of the flow curves can be largely eliminated using a normalisation procedure by plotting $\log \eta'/\log \eta_0$ versus $\log \dot{\gamma}$ [24], so that separate master-curves are obtained at different molecular weights and molecular weights distributions. The diagram shows how this appears for the case of isolated HA and after recombination as compared with the native ox synovial fluid and it is emphasized by the markedly different intrinsic viscosity values.

Furthermore, we examined how far the flow curves of the synovial fluids were altered by the caesium-chloride. For this we used a pathological non-inflammatory synovial fluid sample and dialysed it against 0.01 M phosphate buffer, pH 7.3. A portion of this was adjusted to a density of $\rho = 1.63$ g/ml by the addition of caesium chloride and after a reaction time of about 2 hours dialysed against the same phosphate buffer to remove the caesium chloride.

A comparison of these two samples showed that an irreversible alteration in the flow curves and also in the normal forces was caused by the caesium chloride. The different positions of the master curves suggest a change in conformation and/or alteration of the distribution of molecular weights (fig. 3).

For this reason the caesium chloride-density-gradient ultracentrifugation is probably ruled out as an isolation method because of the irreversible changes in the viscoelasticity that take place.

In searching for another method, which does not degrade HA during isolation, we came across the isopycnic ultracentrifugation with metrizamide. This cyclic iodine compound had the advantage of not being ionic. The separation method using metrizamide gradients is relatively new and was first described by a number of authors in 1973 [8, 9, 13–15].

In order to examine the influence of metrizamide as compared with caesium chloride and sodium chloride on the flow curves, a non-inflammatory synovial fluid sample was used. It was mixed with the above mentioned substances in equal portions and at different densities. All the samples shown in figure 4 were finally dialysed with a phosphate buffer in order to obtain the same pH values. Whereas the sample treated with metrizamide show no substantial decrease in the flow curve as compared with the native synovial fluid sample, the caesium chloride and sodium chloride samples demonstrate the same markedly lower positions. Besides this the resulting master curves for native synovial fluid and that with

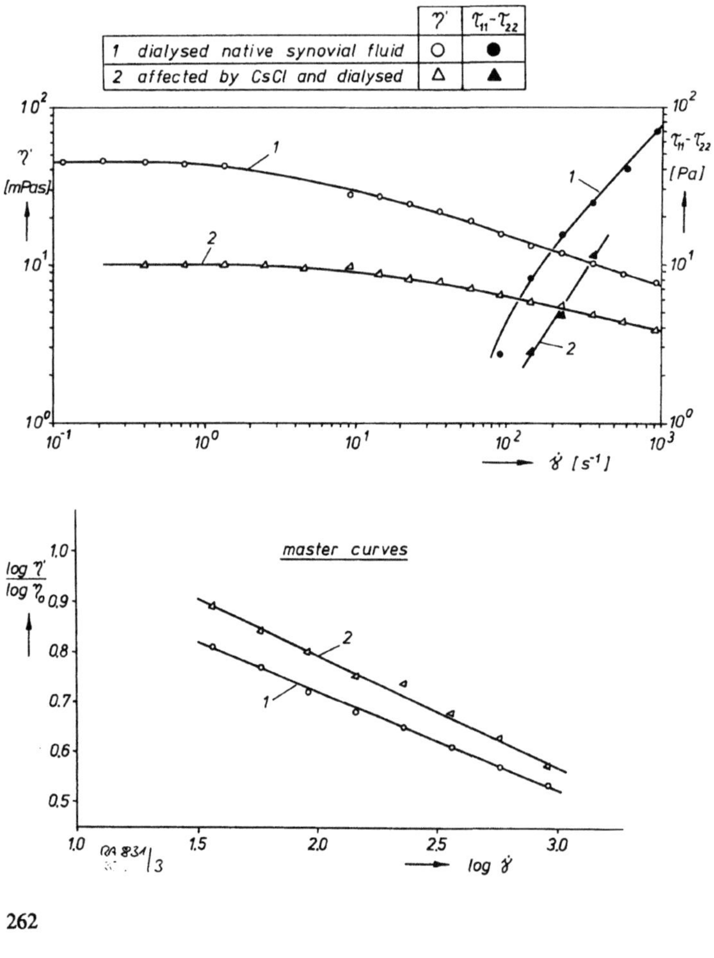

Fig. 3. Influence of CsCl on flow curve, normal forces and master-curve of a human pathological synovial fluid sample ($\vartheta = 37\,°C$)

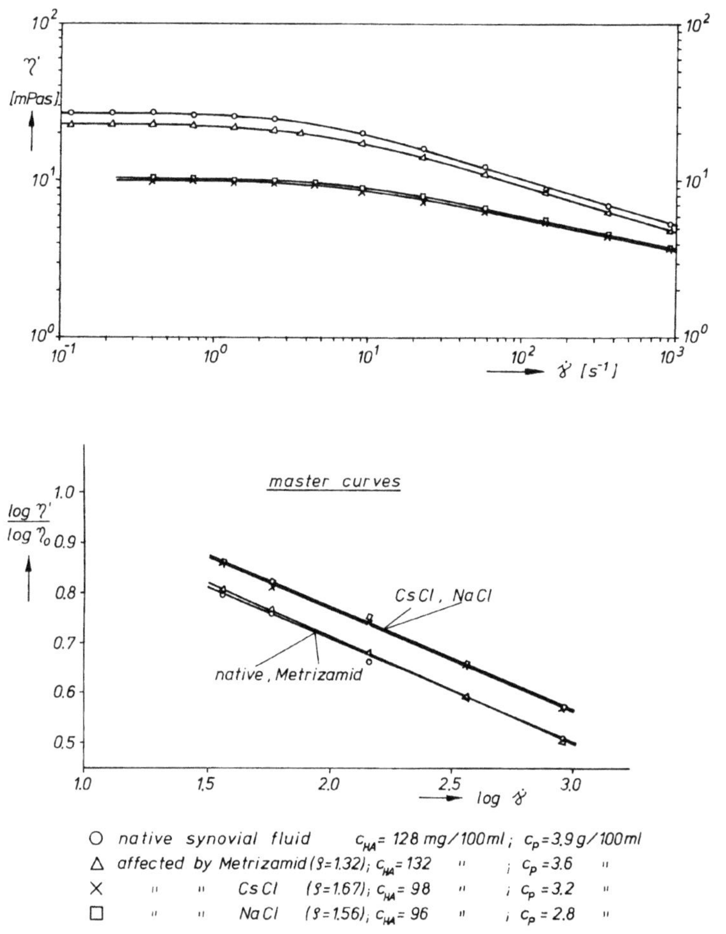

Fig. 4. Influence of Metrizamide, CsCl and NaCl on flow curves and master-curves of a human synovial fluid sample (non-inflammatory) (ϑ = 37 °C)

metrizamide were identical and also the samples treated with caesium chloride and sodium chloride were identical but different from the others. A similar picture was observed when an inflammatory sample was used and treated in the method previously described.

Thus metrizamide appears to offer a gradient material which is suitable for a mild separation of HA because of its non-ionic character.

In order to test whether HA could be separated from the proteins an inflammatory synovial fluid was employed. It was dialysed for 24 h against a 0.01 M phosphate buffer, pH 7.3 at 4 °C and centrifuged at 3500 rpm for 10 minutes to remove the fibrin coagulate. The undiluted, dialysed synovial fluid was then regulated to a density of 1.22 g/ml by the addition of metrizamide and centrifuged for about 70 hours at 4 °C in a fixed angle rotor, model Ti 70.1 of a preparation ultracentrifuge (manufactured by Beckmann) at 40000 rpm ≙ 100000 g. The total volume of 28 ml was divided between two centrifuge tubes. Finally the contents of the tube was divided into eight fractions, the density was determined refractometrically ($\rho_{5^\circ} = 3.453\, \eta_{20^\circ} - 3.601$) as proposed by Birnie and Rickwood [3] (η_{20° is the refractive index at 20 °C), and in each case 0.1 ml was taken for the HA determination by the uronic acid method of Bitter and Muir. The remainder of each fraction was dialysed for 70 h at 4 °C against 0.01 M phosphate buffer, pH 7.3, to remove metrizamide. A protein determination of each fraction followed using the Biuret method of Mehl.

This first isopycnic ultracentrifugation attempt with metrizamide did indeed show a well formed density gradient (fig. 5), but an optimum separation of HA from proteins did not occur because the proteins were distributed over the entire gradients because of interactions with metrizamide. Hüttermann and Wendlberger [6] amongst others described this phenomenon in 1976. They used heavy water (D_2O) as the solvent instead of light water (H_2O) and they were thus able to avoid the reversible formation of metrizamide-protein complexes.

We utilized this information in a second centrifugation experiment in which we used the same procedure

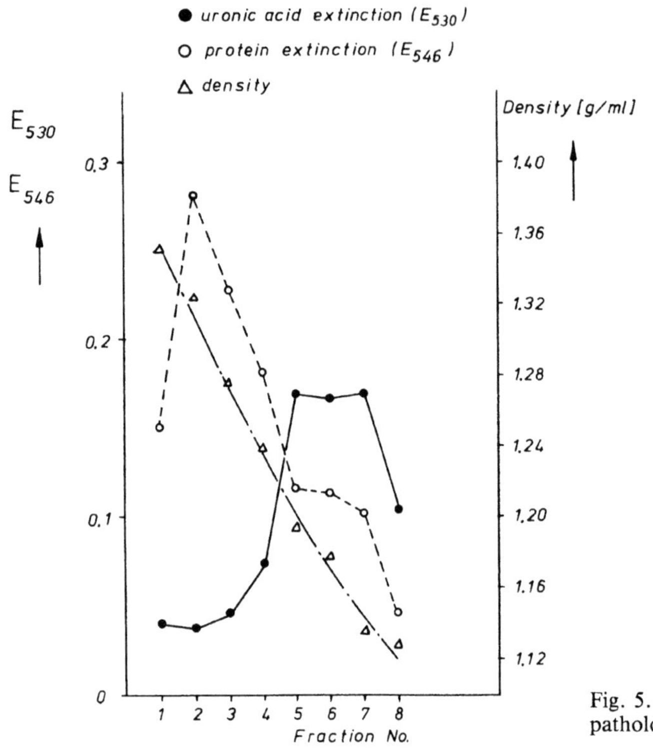

Fig. 5. Metrizamide density-gradient ultracentrifugation of pathological human synovial fluid

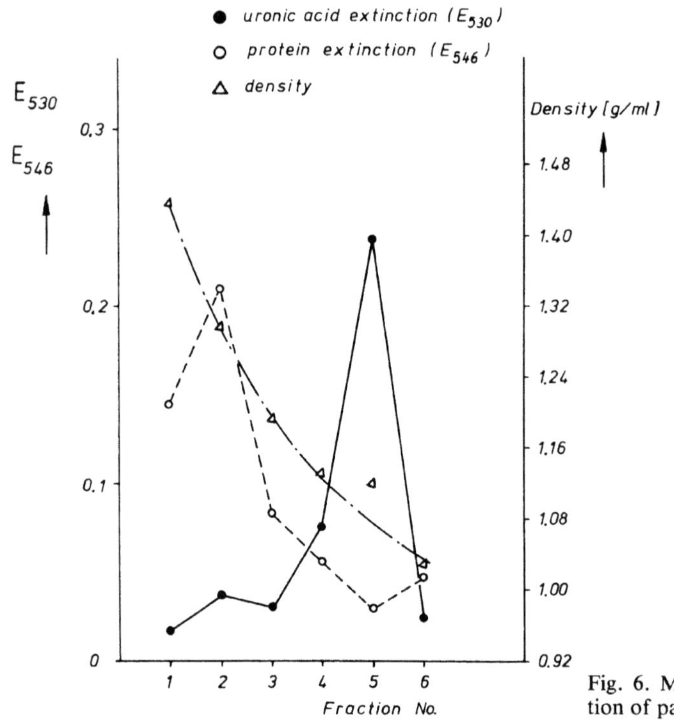

Fig. 6. Metrizamide/D_2O density-gradient ultracentrifugation of pathological human synovial fluid

as for the first, however, we diluted the synovial fluid sample with heavy water in the ratio 4:1.

This method led to a significantly better separation of macromolecules. Hyaluronic acid and proteins appear in two distinct peaks in the diagram (fig. 6), the effect of dilution has probably being an important reason for this.

3. Conclusions

On the basis of this investigation it could be shown that CsCl causes irreversible changes in the viscoelasticity of synovial fluid and hyaluronic acid-protein complex. With metrizamide as a gradient medium it appears possible to use an isopycnic ultracentrifugation for mild isolation of hyaluronic acid from synovial fluids without altering their rheological properties.

Further investigations are necessary to achieve the best separation of the HA. Recombination experiments will reveal the extent to which hyaluronate-protein complexes are responsible for the viscoelasticity of normal and pathological synovial fluids.

Acknowledgement

The authors wish to express their thanks to Deutsche Forschungsgemeinschaft, SFB 54, for supporting this work.

References

1. Altmann, S., H. Zeidler, R. Gaffga, W. M. Kulicke, Rheol. Acta **19**, 642 (1980).
2. Balazs, E. A., L. Sundblad, Acta Soc. med. Upsaliensis **64**, 137 (1959).
3. Birnie, G. D., D. Rickwood, in: D. Rickwood (ed.), Biological Separations in Iodinated Density-Gradient Media, p. 193, Information Retrieval Ltd. (London 1976).
4. Blumberg, B. S., A. G. Ogston, Biochem. J. **66**, 342 (1957).
5. How, M. J., V. J. W. Long, D. R. Stanworth, Biochim. Biophys. Acta **194**, 81 (1969).
6. Hüttermann, A., G. Wendlberger, in: D. Rickwood (ed.), Biological Separations in Iodinated Density-Gradient Media, p. 25, Information Retrieval Ltd. (London 1976).
7. Johnston, J. P., Biochem. J. **59**, 626 (1955).
8. Mathias, A. P., C. A. Wynter, FEBS Lett. **33**, 18 (1973).
9. Mullock, B. M., R. H. Hinton, Trans. Biochem. Soc. **1**, 579 (1973).
10. Myers, R. R., S. Negami, R. K. White, Biorheol. **3**, 197 (1966).
11. Ogston, A. G., J. E. Stanier, Biochem. J. **46**, 364 (1950).
12. Preston, B. N., M. Davies, A. G. Ogston, Biochem. J. **96**, 449 (1965).
13. Rickwood, D., A. Hell, G. D. Birnie, FEBS Lett. **33**, 221 (1973).
14. Rickwood, D., G. D. Birnie, FEBS Lett. **50**, 102 (1975).
15. Rickwood, D., in: G. D. Birnie, D. Rickwood (eds.), Centrifugal Separations in Molecular and Cell Biology, p. 219 (London 1978).
16. Ropes, M. W., W. Bauer, Synovial fluid changes in joint diseases, Harward University Press (Cambridge 1953).
17. Sandson, J., D. Hamerman, J. Clin. Invest. **41**, 1817 (1962).
18. Scher, J., D. Hamerman, Biochem. J. **126**, 1073 (1972).
19. Silpananta, P., J. R. Dunstone, A. G. Ogstone, Biochem. J. **104**, 404 (1967).
20. Silpananta, P., J. R. Dunstone, A. G. Ogstone, Biochem. J. **109**, 43 (1968).
21. Sundblad, L., Acta Soc. med. Upsaliensis **58**, 113 (1953).
22. Thurston, G. B., H. Greiling, Rheol. Acta **17**, 433 (1978).
23. Wilkins, J., J. Clin. Invest. **41**, 1817 (1962).
24. Zeidler, H., S. Altmann, B. John, R. Gaffga, W. M. Kulicke, Rheol. Acta **18**, 151 (1979).

(Received June 16, 1982)

Authors' address:

Ing. grad. S. Altmann, Prof. Dr. med. H. Zeidler
Abteilung für Krankheiten der Bewegungsorgane und des Stoffwechsels des Zentrums Innere Medizin und Dermatologie
Medizinische Hochschule Hannover
Karl-Wiechert-Allee 9
D-3000 Hannover

Viscosity of concentrated protein solutions

J. Lefebvre

Institut National de la Recherche Agronomique (I.N.R.A.), Laboratoire de Biochimie et Technologie des Protéines, Nantes (France)

Abstract: The viscosity of solutions of four proteins (Bovine Serum Albumin, Ovalbumin, α_{s-1} Casein, Lysozyme), brought to the random coil conformation, has been measured over a large concentration range extending into the entanglement region. A master curve is obtained in the dilute and semi-dilute regions with the reduced variables $\tilde{\eta} = \eta_{sp}/[\eta]c$ and $\tilde{c} = c/\gamma$ of Simha and Utracki.

By using Graessley's expression for the polymer coil expansion at a given concentration in the semi-dilute region ($c^* \leqslant c \leqslant c^{**}$), a simple equation is established giving the relative viscosity η_r as a function of concentration c: for $c^* \leqslant c \leqslant c^{**}$, $\ln \eta_r = 2a[\eta]c^*(c/c^*)^{1/2a} - (2a - 1)[\eta]c^*$; where c^* is the incipient overlap concentration, $[\eta]$ the intrinsic viscosity, and a the Mark-Houwink exponent for the polymer-solvent considered.

This equation fits well the experimental results. The adjustment yields for the parameter a values which are comprised between 0.6 and 0.7, as expected, for Bovine Serum Albumin and Ovalbumin, but very close to 0.5 for α_{s-1} Casein and Lysozyme. This can be explained by the fact that the molecular weights of the two latter proteins are lower than, or very close, the critical molecular weight; the critical molecular weight is estimated to be about 20000.

Key words: Viscosity, concentrated solution, protein

1. Introduction

In food media, and generally speaking in biological systems, proteins when present as solutions, are found in most cases at rather high concentrations. Yet, our knowledge on the properties of proteins in concentrated solutions is still very limited. One of the reasons is that the choice of the methods applicable is much more limited than in the case of dilute solutions.

Among the methods which are suitable in order to gain informations on the conformation and interactions of proteins in concentrated media, viscometry has been mostly used. However, the few papers published on this matter [1–6] treat only of proteins (generally globular) in their "native" state (we consider here only the case of well defined dissolved pure proteins). The results are accordingly interpreted on the basis of models derived from that of the hard quasispherical particles dispersion [4, 6].

During food processing, proteins are submitted to conditions which can be well beyond the stability domain of the "native" structure. On another hand, food systems (and in general, biological systems) are often polyphasic; proteins are adsorbed on the interfaces; this process is concomitant with extensive conformational changes and leads to adsorption layers which can be viewed as concentrated solutions of denatured proteins [7, 8]. The properties of concentrated solutions of proteins in their "denatured" states are thus of special interest in food science and technology.

The different kinds of denaturation processes result in protein conformations which are, as a rule, more expanded and less rigid than the native ones. In many cases, they are not very well known, and at all events every protein is singular in this respect. Nevertheless, the classical studies of Tanford and his coworkers [9] have shown that all proteins can be brought to the same fully denatured state, which is the random coil configuration; this state can be reached by using as denaturing agents urea or guanidine hydrochloride (GuCl) at sufficient concentrations, when protein crosslinks (disulfide bonds) are disrupted by reducing agents.

We report here results on the viscosity of solutions of some proteins in the random coil state up to high concentrations and examine them by the light of polymer science.

2. Materials and methods

2.1. Materials

Bovine Serum Albumin (B.S.A) (fraction V), Ovalbumin (grade V) and Lysozyme (from egg-white, grade I) were purchased as lyophilized powders from Sigma Chemical Company. α_{S-1} Casein was a gift from our colleagues of the Dairy research. B.S.A. was defatted with pentane at 4 °C before use.

2.2. Obtention of the proteins in the random coil conformation

B.S.A., Ovalbumin and Lysozyme are submitted successively to reduction and alkylation as described in [11], in order to disrupt disulfide crosslinks and to prevent their subsequent re-formation. α_{S-1}-casein is devoided of disulfide bonds and of SH-groups (table 1). The uncrosslinked polypeptidic chains are dissolved in 6 M GuCl-0.1 M buffer (acetate or phosphate), where they are in the random coil conformation [9]. α_{S-1} casein is in fact already randomly coiled in the native state [9, 10]. Some literature data on the proteins considered, in their native and random coil states, are summarized in table 1.

2.3. Viscosity measurements

A LS-30 (Contraves, Zürich) low-shear coaxial cylinders viscometer was used. For each protein solution, the flow curve was recorded at 25 °C for shear rates from 0 to 128.5 s^{-1}, excepted in the case of the most viscous samples for which the range was 0 – 1.285 s^{-1}.

3. Results and discussion

We have examined the proteins listed above in various conformational states obtained in different solvents (acetate or phosphate buffer, 6 M LiBr, Chloroethanol – HCl, 6 M GuCl) with or without previous reduction – alkylation. Figure 1 gives some examples of the relative viscosity η_r vs. concentration c curves for Lysozyme, showing how they range between that of native protein (quasispherical hard particle) and that of the random coiled one, and pointing out the large effect of "denaturation" on viscosity. In all the experiments we report here, the flow behaviour was newtonian in the shear rate ranges considered.

We are going now to limit the discussion to the case of randomly coiled protein solutions only. Ln η_r – ln c curves (fig. 2) extend into the high concentrations region (linear part of the curves) where it is usually considered that entanglements take place [13]. The concentration c_e at the onset of straight line behaviour is compared in table 2 to its theoretical value c_{cr} (assuming that an uniform segment density is reached) according Cornet [14]. Using the reduced variables of Simha and Utracki [15, 16]: $\tilde{\eta} = \eta_{sp}/[\eta]c$ and $\tilde{c} = c/\gamma$ with the curve of α_{S-1}-Casein as reference ($\gamma = 1$), a master curve (fig. 3) is obtained which fits the experimental points reasonably up to

Table 1. Molecular characteristics of the proteins studied

	Native state				Random coil state[d])	
	$M \cdot 10^{-3}$ [a])	n [b])	$S-S$/mol [c])	$[\eta]$ (ml/g)	n	$[\eta]$ (ml/g)
B.S.A. [9]	69	605	17	3.7	605	52.2
Ovalbumin [12]	43	346	1	3.9	346	31.1
α_{S-1} Casein [10]	23.6	199	0	19.2[e])	199	19.2
Lysozyme [9]	14.3	129	4	2.7	129	17.1

[a]) Molecular weight, [b]) Number of residues per molecule, [c]) Number of disulfide bonds per molecule, [d]) Disulfide bonds reduced (in 6 M GuCl at 25 °C), [e]) Completely dissociated (6 M GuCl at 25 °C).

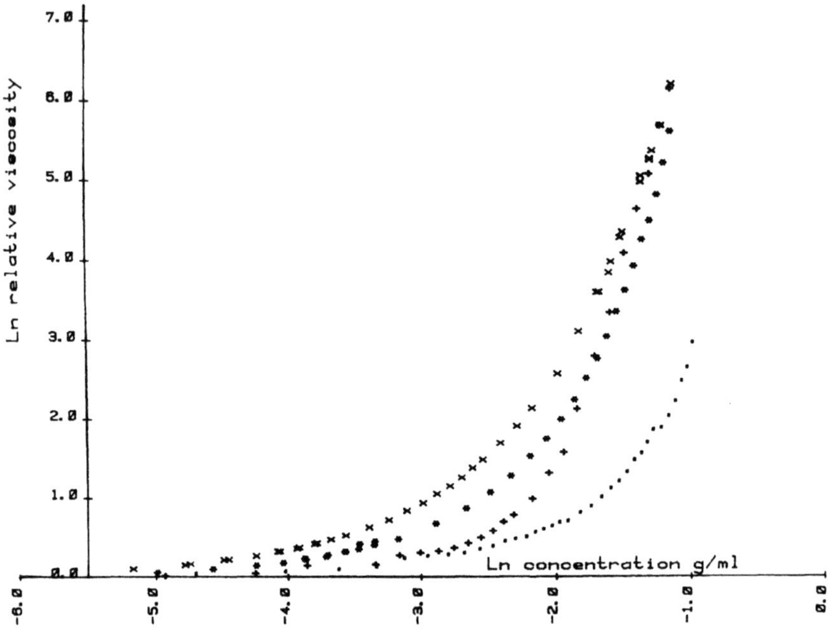

Fig. 1. Effect of protein conformation on solution viscosity: viscosity of Lysozyme solutions at 25 °C.
■ Native state: acetate, pH 4.8; + LiBr denatured: 6 M LiBr, * Crosslinked random coil: 6 M GuCl, × Random coil conformation: reduced-alkylated, 6 M GuCl

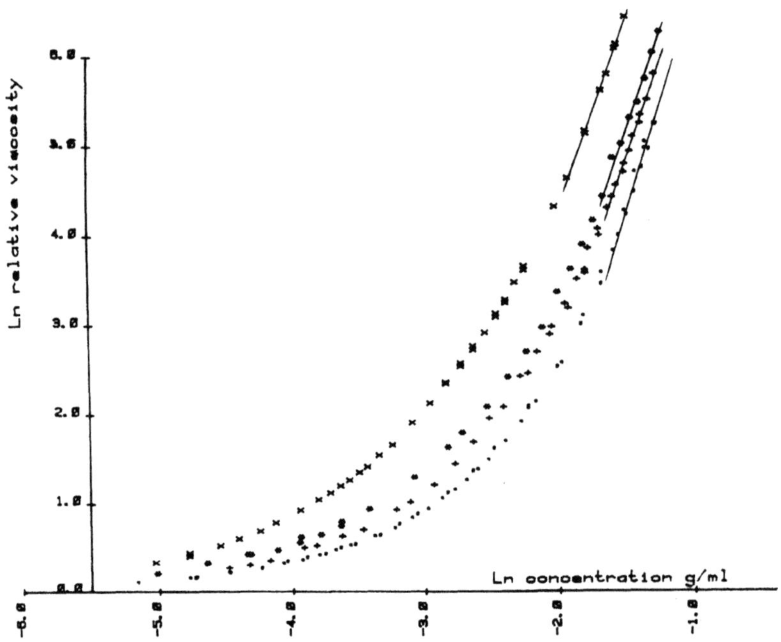

Fig. 2. Viscosity of randomly coiled proteins solutions (reduced-alkylated proteins in 6 M GuCl, 25 °C).
■ Lysozyme; + α_{S-1} Casein; * Ovalbumin; × Bovine Serum Albumin

Table 2. Randomly coiled proteins (in GuCl 6 M at 25 °C): Experimental results

	$[\eta]$ (ml/g)	c_e (g/ml)[a])	c_e/c_{cr}[a])	γ[a])	c_e/γ	ε_1[b])
B.S.A.	51.4	0.12	0.67	0.66	0.18	4.21
Ovalbumin	32.6	0.17	0.65	0.956	0.18	4.32
α_{S-1} Casein	23.4	0.18	0.56	1	0.18	4.41
Lysozyme	18.4	0.20	0.52	1.122	0.18	4.66

[a]) See text, [b]) Slope of the $\ln\eta_r - \ln c$ curve in the straight line region ($c \geqslant c_e$).

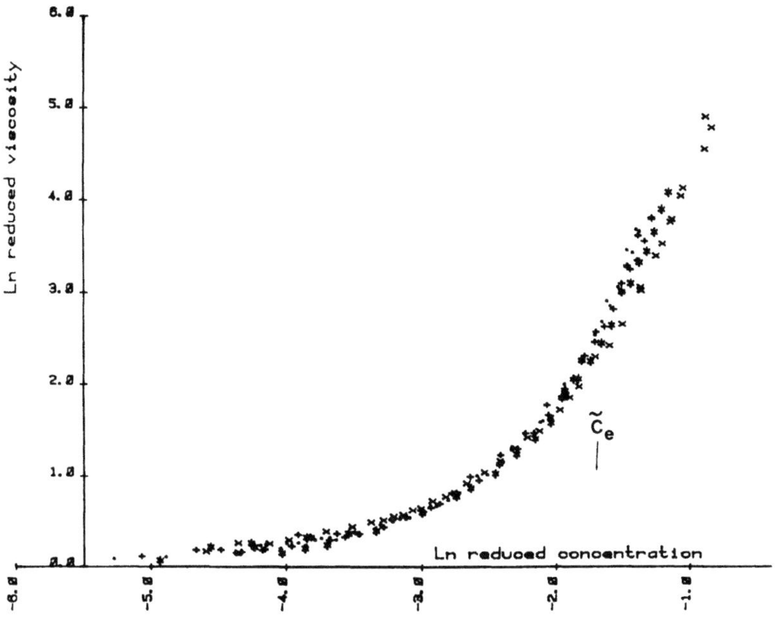

Fig. 3. Reduced viscosity: concentration curve for randomly coiled proteins (6 M GuCl, 25 °C). Reduced viscosity: $\eta_{sp}/c[\eta]$; reduced concentration: c/γ. Legend as in figure 2 – values of γ: see table 2

$\tilde{c} \simeq \tilde{c}_e(\tilde{c}_e = c_e/\gamma)$; table 2 shows that $\tilde{c}_e = 0.18$ for all four proteins and gives the values of γ used for the adjustment. For $\tilde{c} > c_e$, there is a progressive departure as pointed out by Simha and Utracki [15, 16] in the case of polymer samples with relatively small molecular weights.

Long ago, Simha and coworkers [17] introduced the notion of an "effective intrinsic viscosity" at finite concentration, defined as $[\eta(c)] = \dfrac{d(\ln\eta_r)}{dc} = \dfrac{1}{\eta_r}\dfrac{d\eta_r}{dc}$, in order to reflect the contraction of the coil volume in the semi-dilute region. The idea has been resumed recently [18] to give an expression for the Huggins constant. We have used this notion to interpret our results on the whole range of concentrations:

– In the dilute region ($c \leqslant c^*$; where c^* is the incipient overlap concentration), $[\eta(c)] = [\eta]$ i.e. the intrinsic viscosity, and thus

$$\ln\eta_r = [\eta]c^*(c/c^*)\,. \tag{1}$$

– In the concentrated region ($c \geqslant c^{**}$), $[\eta(c)] = [\eta]_\theta$, i.e. the intrinsic viscosity of the unperturbed chain, and thus

$$\ln\eta_r = [\eta]_\theta c + cst\,. \tag{2}$$

– In the semi-dilute region ($c^* \leqslant c \leqslant c^{**}$), $[\eta(c)]$ is a decreasing function of c. The problem is to give a mathematical form of this function. We met with a simple solution of it.

Table 3. Randomly coiled proteins (in 6 M GuCl at 25 °C): Fitting of the theoretical equations on the results (see text)

	c^* (g/ml)$^{a)}$	c^{**} (g/ml)$^{b)}$	Dilute region ($c \leqslant c^*$)	semi-dilute region ($c^* \leqslant c \leqslant c^{**}$)		concentrated region ($c \geqslant c_{e'}$)
			$[\eta]c^{*c)}$	$a^{d)}$	mean dev. (%)$^{d)}$	$[\eta]_\theta{}^{e)}$
B.S.A.	0.0125	0.075	0.64 ± 0.01	0.653	0.70	22.53
Ovalbumin	0.0196	0.078		0.675	1.50	19.22
α_{S-1}-Casein	0.0274	0.055	0.657 ± 0.007	0.509	0.40	19.72
Lysozyme	0.0347	0.080		0.503	0.50	19.89

a) From eq. (4′) with our values for $[\eta]$, b) from eq. (4′) with $\alpha(0)$ from $L_\theta^2 = 70\,n \cdot 10^{-16}$ [9] and $[L(0)]^3 = M[\eta]/\phi$, c) from the slope of eq. (1), d) adjustment of eq. (7) on the results, e) from the slope of eq. (2).

In a recent paper [19], Graessley established an expression for the polymer coil expansion coefficient $\alpha(c)$ in a given solvent, at concentration c in the semi-dilute region:

$$[\alpha(c)]^3 = ([\eta]/[\eta]_\theta)_{M_s}; \tag{3}$$

the sub-index M_s means that $[\eta]$ and $[\eta]_\theta$ are relative to the molecular weight M_s, corresponding to the screening distance beyond which interactions between units along the chain are screened out by the other chains. For the boundaries of the semi-dilute region, he gives $c^* = 6^{3/2}\ M/8\ N_a L^3(0)$ (N_a Avogadro's number; $L^2(0)$ mean square end-to-end chain length at zero concentration) and

$$c^{**} = [\alpha(0)]^8; \tag{4}$$

if we take for Flory's constant $\phi = 2.1 \cdot 10^{23}$ C.G.S. [9],

$$c^* = 0.64/[\eta]\,. \tag{4'}$$

From eq. (3) a straightforward derivation yields

$$[\alpha(c)/\alpha(0)]^3 = (M_s/M)^{\frac{2a-1}{2}} \tag{5}$$

where a is the Mark-Houwink exponent for the polymer-solvent system considered. Considering that the concentration which corresponds to a screening length of M_s is the overlap concentration for chains with molecular weight M_s [19],

$$c = 0.64/[\eta]_{M_s} = 0.64/KM_s^a\,,$$

eq. (5) gives

$$[\alpha(c)/\alpha(0)]^3 = (c^*/c)^{\frac{a-1/2}{a}}\,. \tag{6}$$

Now, we write $[\alpha(c)/\alpha(0)]^3 = [\eta(c)/\eta(0)]$, then eq. (6) becomes

$$(1/\eta_r)(d\eta_r/dc) = [\eta](c^*/c)^{\frac{a-1/2}{a}}$$

and so

$$\ln\eta_r = 2a[\eta]c^*(c/c^*)^{1/2a} - (2a-1)[\eta]c^*\,. \tag{7}$$

We have adjusted eq. (7) on our experimental ($\ln\eta_r - c/c^*$) data by the least-squares method. The results are given in table 3, together with those relative to the dilute [eq. (1)] and concentrated [eq. (2)] regions. The fit is rather good. The main fact to be pointed out is that (whereas the values we find for a in the case of B.S.A. and ovalbumin are not far from that derived by Tanford [9] from the linear $\log[\eta] - \log M$ plot he obtains with a large set of randomy coiled proteins the molecular weights of which encompass ours) a is very close to 0.5 (the value for unperturbed chains) for α_{S-1} casein and Lysozyme. The reason is to be sought for in the relatively low molecular weights of these last two proteins. Indeed, resuming Tanford's data, we found that they fall on two different straight lines in the $1/[\eta]$ vs. $1/M^{1/2}$ plot of Dondos and Benoit [20], one for the lower molecular weights and the other for the higher ones, which intersect at $M \simeq 20000$, value which can be thus taken as the critical molecular weight for proteins.

The fact that $[\eta]_\theta$ (table 3) is very near to $[\eta]$ for Lysozyme and α_{S-1} Casein is in agreement with the values for a.

References

1. Pradipasena, P., C. K. Rha, Polymer Eng. Sci. **17**, 861 (1977).
2. Pradipasena, P., C. K. Rha, J. Texture Studies **8**, 311 (1977).

3. Ross, P. D., A. P. Minton, Biochem. Biophys. Research Commun. **76**, 971 (1977).
4. Olteanu, M., V. Trandafir, M. Leca, I. Mindru, I. Ionescu-Bujor, Mat. Plastice **16**, 214 (1979).
5. Giordano, R., A. Salleo, S. Salleo, F. Wanderlingh, Phys. Letters A **70A**, 64 (1979).
6. Menjivar, J. A., C. K. Rha, in: G. Astarita, G. Marrucci, L. Nicolais (eds.), Rheology, Proc. VIIIth Intern. Congr. Rheology, Naples 1980. Vol. 2, p. 293, Plenum Press (New York 1980).
7. MacRitchie, F., Adv. Protein Chem. **32**, 283 (1978).
8. Phillips, M. C., Chem. Ind. **5**, 170 (1977).
9. Tanford, C., Adv. Protein Chem. **23**, 121 (1968).
10. Swaisgood, H. E., Crit. Rev. Food Technol. **3**, 375 (1973).
11. Lefebvre, J., N. Crozet, J. Gueguen et al., Ann. Technol. agric **28**, 153 (1979).
12. Ahmad, F., A. Salahuddin, Biochem. **15**, 5168 (1976).
13. Porter, R. S., J. F. Johnson, Chem. Rev. **66**, 1 (1966).
14. Cornet, C. F., Polymer **6**, 373 (1965).
15. Simha, R., L. Utracki, J. Polymer Sci. A-2, **5**, 853 (1967).
16. Simha, R., L. Utracki, Rheol. Acta **12**, 455 (1973).
17. Weissberg, S. G., R. Simha, S. Rothman, J. Res. Nat. Bureau Standards **47**, 298 (1951).
18. Irzhak, V. I., Intern. J. Polymeric Mater. **8**, 201 (1980).
19. Graessley, W. W., Polymer **21**, 258 (1980).
20. Dondos, A., H. Benoit, Polymer **18**, 1161 (1977).

(Received April 15, 1982)

Author's address:

Dr. J. Lefebvre
Institut National de la Recherche Agronomique
Laboratoire de Biochemie et Technologie des Proteines
F-44072 Nantes Cedex

PAPERS SUBMITTED BUT NOT PRESENTED

Viscoelastic properties of laminated bodies

M. Dahan and M. Predeleanu

Laboratoire de Mécanique Appliquée, Faculté des Sciences, Besancon and
Laboratoire de Mécanique et Technologie, Université de Paris VI

Abstract: In this paper we define a class of two-phase laminate composites for which the constituent elements are either linearly elastic or exhibit viscoelastic effects. We determine the viscoelastic behaviour of the equivalent homogeneous transversely isotropic medium by a single time-independent memory function.

Key words: Anisotropy, composite, homogeneization, viscoelasticity

A stratified medium consisting of a large number of alternating plane, parallel layers of two homogeneous, isotropic viscoelastic materials is considered. The thicknesses of the stiff reinforcing layers and the soft matrix layers are denoted h_1 and h_2 respectively.

The viscoelastic behaviour is defined, generally, by Riemann-Stieltjes convolutions:

$$s_{ij} = 2 \int_{-\infty}^{t} \mu_\alpha(t-\tau) de_{ij}(\tau),$$

$$\sigma_{ii} = 3 \int_{-\infty}^{t} K_\alpha(t-\tau) d\varepsilon_{ii}(\tau) \tag{1}$$

where μ_α and K_α are the relaxation moduli in shear and in dilatation respectively, $\alpha = 1$ corresponds to the fiber behaviour and $\alpha = 2$ corresponds to the matrix behaviour.

The corresponding complex moduli under harmonically varying strain $\tilde{\varepsilon}_{ij} e^{-i\omega t}$ will be noted by a tilde, like the stress and strain amplitudes. Thus we have:

$$\tilde{s}_{ij} = \tilde{\mu}_\alpha(\omega)\tilde{e}_{ij}, \quad \tilde{\sigma}_{ii} = 3\tilde{K}_\alpha(\omega)\tilde{\varepsilon}_{ii}, \quad \alpha = 1, 2 \tag{2}$$

where ω is real circular frequency.

For the elastic behaviour, $\tilde{\mu}_\alpha$ and $\tilde{K}_\alpha$ are frequence-independent and real numbers.

The complex moduli, corresponding to the viscoelastic behaviour, are complex-valued functions of ω and we shall note by superscript R the real part and by superscript I the imaginary one. Thus:

$$\tilde{\mu}_\alpha(\omega) = \tilde{\mu}^R_\alpha(\omega) + i\tilde{\mu}^I_\alpha(\omega),$$

$$\tilde{K}_\alpha(\omega) = \tilde{K}^R_\alpha(\omega) + i\tilde{K}^I_\alpha(\omega). \tag{3}$$

As is known, the relations between $\tilde{\mu}_\alpha$, $\tilde{K}_\alpha$, Young's moduli $\tilde{E}_\alpha$, Poisson's coefficients $\tilde{\nu}_\alpha$ and Lamé's moduli $\tilde{\lambda}_\alpha$ are the same as in linear elastic theory. For instance, we have:

$$\tilde{\mu}_\alpha = \tilde{E}_\alpha/2(1+\tilde{\nu}_\alpha), \quad \tilde{K}_\alpha = \tilde{E}_\alpha/3(1-2\tilde{\nu}_\alpha). \tag{4}$$

It is to note that if Poisson's ratio is time-independent, the viscoelastic behaviour in shear and dilatation is similar. This hypothesis will be adopted in this paper.

By using a continuum theory of a medium with microstructure, it is possible to define macroscopically an equivalent homogeneous solid with transverse isotropy characterized by five independent moduli. Thus for viscoelastic behaviour, the constitutive relations can be defined by means of the complex compliances $\tilde{A}_{ij}$:

$$\tilde{\varepsilon}_{11} = \tilde{A}_{11}(\omega)\tilde{\sigma}_{11} + \tilde{A}_{12}(\omega)\tilde{\sigma}_{22} + \tilde{A}_{13}(\omega)\tilde{\sigma}_{33},$$

$$\tilde{\varepsilon}_{22} = \tilde{A}_{12}(\omega)\tilde{\sigma}_{11} + \tilde{A}_{11}(\omega)\tilde{\sigma}_{22} + \tilde{A}_{13}(\omega)\tilde{\sigma}_{33},$$

$$\tilde{\varepsilon}_{33} = \tilde{A}_{13}(\omega)\tilde{\sigma}_{11} + \tilde{A}_{13}(\omega)\tilde{\sigma}_{22} + \tilde{A}_{33}(\omega)\tilde{\sigma}_{33},$$

$$\tilde{\varepsilon}_{13} = \tilde{A}_{44}(\omega)\tilde{\sigma}_{13}, \quad \tilde{\varepsilon}_{23} = \tilde{A}_{44}(\omega)\tilde{\sigma}_{23},$$

$$\tilde{\varepsilon}_{12} = \tilde{A}_{66}(\omega)\tilde{\sigma}_{12}, \quad \text{where} \quad \tilde{A}_{66} = 2(\tilde{A}_{11} - \tilde{A}_{12}). \tag{5}$$

Different methods are proposed to find the relations between the characteristics of the constituents and the "effective moduli" A_{ij} [1, 2]. Thus, we obtain [3] by a statical approach the following relations:

$$\tilde{A}_{12}(\omega)/\tilde{A}_{11}(\omega) = -\{(1-t)\nu_1\gamma(\omega)/(1-\nu_1^2)+t\nu_1/(1-\nu_2^2)\}/\{(1-t)\gamma(\omega)/(1-\nu_1^2) + t/(1-\nu_2^2)\}\cdot\tilde{E}_2(\omega),$$

$$\tilde{A}_{11}(\omega) = 1/[1-\tilde{A}_{12}^2(\omega)/\tilde{A}_{11}^2(\omega)]\{(1-t)\gamma(\omega)/(1-\nu_1^2) + t/(1-\nu_2^2)\}\cdot\tilde{E}_2(\omega),$$

$$\tilde{A}_{33}(\omega) = 2\{(1-t)\nu_1/(1-\nu_1) + t\nu_2/(1-\nu_2)\}^2/\{(1-t)\gamma(\omega)/(1-\nu_1)+t/(1-\nu_2)\}\cdot\tilde{E}_2(\omega) + (1-t)/\gamma(\omega)\cdot[1-2\nu_1^2/(1-\nu_1)] + t[1-2\nu_2^2/(1-\nu^2)]/\tilde{E}_2(\omega),$$

$$\tilde{A}_{13}(\omega) = -[\tilde{A}_{11}(\omega) + \tilde{A}_{12}(\omega)]\cdot\{(1-t)\nu_1/(1-\nu_1) + t\nu_2/(1-\nu_2)\},$$

$$\tilde{A}_{44}(\omega) = \{2(1-t)(1+\nu_1)/\gamma(\omega) + 2t(1+\nu_2)\}/\tilde{E}_2(\omega) \tag{6}$$

with the notations

$$\gamma(\omega) = \tilde{E}_1(\omega)/\tilde{E}_2(\omega), \quad t = h_2/(h_1+h_2). \tag{7}$$

Generally, $\tilde{E}_\alpha^R$ is equivalent to E_α [4] and for the elastic materials we have the equality $\tilde{E}_\alpha^R = E_\alpha$. In the composite materials, the ratio of Young's moduli of the reinforcing material and the matrix is important (see for instance [5, 6]: $E_1/E_2 \sim 40$). Then we can assume $\tilde{E}_1^R(\omega) \gg \tilde{E}_2^R(\omega)$ and consider the limited development of the complex compliances $\tilde{A}_{ij}$ in comparison with $\tilde{E}_2^R(\omega)/\tilde{E}_1^R(\omega)$. We get from eqs. (6):

$$\tilde{A}_{11}(\omega) = \frac{1}{(1-t)\gamma_m}\cdot\frac{1}{\tilde{E}_2(\omega)},$$

$$\tilde{A}_{12}(\omega) = \frac{-\nu_1}{(1-t)\gamma_m}\cdot\frac{1}{\tilde{E}_2(\omega)},$$

$$\tilde{A}_{13}(\omega) = -\frac{(1-t)(1-\nu_2)\nu_1 + (1-\nu_1)\nu_2 t}{(1-t)(1-\nu_2)\gamma_m}\cdot\frac{1}{\tilde{E}_2(\omega)},$$

$$\tilde{A}_{33}(\omega) = \left\{t\left(1-\frac{2\nu_2^2}{1-\nu_2}\right) + \frac{1}{\gamma_m}\left[2\frac{1-\nu_1}{1-t}\cdot\left(\frac{1-t}{1-\nu_1}\nu_2 + \frac{t\nu_1}{1-\nu_2}\right)^2 + (1-t)\left(1-\frac{2\nu_1^2}{1-\nu_1}\right)\right]\right\}\cdot\frac{1}{\tilde{E}_2(\omega)},$$

$$\tilde{A}_{44}(\omega) = \left\{2t(1+\nu_2) + \frac{2}{\gamma_m}(1-t)(1+\nu_1)\right\}\cdot\frac{1}{\tilde{E}_2(\omega)}, \tag{8}$$

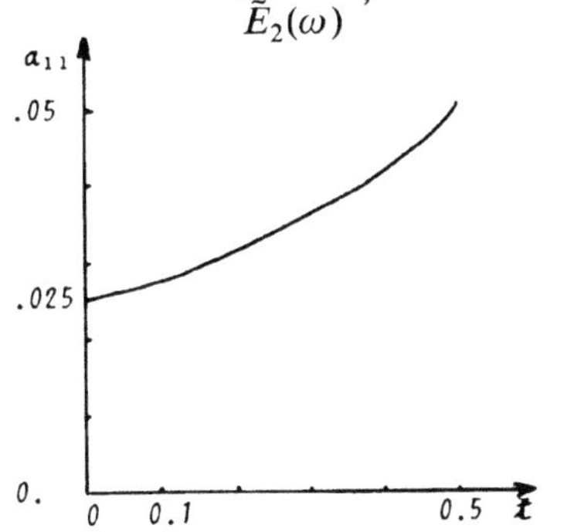

Fig. 1. Variation of a_{11}

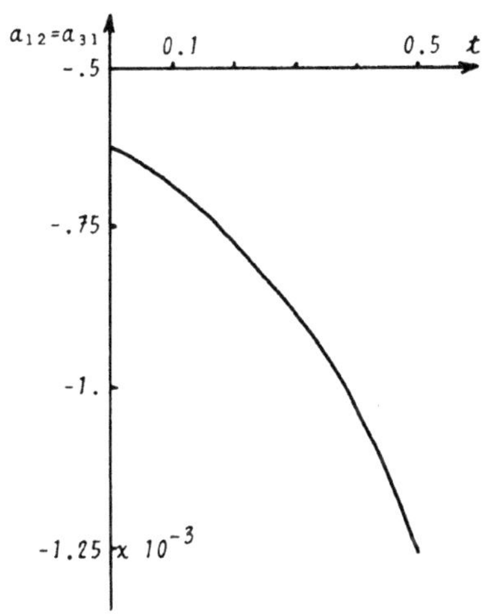

Fig. 2. Variation of a_{12} and a_{13}

where γ_m is an average value of the ratio $\tilde{E}_1^R(\omega)/\tilde{E}_2^R(\omega)$.

The complex compliances are the products of constants a_{ij} by the inverse of matrix Young's modulus, i.e. $\tilde{A}_{ij}(\omega) = a_{ij}/\tilde{E}_2(\omega)$.

In figures 1 and 2, we get the variations of the coefficients a_{11}, a_{12}, a_{13} in respect to the parameter t and when $\nu_1 = \nu_2 = 0.25$. For these values of Poisson's ratio, a_{33} and a_{44} are linearly dependent on t.

References

1. Harrison, W. J., C. M. Gerrard, J. Soil Mechs. Found. Div. ASCE **98**, 1325–1345 (1972).
2. Dahan, M., M. Predeleanu, The Hertz Problem for an axisymmetrical indenter and a viscoelastic anisotropic composite material. IV National and First International Meeting on Composite Materials, Milan 19–21 November 1980.
3. Dahan, M., Problèmes de contact dans la théorie de l'élasticité transversalement isotrope. Thesis University Paris VI, 1980.
4. Ferry, J., Viscoelastic properties of polymers. John Wiley (New York 1961).
5. Hlavacek, M., Acta Tech. CSAV **24**, 47–67 (1979).
6. Hashin, Z., Int. J. Solids Structures **6**, 797–807 (1970).

(Received April 16, 1982)

Authors' addresses:

Dr. M. Dahan
Laboratoire de Mécanique Appliquée
Faculté des Sciences
Université de Franche-Comté
F-25030 Besançon

Prof. Dr. M. Predeleanu
Laboratoire de Mécanique et Technologie
Université de Paris VI-ENSET
F-94230 Cachan

Rheologica Acta Rheol. Acta **21**, 629–631 (1982)

Induction period in curing of oligomers at high shear rates

A. Ya. Malkin and V. P. Beghishev

Research Institute of Plastics, Moscow, and Institute of Continuum Mechanics, Academy of Sciences, Perm (U.S.S.R)

Key words: Curing, induction period, high shear rate, heat generation, oligomer

The processing of oligomers and oligomer-filled compounds into end articles is based on the ability of these materials to flow during a definite time, which may be called the induction period t^*. For $t > t^*$ their viscosities increase rapidly and they lose this ability. The parameter t^* and its temperature dependence are the main technological properties of curing compounds and the determination of t^* has a general significance for different methods of processing (and applying) various polymers.

The magnitude of t^* can be measured using a rotational device by observing the variation of viscosity with time. If measurements are taken at low shear rates we find a value of t_0^* independent of shear rate. However, if sufficiently high shear rates are used a decrease in t^* with increasing $\dot{\gamma}$ up to the complete disappearance of the induction period is observed.

The dependence of t^* on $\dot{\gamma}$ has a direct technological significance in the processing of curing compounds, since in real processes shear rates of the order of hundreds of sec^{-1} are encountered. At the same time the question arises as to the nature of this phenomenon and the methods to be used for analysing it. In this work we consider the effect of heat generation resulting from intensive deformation on the duration of the induction period. Comparison of the results of such calculations with the known experimental data makes it possible to estimate whether this effect is of a primary or secondary importance.

The rate of heat generation in shear flow is determined by the product of stress τ by shear rate $\dot{\gamma}$. The variation of temperature T with time t is expressed by the equation:

$$\frac{dT}{dt} = \frac{\tau\dot{\gamma}}{\rho C}, \tag{1}$$

where C is the heat capacity and ρ is the density.

Since the experiments were carried out at $\dot{\gamma}$ = const, then taking the temperature dependence of viscosity (or τ) into account, we can write eq. (1) in the following way:

$$\frac{dT}{dt} = \frac{A\dot{\gamma}}{\rho C}\exp(U/RT) \tag{2}$$

where A is a preexponential factor, U the activation energy of viscous flow, and R the gas constant.

The formulation of a nonisothermal curing criterion occupies the central place in the problem under consideration. By analogy with the well-known Bailey rupture criterion, which sums up the contribution of the prehistory of loading, we introduce a nonisothermal curing criterion, which sums up the contributions of curing processes at different temperatures. Such a criterion is of the form:

$$B\int_0^{t_n}\exp\left[\frac{E}{RT(t)}\right]dt = 1\,, \tag{3}$$

where E is the activation energy of the processes that result in the curing of a polymer and B is the normalizing constant which can be expressed in terms of t_0^* (the zero-shear induction period at the initial temperature T_0):

$$B = (t_0)^{-1}\exp\left(\frac{E}{RT_0}\right). \tag{4}$$

Eqs. (2) and (4) make it possible to solve the stated problem, i.e. to calculate t^* as a function of $\dot{\gamma}$. It is useful to write these equations in terms of dimensionless variables. Let us denote dimensionless temperature by $\theta = \frac{E}{RT^2}(T - T_0)$, dimensionless time by

821

$\tilde{t} = t/t_0^*$ and the dimensionless induction period by $\tilde{t}^* = t_n^*/t_0^*$. Let us also introduce the parameters $\beta = RT_0/E$ and $\xi = U/E$ and dimensionless shear rate

$$\delta = \frac{E}{RT_0^2}\,\frac{A\dot{\gamma}t_0^*}{\rho C}\exp(U/RT_0)\,.$$

Then eqs. (2) and (3) can be written in the following form:

$$\frac{d\theta}{dt} = \delta\exp\left(-\xi\frac{\theta}{1+\beta\theta}\right), \tag{5}$$

$$\int\limits_0^{\tilde{t}^*}\exp\left(\frac{\theta}{1+\beta\theta}\right)dt = 1\,. \tag{6}$$

Figure 1 shows the calculated dependences of $\tilde{t}^*$ on δ with δ varying from 0.01 to 100 for different values of the parameter ξ (from 0 to 1.8) and $\beta = 0.03$. The effect of the parameter ξ on the variations in $\tilde{t}^*$ with δ is important only at high shear rates.

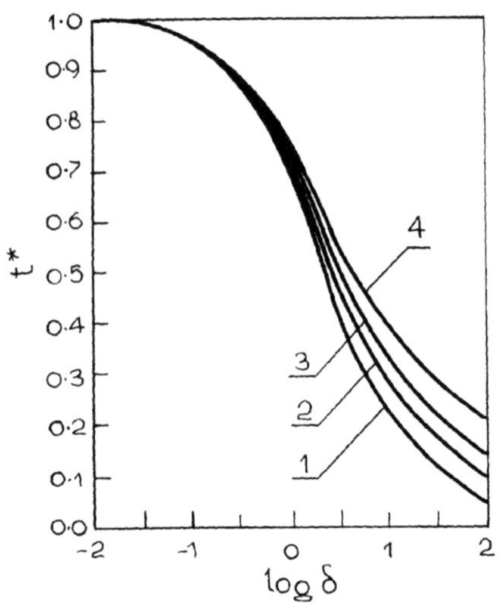

Fig. 1. The variation of the dimensionless induction period t with the dimensionless shear rate δ at $\beta = 0.03$. 1) $\xi = 0$; 2) $\xi = 0.6$; 3) $\xi = 1.2$; 4) $\xi = 1.8$

The graphs shown in figure 1 give the general solution of the stated problem. It is also of interest to consider in greater detail the situation when $\xi = 0$. In this case the shear stress at $t < t^*$ is almost constant in time. This means that we can neglect the dependence of the viscosity on temperature due to the relatively low values of the activation energy of viscous flow. In this case the temperature must rise linearly with time:

$$T = T_0 + \frac{\tau_0\dot{\gamma}}{C\rho}t\,, \tag{7}$$

where $\tau_0 = A\exp(U/RT_0)$ is the shear stress at the initial temperature T_0. Then eq. (3) takes the form:

$$\int\limits_0^{\tilde{t}^*} e^{\delta\tilde{t}}d\tilde{t} = 1 \tag{8}$$

or, after the integration we obtain the following final formula for the $\tilde{t}^*$ vs. Γ dependence:

$$\tilde{t}^* = \frac{1}{\Gamma}\ln(1+\Gamma) \tag{9}$$

where the dimensionless rate is

$$\Gamma = \frac{E}{RT_0^2}\,\frac{\tau_0\dot{\gamma}t^*}{c\rho}\,. \tag{10}$$

Let us compare the results of calculations with the experimental data for curing phenolic resin-based compounds from [1] and presented in figure 2.

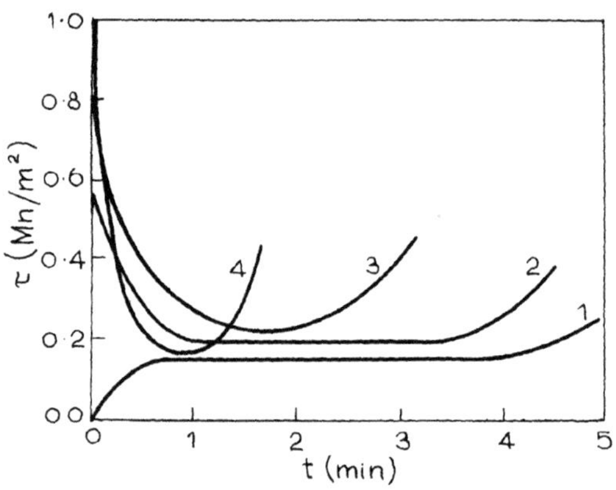

Fig. 2. Real plastograms at the following shear rates: 1) 0.015; 2) 0.95; 3) 3.75; 4) 15 s^{-1} (According to [1])

Suppose that the zero-shear value of t_0^* is approximately 240 s (4 min) at $T_0 = 373$ K and $\tau_0 = 2 \cdot 10^6$ dyn/cm^2 (0.2 MPa), $C = 0.5$ cal/g · deg (2 kJ/kgK), $\rho = 1 \cdot 10^3$ kg/m^3 and $E = 10$ kcal/mol (41.8 kJ/mol). Then

$$\delta = 0.72\,\dot{\gamma}\,.$$

At the low shear rates usually used as standard in such tests, $\delta \approx 0.01$ and it follows directly from figure 1 that this corresponds to $t^* \approx 1$, i.e. to the zero-shear value of the induction period.

The maximum shear rate used in [1] is 15 s^{-1} giving $\delta = 10.8$, which corresponds to $\tilde{t}^* = 0.2$. This means that the induction period is reduced to $\frac{1}{5}$ of its previous value, i.e. to 0.8 min. Though the cited numbers should be treated as estimates, the value of the induction period t^* obtained in these calculations nevertheless almost coincides with the experimental results given in figure 2 (data from [1]).

This means that the observed effect of the shortening in the induction period with increasing shear rate can be completely explained by the nonisothermal nature of the flow of the reactive compounds at high shear rates.

References

1. Sokolov, A. D., Plastmassy Plastics (in Russian) No. 7, 43 (1979).

(Received April 16, 1982)

Authors' addresses:

Prof. Dr. A. Ya. Malkin
Research Institute of Plastics
Perovskii pr. 35
USSR-111112 Moscow E-112

Dr. V. P. Beghishev
Institute of Continuum Mechanics
Academy of Science
ul. Koroleva 1
USSR-614013 Perm

Eine neue Methode zur Auswertung von Plasti-Corder-Plastogrammen

J. Černoch und J. Tluchoř

Ingenieurtechnisches Institut Neratovice des Konzerns Chemopetrol Prag, und Chemopetrol k.p. Spolana, Neratovice (ČSSR)

Abstract: A new method for the evaluation of plastograms using the Brabender Plasti-Corder is proposed, which allows the whole measured curve to be considered. The applicability of this method is examined, by looking at the influence of various slip agents on the rheological properties of PVC blends.

Key words: Plastogram, Brabender Plasti-Corder, PVC blend, slip agent

1. Einleitung

Eines der am meisten verwendeten Laborgeräte zur Beurteilung der rheologischen Eigenschaften von Plasten ist der Plasti-Corder der Firma Brabender, mit dem die Gelatinierung, die Plastizierung, die Absorption von Weichmachern und andere Eigenschaften gemessen werden können. Zur Auswertung der Plastogramme (Abhängigkeit des Deformationswiderstands von der Knetzeit und der Temperatur) wurden verschiedene Methoden vorgeschlagen. Die Ansichten der einzelnen Autoren darüber, welcher Teil der Kurve zur Beurteilung der rheologischen Eigenschaften ausgewertet werden sollte, sind jedoch unterschiedlich.

Bei einem Vergleich mancher Plastogramme kann man feststellen, daß die Kurven zwar einen unterschiedlichen Verlauf haben, daß dieser jedoch mit Hilfe der bisher verwendeten charakteristischen Werte wie z. B. der Plastizierzeit, des maximalen Drehmoments, des Gleichgewichtsdrehmoments oder verschiedener Verhältniszahlen der einzelnen Parameter nicht genügend genau charakterisiert werden kann. Die Ursache liegt darin, daß es sich bei den erwähnten charakteristischen Werten im Grunde genommen um Punktparameter handelt, die nicht den gesamten Kurvenverlauf charakterisieren.

Das angedeutete Problem ließe sich mit gewissen Schwierigkeiten mit Methoden der numerischen Mathematik lösen und zwar durch Aufsuchen einer „geeigneten Approximation" der gemessenen Abhängigkeit des Drehmoments M von der Zeit t, die ganz allgemein in der Form $M = f(t)$ geschrieben werden kann. Der Term „geeignete Approximation" ist jedoch sehr subjektiv, und daher wird bei der folgenden Auswertung der Plastogramme einer der Statistik entlehnten Methode der Vorzug gegeben.

2. Theoretische Grundlagen

Aus der Theorie der mathematischen Statistik ist bekannt, daß die Wahrscheinlichkeitsverteilung einer zufälligen Variablen außer durch die analytische Beschreibung der Verteilungsfunktion selbst auch durch ihre statistischen Momente charakterisiert werden kann. Praktische Bedeutung haben die allgemeinen und die zentralen Momente der Verteilungsfunktion. Die allgemeinen Momente sind (für eine mit x bezeichnete zufällige Variable) folgendermaßen definiert:

$$m_k = E[x^k] = \int_{-\infty}^{\infty} x^k f(x)\,dx\,. \qquad (1)$$

Hierbei ist m_k das allgemeine Moment k-ten Grades, E der Operator des Mittelwerts, in unserem Falle der k-ten Potenz der zufälligen Variablen x. Im Integral auf der rechten Seite von Gl. (1) bedeutet x^k die k-te Potenz des Wertes der zufälligen Variablen x und $f(x)$ ihre Verteilungsfunktion.

Die zentralen Momente der Verteilungsdichte der zufälligen Variablen x sind dagegen folgendermaßen definiert:

$$m_k' = E[(x - m_1)^k] = \int_{-\infty}^{\infty} (x - m_1)^k f(x)\,dx\,. \qquad (2)$$

m_1 ist das erste allgemeine Moment, d. h. der Mittelwert der Verteilung. Das zweite zentrale Moment (m_2') stellt die Streuung der zufälligen Variablen dar. Im Falle einer symmetrischen Wahrscheinlichkeitsverteilung sind alle ungeraden zentralen Momente gleich Null. Aus dem Nichtverschwinden von ungeraden zentralen Momenten kann daher auf die Asymmetrie der Verteilung geschlossen werden. So wird im folgenden aus m_3' ein dimensionsloser Asymmetriekoeffizient k_3 gebildet, s. Gl. (10). Entsprechend wird aus m_4' ein dimensionsloser Exzeßkoeffizient k_4 gebildet, s. Gl. (11), der als Maß für die Spitzheit der Wahrscheinlichkeitsverteilung dient. Diese beiden Koeffizienten charakterisieren also die Abweichung der Verteilungsfunktion vom Verlauf einer Normalverteilung.

Die Wahrscheinlichkeitsverteilung $f(x)$ einer zufälligen Variablen x muß nun folgende Bedingungen erfüllen:

P1: $f(x)$ muß eine stückweise stetige Funktion darstellen.

P2: Für $x \in (-\infty, \infty)$ muß $f(x) \geqslant 0$ sein.

P3: $\int_{-\infty}^{\infty} f(x)\,dx = 1.$

3. Methode

Das bei der Messung mit dem Brabender Plasti-Corder aufgenommene Drehmoment M ist im allgemeinen eine Funktion der Zeit t und der Temperatur T. Eine Anzahl von Parametern müssen bei der Messung konstant gehalten werden (z. B. die Umdrehungen der Maschine, der Knetkammertyp, die Einwaage der Mischung u. a.). Für eine gegebene Mischung und konstante Plastizierbedingungen kann geschrieben werden:

$$M = \phi(t, T), \quad T = \psi(t)$$

und daraus dann weiter:

$$M = \phi[t, \psi(t)] = \varphi(t).$$

Um bei der Beschreibung der Funktion $M = \varphi(t)$ statistische Methoden anwenden zu können, muß sie jedoch so transformiert werden, daß sie die an die Verteilungsfunktion gestellten Bedingungen P1 bis P3 erfüllt.

Da das Plastogramm im abgeschlossenen Intervall $t_0 - t_n$ ausgewertet werden soll, machen wir in Übereinstimmung mit der Bedingung P2 die Voraussetzung:

$$\int_{-\infty}^{t_0} f(t)\,dt = 0, \tag{3}$$

$$\int_{t_n}^{\infty} f(t)\,dt = 0. \tag{4}$$

Nach P3 muß dann

$$\int_{t_0}^{t_n} f(t)\,dt = 1 \tag{5}$$

sein. Dies führt zu der Identifikation

$$f(t) = \frac{\varphi(t)}{J}, \tag{6}$$

wobei

$$J = \int_{t_0}^{t_n} \varphi(t)\,dt \tag{7}$$

gilt. Für äquidistante Punkte des Plastogramms, d. h. bei Ablesung der Werte M_i in gleichen Zeitintervallen Δt gilt entsprechend:

$$f(t_i) = \frac{\varphi(t_i)}{J} \tag{8}$$

mit

$$J = \sum_{i=0}^{n} \varphi(t_i)\,\Delta t. \tag{9}$$

4. Auswertegleichungen

Der Asymmetriekoeffizient sei gegeben durch den Ausdruck:

$$k_3 = \frac{m_3'}{(\sqrt{m_2'})^3} \tag{10}$$

und der Exzeßkoeffizient durch die Beziehung:

$$k_4 = \frac{m_4'}{(\sqrt{m_2'})^4}. \tag{11}$$

Nach Umformung erhält man daraus:

$$k_3 = \frac{m_3 - 3m_1m_2 + 2m_1^3}{(\sqrt{m_2 - m_1^2})^3}, \tag{12}$$

$$k_4 = \frac{m_4 - 4m_1m_3 + 6m_1^2m_2 - 3m_1^4}{(\sqrt{m_2 - m_1^2})^4}, \tag{13}$$

Tab. 1. Abgelesene und berechnete Parameter der Plastogramme.

	Mk_1 [Nm]	Mk_2 [Nm]	Mk_3 [Nm]	t_1 [min]	t_2 [min]	P [Nm/min]	L [Nm/min]	F [Nm/min]	k_3	k_4	K
L-GH 3	18,5	33,5	17,5	0,3	0,6	50,0	55,7	100,4	0,41	1,85	0,76
L-G 10	18,5	36,0	19,3	0,1	1,0	19,4	36,0	69,8	0,41	1,83	0,75
L-G 60	12,7	32,0	19,5	0,25	1,5	15,8	21,7	55,6	0,32	1,81	0,58
L-GH-2	12,0	29,5	19,2	0,5	2,1	16,9	14,0	34,3	0,15	1,79	0,27
L-G-72	9,2	24,9	19,2	1,1	3,3	7,1	7,6	20,5	−0,03	1,97	−0,06
L-G-47	7,3	20,1	16,0	2,0	5,8	3,4	3,5	9,5	−0,41	2,27	−0,93
Wachs E	7,2	19,5	17,0	2,2	7,5	2,3	2,6	7,0	−0,52	2,28	−1,19
L-G-21	5,5	15,4	14,8	2,9	10,3	1,5	1,5	4,2	−0,58	2,39	−1,39

mit

$$m_k = \frac{\sum_{i=0}^{n} t_i^k M_i}{\sum_{i=0}^{n} M_i} \qquad (14)$$

Für die Anwendung der Gln. (12 – 14) muß aus der gemessenen Beziehung $M = \varphi(t)$ im gewählten Bereich (t_0, t_n) in äquidistanten Punkten t_i der entsprechende Wert M_i abgelesen werden, und mit Hilfe der früher angeführten Gleichungen läßt sich dann der Wert der allgemeinen Momente ausrechnen.

5. Experimente

Nach der angeführten Methode wurden die Plastogramme einer Reihe PVC-hart-Proben ausgewertet, die je 1,5 Masseteile verschiedener Gleitmittel (Firma Henkel) enthielten. Die verwendeten Gleitmittel sowie auch die Meßbedingungen waren die gleichen, wie in einer früheren Untersuchung [1] mitgeteilt. In Tabelle 1 sind der Anschaulichkeit halber zuerst die Werte aufgeführt, die aus den Plastogrammen abgelesen und berechnet wurden, und weiter die berechneten Werte des Asymmetriekoeffizienten k_3, des Exzeßkoeffizienten k_1 sowie des Produkts dieser beiden Koeffizienten $K = k_3 \cdot k_4$.

Da in der früheren Arbeit [1] schon ausführlich über den Einfluß der inneren und äußeren Gleitmittel auf die aus dem Plastogramm abgelesenen einzelnen Punktparameter gesprochen worden ist, beschränken wir uns hier nur auf den Asymmetrie- und den Exzeßkoeffizienten sowie auf deren Produkt K. Bei Gleitmitteln mit vorwiegend innerer Gleitwirkung (diese sind im oberen Teil der Tabelle 1 angeführt) sind die Werte des Asymmetriekoeffizienten k_3 höher als 0,3. Bei Gleitmitteln mit sogenannter kombinierter Wirkung liegen die Werte k_3 ungefähr im Bereich von −0,3 bis +0,3, und bei Gleitmitteln mit vorwiegend äußerer Gleitwirkung sind die k_3-Werte kleiner als −0,3. Die Werte des Exzeßkoeffizienten ändern sich von 1,8 bis 2,4, wobei sich die Mittel mit sogenannter kombinierter Gleitwirkung durch die kleinsten Werte auszeichnen. Weiter wurde gefunden, daß auch das Produkt der Koeffizienten K sehr gut die Gleitwirkung der einzelnen Gleitmittel charakterisiert. Dies kann wie folgt präzisiert werden: je höher der Wert des Produktes K ist, desto ausgeprägter ist die innere Gleitwirkung des geprüften Mittels, und umgekehrt, je niedriger dieser Wert ausfällt, um so stärker zeichnet sich das Gleitmittel durch eine äußere Gleitwirkung aus.

Was soeben über die Koeffizienten k_3 und k_4 sowie über deren Produkt K gesagt wurde, stimmt voll mit den aus den Plastogrammen abgelesenen Parametern der Tabelle 1 überein und bestätigt insofern die allgemeine Brauchbarkeit der vorgeschlagenen Methode. Besonders geeignet ist diese neue Methode in den Fällen, bei denen sich die Plastogramme durch einen unterschiedlichen Verlauf auszeichnen, die einzelnen abgelesenen Parameter (Drehmoment, Plastizierzeit, u. a.) dagegen praktisch gleiche Werte ergeben.

Literatur

1. Černoch, J., Z. Štíhel, J. Tluchoř, Plaste und Kautschuk **27**, 621 (1980); **28**, 200, 270 (1981).

(Eingegangen am 16. April 1982)

Anschrift der Verfasser:

Dipl.-Ing. J. Černoch
Ingenieurtechnisches Institut Neratovice
Chemopetrol-Konzern Prag (ČSSR)

Dipl.-Ing. J. Tluchoř
Chemopetrol k.p. Spolana
Neratovice (ČSSR)

Rheologica Acta Rheol. Acta **21**, 635 – 636 (1982)

Rheological properties of vinyl chloride polymer melts

M. Kozłowski

Institute of Organic Technology and Plastics, Technical University Wrocław (Poland)

Key words: Extrusiometry, melt viscosity, melt elasticity, activation energy, PVC

The rheological investigations of polymer melts have as a practical aim the determination of the processing behaviour of plastics. The knowledge of the correlations between rheological properties and processing conditions, construction parameters of processing machines and their equipment, polymer morphology and mechanical properties of end products has been required for the optimisation of the polymer processing technology [1]. These relationships are also useful when "tailor-made" plastics are required.

In this paper the processing behaviour of PVC as well as of vinyl chloride-vinyl acetate and vinyl chloride-propylene copolymers are presented. The investigated polymers can be used for two-component PVC systems, exhibiting improved processing properties [2]. The results were obtained using extrusiometry, a method most similar to the practical processing by extrusion molding [3 – 5].

A Göttfert extrusiometer equipped with a screw of 30 mm diameter and 20 D length and compression ratio 1 : 1 was used. The dimensions of the slit rheometric die were as follows: L = 90 mm, B = 20 mm, H = 2 mm; the temperature of the die varied from 453 to 473 K. The measurements were made with various screw speeds (5 – 50 min^{-1}), recording the torque, the inlet and outlet pressures in the die and the extrusion yield.

The exponent m of the Ostwald-de Waele power-law relation, wall shear stress τ, true shear rate $\dot{\gamma}$, melt viscosity η and activation energy of viscous flow E_y were obtained in the usual manner. Die swell behaviour was determined as a ratio β of the cross section of the extruded profile to that of the die.

A direct proportionality between extrusion yield and screw speed was obtained for all investigated polymers. This indicates the correct choice of the screw and die, indicating the correct interpretation of the extrusiometry results.

The characteristics of the rheological behaviour are given in table 1 and figure 1. The fundamental influence of the molecular weight of the investigated

Table 1. Rheological characteristic of vinyl chloride polymers melts.

Polymer	Melt viscosity*) η_{100} [kPa · s]	Exponent m	Activation energy E_y [kJ/mol]
PVC	2.05	3.9	78.7
VC/P 55/7	1.25	1.8	75.3
VC/P 49/5	0.66	1.1	73.4
VC/VA 59/10	1.70	2.2	55.9
VC/VA 49/14	0.56	1.6	35.7

*) at shear rate $\dot{\gamma}$ = 100 s^{-1}

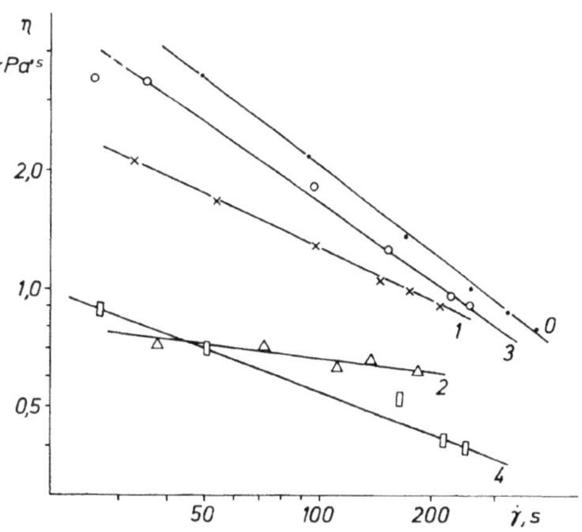

Fig. 1. Melt viscosity versus shear rate for PVC and copolymers (463 K) 0 – PVC: 1 – VC/P 55/7; 2 – VC/P 49/5; 3 – VC/VA 59/10; 4 – VC/VA 4914

810

polymers on the rheological properties of their melts is shown, because the molecular weight distributions were almost all the same. Similar values η_{100} for VC/VA 49/14 and VC/P 49/5 suggest that the type and content of the comonomer has a rather small influence on the melt viscosity. Obviously it is less for all copolymers than for PVC, as a result of the weakness of the secondary forces between macromolecules. The flow behaviour represented by the non-Newtonian index m is different, as well. The copolymer melts, especially those having a lower molecular weight, are more Newtonian than PVC. Observations with a polariscope and methylen chloride-test showed that copolymers have smaller internal stresses. As a result, the anisotropy of the copolymer extrudates is less than that of PVC.

The activation energy differences for PVC and VC/P copolymers are rather small – only an influence of the molecular weight on E_γ has been observed. The markedly smaller values E_γ for the VC/VA copolymers may be connected with their smaller polarity and higher intermolecular distances than for PVC, which makes a better mobility of the VC/VA chains possible. Powell and Khanna [6] suggested that the decrease of E_γ value for VC/VA copolymers may be due to their higher degree of branching.

Comparing the relationships of the extrusion yield and torque versus screw speed, different parts of the energy absorbed by the investigated polymers melts have been found. The elastically stored energy has been determined by the die swell ratio β. It rises as the shear rate increases – PVC exhibits higher melt elasticity than copolymers.

It was shown that lowering of the shear rate values at the melt flow of copolymers through the die was due to lubricating action of propylene and vinyl acetate units in the polymer chains. This conclusion was confirmed by evaluation of the surface quality of extruded profiles and observations of the melt fracture. The critical shear rate at which flow instability or turbulent flow occurred was higher for copolymers than for PVC.

References

1. Han, C. D., "Rheology in Polymer Processing" Academic Press (New York 1976).
2. Kozłowski, M., W. Laskawski, Angew. Makromol. Chem. **91**, 1, 17, 29 (1980).
3. Hufnagel, W., Z. Doboczky, Plastverarbeiter **17**, 61, 135, 211 (1966).
4. Rothenpieler, A., Kunststoff-Rdsch. **19**, 613 (1972).
5. Barth, Hj., Kunststoffe **69**, 370 (1979).
6. Powell, E., S. K. Khanna, J. Appl. Polymer Sci. **16**, 2013 (1972).

(Received April 16, 1982)

Author's address:

Dr. M. Kozłowski
Institute of Organic Technology and Plastics
Technical University Wrocław
ul. Wyb. Wyspiańskiego 27
PL-50-370 Wrocław

Rheologica Acta Rheol. Acta **21**, 637 – 638 (1982)

Rheologisches Verhalten von mehrphasigen Schmelzen aus Polyvinylchlorid und Polyurethan

J. Pigłowski

Institut für Organische Technologie und Kunststoffe, Technische Universität Wrocław (V.R. Polen)

Key words: Melt viscosity, activation energy, surface tension, PVC/PU system

Der Schlagzähigkeit von Hart-PVC wird durch Herstellung eines Mehrphasensystems, bestehend aus PVC und einer Elastomerkomponente, erhöht. Dabei ist die Verteilung der Kautschukphase als Netzwerk oder in Form dispergierter Kugeln verschiedener Größe möglich [1]. Zu solchen Kompositionen gehört auch das System PVC/Polyurethan [2]. Im folgenden wird das Fließverhalten von PVC/PU-Systemen in Abhängigkeit von der Morphologie und von der Oberflächenspannung der geschmolzenen Polymere diskutiert.

Die Ermittlung des Fließverhaltens wurde mit dem Göttfert-Extruder und dem Göttfert-Wiedemann-Hochdruck-Kapillar-Rheometer durchgeführt. Die Berechnung der scheinbaren Schubspannung τ_a und der scheinbaren Schergeschwindigkeit D erfolgte nach der bekannten Formel [3]. Zur Darstellung der Fließkurve wurden die vom Rheometer ermittelten Werte als Funktion $\log D = f(\log \tau_a)$, d. h. in doppeltlogarithmischem Maßstab dargestellt. Mit dem Göttfert-Extruder wurden Druck-Durchsatzkurven $G = f(\Delta p)$ aufgenommen. Der Fließexponent m wurde rechnerisch mittels der Gleichungen $m = \Delta(\log D)/\Delta(\log \tau_a)$ bzw. $m = \Delta(\log G)/\Delta(\log \Delta p)$ entsprechend für Rheometer und Extruder ermittelt. Für die rheologischen Untersuchungen wurden die unter definierten Bedingungen hergestellten Walzfelle direkt verwendet. Es wurde der Einfluß von drei Polyurethantypen im Konzentrationsbereich von 5 bis 15 Gew.-% auf das Fließverhalten von PVC untersucht.

Aus den Rheometer-Messungen läßt sich erkennen, daß die PVC/PU-Systeme stark vom Newtonschen Reibungsgesetz abweichen, denn die m-Werte liegen in einem Bereich von 2,4 bis 4,6. Wenn noch Temperatureinflüsse auf das Fließverhalten mitberücksichtigt werden müssen, so ist mit dem Faktor $e^{-E/RT}$ zu multiplizieren, in dem E eine Aktivierungsenergie für die sich beim Fließvorgang abspielenden Platzwechselvorgänge darstellt. Es wurde festgestellt, daß die Aktivierungsenergie von PVC und PVC/PU-Systemen praktisch unabhängig von der Zusammensetzung der Mischungen und dem verwendeten Polyurethantyp zwischen $3{,}9 \cdot 10^4$ und $7{,}9 \cdot 10^4$ J/mol streut. Die Auswertung der Fließexponenten- und Aktivierungsenergiewerte zeigt, daß zwischen diesen Größen keine Korrelation besteht. Die scheinbare Schmelzviskosität $\eta_{D=10\,s^{-1}}$ der Kompositionen sinkt mit zunehmendem Kautschuk-Gehalt, wie Abbildung 1 zeigt.

Im Gegensatz zu den Rheometer-Messungen weisen die Extruder-Meßwerte eine Korrelation auf. Diese Tatsache kann in der Weise interpretiert werden, daß

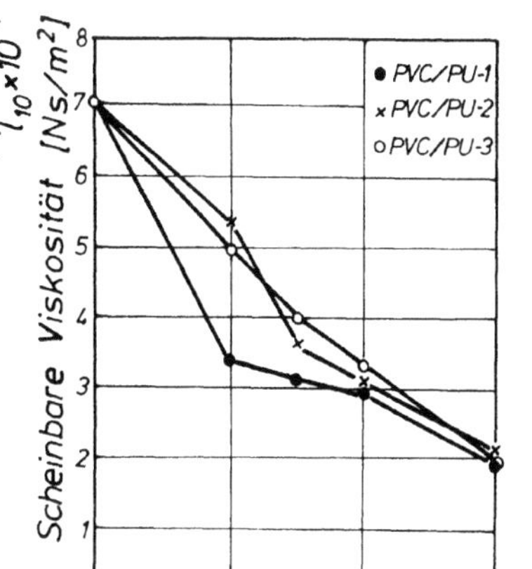

Abb. 1. Konzentrationsabhängigkeit der scheinbaren Viskosität von PVC/PU-Schmelzen

787

hier unterschiedliche Scherenergien umgesetzt werden, die die morphologischen Verhältnisse (Homogenität) beeinflussen. So ist mit steigender Polyurethan-Konzentration eine Verringerung des Fließexponenten zu beobachten. Für die Mehrzahl der Kompositionen nimmt der Fließexponent mit der Temperatur ab. Im Vergleich zu den Rheometer-Meßwerten sind die Fließexponenten und Aktivierungsenergien kleiner, liegen aber in derselben Größenordnung (m = 1,6 – 4,0, $E = 4 \cdot 10^4$ J/mol). Auch die entsprechenden mit dem Brabender-Plastograph erhaltenen Meßkneter-Ergebnisse deuten darauf hin, daß der Polyurethan-Zusatz die Verarbeitungsbedingungen erleichtert.

Die Einzelkomponenten des schlagzähen PVC/PU-Systems haben deutlich unterschiedliche Schmelzviskositäten. So kann durch die scheinbare Schmelzviskosität des Systems auch dessen Morphologie korreliert werden. Aus elektronenmikroskopischen Aufnahmen läßt sich erkennen, daß sich während der Verarbeitung ein Kautschuknetzwerk ausbildet. Liegen PVC-Partikel und damit im Schlagzähverbund eine Netzmorphologie vor, so bildet die Weichphase die Außenphase. Da diese die niedriger viskose Komponente darstellt, resultiert ein leicht fließendes Zweiphasen-System. Die Unterschiede in der Schmelzviskosität sind auch auf die Grenzflächenspannung zwischen den Polymeren mit zurückzuführen. Die Oberflächenspannungswerte, die mit der Methode des hängenden Tropfens für die geschmolzenen Polymere erhalten wurden, zeigen, daß die Oberflächenspannung von PU-1 um 10 mN/m kleiner ist als die entsprechenden Werte für PU-2 und PU-3.

Die Ergebnisse der Untersuchungen an PVC/PU-Systemen zeigen somit, daß ihre rheologischen Eigenschaften im untersuchten Schubspannungs- und Schergeschwindigkeitsbereich auf die Zweiphasen-Morphologie, die Konzentrations- und die Viskositätsverhältnisse beider Polymeren in der Schmelze, sowie auf die Adhäsion der beiden Phasen zurückzuführen sind.

Literatur

1. Fleischer, D., H. Scherer, J. Brandrup, Angew. Makromol. Chem. **58/59**, 121 (1977).
2. Pigłowski, J., W. Laskawski, Angew. Makromol. Chem. **82**, 157 (1979).
3. Ryborz, H., B. Bremer, Plastverarbeiter **32**, 193 (1981).

(Eingegangen am 16. April 1982)

Anschrift des Verfassers:

Dr. J. Pigłowski
Institut für Organische Technologie und Kunststoffe
Technische Universität Wrocław
ul. Wyb. Wyspiańskiego 27
PL-50-370 Wrocław

Rheologica Acta Rheol. Acta 21, 639 – 640 (1982)

Diffusion of macromolecules in non-uniform velocity gradients

R. Drouot

Université P. et M. Curie, Laboratoire Mécanique Théorique associé au C.N.R.S., Paris

Key words: Non-homogeneous flow, macromolecule, diffusion

The purpose of this communication is the study of the phenomenon of diffusion of macromolecules in non-homogeneous flows. Non-homogeneous flow is one in which the velocity gradients vary spatially over the flow domain of interest. A kinetic approach of this problem has been made by Bird [1] and Aubert and Tirrell [2]. These last two authors built a theory using bead-spring-type macromolecular models (linear elastic dumbell and Rouse models). They have shown that the macromolecular solute does not move with the local center of mass-solvent-velocity in non-homogeneous flow. Particularly they have shown that migration along streamline occurs in Poiseuille flow and migration across streamline occurs in Couette flow. They have calculated the concentration profiles in Couette flow. These theoretical results (for concentration profiles) are in good agreement with experimental data obtained by Dill and Zimm [3] for D.N.A. molecules.

Our purpose is to propose a continuum thermodynamical theory which allows for the same diffusion effect as that of Bird and Aubert and Tirrell.

The expression for the dumbell model is:

$$\langle v_i \rangle = \tfrac{1}{2} \langle \boldsymbol{R}_k \boldsymbol{R}_l \rangle v_{i,kl}$$

where vector $\boldsymbol{R}$ is the vector that joins end points of the dumbell.

First we introduce a vectorial internal variable $\boldsymbol{h}$ which accounts for the microstructure.

Several physical interpretations can be given to such a variable. It can be related to $\boldsymbol{R}$ or to the principal axis of a deformation tensor as defined by Lhuiller and Ouibrahim [4].

We apply the general framework of internal variable theory as outlined in Maugin [5]. In a continuum theory involving so-called internal variables, it is assumed that the usual balance laws of mechanics are not affected by the presence of the internal variables. These variables have a time evolution equation which we need to establish.

We suppose that the fluid is incompressible, and that the free energy density ψ is function of the concentration c, the temperature θ and of $\boldsymbol{h}$. Using objectivity of the different quantities, we obtain an expression for Clausius-Duhem inequality

$$-\boldsymbol{J} \cdot \nabla \left(\frac{\partial \psi}{\partial c} \right) - \rho \frac{\partial \psi}{\partial \boldsymbol{h}} \cdot D^J \boldsymbol{h} + \sigma_{ij} \boldsymbol{D}_{ij} - \boldsymbol{\Phi} \cdot \nabla \theta \geqslant 0$$

where

– $\nabla \left(\dfrac{\partial \psi}{\partial c} \right) = \nabla \mu$ is related to the usual chemical potential,

– $\boldsymbol{J}$ is the flux diffusion,

– $\boldsymbol{\Phi}$ is the new entropy flux vector defined by

$$\boldsymbol{\Phi} = -\frac{1}{\theta} \left(\frac{\partial \psi}{\partial c} \right) \boldsymbol{J} - \frac{q}{\theta},$$

– q is the usual heat-flux vector,

– σ is the symmetric Cauchy stress tensor.

Onsager relations give constitutive equations for σ, $\boldsymbol{J}$, $D^J \boldsymbol{h}$, $\boldsymbol{\Phi}$. Here we consider only isothermic transformation. These constitutive equations are similar to those proposed by Abhiraman and George [6], Kaloni [7] and Ericksen [8] in fluid having directional properties described by a so-called "director" field of variable magnitude. However, these authors did not consider any thermodynamical restrictions so that they are not able to establish any relations between the various material coefficients they have introduced in the time-evolution equation for $\boldsymbol{h}$ and in the expression of the stress tensor. On the basis of this first model we obtain a diffusion flux which is proportional to the chemical potential but we cannot deduce the same form as that obtained by Aubert and Tirrell. In a second part we introduce a second-order gradient theory (see Germain [9] for this notion). The thermodynamical description becomes much more involved while it allows us to reach the expected result. We add in the previous set of independent variables the gradient of ρ and c. The equations of balance of mass are unchanged and we consider only the balance of energy and the Clausius-Duhem inequality. The energy balance is modified as follows

$$\rho \frac{d\varepsilon}{dt} + q_{i,i} - \sigma_{ij} D_{ij} - \mu_{ij} K_{ij} - \mu_{ijk} K_{ijk} = \rho r$$

where

$$K_{ij} = -\tfrac{1}{2} \varepsilon_{ipq} v_{p,qj} ,$$

$$K_{ijk} = \tfrac{1}{3}(v_{i,jk} + v_{j,ik} + v_{k,ij}) .$$

The entropy inequality can be written in the following form:

$$-\boldsymbol{J} \cdot \nabla \tilde{\mu} - \rho \frac{\partial \psi}{\partial \boldsymbol{h}} \cdot D^J \boldsymbol{h} + \tilde{\sigma}_{ij} D_{ij} + \mu_{ik} K_{ij}$$

$$+ \mu_{ijk} K_{ijk} - \boldsymbol{\Phi} \cdot \nabla \theta \geqslant 0 ,$$

quantities such as $\boldsymbol{\Phi}$ extra flux entropy vector, $\tilde{\mu}$ et $\tilde{\sigma}$ have to be redefined.

Onsager relations give us new constitutive equations and we are essentially interested in the formulation of the diffusion-flux vector $\boldsymbol{J}$. As compared to the simpler approach in part one, this flux will include a contribution from the new thermodynamical forces K_{ijk} an K_{ij}: For example the contribution from K_{ijk} is:

$$\underset{1}{J_i} = \alpha v_{i,kk} + \alpha_g (v_{i,kl} d_k d_l + 2 v_{j,li} d_j d_l)$$

$$+ d_i (\alpha_{10} d_j v_{j,kk} + \alpha^0 v_{j,kl} d_k d_l d_j)$$

where $\boldsymbol{d}$ is the unit vector of $\boldsymbol{h}$.

The first term is an isotropic term which exists in the absence of any variable $\boldsymbol{h}$.

The term $\alpha_g v_{i,kl} d_k d_l$ has the same structure as the term of Aubert and Tirrell. We have also additional terms that involve $\boldsymbol{d}$ and are not given in the approach of Aubert and Tirrell. We can see that such terms may be responsible for migration of macromolecules across streamlines.

In this work we have shown how a second-gradient theory allows one to recover the same results as those of Aubert and Tirrell if a vectorial internal variable is introduced.

References

1. Byron Bird, R., J. Non-Newtonian Fluid Mech. **5**, 1–12 (1972).
2. Aubert, J. H., M. Tirrell, J. Chem. Phys. **72**, 2694–2701 (1980); **73**, 4103–4112 (1980).
3. Dill, K. A., B. H. Zimm, Nucleic Acids Research **7**, 735–749 (1979); Dill, K. A., Biophysical Chemistry **10**, 327–334 (1979).
4. Lhuillier, D., A. Ouibrahim, J. de Mécanique **19**, 725–741 (1980).
5. Maugin, G. A., J. de Mécanique **18**, 541–563 (1979).
6. Abhiraman, A. S., W. George, J. Polym. Sci. **18**, 127–156 (1980).
7. Kaloni, P. N., Int. J. Engng. Sci. **3**, 515–532 (1965).
8. Ericksen, J. L., Arch. Rational Mech. Analysis **4**, 231–237 (1960).
9. Germain, P., J. de Mécanique **12**, 234–274 (1973).

(Received April 16, 1982)

Author's address:

Dr. R. Drouot
Laboratoire Mécanique Théorique
Université P. et M. Curie
Tour 66-4, place Jussieu
F-75230 Paris Cedex 05

Rheologica Acta Rheol. Acta **21**, 641 – 643 (1982)

Zur Modellierung und experimentellen Untersuchung von Dehnströmungsprozessen am Beispiel von hochpolymeren Lösungen

E.-O. Reher, R. Kärmer und R. Lipp

Sektion Verfahrenstechnik der Technischen Hochschule „Carl Schorlemmer", Leuna-Merseburg (DDR)

Zusammenfassung: Unter Benutzung einer verallgemeinerten Rheologischen Zustandsgleichung nach Jeffrey wird ein Modell für verschiedene Formen von Dehnströmungen erarbeitet.

Es wird gezeigt, daß Ergebnisse dieser Modellierung als Grundlage für die Auswertung von experimentell ermittelten Dehnviskositäten von hochpolymeren Lösungen dienen können.

Abstract: By using a generalized rheological equation of Jeffrey a model for various types of elongational flow is given.

It is shown that the results of this modelling can be used as a basis for the interpretation of elongational viscosities of high-polymeric solutions obtained by experiment.

Key words: Elongational flow, viscometry, polymer solution, Jeffrey model

1. Einführung

In der Literatur existieren eine Vielzahl von Arbeiten, in denen unterschiedlichste rheologische Zustandsgleichungen viskoelastischer Flüssigkeiten auf ihre Eignung zur Widerspiegelung von Dehnströmungen untersucht werden. In [1] wird dabei auf die besondere Bedeutung der Fließgleichung nach Jeffrey hingewiesen, die für unterschiedliche kinematische Bedingungen und unter Berücksichtigung verschiedenartigsten Stoffverhaltens gestattet, zeitliche und örtliche Verläufe von Spannungen und Deformationsgeschwindigkeiten zu modellieren.

2. Dehnströmung einer Jeffrey-Flüssigkeit

Den technisch interessantesten Fall einer Dehnströmung stellt die sog. „Spinnströmung" dar (Abb. 1), die im Vergleich zu anderen, meist nur für die Viskosimetrie wichtigen Strömungssituationen dadurch charakterisiert werden kann, daß im gedehnten Freistrahl keine zeitlichen, sondern nur örtliche Veränderungen von Geschwindigkeiten, Deformationsgeschwindigkeiten und Spannungen auftreten. Die am

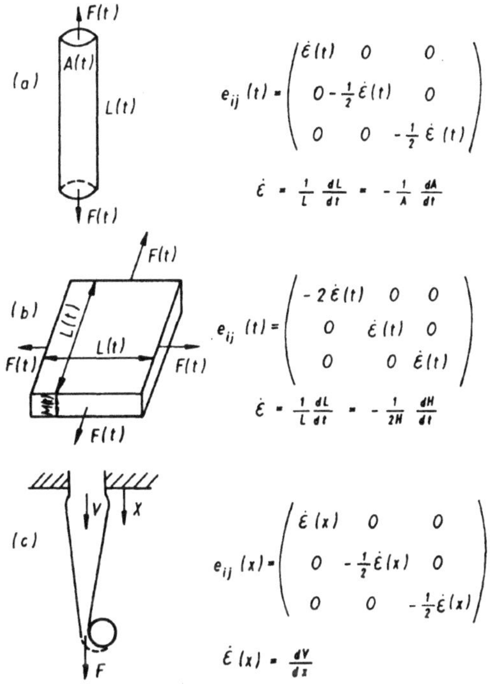

Abb. 1. Basis-Dehnströmungen: (a) einachsiges Strecken (uniaxial stretching), (b) zweiachsiges Strecken (biaxial stretching), (c) Spinndehnung

häufigsten zur Anwendung kommenden Spezialfälle einer Spinnströmung sind die einachsige Dehnung eines aus einer Kreislochdüse extrudierten runden Freistrahls und die Verstreckung eines seitlich festgehaltenen und damit auf eine konstant bleibende Breite begrenzten ebenen Freistrahls, der aus einer Breitschlitzdüse extrudiert wird.

Die kinematischen Bedingungen beider Strömungstypen sind durch den Deformationsgeschwindigkeitstensor

$$e_{ij} = \dot{\varepsilon}(x_1) \begin{pmatrix} 1 & 0 & 0 \\ 0 & -\frac{1}{2}\alpha & 0 \\ 0 & 0 & (\frac{1}{2}\alpha - 1) \end{pmatrix}$$

mit $\dot{\varepsilon} = \partial v_1/\partial x_1$ hinreichend charakterisiert, wobei die Größe α für die einachsige Spinnströmung den Wert $\alpha = 1$ und für die begrenzte zweiachsige Spinnströmung den Wert $\alpha = 2$ annimmt. Aus einer Kräftebilanz an der freien Oberfläche des Spinnstrahls folgt für die Gesamtspannungen T_{ii}

$$T_{22} = -p + \tau_{22} = 0\,,$$

$$T_{11} + (\alpha - 1)\tau_{33} = -3p\,,$$

$$T_{11} = \tau_{11} - \tau_{22} = T\,,$$

$$T_{33}(\alpha - 1) = -(\tau_{22} - \tau_{33}) = -T_3\,.$$

Unter Benutzung einer verallgemeinerten Zustandsgleichung nach Jeffrey

$$\tau_{ij} + \lambda_1 \frac{\mathbb{D}\tau_{ij}}{\mathbb{D}t} = \eta e_{ij} + \lambda_2 \frac{\mathbb{D}e_{ij}}{\mathbb{D}t}\,,$$

worin $\mathbb{D}/\mathbb{D}t$ eine objektive Zeitableitung bedeutet, die noch einen Parameter a enthält, ergibt sich folgende Spannungsbilanz:

$$\begin{aligned} &\lambda_1^2 v^2 T'' \\ &\quad + \lambda_1 v T' \left(2 + \lambda_1 v' - \lambda_1 \frac{v v''}{v'} + (\alpha - 2)\lambda_1 a v'\right) \\ &\quad + T\left(1 + (\alpha - 2)\lambda_1 a v' - 2\alpha\lambda_1^2 a^2 v'^2 - \lambda_1 \frac{v v''}{v'}\right) \\ &= (\alpha + 2)\eta_0 v' \Bigg[1 + a(\alpha - 2)\lambda_2 v' + \alpha a^2 \lambda_1 \lambda_2 v'^2 \\ &\quad + \lambda_2 \frac{v v''}{v'} + \lambda_1 \lambda_2 \left(\frac{v'' v}{v'}\right) + a(\alpha - 2)\lambda_1 \lambda_2 v v'' \\ &\quad + \lambda_1 \lambda_2 \frac{v^2 v'''}{v'}\Bigg]\,, \end{aligned} \tag{1}$$

$$\begin{aligned} &\lambda_1^2 v^2 T_3'' + \lambda_1 v T_3' \left(2 + 2a\lambda_1 v' - \lambda_1 \frac{v v''}{v'}\right) \\ &\quad + T_3 \left(1 + 2a\lambda_1 v' - \lambda_1 \frac{v v''}{v'}\right) \\ &\quad = -2\eta_0 v' \Bigg[1 + \lambda_2 \frac{v v''}{v'} + 2a\lambda_2 v' \\ &\quad + \lambda_1 \lambda_2 \frac{v^2 v'''}{v'} + 2a\lambda_1 \lambda_2 v v'' \\ &\quad - \lambda_1 \lambda_2 \frac{v^2 v'''}{v'}\Bigg]\,. \end{aligned} \tag{2}$$

In [1] wird gezeigt, daß mit diesen Basisgleichungen schon eine umfangreiche Analyse der Jeffrey-Fließgleichung durchgeführt werden kann.

Zur Modellierung der Spinnströmung werden noch der Impulssatz in Fließrichtung x

$$\rho v \frac{dv}{dx} = \rho g + \frac{1}{A} \frac{dF}{dx} \tag{3}$$

und die Kontinuitätsbedingung

$$v(x) A(x) = \text{const} = Q \tag{4}$$

benötigt. Die Größen F, A und Q stellen dabei die in normaler Richtung am Strahl angreifende Kraft, den Strahlquerschnitt und den Durchsatz dar.

Eine Zusammenfassung der Gln. (1 – 4) zur sog. „Bewegungsgleichung" würde den üblichen Weg zur Berechnung des Strömungsproblems darstellen. Obwohl das Aufstellen dieser Bewegungsgleichung auch im vorliegenden Fall möglich wäre, soll darauf verzichtet und die Lösung der Aufgabe durch eine direkte numerische Behandlung des vorliegenden Differentialgleichungssystems nach einem modifizierten Runge-Kutta-Verfahren erhalten werden.

3. Dehnviskosimetrie hochpolymerer Lösungen

Da Dehnströmungen in zunehmendem Maße bei vielen Formgebungsprozessen in der Verarbeitungstechnik eine entscheidende Rolle spielen, wird in der Literatur übereinstimmend die rheologische Charakterisierung des Materialverhaltens der zu verarbeitenden Fluide bei Dehnbelastung als unbedingte Notwendigkeit dargestellt. Arbeiten zu dieser Problematik werden mit der Zielstellung verfolgt, analog zur einfachen Scherströmung die Materialfunktion der Dehnströmung zu definieren und deren exakte meßtechnische Erfassung in geeigneten Rheometerströmungen zu gewährleisten. Gegenwärtig bereitet dabei die rheologische Charakterisierung des Dehnverhaltens von hochpolymeren Lösungen die im Vergleich zu anderen Stoffgruppen größten Schwierigkeiten.

Als Rheometerströmungen mit Dehnkinematik können die „einachsige Dehnung" (uniaxial elongation), die „begrenzte zweiachsige Dehnung" (strip biaxial extension) und die „zweiachsige gleichgerichtete Dehnung" (equal biaxial elongation) definiert werden [2], die alle der Anforderung genügen, daß der Deformationsgeschwindigkeitstensor *keine* Funktion der Zeit t und der Fließrichtung x ist.

In [3] wird dargelegt, daß eine praktische Realisierung dieser Strömung nicht möglich ist und als Basis-Dehnströmung für entsprechende Rheometer nur die in Abbildung 1 dargestellten Strömungstypen zur Anwendung kommen können. Dehnviskosimeter für hochpolymere Lösungen basieren zumeist auf der Spinnströmung und genügen daher nur unzureichend der Forderung nach stationären Strömungsverhältnissen.

Im Bestreben, Möglichkeiten zur Korrektur bzw. zu einer exakteren Interpretation der im physikalischen Sinne nur scheinbaren Meßergebnisse zu erarbeiten, wird in [3] die Vorstellung vertreten, daß sich der Spannungszustand an einer beliebigen Stelle des Spinnstrahles aus einem stationären, allein von der Dehngeschwindigkeit abhängigen Anteil und einem instationären Betrag zusammensetzt, der als Relaxationseffekt auf die in Fließrichtung wechselnde Belastung zurückzuführen ist. Darauf aufbauend wird in [3] eine Möglichkeit zur Korrektur des Einflusses von Gedächtnis- bzw. Relaxationseffekten vorgeschlagen.

Im Gegensatz dazu soll an dieser Stelle eine Methode vorgestellt werden, ohne Approximation auf stationäre Zustände direkt von den Versuchsbedingungen unabhängige Stoffwerte zu ermitteln. Die experimentell bestimmten Materialfunktionen werden dabei nur als scheinbare Größen angesehen, aus denen durch Vergleich und Anpassung mit theoretischen Berechnungen der instationären Spinnströmung die in der verwendeten rheologischen Zustandsgleichung definierten Stoffparameter bestimmt werden können.

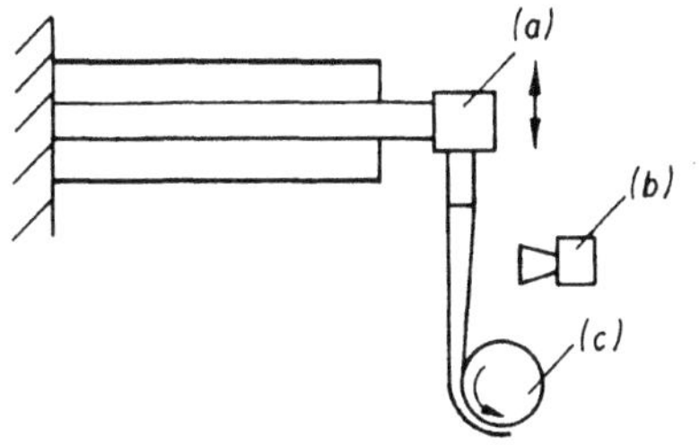

Abb. 2. Spinnwaage: (a) elastisch angeordnete Düse, (b) Kamera, (c) Abzugswalze

Als experimentelle Basis dient das in [3] beschriebene und in Abbildung 2 schematisch dargestellte Dehnviskosimeter, das eine brauchbare gerätetechnische Realisierung einer Spinnströmung darstellt. Als Meßgrößen fungieren die Kraft F_0 am Austritt des Strahles aus der Düse und die fotografisch ermittelte Veränderung der Fläche über der Länge des durch eine Walze abgezogenen Flüssigkeitsstrahles, woraus die Verläufe der Dehngeschwindigkeit $\dot{\varepsilon}$ und der Dehnspannung T bestimmt werden können.

Durch Anwendung eines numerischen Rechenverfahrens, das auf einer Modifikation der „Methode der kleinsten Fehlerquadrate" beruht, werden die im Differentialgleichungssystem (1 – 4) enthaltenen Stoffparameter so variiert, daß eine maximale Angleichung der theoretisch ermittelten Verläufe mit den experimentellen Werten erreicht wird.

Bei Annahme eines Stoffverhaltens nach dem „lower convected Maxwell-Modell" ($a = -1$, $\lambda_2 = 0$) erhält man z.B. für eine 2%ige PEO-Lösung in Wasser die Stoffwerte $\lambda_1 = 0{,}04$ s und $\eta = 23$ Pa s.

Literatur

1. Lipp, R., J. Bergmann, E.-O. Reher, Plaste und Kautschuk **28**, 272 (1981).
2. Petri, C. J. S., Elongational Flow, Verlag Pitman (London 1979).
3. Reher, E.-O., R. Kärmer, in: G. Astarita, G. Marrucci, L. Nicolais (eds.), Rheology; Proc. VIII. Intern. Congr. Rheology Naples Sept. 1 – 5, 1980, Plenum Press (New York 1980), Vol. 2, pp. 371.

(Eingegangen am 16. April 1982)

Anschrift der Verfasser:

Prof. Dr. E.-O. Reher, Dr. R. Kärmer, Dr. R. Lipp
Sektion Verfahrenstechnik
Technische Hochschule „Carl Schorlemmer"
DDR-4200 Merseburg

Rheologica Acta Rheol. Acta **21**, 644 – 646 (1982)

Zum rheologischen Verhalten konzentrierter Dispersionen organischer Polymere

H. Zecha, K. Wulf und H. Schlacken

Akademie der Wissenschaften der DDR, Institut für Polymerenchemie Teltow-Seehof (DDR)

Zusammenfassung: Die Abhängigkeit der scheinbaren Scherviskosität von den Parametern der Teilchengrößenverteilung wird an monodispersen und polydispersen PVC-E-Suspensionen untersucht. Unterschiede im rheologischen Verhalten werden bei gleicher Größenverteilung auf die Eigenschaften der Grenzfläche zwischen disperser und fluider Phase zurückgeführt. Deren Schichtdicke bestimmt sich aus der Abweichung vom „Einstein-Verhalten". Zur Beschreibung des Fließverhaltens konzentrierter PVC-Weichmacher-Suspensionen können verschiedene Modell-Theorien herangezogen werden.

Abstract: The dependence of the apparent shear viscosity on parameters of the particle size distribution is investigated on mono- and polydisperse PVC-E-suspensions. For equal size distributions, differences in the rheological behaviour are reduced to properties of the interface between disperse and fluid phases. The thickness of the interface is estimated from the deviation of the "Einstein behaviour". The flow behaviour of concentrated PVC-plasticizer suspensions can be described by various model theories.

Key words: PVC-plastisol, particle size distribution, viscosity, interface, stratified-flow model

Es ist bekannt, daß für monodisperse Suspensionen mit Teilchendurchmessern $x \gg 1$ µm die absolute Teilchengröße keinen Einfluß auf die Scherviskosität hat. Für $x \lesssim 1$ µm hingegen nimmt die scheinbare Viskosität mit abnehmender Teilchengröße bei niedrigen Schergeschwindigkeiten zu, während sich bei hohen Schergeschwindigkeiten die Viskositätswerte angleichen. Mischungen von monodispersen Systemen haben geringere Viskositäten; mit zunehmender Polydispersität nimmt die Viskosität ab.

In Tabelle 1 werden experimentell ermittelte Korrelationen zwischen Teilchengrößenverteilungsparametern und scheinbaren Scherviskositäten für verpastete sprühgetrocknete PVC-Emulsionspolymerisate [1] angegeben. Solche polydispersen Systeme enthalten Teilchen im Größenbereich 0,1 µm $\leqslant x \leqslant$ 60 µm. Für $D = 0{,}5$ s^{-1} werden die bekannten Aussagen bestätigt, für $D = 50^{-1}$ ergeben sich davon abweichende Resultate. Wie der Vergleich der Ergebnisse für die beiden Schergeschwindigkeiten ausweist, tritt dilatantes Fließen beim PVC-Typ 1 erst bei größeren Teilchengrößenverteilungsparametern auf als beim PVC-Typ 2. Die Typen 1 und 2 unterscheiden sich hinsichtlich der bei der Polymerisation eingesetzten Emulgatorsysteme. Dies führt zu der Frage, welche Eigen-

Tab. 1. Abhängigkeit der scheinbaren Viskosität η_s in Pa s vom Teilchengrößenmittelwert $\bar{x}_2$ (in µm) bzw. der Polydispersität σ_2 (in µm) von der Schergeschwindigkeit D bei $D = 0{,}5$ s^{-1} und $D = 50$ s^{-1} für zwei PVC-E-Typen (1 und 2)

$\bar{x}_2$; σ_2	$D = 0{,}5$ s^{-1}				$D = 50$ s^{-1}	
	$\eta_s(\bar{x}_2)$		$\eta_s(\sigma_2)$		$\eta_s(\bar{x}_2)$	$\eta_s(\sigma_2)$
	1	2	1	2	1/2	1/2
2	250	5,5	–	–	1,5	–
4	92	3,8	280	6,5	4,0	0,8
6	42	2,8	120	4,3	6,5	2,5
8	22	2,0	55	3,0	8,7	4,7
10	12	1,6	25	2,2	10,8	7,4
12	7,3	1,3	12	1,6	12,2	11,0
14	4,8	–	8	1,3	12,8	–
16	3,4	–	2,5	–	–	–

schaften der Grenzfläche zwischen disperser Phase und Weichmacher das Scherfließverhalten beeinflussen. Hierzu wurden an verschiedenen quasimonodispersen verdünnten PVC-Weichmacher-Dispersionen aus den Abweichungen vom „Einstein-Verhalten" Grenzflächenschichtdicken Δr in der Größenordnung von 50 nm bei $D \approx 10\ s^{-1}$ unabhängig von der Teilchengröße bestimmt. Es wird angenommen [2], daß Grenzflächenschichten aus solvatisierten Makromolekülabschnitten bestehen. Man kann darüber hinaus auch echte Solvatschichten oder den Einfluß von Ladungen [3] nicht ausschließen. Extraktion von löslichen Bestandteilen aus den PVC-Pulvern mit Methanol führte zu einer Abnahme der Grenzflächenschichtdicke auf etwa die Hälfte. Aus der für die Mooney-Gleichung

$$\eta_r = \exp\frac{k\phi}{1 - s\phi} \qquad (1)$$

üblichen Auftragung wurden für die polydispersen Materialien $0{,}1\ \mu m \leqslant x \leqslant 60\ \mu m$ bei $D = 10\ s^{-1}$ Werte für $k = 3{,}3$ und $1/s = 0{,}83$ erhalten, wobei mit zunehmendem Schergefälle k abnimmt und $1/s$ noch ansteigt und umgekehrt. Für ein monodisperses System mit $x = 210$ nm ergaben sich für verschiedene Schergeschwindigkeiten hingegen die in Abbildung 1 dargestellten Ergebnisse. Für $D = 10\ s^{-1}$ wurde der Schnittpunkt mit der Ordinate unter der Annahme von $\Delta r = 50$ nm berechnet. Die eingezeichnete Gerade schneidet die Abszisse im Wert $1/s = 0{,}23$. Unter Berücksichtigung von Δr bestimmt man hieraus eine effektive Volumenkonzentration $\phi = 0{,}739$, womit eine dichteste Kugelpackung charakterisiert ist. Das Ansteigen der Werte $\phi/\ln\eta_r$ bei Konzentrationen $\phi \gtrsim 0{,}2$ wird zunächst als Abnahme von Δr mit entsprechender Abnahme von k und Zunahme von $1/s$ interpretiert. Gleichermaßen wirkt sich offensichtlich die Erhöhung des Schergefälles aus. Ursächlich hierfür kann die jeweils zunehmende gegenseitige Beeinflussung der Teilchen sein.

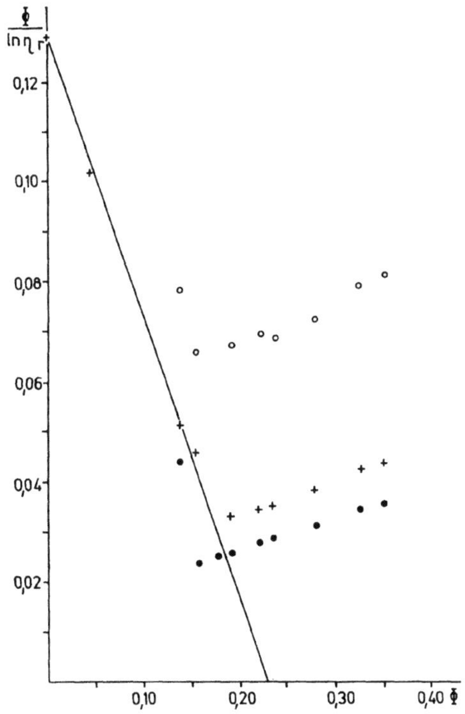

Abb. 1. Mooney-Auftragung für den Zusammenhang zwischen relativer Viskosität und Volumenkonzentration. ●: $D = 0{,}1\ s^{-1}$, +: $D = 10\ s^{-1}$ ($k = 8{,}04$; $s = 0{,}23$), ○: $D = 100\ s^{-1}$

Die monodispersen PVC-E-Pasten (200 nm ... x ... 500 nm) verhielten sich ausschließlich strukturviskos, während polydisperse, aus sprühgetrockneten Pulvern hergestellte Pasten im niederen Beanspruchungsbereich $D \approx (10^0 \ldots 10^1)\ s^{-1}$ mitunter auch Dilatanz zeigten. Diese Dilatanz wird durch eine Temperung des Pulvers vor dem Verpasten, die den Kornzerfall [1] einschränkt, noch verstärkt. Mit wachsender Konzentration verschiebt sich das „Dilatanzmaximum" bekanntlich zu niederen Beanspruchungen. Typisch für Mikrosuspensionspolymerisate und Teilchengrößen $x \gtrsim 500$ nm ist ein Dilatanzmaximum im Bereich $D \approx (10^2 \ldots 10^3)\ s^{-1}$, das sich mit wachsender Konzentration und bei Monodispersität ausprägt, zu kleineren D-Werten verschiebt und als Fließinstabilität auftreten kann [4]. Dies wird durch Zumischung von groben Teilchenfraktionen unterdrückt [5]. Experimentell ermittelte Fließkurven von PVC-E-Pasten ließen sich oberhalb einer Schergeschwindigkeit D_0 in einem gewissen Schergeschwindigkeitsbereich durch das Fließgesetz

$$\tau = \tau_B + \eta_{Pl} D \qquad (2)$$

beschreiben.

Unterhalb D_0 kann die scheinbare Viskosität stetig zunehmen oder Extremwerte und damit einen Dilatanzabschnitt durchlaufen. Auch eine echte Fließgrenze kann auftreten.

Das strukturviskose oder dilatante Fließen konzentrierter Dispersionen wird durch verschiedene Modellverstellungen bzw. -theorien gedeutet, deren Allgemeingültigkeit jedoch begrenzt ist. Firth, van de Ven und Hunter sowie Frayne [6] beschreiben ein Fließverhalten nach Gl. (2) für $D \geqslant D_0$ durch ein elastisches Flockenmodell. Die Modellparameter τ_B, D_0 und η_{Pl} können für geflockte Systeme mit Primär- und Sekundärflocken auf die mikrophysikalischen Parameter der Dispersion (Partikelgröße, elektrostatische und Dispersionswechselwirkungskräfte) zurückgeführt werden. Eine Anwendung des Modells auf konzentrierte PVC-Weichmacher-Dispersionen erfordert Kenntnisse über die Mechanismen von Aggregations- und Desaggregationsvorgängen und die zugehörigen Energien. Da es hieran bisher mangelt, sind auch andere Vorstellungen, die zur Interpretation der Strukturviskosität herangezogen werden, wie das „dynamische Teilchennetzwerk", hypothetisch.

An monodispersen, flockungsstabilen Dispersionen ist die Ausbildung einer Schichtströmung beobachtet worden [4, 7]. Aus dem Experiment folgt [7], daß sich mit zunehmendem Schergradienten der Abstand zwischen den Schichten $2(d''-d')$ vergrößert und der Abstand der Teilchen (Teilchenradius r) in den Teilchenschichten $2d'$ abnimmt. Da gleichzeitig die Teilchenrotation eingeschränkt wird, muß der Abfall der Schergeschwindigkeit D in der Dispersion (η_r) über den Viskosimeterspalt h für Teilchenschichten (g_2, η_{rs}) kleiner sein als in den teilchenfreien Schichten (g_1, η_0). Für die Schubspannung und die Schergeschwindigkeit gelten [7]:

$$\tau = \eta_0 \cdot \eta_r \cdot D = \eta_0 \cdot g_1 = \eta_0 \cdot \eta_{rs} \cdot g_2, \quad (3)$$

$$D \cdot h = 2N \cdot g_1(d'' - d') + 2Ng_2(r + d'). \quad (4)$$

Man erhält aus Gl. (3) und (4), einigen geometrischen Überlegungen sowie $h = 2N(r + d'')$:

$$\eta_r = \frac{1}{1 - K\left(1 + \dfrac{d'}{r}\right)^3 \left(1 - \dfrac{1}{\eta_{rs}}\right)\phi}. \quad (5)$$

$K = 6/\pi$ bzw. $(3 \cdot \sqrt{3})/\pi$ für kubische bzw. hexagonale Anordnung der Teilchen in den Schichten. Führt man z. B. für η_{rs} Gl. (1) unter Berücksichtigung der geometrisch festgelegten Volumenkonzentration in den Schichten bei kubischer Teilchenanordnung mit $k = 2{,}5$ und $s = 6/\pi$ ein:

$$\eta_{rs} = \exp\frac{6}{\pi}\left[\frac{2{,}5}{(1 + d'/r)^3 - 1}\right],$$

so kann man zeigen, daß die experimentell beobachteten Abstandsveränderungen bei wachsender Schergeschwindigkeit zur Abnahme von η_r bei Zunahme von η_{rs} führen. Das Aufbrechen der Schichtströmung bei einer kritischen Schergeschwindigkeit g_1 bewirkt nach [4] Dilatanz.

Moderne hydrodynamische Theorien bieten einen passenden Rahmen für ein theoretisches Verständnis rheologischer Phänomene, auch wenn der gegenwärtige Kenntnisstand noch keine Beschreibung realer Systeme ermöglicht. Wegen seiner Allgemeingültigkeit scheint das auf Vorstellungen der Vielteilchenstreutheorie basierende Konzept von Freed und Muthukumar [8] hierzu besonders geeignet.

Danksagung

Dem Kombinat VEB Chemische Werke BUNA, Schkopau (DDR) danken wir für die Bereitstellung von Probenmaterialien und die großzügige Unterstützung unserer Arbeiten. Unseren Kolleginnen Marlene Versäumer, Erika Linow, Kristina Blümel und Brigitte Beuster gilt Dank für sorgfältige Experimentalarbeiten.

Literatur

1. Bischof, C., W. Possart, K. Wulf, H. Zecha, Kunststoffe **70**, 92–97 (1980); Zecha, H., I. Zenke, H.-J. Purz, C. Bischof, Angew. Makromol. Chem. **104**, 169–188 (1982).
2. Willey, S. J., C. W. Macosco, Trans. Soc. Rheol. **22**, 325 (1978).
3. Berezov, L. V., F. D. Ovcharenko, Yu. A. Merinov, Kolloidnij Shurnal, Moskau **44**, 10–16 (1982).
4. Hoffmann, R. L., Trans. Soc. Rheol. **16**, 155–173 (1972).
5. Weinhold, G., K.-P. Paul, Kunststoffe **71**, 436–442 (1981).
6. van de Ven, T. G. M., R. J. Hunter, Rheol. Acta **16**, 534–543 (1977).
7. Schmutzler, K., Promotion B, Technische Hochschule Leuna-Merseburg 1980.
8. Freed, K. F., M. Muthukumar, J. Chem. Phys. **69**, 2657–2671 (1978).

(Eingegangen am 16. April 1982)

Anschrift der Verfasser:

Dr. H. Zecha, Dr. K. Wulf, H. Schlacken
Akademie der Wissenschaften der DDR
Institut für Polymerenchemie
Kantstraße 55
DDR-1530 Teltow-Seehof

Positive and negative flow and viscosity of clay suspensions

Z. Sobotka

Key words: Positive and negative flow, viscosity, non-Newtonian liquid, transformed flow tensor, clay suspension

1. Introduction

The author introduces the notion of "positive" and "negative" viscosities for the positive and negative flow of liquids. In positive flow, with higher value of viscosity, the liquids with structure resist movement more than in negative flow which is characterized by lower viscosities.

The different resistance to the flow of liquids in two opposite directions through a pipe could be caused by the orientated roughness of the interior surface of the pipe.

Using the idea of an orientated roughness of the bottom and of the walls it is possible to explain the preferred sense of the eddies above the outlets of vessels or reservoirs. The occurrence of eddies is conditional on differences between the positive and negative shears in liquids. If there were no differences between the shears, the liquid would flow out of the vessels in an axially symmetric manner. Such a form of outflow also occurs in very viscous liquids and at low velocities which may correspond to low pressures in the bottom region.

2. Theoretical background

The stress tensor of liquids is related to the strain rate tensor or rather to the flow tensor, which in terms of the velocity vector v_i is given by

$$e_{ij} = \frac{1}{2}\left(\frac{\partial v_i}{\partial x_j} + \frac{\partial v_j}{\partial x_i}\right), \tag{1}$$

where x_i $(i = 1, 2, 3)$ denote orthogonal Cartesian coordinates of a typical particle at the time t.

The orientated structure of liquids involves the anisotropy. For the analysis of anisotropic liquids, the author introduces the transformed flow tensor which has the same principal axes as the stress tensor and which is defined by

$$\beta_{ij} = b_{ijkl} e_{kl}, \tag{2}$$

where b_{ijkl} is the dimensionless fourth-rank tensor of anisotropy [5].

In order to express the rheological behaviour of general non-Newtonian liquids in a simplified form, the author introduces further the equivalent stress tensor given by

$$\bar{\sigma}_{ij} = a_0(t)\sigma_{ij} + a_1(t)\frac{D\sigma_{ij}}{Dt} + a_2(t)\frac{D^2\sigma_{ij}}{Dt} + \cdots + a_M(t)\frac{D^M\sigma_{ij}}{Dt^M} + a_{-1}(t)\int_{t_0}^{t} K(t-\tau)\frac{\partial \chi_k(\tau)}{\partial x_i(t)} \cdot \frac{\partial \chi_l(\tau)}{\partial x_j(t)}\sigma_{kl}(\tau)\,d\tau, \tag{3}$$

where

$$\frac{D\sigma_{ij}}{Dt} = \frac{\partial \sigma_{ij}}{\partial t} + v_\lambda \frac{\partial \sigma_{ij}}{\partial x_\lambda} + \omega_{i\lambda}\sigma_{\lambda j} + \omega_{j\lambda}\sigma_{i\lambda} \tag{4}$$

is the co-rotational derivative of the stress tensor according to Oldroyd [2],

$$\omega_{ij} = \frac{1}{2}\left(\frac{\partial v_i}{\partial x_j} - \frac{\partial v_j}{\partial x_i}\right) \tag{5}$$

is the vorticity tensor and $\chi_k(\tau)$ in the Oldroyd's convected integral denote the position at time τ of the element which is instantaneously in the point $x_k(t)$ at time t.

The equivalent transformed flow tensor is defined in an analogous manner:

$$\bar{\beta}_{ij} = b_0(t)\beta_{ij} + b_1(t)\frac{D\beta_{ij}}{Dt} + b_2(t)\frac{D^2\beta_{ij}}{Dt^2} + \cdots$$

$$+ b_N(t)\frac{D^N\beta_{ij}}{Dt^N} + b_{-1}(t)\int_{t_0}^{t} L(t-\tau)\frac{\partial\chi_k(\tau)}{\partial\chi_i(t)}$$

$$\cdot \frac{\partial\chi_l(\tau)}{\partial x_j(t)}\beta_{kl}(\tau)\,d\tau\,, \tag{6}$$

where D/Dt denotes again the co-rotational derivative.

The coefficients $a_K(t)$ and $b_L(t)$ in eqs. (3) and (6) may be functions of time or constants, respectively.

The equivalent stress tensor can be expressed by an isotropic tensorial function of the equivalent transformed flow tensor

$$\bar{\sigma}_{ij} = f_{ij}(\bar{\beta}_{kl})\,, \tag{7}$$

which may be, according to rules of the tensorial algebra [3], expanded as follows:

$$\bar{\sigma}_{ij} = \phi_0\delta_{ij} + \phi_1\bar{\beta}_{ij} + \phi_2\bar{\beta}_{i\lambda}\bar{\beta}_{\lambda i}\,, \tag{8}$$

where ϕ_0, ϕ_1 and ϕ_2 are scalar functions of invariants and δ_{ij} is the Kronecker delta.

Expressing eq. (7) in terms of principal values

$$\bar{\sigma}_{\mathrm{I}} = f(\bar{\beta}_{\mathrm{I}})\,,\quad \bar{\sigma}_{\mathrm{II}} = f(\bar{\beta}_{\mathrm{II}})\,,\quad \bar{\sigma}_{\mathrm{III}} = f(\bar{\beta}_{\mathrm{III}}) \tag{9}$$

and writing eq. (8) for all three principal values of the transformed flow tensor, we obtain the set of three equations

$$f(\bar{\beta}_{\mathrm{I}}) = \phi_0 + \phi_1\bar{\beta}_{\mathrm{I}} + \phi_2\bar{\beta}_{\mathrm{I}}^2\,, \tag{10}$$

$$f(\bar{\beta}_{\mathrm{II}}) = \phi_0 + \phi_1\bar{\beta}_{\mathrm{II}} + \phi_2\bar{\beta}_{\mathrm{II}}^2\,, \tag{11}$$

$$f(\bar{\beta}_{\mathrm{III}}) = \phi_0 + \phi_1\bar{\beta}_{\mathrm{III}} + \phi_2\bar{\beta}_{\mathrm{III}}^2 \tag{12}$$

for determining the scalar functions ϕ_0, ϕ_1 and ϕ_2. Substituting them into eq. (8), we obtain the tensorial constitutive equation with one material scalar function $f(\bar{\beta}_K)$ only. This equation can, after some rearrangements, be written as follows:

$$\bar{\sigma}_{ij} = \frac{\bar{\beta}_{\mathrm{II}}\bar{\beta}_{\mathrm{III}}\delta_{ij} - (\bar{\beta}_{\mathrm{II}} + \bar{\beta}_{\mathrm{III}})\bar{\beta}_{ij} + \bar{\beta}_{i\lambda}\bar{\beta}_{\lambda j}}{(\bar{\beta}_{\mathrm{I}} - \bar{\beta}_{\mathrm{II}})(\bar{\beta}_{\mathrm{I}} - \bar{\beta}_{\mathrm{III}})} f(\bar{\beta}_{\mathrm{I}})$$

$$+ \frac{\bar{\beta}_{\mathrm{I}}\bar{\beta}_{\mathrm{III}}\delta_{ij} - (\bar{\beta}_{\mathrm{I}} + \bar{\beta}_{\mathrm{III}})\bar{\beta}_{ij} + \bar{\beta}_{i\lambda}\bar{\beta}_{\lambda j}}{(\bar{\beta}_{\mathrm{II}} - \bar{\beta}_{\mathrm{I}})(\bar{\beta}_{\mathrm{II}} - \bar{\beta}_{\mathrm{III}})} f(\bar{\beta}_{\mathrm{II}})$$

$$+ \frac{\bar{\beta}_{\mathrm{I}}\bar{\beta}_{\mathrm{II}}\delta_{ij} - (\bar{\beta}_{\mathrm{I}} + \bar{\beta}_{\mathrm{II}})\bar{\beta}_{ij} + \bar{\beta}_{i\lambda}\bar{\beta}_{\lambda j}}{(\bar{\beta}_{\mathrm{III}} - \bar{\beta}_{\mathrm{I}})(\bar{\beta}_{\mathrm{III}} - \bar{\beta}_{\mathrm{II}})} f(\bar{\beta}_{\mathrm{III}})\,. \tag{13}$$

This relation can also be obtained on the basis of the Sylvester formula for matrix functions.

Table 1. "Positive" and "negative" shear viscosities of suspensions of clay in water.

Clay/water	1.50				1.75			
No.	Shear stress [Pa]		Viscosity [Pa s]		Shear stress [Pa]		Viscosity [Pa s]	
	Pos.	Neg.	Pos.	Neg.	Pos.	Neg.	Pos.	Neg.
1	–	–	–	–	166	104	502.4	316.3
2	98	–	181.0	–	169	117	284.3	196.4
3	117	86	129.6	95.5	175	126	176.8	127.1
4	141	98	87.2	60.7	209	141	117.2	79.3
5	190	117	70.5	43.2	239	160	80.6	53.7
6	221	154	45.5	31.6	295	190	55.1	35.6
7	233	166	28.8	20.5	375	221	42.0	24.8
8	245	175	16.8	12.0	461	252	28.7	15.7
9	258	178	10.6	7.3	479	270	17.9	10.1
10	276	184	6.3	4.2	534	282	11.1	5.9
11	289	190	4.0	2.6	571	301	7.1	3.8
12	307	209	2.3	1.6	596	332	4.1	2.3

If the material function is a nonsymmetrical one with respect to the origin, i. e. if $f(\bar{\beta}_K) \neq -f(-\bar{\beta}_K)$, we have different flow rules in two opposite directions which are characterized by different values of positive and negative viscosities. For the equivalent positive and negative shear stresses in the plane 12, we have

$$\bar{\sigma}_{12} = \frac{\bar{\beta}_{12}(\bar{\beta}_{11} + \bar{\beta}_{22}) + \bar{\beta}_{13}\bar{\beta}_{32} - (\bar{\beta}_{II} + \bar{\beta}_{III})\bar{\beta}_{12}}{(\bar{\beta}_I - \bar{\beta}_{II})(\bar{\beta}_I - \bar{\beta}_{III})} f(\bar{\beta}_I)$$
$$+ \frac{\bar{\beta}_{12}(\bar{\beta}_{11} + \bar{\beta}_{22}) + \bar{\beta}_{13}\bar{\beta}_{32} - (\bar{\beta}_I + \bar{\beta}_{III})\bar{\beta}_{12}}{(\bar{\beta}_{II} - \bar{\beta}_I)(\bar{\beta}_{II} - \bar{\beta}_{III})} f(\bar{\beta}_{II})$$
$$+ \frac{\bar{\beta}_{12}(\bar{\beta}_{11} + \bar{\beta}_{22}) + \bar{\beta}_{13}\bar{\beta}_{32} - (\bar{\beta}_I + \bar{\beta}_{II})\bar{\beta}_{12}}{(\bar{\beta}_{III} - \bar{\beta}_I)(\bar{\beta}_{III} - \bar{\beta}_{II})} f(\bar{\beta}_{III}) . \quad (14)$$

3. Experimental verification

The author has confirmed the existence of the "positive" and "negative" viscosities by tests of various suspensions of clay as well as those of clay and cement in water.

These suspensions were stirred at 1700 r.p.m. during the first 5 minutes and then at 600 r.p.m. for three and a half hours in the rotation stirrer before being tested at 12 velocities in the rotational viscometer Rheotest. If the moving cylinder of the viscometer rotated in the opposite sense to that of the stirrer, the suspension exhibited higher viscosities than when the cylinder and stirrer were rotated in the same direction. The differences between the positive and negative viscosities are due to the orientated structure of suspensions, which has been formed in the course of stirring. They increase with the density of suspensions.

Illustrative examples of test results are listed in table 1.

References

1. Mandel, J., Introduction à la mécanique des milieux continus (Varsovie 1974).
2. Oldroyd, J. G., Non-Linear Stress, Rate of Strain Relations at Finite Rates of Shear in So-Called Linear Elastico-Viscous Liquids. In: Second-Order Effects in Elasticity, Plasticity and Fluid Dynamics. Ed. Reiner, Abir. Jerusalem, Oxford (1964).
3. Reiner, M., Am. J. Math. **70**, 433–446 (1948).
4. Sobotka, Z., ZAMM **49**, 25–32 (1969).
5. Sobotka, Z., The Rheological Behaviour of Complex Viscoelastic Solids, in: D. Abir (ed.), Contributions to Mechanics, Pergamon Press (Oxford 1969).
6. Sobotka, Z., J. Franklin Inst. **308**, 423–446 (1979).

(Received April 16, 1982)

Author's address:

Dr. Z. Sobotka
Švédská 10
Praha 5 (Czechoslovakia)

Rheologica Acta Rheol. Acta **21**, 650–652 (1982)

The unified description of viscoelastic and thixotropic properties of human blood

P. Říha

Institute of Hydrodynamics, Czechoslovak Academy of Sciences, Prague

Abstract: A modified molecular network theory is used for the description of human blood having both viscoelastic and thixotropic properties. These diverse rheological properties are differentiated in the theoretical description through the material constants representing quantitatively the rate of both mechanisms of the rheological behaviour.

Key words: Human blood, viscoelasticity, thixotropy

1. Introduction

Generally, the viscoelasticity and thixotropy of fluids are connected with the effects of the fluids' inner structure on their rheological behaviour [1, 2]. The inner structure is usually formed by dispersed particles or by molecules having intermolecular interactions. In the absence of flow, either molecules or particles link together. Under shear the links are broken by hydrodynamic forces. The increase of shear rate leads to the transient reduction in size and in number of the associated structural units which corresponds to smaller resistance against flow.

The very similar molecular or microscopic processes associated with viscoelasticity and thixotropy provide the basis for a unifying concept of these diverse rheological properties. Therefore, it can be assumed that the unifying concept leading to a theory permitting quantitative analysis may result from theories which are successful in describing the mechanism governing either viscoelasticity or thixotropy.

2. Theory

For the description of fluids having both thixotropic and viscoelastic properties, we shall use Lodge's molecular network theory in the modification developed by Carreau [1] and De Kee and Carreau [3]. The modification includes the dependence of the memory function on the second invariant of the rate of deformation tensor through the functions f_p and g_p. With respect to the network theory, f_p represents the dependence of the rates of creation of junctions on the rate of deformation, whereas g_p is the inverse dependence of the probabilities of loss of junctions.

No information as to the form of the functions f_p and g_p can be obtained from the network theory. For our purposes we shall assume that the functions f_p and g_p may be in the form

$$f_p = f_{pv} f_{pt}, \quad g_p = g_{pv} g_{pt}, \tag{1}$$

where f_{pv} and g_{pv} represent the mechanism of viscoelastic behaviour and f_{pt} and g_{pt} represent the mechanism of thixotropic behaviour of fluids.

According to [3], we propose the following exponential relations for f_p and g_p, namely

$$f_{pv} = \exp[-(\tfrac{1}{2}\,\mathrm{II})^{1/2} v_p (3 - 2c_1)], \tag{2a}$$

$$g_{pv} = \exp[-(\tfrac{1}{2}\,\mathrm{II})^{1/2} v_p (c_1 - 1)], \tag{2b}$$

$$f_{pt} = \exp[-(\tfrac{1}{2}\,\mathrm{II})^{1/2} t_p (3 - 2c_2)], \tag{2c}$$

$$g_{pt} = \exp[-(\tfrac{1}{2}\,\mathrm{II})^{1/2} t_p (c_2 - 1)]. \tag{2d}$$

The constants v_p and t_p have the units of time, whereas c_1 and c_2 are dimensionless parameters. II denotes the second invariant of the rate of deformation tensor.

Using eqs. (1) and (2), the following relation may be obtained for the non-Newtonian shear viscosity [3]

$$\eta = \sum_{p=1}^{k} \eta_p \exp[-(v_p + t_p)\dot{\gamma}] + \eta_\infty . \qquad (3)$$

Similarly, the stress relaxation experiment yields the following time dependence of shear viscosity [3]

$$\eta_r = \sum_{p=1}^{k} \eta_p \exp[-(v_p + t_p)\dot{\gamma}_0] \exp[-t/(\lambda_p + \tau_p)] + \eta_\infty[1 - h(t)] . \qquad (4)$$

The constants η_p have the units of viscosity. λ_p and τ_p have the units of time and represent the contribution of viscoelasticity and thixotropy, respectively. η_∞ denotes the limiting shear viscosity, $h(t)$ is the unit step function: $h(t) = 0$ for $t \leqslant 0$ and $h(t) = 1$ for $t > 0$. $\dot{\gamma}_0$ is the initial constant shear rate.

3. Experimental

An example of fluid with inner structure having both viscoelastic [4, 5] and thixotropic [5, 6] properties is human blood. The inner structure of blood is formed by red blood cells (RBCs). According to experiments [7], RBC aggregation and RBC deformation are two major mechanisms affecting the rheological behaviour.

The measurements of rheological properties of human blood were carried out with the Contraves LS 30 viscometer and with the Weissenberg Rheogoniometer (Model R18). The experimental arrangement and further details are given in [4]. All studies were performed at a temperature of 37 °C.

Fresh blood samples were drawn from normal human subjects, with citrate solution as the anticoagulant. Hematocrits were determined with a micro-hematocrit centrifuge TH-11 (Fa. Janetzki) after 1 min centrifugation.

4. Results

Figure 1 shows the comparison of experimental results with theoretical predictions for the shear rate dependence of shear viscosity and frequency dependence of the viscous component of the complex viscosity.

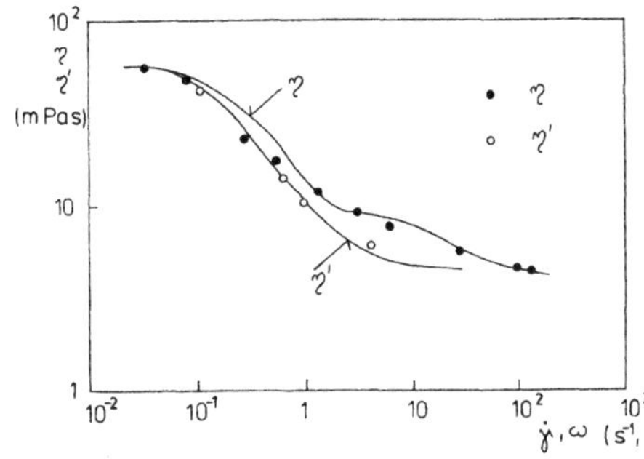

Fig. 1. Shear viscosity η and viscous component η' of the complex viscosity for normal human blood with citrate solution as anticoagulant, the hematocrit being 0.39. Parameters: η_0 = 58 mPas, η_1 = 48.3, η_2 = 5.2, η_∞ = 4.5, $(v_1 + t_1)$ = 2.5 s, $(v_2 + t_2)$ = 0.047, λ_1 = 5 s, λ_2 = 0.54

The material constants were determined by following the procedure suggested by De Kee and Carreau [3]. Briefly stated, the zero-shear viscosity η_0 is read directly from the measurements. The constants η_p and $(v_p + t_p)$ can be estimated from the intercepts and the slopes of the curve of the viscosity data. The constants λ_p can be evaluated by fitting the equation for the viscous component η' of the complex viscosity [3] to the experimentally obtained plot η' versus ω.

As follows from eqs. (1 – 4), the effect of the viscoelastic and thixotropic mechanisms on the rheological behaviour is differentiated in the theoretical description through the material constants v_p, t_p and λ_p, τ_p. The values of $(\lambda_p + \tau_p)$ may be determined from the plot of the shear stress relaxation data shown in figure 2. Since the values of λ_p connected with the viscoelastic contribution to the fluid behaviour were evaluated earlier from the frequency dependence of η', the time constants τ_p connected

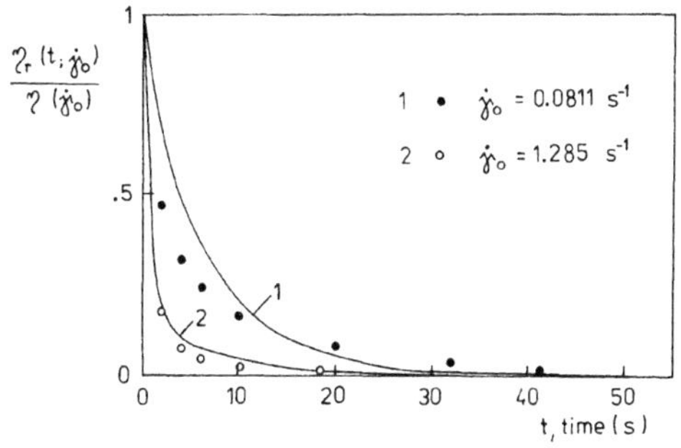

Fig. 2. Shear stress relaxation function for normal human blood with citrate solution as anticoagulant, the hematocrit being 0.39. Parameters: η_0 = 58 mPas, η_1 = 48.3, η_2 = 5.2, η_∞ = 4.5, $(v_1 + t_1)$ = 2.5 s, $(v_2 + t_2)$ = 0.047, $(\lambda_1 + \tau_1)$ = 7 s, $(\lambda_2 + \tau_2)$ = 0.76

with the effect of the thixotropic mechanism can be calculated.

However, the steady and oscillatory shearing flow data as well as the stress relaxation data obtained here by rotational viscometers do not enable us to determine the values of the material constants v_p and t_p. Looking for an appropriate flow experiment for determining these constants, it appears that a capillary viscometer with oscillatory flow developed by Thurston [5] could be used for this purpose.

The abovementioned type of measuring device is capable of determining apart from the same rheological characteristics that can be obtained by flow experiments in rotational viscometers, the shear rate dependence of the viscoelasticity at a fixed frequency. Comparisons of these data and the predictions supplemented by the sums $(v_p + t_p)$ evaluated from the steady shearing flow measurements yield the values of the time constants v_p and t_p.

5. Conclusions

The results presented herein demonstrate that the modified molecular theory is able to predict the functions in figures 1 and 2 reasonably well and to assess to some extent the effects of the viscoelastic and thixotropic mechanisms on the rheological behaviour. At the same time the presented results demonstrate that further studies are required to confirm the full suitability of the proposed concept, especially with respect to the interpretation of the results in connection with the behaviour and the properties of the inner structure.

References

1. Carreau, P. J., Trans. Soc. Rheol. **16**, 99 (1972).
2. Mewis, J., J. Non-Newtonian Fluid Mech. **6**, 1 (1979).
3. De Kee, D., P. J. Carreau, J. Non-Newtonian Fluid Mech. **6**, 127 (1979).
4. Říha, P., in: G. Astarita, G. Marrucci, L. Nicolais (eds.), Rheology, Vol. 3: Applications, p. 575, Plenum Press (New York 1980).
5. Thurston, G. B., Biorheology **16**, 149 (1979).
6. Huang, C. R., et al., Biorheology **12**, 279 (1975).
7. Chien, S., et al., Biorheology **12**, 341 (1975).

(Received April 16, 1982)

Author's address:

Dr. P. Říha
Institute of Hydrodynamics
Czechoslovak Academy of Sciences
Podbabská 13
CSSR-16612 Prague 6

Rheologica Acta Rheol. Acta **21**, 653 – 655 (1982)

Influence of a chemical modification on the formation of protein gels

K. D. Schwenke and H. Schmandke

Central Institute of Nutrition, Potsdam-Rehbrücke, Research Centre for Molecular Biology and Medicine, Academy of Sciences of the GDR

Abstract: Covalent cross-linking of casein by dialdehyde starch and heat-denaturation of acetylated faba-bean proteins result in a formation of gels, the properties of which were studied by measuring the creep behavior.

Key words: Protein gel, acetylation, creep, dialdehyde starch, shear modulus

1. Introduction

As widely distributed physical states among foodstuff systems, gel-like structures are of great importance for the processing of unconventional food-proteins into new types of products. As only a few proteins exhibit optimal properties for processing technology (functional properties) when in their natural condition, effort must be made to improve them by physico-chemical or chemical modification [1]. In the work submitted here, two possibilities of influencing the gel-forming capacity of food-proteins are put forward, the covalent cross-linking of the protein molecules by reaction with a polycarbonyl compound and the blocking of the basic amino-acid residues of the protein by acetylation.

2. Thermo-reversible and thermo-irreversible gels by the reaction of casein with dialdehyde starch (DAS)

Dialdehyde starch reacts with casein in weak acidic, neutral or alkaline solution blocking of basic amino-acid residues (lysine, histidine, arginine) and inter-molecular cross-linking. Under suitable conditions spontaneous gelation also takes place. The properties of the gels formed can be regulated by the quantity of carbonyl reagent, the concentration of protein and the pH value of the system.

The first stage of gel-formation is characterised by an association of covalently connected protein molecules. Between 10 and 20% of the basic amino-acid residues are irreversibly blocked. A thoroughly united covalent spatial network does not come into being; the resulting gels fuse.

Through an increase of the amino-acid blocking (20 – 30% of the basic residues) and the covalent cross-linking of the protein, a raising of the quantity of DAS causes the transition to a non-fusible system. Its final condition is marked by a cohesive covalent spatial network.

The investigation of the creep of the gel during a stress-phase and after a release-phase in Tolstoy's apparatus [2, 3] produced the following results:

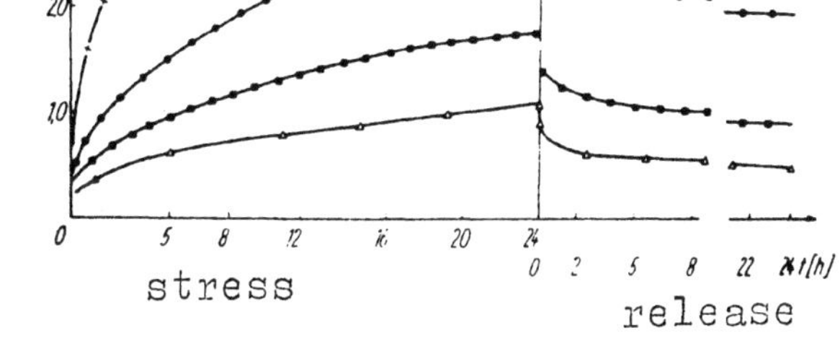

Fig. 1. Creep of a gel from modified casein (10% casein, 1% DAS, pH 8.4) during a 2-week storage, according to [4]. ×--× 1 d, ●--● 3 d, ■--■ 8 d, △--△ 15 d; J = creep compliance

1. Total compliance decrease during a two-week storage (fig. 1).

2. At the beginning of gelation (1st – 3rd day) a decrease of the elastic component of compliance ensues, owing to progressive cross-linking even during the 24-hour stress-phase of the experiment.

3. The double-logarithmic plot of the change in the elastic component against time, according to Ferry, shows that the system tends towards a normal rubber-elastic condition.

4. The change in the elastic component in standard gels (10% casein, 2.5% DAS) is completed at pH 8.0 after 7-day storage (still earlier at higher pH values) [4, 5]. The decrease of the total compliance thereupon follows according to the changes in the plastic component (fig. 2).

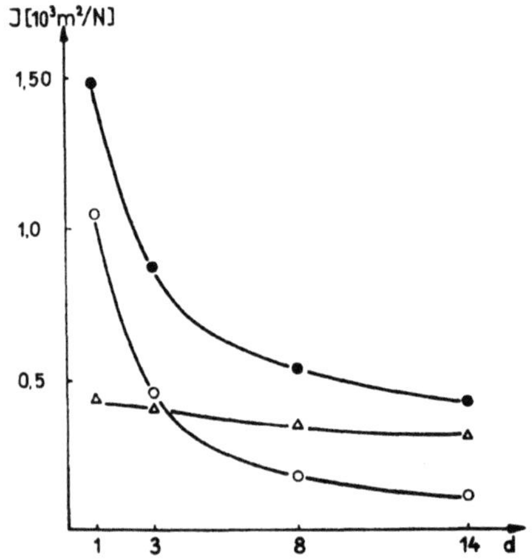

Fig. 2. Dependence of compliance (●--●) and of the plastic (○--○) and elastic (△--△) component on time; System: 10% casein, 2.5% DAS, ph 8.0

In conformity with these results, the decrease of absorptive capacity and the sol fraction as well as the increase of irreversible amino-acid blocking (especially of lysine) is completed within 7 days. Hence a connection between the covalent cross-linking and the irreversible blocking of lysine may be deduced. Applying the Flory Relation between shear-modulus and network nodal density gives a result, allowing for network defects, of 3 to 4 net nodes per polypeptide chain. For interpreting the change in the plastic component, which proceeds parallel with the increase of the reversibly blocked lysine residues (carbinolamine and Schiff's bases) component, it may be assumed that this corresponds to the condition of the non-covalently connected associate in the spatial network. Indirect evidence for this are the changes in the creep of the gel in the presence of a detergent, which weakens or dissolves the inter-molecular links between the associates. Thus dodecylsulphate changes the elastic component of compliance, conditioned by covalent links, only a little, but raises the total compliance and its plastic component.

3. Thermotropic gels from acetylated faba-bean proteins

The blocking of basic residues by acylation raises the negative net charge and disturbs the electrostatic equilibrium between the polypeptide chains of a protein. An acylation consequently facilitates the decomposition of the structure of the native protein. Thus with a growing level of acetylation of faba-bean protein the heat and temperature of the denaturing fall [6]. With a high degree of substitution the negative charge of the protein predominates, whereby

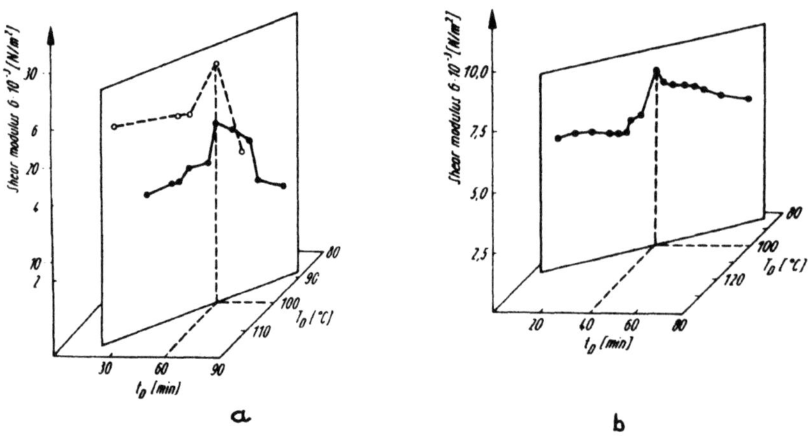

Fig. 3. Shear-modulus of faba-bean proteins (pH 7), unmodified (a) and 78% acetylated (b), and its dependence on heating time and temperature, according to [9].
●--● 17.6% protein, ○--○ 24.9% protein

The relative changes $\Delta G'/G'$ and $\Delta G''/G''$ which occur at the clearing temperature, decrease considerably with increasing spacer length. The polymer having spacer length $m = 3$, exhibits a nematic phase whereas the other ones are smectic. As the smectic phase is of higher order, a greater step of G' and G'' at T_c was expected with spacer lengths 4 and 6. The experimental results show the opposite behaviour.

What we have to conclude from this fact is that the rheological properties are largely dominated by the properties of the polymer backbone. The mesogenic side groups with their specific transition can only act as a small disturbance, and their influence is the smaller the weaker the coupling between the mesogenic groups and the polymer main chain is.

This view is supported by the fact that G' and G'' are in the usual order of magnitude found with common polymers, in the glassy state and in the rubbery region as well.

Frequency dependent measurement of G' and G'' were also carried out in the frequency range from 10^{-3} to 1 Hz. A specific influence of the phase transition on the data could not be detected. The curves below and above T_c appeared to be similar in shape.

3.2. Thermoelastic and photoelastic behaviour

Figure 3 shows the results of an experiment which is quite usual in rubber elasticity. The sample is stretched to a certain extent and then kept at this length. The retractive force divided by the cross-rectional area of the undeformed sample, σ^0, is measured as function of temperature. In the upper temperature range, $T > T_c$, we find the typical behaviour of rubbers. At low elongation, σ^0 decreases with increasing temperature. This is because of the thermal extension of the sample. At a certain elongation, in our case at about 12%, σ^0 is independent of T. This is the so-called thermoelastic inversion point. At still higher elongation – we could not go so far because the sample broke – one should find an increase of σ^0 with rising temperature.

The clearing temperature of this sample is 283 K. At this point, the retractive force decreases substantially, even going to some negative values. Although the specimen had been stretched at higher temperature, now the tensile force changes into a compressive force. What we observe in the experiment is that the sample, a thin rubber strip, buckles when we go from higher to lower temperatures. This is an absolutely unusual and surprising behaviour in the rubbery region and has to be due to the ordering of the mesogenic side groups.

In the undeformed state above T_c, the network is isotropic, the statistical segments being randomly distributed. When the sample is stretched, the statistical segments and hence the mesogenic groups have an anisotropic distribution, but there is not yet a correlation among the mesogenic groups. When cooling down below T_c at this fixed length, the mesogenic groups become anisotropically ordered in the liquid crystalline state. As the network is already deformed uniaxially, the mesogenic groups adapt this macroscopic orientation, yielding a macroscopically ordered liquid crystal.

If now the force is removed (still below T_c), the sample will not move all the way back to its initial

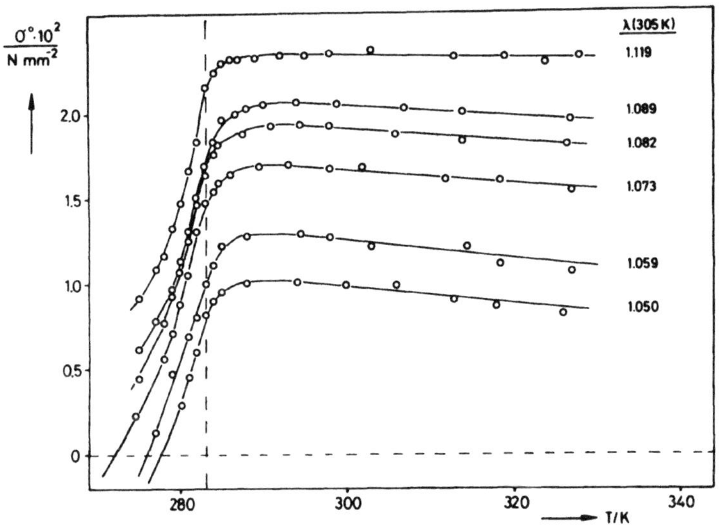

Fig. 3. Reduced stress σ^0 at constant sample length as function of temperature for a liquid crystalline elastomer with spacer length $m = 3$. The deformation was applied at the highest temperature. Different stretching ratios are indicated

List of Contributed Papers of which no Short Versions are Included

1. U. Akbay, St. Sponagel, Institut für Mechanik, Technische Hochschule Darmstadt (F.R.G.):
Nichtlineare Erscheinungen beim Strömen viskoelastischer Flüssigkeiten.

2. J. Aleman, J. P. Lesbats, R. Legros, Istituto de Plasticos y Caucho, Madrid (Spain):
Longitudinal bulk flow of epoxide prepolymers. (cf. J. P. Lesbats, R. Legros, J. Aleman, J. Polym. Sci., Polym. Chem. Ed. **20**, 1971 – 1984 (1982)).

3. H. J. Booß, Siemens AG, Berlin (F.R.G.):
Homogenitätsprüfung von Polymer-Gemischen mit dem Dehnungstester.
(Vorgesehen zur Veröffentlichung in Kautschuk + Gummi, Kunststoffe).

4. O. S. Brüller, B. Reichelt, Universität Duisburg (GH), FB7-Werkstofftechnik, Duisburg (F.R.G.):
On the computation of the principal normal stress difference of polymer melts in circular and slit dies.

5. L. Choplin, P. J. Carreau, Chemical Engineering Department, Laval University, Quebec (Canada):
End effects: a new interpretation.
(Intended for publication in Rheol. Acta).

6. D. Hadjistamov, Ciba-Geigy AG, Basel (Switzerland):
Viskoelastisches Verhalten von Epoxidharzen und gefüllten Formulierungen auf deren Basis.

7. J. Hutton, B. Jones, Shell Research Ltd., Thornton Research Centre, Chester (U.K.):
Effects of isotropic pressure on transitions in viscosity-temperature properties of polymer-containing motor oils.

8. Y. Ivanov, Bulgarian Academy of Sciences, Central Laboratory of Physico-Chemical Mechanics, Sofia (Bulgaria):
On the rheology of short fibers filled liquid polymers.

9. W. M. Jones, I. Rees, University College of Wales, Department of Physics, Aberystwyth, Dyfed (U.K.):
The "stringiness" of dilute polymer solutions.
(cf. Report of the Newcastle Meeting of the British Society of Rheology, to be published in J. Non-Newtonian Fluid Mech.).

10. R. J. J. Jongschaap, Twente Institute of Technology, Department of Applied Physics, Enschede (The Netherlands):
The average stress tensor in systems with interacting particles.
(to be published in Rheol. Acta).

11. Z. Kembłowski, J. Torzecki, Institute of Chemical Engineering, Technical University, Łódź (Poland):
Rheological properties of reactive polymer melts.
(to be published in Rheol. Acta).

12. C. W. Macosko, J. M. Starita, Department of Chemical Engineering and Materials Science, University of Minnesota, Minneapolis (U.S.A.):
New instrumental methods for characterizing polymer melts and solutions.

13. Ch. D. Meier, W. Bulacher, Contraves AG, Dept. IBVA, Zürich (Switzerland):
Tension and relaxation tests with rotational rheometers.

14. P. Michels, K. Kleesiek, H. Greiling, Institut für Klinische Chemie und Pathobiochemie, Technische Hochschule Aachen (F.R.G.):
Korrelation reologischer und molekularer Eigenschaften des Hyaluronats in Synovialflüssigkeiten.
(Vorgesehen zur Veröffentlichung in Rheol. Acta).

15. D. Nienass, Institut für Verfahrenstechnik, Technische Hochschule Aachen (F.R.G.):
Numerical calculation of viscoelastic liquids streaming through a contracted conduit.

16. V. Ribitsch, J. Schurz, F. Rainer, Institut für Physikalische Chemie, Universität Graz (Austria):
The rheology of healthy human synovial fluids. (cf. V. Ribitsch, F. Rainer, J. Schurz, Rheol. Acta **21**, 81 – 89 (1982)).

17. F. R. Schwarzl, Institut für Werkstoffwissenschaften V, Kunststoffe, Universität Erlangen (F.R.G.):
On the mechanical properties of polystyrene.
(intended for publication in Rheol. Acta).

18. J. Ulbrecht, Y. Kawase, State University of New York, Department of Chemical Engineering, Buffalo, N.Y. (U.S.A.):
Influence of walls on the motion of a sphere in non-Newtonian liquids.
(to be published in Rheol. Acta).

Subject Index

GPSR Compliance

The European Union's (EU) General Product Safety Regulation (GPSR) is a set of rules that requires consumer products to be safe and our obligations to ensure this.

If you have any concerns about our products, you can contact us on ProductSafety@springernature.com

In case Publisher is established outside the EU, the EU authorized representative is:

Springer Nature Customer Service Center GmbH
Europaplatz 3
69115 Heidelberg, Germany

Batch number: 10578410

Printed by Printforce, the Netherlands